Advances in the Internet of Things

Many reports estimated that in 2024, the number of Internet of Things (IoT) devices exceeded 18 billion worldwide, with predictions suggesting that it could reach nearly 40 billion by 2033. Despite primarily being consumer devices, a growing number of them will find use in industrial and enterprise applications. This shows the significance of IoT and how it shapes the future. However, to realize its full potential, we must address its emerging challenges and highlight recent applications, advances, and trends, which is the focus of this book.

Security and privacy represent some of the key challenges IoT adopters face. The severity of these issues is exacerbated by the growing number of IoT devices, the expansion of Industry 4.0 (and the emergence of Industry 5.0), and the significant increase in cybersecurity attacks. Considering that ensuring security and privacy is crucial for the successful adoption of IoT, this book dedicates several chapters to these areas. This book also introduces some novel models that improve IoT environments and presents several practical implementations that utilize IoT to demonstrate some of its real-world applications. Furthermore, it examines several emerging technologies that enable the realization of advanced IoT environments. We see most IoT advances in three main areas: the integration of artificial intelligence/machine learning, network technologies, and hardware design. Therefore, this book dedicates several chapters to these areas. Most chapters touch on artificial intelligence/machine learning, emphasizing the significance of these technologies in today's and next-generation applications.

The main objective of this book is to capture the state of the art in IoT and explore some of its emerging challenges, solutions, and technologies. This peer-reviewed book serves as a reference for researchers, academics, practitioners, and graduate-level students.

Qusay F. Hassan, PhD, is an independent researcher and technology evangelist with over 20 years of practical experience in information communication technology (ICT). Throughout his career, he has held various roles in software development, IT services, and teaching. In his previous position as a systems analyst at the U.S. Agency for International Development (USAID), he actively contributed to the digital transformation initiative, helping adopt new technologies and implement innovative solutions across sectors. He received his PhD in computers and information sciences from Mansoura University, Egypt, in 2015. His technical experience and research interests span diverse areas, including software engineering, distributed systems, and big data. He has authored several publications and reviewed many articles and books in these areas. He is also the editor of several well-received academic books, including two volumes on IoT: *Internet of Things A to Z: Technologies and Applications, Second Edition* (Wiley, 2025) and *Internet of Things: Challenges, Advances, and Applications* (CRC Press, 2017). He is a senior member of IEEE.

Advances in the Internet of Things

Challenges, Solutions, and Emerging Technologies

Edited by
Qusay F. Hassan

CRC Press
Taylor & Francis Group
Boca Raton London New York

CRC Press is an imprint of the
Taylor & Francis Group, an informa business

Designed cover image: Shutterstock_1822762331

First edition published 2026
by CRC Press
2385 NW Executive Center Drive, Suite 320, Boca Raton FL 33431

and by CRC Press
4 Park Square, Milton Park, Abingdon, Oxon, OX14 4RN

CRC Press is an imprint of Taylor & Francis Group, LLC

ISBN: 978-1-032-82840-4 (hbk)
ISBN: 978-1-032-82847-3 (pbk)
ISBN: 978-1-003-50663-8 (ebk)

DOI: 10.1201/9781003506638

Typeset in Times
by codeMantra

In memory of Dongol

You are missed every day.

Contents

PART II Novel Solutions and Applications

PART III Emerging Technologies

Preface

The Internet of Things (IoT) is an emerging computing paradigm that extends connectivity beyond traditional information technology (IT) to everyday objects, such as consumer electronics (e.g., home appliances and personal health monitoring devices), roads and cars, energy meters, and industrial machines. In this model, objects have unique identifiers (mostly IP addresses), can communicate with each other, monitor their environments, and react to changes autonomously or with minimal human intervention. This paves the way for innovations that could improve quality of life, resource monitoring, and industrial processes.

Many reports estimated that in 2024, the number of IoT devices exceeded 18 billion worldwide, with predictions suggesting that it could reach nearly 40 billion by 2033. Despite primarily being consumer devices, a growing number of them will find use in industrial and enterprise applications. This highlights the significance of IoT and how it will be shaping the future. However, to realize its full potential, we must address its emerging challenges and highlight recent applications, advances, and trends, which is the focus of this book.

Security and privacy represent some of the key challenges that IoT adopters face. The growing number of IoT devices, the expansion of Industry 4.0 (and the emergence of Industry 5.0), and the significant increase in cybersecurity attacks exacerbate the severity of these challenges. Considering that ensuring security and privacy is crucial for the successful adoption of IoT, this book dedicates several chapters to these areas. This book also introduces novel models that improve IoT environments and presents several practical implementations that demonstrate some of its real-world applications. Furthermore, it examines several emerging technologies that enable the realization of advanced IoT environments. We see most advances in three main areas: the integration of artificial intelligence/machine learning, network technologies, and hardware design. Therefore, this book dedicates several chapters to these areas. Most chapters touch on artificial intelligence/machine learning, emphasizing the significance of these technologies in today's and next-generation applications. This book also highlights the role of cloud computing and edge computing in enabling IoT.

AUDIENCE AND SCOPE

This book is a result of research efforts made by its contributors in various areas of the domain. The book aims to capture the current state of IoT and explore some of its emerging challenges, solutions, and trends. This peer-reviewed book serves as a reference for researchers, academics, practitioners, and graduate-level students.

ORGANIZATION OF THIS BOOK

This book contains 13 chapters authored by a group of international domain experts and researchers. All chapters include sufficient details, examples, illustrations, necessary background information, and references to make them clear and coherent. The editor and reviewers thoroughly reviewed all chapters to ensure clarity and accuracy. Each chapter went through several review/revision cycles to improve it and present it in the best way possible.

The content of this book is organized in four parts, followed by a glossary where key terms discussed throughout the chapters are defined. Because IoT spans diverse disciplines, readers from diverse areas will find some or all chapters of interest. Below is an overview of the book's parts.

Part I: Concepts, Challenges, and Techniques

Part I addresses some of the current and emerging challenges, as well as techniques utilized in mitigating them. This part contains five chapters, many of which discuss security and cybersecurity-related issues. These chapters emphasize how such pressing issues are threatening modern IoT implementations and present various approaches for making them more safe and secure. This part begins with a broad overview of IoT to cover the basics and pace the reader before moving on to discuss various challenges and mitigation measures. The following is an overview of the chapters:

Chapter 1 introduces the reader to IoT. It begins with a brief overview of its history and common definitions and proceeds by presenting the main architectural and reference models proposed. This chapter also gives an overview of key enabling technologies, main application domains, and adoption challenges.

Chapter 2 presents a review of various IoT security and privacy threats and countermeasures, with a special focus on smart home applications. Throughout its sections, this chapter elaborates on relevant issues and solutions and identifies future challenges and research directions. This chapter discusses several important topics, including fingerprinting, traffic morphing, trigger-action platform safety measures, network intrusion detection systems, and user behavior modeling techniques.

Chapter 3 offers an overview of radio frequency fingerprinting techniques for IoT security. This chapter covers both classical and state-of-the-art deep learning-based approaches and their applications in improving security in IoT devices and networks. It also presents relevant emerging techniques and open challenges.

Chapter 4 surveys several blockchain consensus mechanisms for resource-constrained IoT systems. This chapter focuses on the Proof-of-X, Practical Byzantine Fault Tolerance, Directed Acyclic Graph, and Block Alliance Consensus methods. It highlights how these mechanisms differ in their energy efficiency, scalability, and fault tolerance. It also discusses their potential applications in the IoT. This chapter concludes with a performance assessment of the discussed consensus mechanisms.

Chapter 5 introduces a ransomware threat modeling framework for smart cities. The proposed framework addresses attack surfaces, threat actors, attack vectors, attack analyses, and mitigation strategies. This chapter explores common security threats in smart cities with a special focus on ransomware threats. This chapter uses a smart water management system as an example to demonstrate the application of the proposed framework. Nevertheless, the framework is adaptable and can expand to other applications within the broader context of a smart city. This chapter is a valuable resource for stakeholders involved in implementing and securing industrial IoT environments.

Part II: Novel Solutions and Exemplary Applications

Part II consists of four chapters that present novel and innovative IoT solutions. These chapters exemplify the use of IoT in solving real-world problems and how to integrate machine learning in such environments. The following is an overview of the chapters:

Chapter 6 addresses the integration of machine learning into heterogeneous IoT environments with a specific focus on cyber-physical systems. This chapter proposes a novel framework that considers the hardware specification of every processing node in the network and divides the AI inference model accordingly for distributed learning and execution. It also proposes a blockchain-based framework for secure distribution and execution of these AI models. This chapter discusses in detail the overall system architecture, its constituent parts, and how it works. Moreover, it uses power grids as a case study to explain the relevance of the proposed system in real-world applications. This chapter also presents and discusses relevant simulation results.

Chapter 7 demonstrates the integration of IoT and machine learning in environmental monitoring. This chapter presents a novel, low-cost, energy-efficient framework for large-scale landslide monitoring in remote and rural areas. This chapter discusses in detail the implementation of the proposed framework, covering its layers and components, machine learning algorithms used, data flows, and its notification system. This chapter also presents the testing results acquired from deploying the framework in landslide-prone regions of the Himalayas. Results show the efficiency of the presented framework in predicting slope movements ahead of time and how it enables improved hazard mitigation.

Chapter 8 discusses the use of IoT and machine learning in medical asset tracking and monitoring. Specifically, it focuses on monitoring vital medical supplies (blood, vaccines, and other medicines) during transport. This chapter designs a low-cost, power-efficient system that monitors and predicts the internal temperature and relative humidity of host iceboxes. This chapter provides the technical details of the proposed system and covers its architecture, elements, and development. This chapter also evaluates the system and demonstrates its effectiveness in guaranteeing medical supplies reach their destinations in perfect condition.

Chapter 9 tackles the use of IoT in remote patient monitoring by proposing an affordable, portable, and easy-to-use device for heart rate variability assessment. This chapter provides a detailed discussion of the proposed device, covering its hardware components, software implementation, and evaluation in a controlled testing environment. This chapter explains how the proposed device enables elderly and physically challenged patients, and those who live in remote areas, to keep track of heart rate variability at home.

Part III: Emerging Technologies

Part III offers three chapters on some of the emerging technologies in the broader context of IoT. The presented technologies are diverse, which touch on areas related to machine learning, cloud computing, edge computing, smart radio environments, and hardware architecture. The following is an overview of the chapters:

Chapter 10 reviews split learning and synergetic inference as cutting-edge privacy-preserving techniques for distributed machine learning. This chapter explains in depth how such offloading techniques enable collaboration between edge devices and the cloud. This chapter introduces both concepts and covers their evolution, approaches, applications, and optimization, with a specific focus on split learning. This chapter explores common challenges in such cross-device learning environments, including data heterogeneity, privacy, and security.

Chapter 11 introduces the reader to smart radio technology and its applications for next-generation IoT networks. This chapter covers in great detail its system architecture, hardware components, software enablers, and deployment modes. Moreover, this chapter addresses some open problems associated with the technology and explains how they shape emerging trends and research directions toward its practical realization.

Chapter 12 provides a thorough overview of RISC-V, an emerging open standard instruction set architecture that is based on the Reduced Instruction Set Architecture Computer (RISC) design approach. Its low-power characteristics and customizable features position it well for current and future AI and IoT applications. This chapter begins with a summary of the underlying principles behind the Complex Instruction Set Architecture Computer (CISC) and RISC. It then discusses in detail RISC-V, covering adoption catalysts, various technical aspects, vendors and initiatives, global interest, and challenges faced. This chapter explains how RISC-V is well suited for developing both low-end and high-end computing platforms. This chapter concludes with a case study on the use of RISC-V in building accelerated computing platforms for high-performance data center applications, like those used in processing massive IoT data.

Part IV: Case Study

Part IV is the final part of this book, which consists of a non-technical chapter for policymakers and stakeholders involved in planning and managing national IoT projects. This part targets readers with an interest in the broader context of the technology.

Specifically, Chapter 13 explores India's recent effort in implementing IoT. This chapter highlights major government and private projects, enablers, bottlenecks, emerging trends, and challenges. It also addresses the role IoT and ongoing digital transformation are playing in shaping India's diplomacy with its neighboring countries. This chapter serves as a case study illustrating the typical adoption path of IoT in developing countries. This chapter addresses some of the political and socioeconomic aspects of the story.

Acknowledgments

I would like to thank each and every person who contributed to this project and made the publication of this book possible. In particular, I am grateful to the authors for their time, effort, and patience throughout the preparation of this book. I also appreciate the reviewers, whose invaluable efforts and feedback helped improve the quality of the chapters.

Many thanks to CRC Press for publishing this book. Special thanks go to their editorial team, especially Randi Slack for her outstanding support throughout the various stages of this project. I also extend my appreciation to the Codemantra team for their copyediting work on this book.

I am also thankful to my family and friends for their constant support and encouragement, especially Ahmed Emad ElDean Khalifa, Amany Alqatawy, and Andreia Barbu. Lastly, a special thanks to Sanbar and Zoe for bringing me joy and keeping me company for over a year while I worked on this book.

Qusay F. Hassan
Cairo, Egypt
November, 2025

Contributors

Karolina Baras has been an assistant professor at the University of Madeira since 2012 and a researcher at the Madeira Interactive Technologies Institute since 2015. She holds a PhD degree in Technologies and Information Systems obtained in 2012 from the University of Minho, Portugal. Her research interests are in the field of ubiquitous computing in general and in the areas of the Internet of Things, smart cities, and positive technologies in particular.

Adarsha Bhattarai received his BS degree in Electrical and Electronics Engineering in July 2021. He is currently a PhD student in the Department of Electrical and Computer Engineering at the University of Nebraska–Lincoln. His primary research interests include digital signal processing, the Internet of Things, artificial intelligence, and security in wireless networks. He has co-authored a number of academic research papers in the area of AI applications that have been published in professional journals, such as IEEE and Springer Nature.

Arnav Bhavsar is an associate professor in the School of Computing and Electrical Engineering, Indian Institute of Technology (IIT), Mandi, and the Chair of the Indian Knowledge Systems and Mental Health Applications Center at IIT Mandi. He earned his PhD degree from IIT Madras. With expertise spanning computer vision, medical image analysis, deep learning, and machine learning, he has recently dedicated his focus to EEG signal processing and brain decoding using deep learning techniques, as well as object detection and localization in drone imagery. He has been recognized for his contributions, receiving the DST INSPIRE research grant in 2014 and engaging in collaborative projects with industry partners for translational research. Additionally, he serves as a core faculty member of the Technology Innovation Hub (TIH) at IIT Mandi, overseeing the research and development vertical. A prolific author, Dr. Bhavsar regularly publishes papers in prestigious conferences and journals in the fields of computer vision, medical image analysis, and machine learning. He also contributes as a review editor for *Frontiers in Brain Imaging* and *Frontiers in Radiology* and is a sought-after speaker at workshops and summer schools. Additionally, he is an active member of the IEEE community.

Lina Brito received her PhD degree in Distributed Systems and Networks in 2011 from the University of Madeira (UMa), Portugal. She is an assistant professor at the Exact Sciences and Engineering Faculty in the University of Madeira and a researcher at Madeira Interactive Technologies Institute since 2014. Her research interests are in the areas of networking, network management, wireless sensor networks, and IoT.

Ediz Cetin (MIET, MIEEE) is an associate professor in the School of Engineering, Macquarie University, Sydney, New South Wales (NSW), Australia. Prior to this, he worked at the Australian Centre for Space Engineering Research (ACSER), University of New South Wales (UNSW), Sydney, NSW, Australia, and at the University of Westminster, London, UK. He received his BEng (Hons) degree in Control and Computer Engineering and his PhD degree in Unsupervised Adaptive Signal Processing Techniques for Wireless Receivers, both from the University of Westminster, London, UK. His research interests include radio frequency interference detection, mitigation, and geolocalization; fault-tolerant reconfigurable circuits for space applications; the application of machine learning to radio frequency signals; and the design and low-power implementation of digital circuits. To date, he has authored/co-authored more than 80 technical publications and a book chapter, and he holds two patents in the areas of communications and global navigation satellite system receivers. He is a member of the Institution of Engineering and Technology (IET) and the Institute of Electrical and Electronics Engineers (IEEE). He was the Chair of the IET NSW Local Network as well as the IEEE NSW Circuits and Systems, Solid-State Circuits, and Photonics and Electron Devices joint chapter.

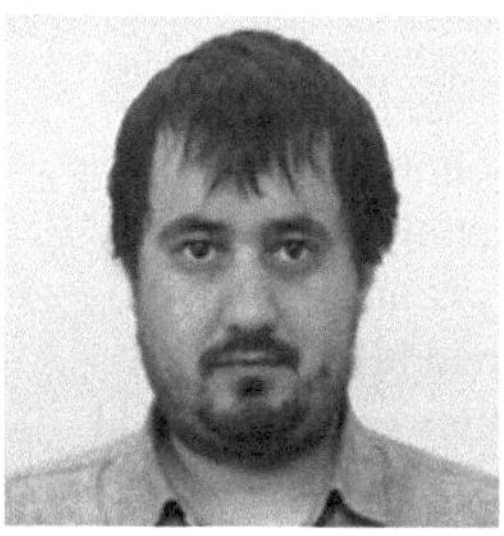

Dimitris Chatzopoulos is an assistant professor in the School of Computer Science at University College Dublin, Ireland. He received his PhD degree in Computer Science and Engineering from the Hong Kong University of Science and Technology and his Diploma and MSc in Computer Engineering and Communications from the University of Thessaly, Greece. He was a visiting researcher at EPFL in 2014 and at Tsinghua University and Cambridge University in 2016. In 2017, he was selected to participate in the 5th Heidelberg Laureate Forum. His research interests include privacy preserving and AI-enabled decentralized applications for mobile and distributed systems.

Varun Dutt works as an associate professor in the School of Computing and Electrical Engineering at the Indian Institute of Technology (IIT) Mandi. He received his PhD in Engineering and Public Policy from Carnegie Mellon University (CMU) in 2011. He also holds several MS degrees from CMU. Before joining IIT Mandi, he was a postdoctoral fellow at the Dynamic Decision-Making Laboratory at Carnegie Mellon University. His research is at the intersection of computer science, economics, and decision-making, focusing on how computational and experimental models could be used for making decisions on managerial, environmental, and socioeconomic issues. He is the associate editor of *Frontiers in Cognitive Science* and a review editor of *Frontiers in Decision Neuroscience*. He is also a senior member of the IEEE.

Akaa Agbaeze Eteng received his BEng degree in Electrical/ Electronic Engineering from the Federal University of Technology Owerri, Nigeria, in 2002; his MEng degree in Telecommunications and Electronics from the University of Port Harcourt, Nigeria, in 2008; and his PhD degree in Electrical Engineering from Universiti Teknologi Malaysia in 2016. He is currently a lecturer in the Department of Electrical/Electronic Engineering, University of Port Harcourt. His research interests include engineered electromagnetic structures, wireless power transfer, electromagnetic energy harvesting, and wirelessly powered communications. He has published research articles on these topics in both international journals and conferences. He has also published several chapters in well-received research and technical books.

Farnaz Farid is working as a lecturer in Cybersecurity and Behavior at Western Sydney University. Prior to that, she worked as an associate lecturer in the School of Computer Science at the University of Sydney. Before moving to Academia, she worked as an IT professional at IBM. She specializes in cybersecurity, network, data science, and web and software systems. She has more than 15 years of research and industry experience in cybersecurity, networking, and distributed systems. She has published 35+ papers in highly esteemed peer-reviewed journals, conferences, and books. Under her successful supervision, students have produced reputable publications. Her current research interest involves various aspects of cybersecurity. She is particularly interested in using artificial intelligence technologies to analyze security, privacy, and usability aspects of the Internet of Things and sensor devices, focusing on data from smart environments, agriculture, water monitoring systems, and healthcare applications.

Aditya Kant Ghising (PhD) is an assistant professor in the Department of Political Science at City College, Kolkata, India. In the recent past, he has served as a guest faculty member in the Department of International Relations at Jadavpur University, India, and in the Faculty of Social Science, Arts and Humanities at Lincoln University College, Malaysia. His research interests include matters pertaining to development, geopolitics, globalization, borders, and borderland studies. Over the years, he has published several academic papers on these themes and has presented his findings in national and international symposia. He has worked as a project assistant for the North-Eastern Development Finance Corporation Limited in Assam, India, with the objective of exploring avenues for investment from Southeast Asia into Northeast India. His teaching areas are international relations, diplomacy, political theory, as well as South and Southeast Asian affairs.

Abbas Husain is currently pursuing his master's degree in the School of Computing and Electrical Engineering at the Indian Institute of Technology, Mandi. He has a bachelor's degree in Computer Science. Before pursuing his master's degree, he worked in the software development industry. His specialty lies in the field of research on slope movement prediction. His areas of interest include IoT and machine learning model development for slope movement prediction.

Yong Jiang (Member, IEEE) received his BS and PhD degrees in Computer Science and Technology from Tsinghua University, Beijing, China, in 1998 and 2002, respectively. He is currently a full professor at the Tsinghua Shenzhen International Graduate School, China. He is also with the Peng Cheng Laboratory in China. His research interests include future network architecture, internet QoS, software-defined networks, and network function virtualization.

Ritu Kapur is a senior project scientist in the Digital Nose Project of the Indian Knowledge System and Mental Health Applications Center (IKSMHA) at the Indian Institute of Technology (IIT) Mandi. She received her PhD degree in Computer Science and Engineering from the Indian Institute of Technology, Ropar, Punjab, India. She has worked as a postdoc and research fellow at the University of Sannio, Italy, and the University of Stuttgart, Germany, respectively. Her current research is focused on Digital Nose for Healthcare, DevOps for cyberphysical systems, software maintenance and evolution, empirical software engineering, and mining software repositories. She has received the DST Accelerate Vigyan Vritika grant, the Microsoft research travel grant, and the ACM CAPS Sigsoft grant for her research. She has served as a program committee member of international conferences in the software engineering field. She serves as a reviewer for software engineering journals such as the *Transactions of Software Engineering and Methodology* (TOSEM), *Transactions of Cloud Computing* (TCC), *Empirical Software Engineering* (EMSE), *Information and Software Technology* (IST), and *Journal of Systems and Software* (JSS).

Akshay Kumar is currently a project associate on the Landslide Monitoring and Early Warning System Development project at IIT, Mandi. He received his master's degree in Computer Applications. Prior to working for IIT, he worked as a software developer and data engineer. His expertise lies in remote sensing techniques, and he has a passion for developing cloud-based early warning systems for landslide-prone areas.

Praveen Kumar is a PhD student in the School of Computing and Electrical Engineering at the Indian Institute of Technology, Mandi. He received his master's degree (by research) in Machine Learning. He also worked in the software development industry. He has published papers and articles in international publications and conferences. His research interests include IoT and early warning and monitoring systems, with a focus on landslides. His expertise also includes developing advanced machine learning models for slope movement prediction.

Yashwant Kumar is a PhD scholar in the IKSMHA Center at the Indian Institute of Technology Mandi. He is developing portable devices for remote health monitoring and management. He earned his bachelor's degree in Electronics and Communication Engineering from the Institution of Engineers (India), Kolkata. His current research is focused on remote HRV assessment for healthcare, biomedical sensing, DevOps, and human-computer interaction.

Bo Li (Member, IEEE) received his BS degree (2008) and MS degree (2010) in Telecommunications Engineering from Xidian University, Xi'an, China, and PhD degree (2023) in Electronics Engineering from Macquarie University, Sydney, Australia. From 2010 to 2016, he was an engineer at the Shanghai Institute of Satellite Engineering, Shanghai, China. From 2022 to 2023, he was with the University of New South Wales (UNSW), Sydney, NSW, Australia. He is currently a postdoctoral research fellow with the School of Engineering, Macquarie University, Sydney. His research interests include cognitive radio, IoT, satellite communications, and information theory, with current emphasis on deep learning for wireless device identification.

Keqin Li received his BS degree in Computer Science from Tsinghua University in 1985 and his PhD degree in Computer Science from the University of Houston in 1990. He is currently a SUNY Distinguished Professor at the State University of New York and also a National Distinguished Professor at Hunan University (China). He has authored or co-authored more than 910 journal articles, book chapters, and refereed conference papers. He received several best paper awards from international conferences, including PDPTA-1996, NAECON-1997IPDPS-2000, ISPA-2016, NPC-2019, ISPA-2019, and CPSCom-2022. He holds nearly 70 patents announced or authorized by the Chinese National Intellectual Property Administration. He is among the world's top five most influential scientists in parallel and distributed computing in terms of both single-year impact and career-long impact, based on a composite indicator of the Scopus citation database. He was a 2017 recipient of the Albert Nelson Marquis Lifetime Achievement Award for being listed in Marquis Who's Who in Science and Engineering, Who's Who in America, Who's Who in the World, and Who's Who in American Education for more than 20 consecutive years. He received the Distinguished Alumnus Award from the Computer Science Department at the University of Houston in 2018. He received the IEEE TCCLD Research Impact Award from the IEEE Technical Committee on Cloud Computing in 2022. He is an AAAS Fellow, an IEEE Fellow, and an AAIA Fellow. He is a member of the SUNY Distinguished Academy. He is a member of Academia Europaea (Academician of the Academy of Europe).

Qing Li (Senior Member, IEEE) received his BS degree in Computer Science and Technology from the Dalian University of Technology, Dalian, China, in 2008, and his PhD degree in Computer Science and Technology from Tsinghua University, Beijing.

Ruoyu Li received his BS degree in Cybersecurity from Huazhong University of Science and Technology, Wuhan, China, in 2017 and his MS degree in Computer Science from Columbia University, New York, USA, in 2019. He is currently a PhD candidate with Tsinghua-Berkeley Shenzhen Institute, Tsinghua University, and also a research intern at Peng Cheng Laboratory, Shenzhen, China. His research interests mainly include AI-assisted network security, IoT security, intrusion/anomaly detection, and in-network security applications.

Yutong Liu is a professor of Radiology at the University of Nebraska Medical Center, specializing in advanced neuroimaging techniques. With a strong focus on developing non-invasive imaging biomarkers, Dr. Liu's research aims to enhance the diagnosis, monitoring, and treatment of neurodegenerative disorders, strokes, and brain tumors. Integrating artificial intelligence (AI) into his work, Dr. Liu leverages AI algorithms to accelerate MRI data acquisition and improve the precision of image data analysis. His innovative approach combines AI with MRI to provide deeper insights into brain metabolism and function, advancing our understanding of neural mechanisms and improving clinical outcomes. He has published extensively in top-tier journals, highlighting his contributions to both the fields of radiology and AI.

Jianhui Lv received his BS degree in Mathematics and Applied Mathematics from the Jilin Institute of Chemical Technology, Jilin, China, in 2012, and his MS and PhD degrees in Computer Science from Northeastern University, Shenyang, China, in 2014 and 2017, respectively. He worked at the Network Technology Lab, Central Research Institute, Huawei Technologies Co. Ltd., Shenzhen, China, as a senior engineer from January 2018 to July 2019. He worked at Tsinghua University as an assistant professor from August 2019 to July 2021. He is currently working as an associate professor at the Pengcheng Laboratory in China. His research interests include computer networks, artificial intelligence, ICN, IoT, bio-inspired networking, evolutionary computation, cloud/edge computing, smart city, healthcare, etc. He has published more than 80 high-quality journal papers and conference papers. He has served as a leader guest editor (LGE) in several international journals (such as *Applied Soft Computing*, *Digital Communications and Networks*, *Expert Systems*, *Wireless Networks*, *International Journal on Artificial Intelligence Tools*, *Mobile Information Systems*, and *Internet Technology Letters*) and as a guest editor in *IEEE TCE*. He also serves as the associate editor of *Internet Technology Letters* (indexed by EI and ESCI), *Journal of Multimedia Information System* (indexed by EI and ESCI), and *International Journal of Swarm Intelligence Research* (indexed by EI and ESCI).

Dongming Peng received his PhD degree in Computer Engineering from Texas A&M University. He is an associate professor in the Electrical and Computer Engineering Department, University of Nebraska–Lincoln. He has published over 100 technical articles in the research areas of digital image processing, multimedia security, wireless sensor networks, cybersecurity, computer architecture, and related topics. He has conducted various research projects in these areas, with special contributions to energy-efficient wireless and secure communications in multimedia sensor networks.

Premchand is an MTech (research) student in the School of Computing and Electrical Engineering (SCEE) at IIT Mandi. He holds a BTech degree in Electronic Instrumentation and Control Engineering. He is currently working as a project associate on the project titled "Developing an AI-Based Cognitive Function Enhancement App by Applying Artificial Intelligence," funded by iHub and HCIF-IIT Mandi. His current research focuses on wearable health monitoring devices.

Priyanka is an MS (by research) student in the School of Computing and Electrical Engineering at the Indian Institute of Technology Mandi. She received her bachelor's degree in Electronics and Communication Engineering. She has an interest in using machine learning for slope movement predictions, communication systems, and the development of IoT devices. Besides, she works as an expert who designs electronic circuits for IoT hardware.

Yan Qiang received his PhD degree in Computer Application Technology from the Taiyuan University of Technology in Taiyuan, China, in 2010. He is currently a professor at the College of Information and Computer, Taiyuan University of Technology. He has authored and co-authored more than 70 professional papers. His research interests include medical image processing and big data.

Surbhi Ranga is an MTech (R) student in the Center for Human-Computer Interaction (CHCI) department at the Indian Institute of Technology Mandi (IIT Mandi). She received a BTech degree in Electrical Engineering from Engineering College, Bikaner. Her current research focuses on the IoT-MediGuard system, which is designed to enhance the reliability and safety of medical supply delivery through IoT technologies and advanced machine-learning models. In addition to her academic pursuits, Surbhi is also a project associate at the Indian Knowledge System and Mental Health Applications (IKSMHA) Center, where she contributes to the IoT-MediGuard project.

Assim Sagahyroon has over 30 years of experience in academia as a faculty and administrator. He is currently the Associate Dean of Undergraduate Affairs in the College of Engineering at the American University of Sharjah, a position he has held since 2019. He also served as the Head of the Computer Science and Engineering Department from 2009 to 2016. Under his supervision, departments were able to secure multiple accreditations by CAA and ABET for their different programs. He oversaw the recent revamp of the college curricula, as well as the development of three new undergraduate programs to be launched in Fall 2024. He participated in numerous university, college, and departmental activities and committees. Professor Sagahyroon received his BSc degree in Electrical Engineering from the University of Khartoum, his master's degree from Northwestern University, and his PhD degree from the University of Arizona with a computer engineering emphasis. Before joining AUS, he held academic posts with Northern Arizona University and California State University, where he was awarded tenure in both affiliations. He was an invited technical reviewer for National Science Foundation programs and served in technical program committees of many international conferences. He has many publications in international conferences and journals. His research interests include innovative applications of emerging technology in the medical field, power consumption of portable electronics, and FPGA-based design.

Chathumi Samaraweera received her BSc degree specializing in Engineering Physics in December 2016. Also, she received her MSc degree in Electrical and Electronic Engineering in July 2022. She is currently a PhD student in the Department of Electrical and Computer Engineering at the University of Nebraska–Lincoln. Her primary research interests include embedded system designs, the Internet of Things, artificial intelligence, and security in wireless sensor networks.

Sahil Sankhyan is a PhD scholar at the Indian Institute of Technology Mandi, specializing in IoT device design. With a master's degree in Electronic Product Design Technology, his research integrates modern IoT technologies and machine learning models for slope movement prediction. He is also interested in traditional techniques like Vrikshaurveda (the science of plant life) and how they affect landslides. His work aims to enhance environmental sustainability and disaster management practices.

Hamid Sharif (Fellow, IEEE) is currently a Charles J. Vranek Professor with the Department of Electrical and Computer Engineering, University of Nebraska–Lincoln (UNL), Lincoln, NE, USA. He is also the Director of the Advanced Telecommunication Engineering Laboratory (TEL), UNL. He has authored or co-authored over 400 research papers in national and international journals and conferences and has been serving on many IEEE and other international journals' editorial boards. He has received a number of research awards and best articles. He is a distinguished lecturer for the IEEE Vehicular Technology Society.

Shivani Sharma is an MTech (research) student in the Indian Knowledge System and Mental Health Applications (IKSMHA) Center at the Indian Institute of Technology (IIT) Mandi. She also works as a project associate-I in the Non-Invasive Glucometer for Diabetes Prediction through Breath Samples Project of the IKSMHA. She completed her BTech in ECE from Jawaharlal Nehru Government Engineering College (JNGEC) Sundernagar, Himachal Pradesh, India. Her current research is focused on the Non-Invasive Glucometer for Diabetes Prediction through Breath Samples and AI Perception Study for Healthcare.

Joana Tirana is a PhD student in the School of Computer Science at University College Dublin. She received her diploma (MSc equivalent) in Electrical and Computer Engineering from the University of Thessaly in Volos, Greece, in 2021. She started her PhD in January 2022 at University College Dublin. Her research area is decentralized AI, focusing on protocols for distributed and collaborative learning like federated learning and finding ways to optimize these techniques by using AI task offloading and split learning.

Kala Venkata Uday is an associate professor with the School of Civil and Environmental Engineering at the Indian Institute of Technology Mandi. He earbned his PhD degree from the Indian Institute of Technology, Bombay, in Geotechnical and Geoenvironmental Engineering. His research interests include geotechnical engineering, slope stabilization, structure retention, environmental geotechnics, geosynthetics, and landslides (early notice, planning, and mitigation).

Tushar Vasudev is an MTech research scholar and researcher at IIT Mandi's School of Mechanical and Material Engineering (SMME), with a focus on intelligent systems and research related to machine learning applications and IoT technologies. He earned his bachelor's degree in Electrical and Communication Engineering from the Manipal Institute of Technology, Sikkim. His current research centers on developing intelligent systems for health monitoring and predictive analytics, with projects like the IoT Smart MediGuard (a smart blood vaccine and medicine monitoring system).

Yingsen Wang received his BS degree in 2019 and PhD degree in 2024 from Taiyuan University of Technology, Taiyuan, China. He is currently employed by the China Electric Power Research Institute Co. Ltd. His research interests are blockchain and distributed consensus mechanisms. He has published many papers in *IEEE Transactions on Vehicular Technology* and *IEEE Transactions on Industry Applications* and served as a reviewer for *IEEE Transactions on Smart Grid*, *IEEE Transactions on Network Science and Engineering*, and other journals. He has published a monograph as the chief editor.

Kaidong Wu received his BS degree in Software Engineering from Jilin University, Changchun, China, in 2023. He is pursuing an MS degree in Computer Technology at Tsinghua Shenzhen International Graduate School, China. His research interests include traffic analysis, AI-assisted network security, and IoT security.

Jingyu Xiao received his BS degree from Wuhan University, Wuhan, China, in 2021. He is currently pursuing his MS degree at the Tsinghua Shenzhen International Graduate School, China. His research interests include anomaly detection, smart homes, and AI for networks.

Dan Zhao received her bachelor's degree in Telecommunications from the Beijing University of Posts and Telecommunications in 2011 and her PhD degree in Electronic Engineering from the Queen Mary University of London in 2015. She was a postdoctoral researcher with the School of Electronic Engineering, Dublin City University. She is currently an assistant researcher at the Peng Cheng Laboratory in China. Her research interests include IoT security, anomaly detection, and in-network intelligence.

Juanjuan Zhao received her PhD degree in Computer Application Technology from the Taiyuan University of Technology, Taiyuan, China, in 2010. She is currently a professor at the School of Computer Science and Technology, Taiyuan University of Technology. Her research interests include medical image processing and deep learning.

Jun Zhao received her BS degree in 2019 and MS degree in 2022 from Taiyuan University of Technology, Taiyuan, China. She is currently pursuing her PhD degree at the College of Computer Science and Technology, Taiyuan University of Technology, Taiyuan, China. Her research interests are industrial vision and image processing.

Qingsong Zou received his BE degree from Southeast University, China, in 2021. He is currently pursuing his PhD degree in Computer Science and Technology at the Tsinghua Shenzhen International Graduate School, China. His research interests include IoT security, network intrusion detection, and encrypted traffic analysis.

Reviewers

John Byabazaire, PhD, University College Dublin, Ireland

Farnaz Farid, PhD, Western Sydney University, Australia

Bernard Fong, PhD, Providence University, Taiwan

Guan Yue Hong, PhD, Western Michigan University, USA

Ahmed Emad ElDean Khalifa, Egypt

Axel Kloth, Abacus Semiconductor Corporation, USA

Dongming Peng, PhD, University of Nebraska - Lincoln, USA

Bruno Miguel Correia Silva, PhD, Instituto de Telecomunicações, University of Beira Interior, Portugal

Jan Sliwa, Bern University of Applied Sciences (Emeritus), Switzerland

Part I

Concepts, Challenges, and Techniques

1 Internet of Things
Architectures, Technologies, Applications, and Challenges[1]

Karolina Baras, Lina Brito and Qusay F. Hassan

1.1 INTRODUCTION

Back in 1989, there were around 100,000 hosts connected to the Internet (Zakon, 2016), and the World Wide Web (WWW) came to life a year later at CERN with the first and only site at the time.[2] Ten years after Tom Berners-Lee unleashed the WWW, a whole new world of possibilities started to emerge when Kevin Ashton, from the Massachusetts Institute of Technology's (MIT) Auto-ID Labs, coined the term *Internet of Things* (IoT) (Ashton, 2009). In the same year, Neil Gershenfeld published his work on things that think, where he envisioned the evolution of the WWW as "things start to use the Net so that people don't need to" (Gershenfeld, 1999). Simultaneously, in Xerox PARC Laboratories in Palo Alto, California, the so-called third era of modern computing was dawning, with Mark Weiser introducing the concept of "ubiquitous computing" in his paper published in *Scientific American* (Weiser, 1991). Tabs, pads, and boards were proposed as the essential building blocks for the computing of the future. Wireless networking and seamless access to shared resources would make the user experience with technology as enjoyable as "a walk in the woods."

In 1999, the number of hosts exceeded 2 million and the number of sites jumped to 4 million (Zakon, 2016). The Institute of Electrical and Electronics Engineers (IEEE) standard 802.11b (Wi-Fi) had just been published, with transmission rates of 11 Mbits/s. GSM was growing fast, but the phones were not at all smart yet. They only allowed making phone calls and sending short messages. GPS signals for civil usage were still degraded with selective availability, and the receivers were heavy, huge, and expensive. The area of wireless sensor networks (WSNs) also emerged in the 1990s with the concept of "smart dust," i.e., a huge amount of tiny devices scattered around an area capable of sensing, recording, and communicating sensed data wirelessly.

At the dawn of the eagerly anticipated twenty-first century, technological growth has accelerated at an unprecedented pace. Although the reports published 10 and 20 years after Weiser's vision showed that not everything turned out to be just as he had imagined, significant changes were introduced in the way we use technology and live with it. Our habits changed, our interaction with technology changed, and the way we grow, play, study, work, and communicate also changed.

In 2005, the International Telecommunications Union (ITU) published its first report on the IoT, noting that

> Machine-to-machine communications and person-to-computer communications will be extended to things, from everyday household objects to sensors monitoring the movement of the Golden Gate Bridge or detecting earth tremors. Everything from tyres to toothbrushes will fall within communications range, heralding the dawn of a new era, one in which today's internet (of data and people) gives way to tomorrow's Internet of Things.
>
> *(ITU-T, 2005)*

DOI: 10.1201/9781003506638-2

Three years later, in 2008, the number of devices connected to the Internet outnumbered the world's population for the first time. The introduction of Internet Protocol version 6 (IPv6)[3] resolved the problem of the exhaustion of IP addresses, which was imminent near the end of the twentieth century. The first international conference on IoT[4] took place in March 2008 to gather industry and academia experts to share their knowledge, experience, and ideas on this emerging concept. In the following years, the number of IoT-related events and conferences grew enormously.

Open-source electronics like Arduino,[5] which reached the market between 2005 and 2008, gave birth to millions of new ideas and projects for home and office automation, education, and leisure. Other examples of single-board computers (SBCs) followed: Raspberry Pi,[6] BeagleBone Black,[7] and so on. Today, one can buy a dozen tiny but fairly powerful computers for less than $100 each, connect them to the Internet and to a plethora of sensors and actuators, collect and analyze gigabytes of data, and make interesting home or office automation projects with real-time visualizations of information generated from the data on the go. Alternatively, one can use remote networks of intelligent devices deployed somewhere else, for example, OneLab.[8]

In 2009,[9] the Commission of the European Communities published a report on the IoT action plan for Europe showing that the IoT had reached a very high level of importance among European politicians, commercial and industry partners, and researchers. Several global standard initiatives were created in recent years to discuss and define IoT-related issues and establish global agreements on standard technologies to be deployed in IoT projects. For example, oneM2M[10] was created in 2012 as a global standard initiative that covers machine-to-machine and IoT technologies, which go from requirements, architecture, and application programming interface (API) specifications to security solutions and interoperability issues.

In 2015, the European Commission created the Alliance for the Internet of Things Innovation, now known as the Alliance for the Internet of Things Edge Computing Innovation (AIOTI)[11] to foster interaction and collaboration between IoT stakeholders. The convergence of cloud computing, the miniaturization and lower cost of sensors and microcontrollers, and the omnipresence of digital connectivity all contributed to making the IoT a reality for years to come. In fact, some sources consider that the four pillars of digital transformation are cloud, mobility, big data, and social networking, and that IoT is based on these.

Statista forecasts that by 2033 there will be over 39 billion "things" connected to the Internet.[12] This number excludes PCs, smartphones, and tablets. Now that the IoT is finally becoming a reality, there is a need for a global understanding of its definition, reference architecture (RA), requirements, and standards. In the following sections, an overview of the current IoT landscape will be given and some of the proposals that are on the table for discussion in several groups, alliances, and consortia focused on IoT will be highlighted. There is at least an agreement on some of the requirements that need to be addressed, but still there is space for improvement and even more collaboration among the stakeholders. For example, unique device identification, system modularity, security, privacy, and low cost are some of the issues that need further discussion and action.

This chapter covers the fundamentals of IoT, its enabling technologies, application domains, and the main challenges that still need to be surpassed.

1.2 DEFINITION OF IOT

There have been several international organizations and research centers involved in the creation of common standards for the IoT. One of the first steps in this process has been to find a common definition. The first definitions of the IoT were tightly coupled to the radio-frequency identification (RFID)-related context of the Auto-ID Labs at MIT, where the term first emerged. As the concept became universal, the definition started to evolve into more general terms. For Kevin Ashton, the meaning of the IoT and the consequences of its implementation in our environments are the following (Ashton, 2009):

> If we had computers that knew everything there was to know about things – using data they gathered without any help from us – we would be able to track and count everything and greatly reduce waste, loss and cost. We would know when things needed replacing, repairing, or recalling, and whether they were fresh or past their best. We need to empower computers with their own means of gathering information, so they can see, hear and smell the world for themselves.

Another definition of the IoT is the following (Atzori et al., 2010):

> The basic idea of this concept is the pervasive presence around us of a variety of things or objects – such as Radio-Frequency IDentification (RFID) tags, sensors, actuators, mobile phones, etc. – which, through unique addressing schemes, are able to interact with each other and cooperate with their neighbors to reach common goals.

The Study Group 20 (SG 20) was created in 2015 as a result of the ten-year experience period that followed the publication of the first ITU report on IoT in 2005 and the findings of the International Telecommunications Union Telecommunication Standardization Sector (ITU-T) Focus Group on Smart Sustainable Cities, which ceased to exist in 2015. In the SG 20 recommendation document Y.2060 (ITU-T, 2012), the following definition is given:

> Internet of things (IoT): A global infrastructure for the information society, enabling advanced services by interconnecting (physical and virtual) things based on existing and evolving interoperable information and communication technologies.
>
> NOTE 1 – Through the exploitation of identification, data capture, processing and communication capabilities, the IoT makes full use of things to offer services to all kinds of applications, whilst ensuring that security and privacy requirements are fulfilled.
>
> NOTE 2 – From a broader perspective, the IoT can be perceived as a vision with technological and societal implications.

The ITU-T document goes on to explain that IoT adds a new dimension ("any thing") to the already existing "any time" and "any place" communication provided by the digital connectivity expansion. In this context, "things" are defined as being physical or virtual identified objects capable of communicating. Physical objects are all kinds of everyday objects that are present in our environments and that can contain sensors, actuators, and communication capabilities. Examples of physical objects are electronic appliances, industrial machinery, and digitally enhanced everyday objects. Virtual objects exist in the information world and can be stored, accessed, and processed. Software is an example of a virtual object.

The European Union (EU) created the IoT European Research Cluster (IERC) as a platform for FP7 (7th Framework Programme for Research and Technological Development) projects on IoT. Currently, the IERC is Working Group 1 (WG 1) of the AIOTI, which was created to establish collaboration and communication between different entities involved in IoT development, standardization, and implementation.

The definition published on the IERC website[13] states that the IoT is

> A dynamic global network infrastructure with self-configuring capabilities based on standard and interoperable communication protocols where physical and virtual "things" have identities, physical attributes, and virtual personalities and use intelligent interfaces, and are seamlessly integrated into the information network.

The International Organization for Standardization/International Electrotechnical Commission (ISO/IEC) Joint Technical Committee 1 (JTC1) was created in 1987 and is responsible for standard development in the information technology (IT) area, having so far published more than 3,000 standards.[14] WG 10 (former SWG 5) is one of JTC1 working groups responsible for IoT-related issues.

In one of the SWG 5 reports published in 2015, the adopted definition for IoT is given in the following terms[15]:

> An infrastructure of interconnected objects, people, systems and information resources together with intelligent services to allow them to process information of the physical and the virtual world and react.

In the Request for Comments (RFC) 7452,[16] which talks about the architectures for networks of smart objects, IoT is defined as follows:

> The term "Internet of Things" (IoT) denotes a trend where a large number of embedded devices employ communication services offered by Internet protocols. Many of these devices, often called "smart objects," are not directly operated by humans but exist as components in buildings or vehicles, or are spread out in the environment.

The IEEE IoT initiative published a document (Minerva et al., 2015) with an overview of the IoT applications and a proposal of a definition in order to start a discussion and to give its community members an opportunity to contribute to the definition of the IoT.[17] The document presents two definitions, one for small-scale scenarios:

> An IoT is a network that connects uniquely identifiable "Things" to the Internet. The "Things" have sensing/actuation and potential programmability capabilities. Through the exploitation of unique identification and sensing, information about the "Thing" can be collected and the state of the "Thing" can be changed from anywhere, anytime, by anything.

The other is for large-scale scenarios:

> Internet of Things envisions a self-configuring, adaptive, complex network that interconnects "things" to the Internet through the use of standard communication protocols. The interconnected things have physical or virtual representation in the digital world, sensing/actuation capability, a programmability feature and are uniquely identifiable. The representation contains information including the thing's identity, status, location or any other business, social or privately relevant information. The things offer services, with or without human intervention, through the exploitation of unique identification, data capture and communication, and actuation capability. The service is exploited through the use of intelligent interfaces and is made available anywhere, anytime, and for anything taking security into consideration.

A fairly complete collection of IoT definitions can be found in Minoli (2013), in which the definitions are organized into two categories: those that define IoT as a concept and those that define IoT as an infrastructure. The following definition tries to encompass both the concept and the infrastructure and can be found in the above-cited reference:

> A broadly-deployed aggregate computing/communication application and/or application-consumption system that is deployed over a local (L-IoT), metropolitan (M-IoT), regional (R-IoT), national (N-IoT), or global (G-IoT) geography, consisting of (i) dispersed instrumented objects ("things") with embedded one- or two-way communications and some (or, at times, no) computing capabilities, (ii) where objects are reachable over a variety of wireless or wired local area and/or wide area networks, and (iii) whose inbound data and/or outbound commands are pipelined to or issued by a(n application) system with a (high) degree of (human or computer-based) intelligence.

Although the wording may be slightly different, it seems that there are several touching points among the definitions. For example, the IoT is made of (physical and virtual) objects that are uniquely identifiable, that are able to capture their context (sensors), and that are able to transmit and/or receive data over the Internet and, in the case of actuators, are able to change their own state or the state of their surroundings – all of which should ideally be done without or with very little direct human intervention.

1.3 PROPOSED ARCHITECTURES AND REFERENCE MODELS

The end goal of IoT systems is to achieve a synergy between different systems, meaning that they should interoperate and communicate automatically to provide innovative services to the users. Therefore, standardization is needed to ensure that IoT platforms will allow distinct systems to reliably interoperate.

While it is expected that the IoT will positively revolutionize all the different sectors of the economy in society, it will also produce a very large amount of data. This not only brings new challenges regarding the management, processing, and transmission of data, but above all, it also brings new concerns regarding data security. So, on top of standardization for interoperability, security standards are also needed to protect the individuals, businesses, and governments that will use the IoT systems.

Recently, several attempts were made – and are still being made – to develop RAs by either standards organizations or industries and universities. However, standardization is difficult to achieve in the real world. RAs are vital for standardization, as they define guidelines that can be used when planning the implementation of an IoT system (Weyrich and Ebert, 2016).

Therefore, with the intention of making interoperability between different IoT systems possible, several attempts have been made in recent years to create reference layered models for IoT (Bassi et al., 2013; Bauer et al., 2013; Gubbi et al., 2013). Several standards development organizations (SDOs) are also engaged in this process, as will be described below (Jakobs, 2018).

oneM2M specifications focus on the creation of a framework to support applications and services, such as smart grids, connected cars, home automation, public safety, and health. During a workshop organized by the European Telecommunications Standards Institute (ETSI), which took place in November 2016, the European Commission highlighted the need for an open common RA for IoT, enabling the integration of different services, for the specific case of smart city applications. In fact, this is of critical importance not only to smart cities but also to all areas of application of IoT technologies. In addition to oneM2M standardization activities, ETSI has also created a working group on sustainable digital multiservice cities, specifically for the case of smart city projects (Antipolis, 2016).

The IEEE has produced numerous standards[18] that relate to several areas of IoT systems and has many ongoing projects to develop new standards also related to the IoT. Among all the standards, projects, and events promoted by IEEE, two important initiatives need to be emphasized: the IEEE Standards Association (IEEE-SA) engaged participants in key regions of the world to create the IoT Ecosystem Study, which encompasses three main areas – market, technology, and standards – but also examines the role of academia and research and the importance of user acceptance; and the IEEE P2413 Working Group is focusing on the creation of a standard architecture for IoT, the "IoT Architecture." The resulting draft standard basically defines an architectural framework for the IoT: it describes different IoT domains, gives definitions of IoT domain abstractions, and identifies commonalities between different IoT domains.[19]

The GSM Association (GSMA)[20] has gathered nearly 800 mobile operators and 300 companies worldwide to address four areas of the mobile industry: "Personal Data (enabling trust through digital identity), Connected Living (bringing the IoT to life), Network 2020 (the future of mobile communications), and Digital Commerce (streamlining interactions and transactions)" (GSMA, 2016). In August 2015, the GSMA established a new project named Mobile IoT Initiative,[21] supported by a group of 26 of the world's leading mobile operators, equipment manufacturers, and module and infrastructure companies, to address the use of low-power wide area (LPWA) solutions in the licensed spectrum.

The GSMA Connected Living Programme (LP) is working with mobile operators to speed up the delivery of IoT solutions that exploit connectivity in innovative ways. In February 2016, the Connected LP also published new guidelines designed to promote the secure development and deployment of services in the IoT market. The result is the document entitled "GSMA IoT Security Guidelines," developed in conjunction with the mobile industry, which offers IoT service providers practical recommendations on handling common security and data privacy threats associated with IoT services (GSMA, 2016).

For 2017, the GSMA Connected LP focuses on four new goals: (1) Mobile IoT, which mainly addresses increasing market awareness and support for licensed spectrum LPWA solutions; (2) completing the technical specification of the Consumer Remote SIM Provisioning; (3) positioning operators as key partners within the IoT big data market through the delivery of data sets and APIs; and (4) supporting operators in the provision of services that enable smart cities (GSMA, 2016). It is important to be aware of these goals to get an idea of the directions in which efforts are being made and of what is happening at the moment in the area.

ITU also created the Internet of Things Global Standards Initiative (IoT-GSI),[22] which worked in detailing the requirements for developing the standards that are necessary to enable the deployment of IoT on a global scale, taking into account the work done by other SDOs. In July 2015, this group decided to create the SG 20, which focuses on "IoT and its applications, including smart cities and communities." Therefore, all activities conducted by the IoT-GSI were transferred to SG 20,[23] which has produced around hundreds of related documents.

Both Industrial Internet Reference Architecture (IIRA)[24] and Reference Architecture Model for Industrie 4.0 (RAMI 4.0) were developed, focusing on taking advantage of IoT to increase the efficiency of the industrial processes, either improving manufacturing itself or making the supply chain from the suppliers to the customers more effective.

Sensor Network Reference Architecture (SNRA),[25] in turn, provides a general overview of the characteristics of a sensor network and the organization of the entities that comprise such a network. It also describes the general requirements that are identified for sensor networks, which relate to IoT systems since sensor networks are used by IoT systems as a tool for collecting data. A working group created by ISO/IEC,[26] involving industry and commerce, academic and research bodies, and government, is working on the development of an RA for IoT (IoT RA) that aims to describe the characteristics and aspects of IoT systems, define the IoT domains, describe the RM of IoT systems, and describe the interoperability of IoT entities.[27] IoT RA intends to become the common reference for all the RAs already proposed by other organizations, including Internet of Things – Architecture (IoT-A) and the standards developed by ITU (WG 10, 2015).

Since there is no universally accepted definition of IoT, different groups have developed different approaches according to the domain in which they are active (ISO/IEC, 2014).

Additionally, GS1,[28] a nonprofit organization working in the area of barcoding standardization, claims that the evolution of standards for IoT has followed a path where no standards existed to a situation where too many standards are available, leading to difficult choices to be made when designing IoT applications (ISO/IEC, 2014). These difficulties may be accentuated by manufacturers who try to protect their products and solutions and are not necessarily interested in adopting open standards or ensuring interoperable solutions.

In 2013, a consortium involving industry and university partners, like Alcatel-Lucent, IBM, NEC Siemens, Sapienza University of Rome, and the University of Surrey, created an architectural reference model (ARM) for IoT, named IoT-A (Bassi et al., 2013). For the partners, achieving interoperability between solutions across various platforms could only be ensured through interoperability both at the communication level and at the service level. At the end of the project, funded by the EU, the benefits of the developed architecture were demonstrated through the implementation of real-life use cases.

In the following sections, some of the more relevant RAs are briefly described.

1.3.1 IoT-A

The IoT-A project is no longer active. However, IoT-A is described here since it is being used as a basis for developing other architectures, such as the IoT RA or Reference IoT Layered Architecture (RILA), which are also discussed in the following sections.

The IoT-A ARM was created in order to achieve interoperability between different IoT systems (Bassi et al., 2013). The IoT ARM is defined to be abstract so that it can be used as a reference for generating concrete system architectures. It consists of an RM and an RA.

The RM, presented in Figure 1.1, provides a common understanding of the IoT domain by modeling its concepts and their relationships. Similar to the Open Systems Interconnection (OSI) model, the IoT RM by itself does not specify the technical particularities of an IoT system.

The *domain model* considers a top-level description of the concepts and entities (physical entities, devices, resources, and services) that represent particular aspects of the IoT domain and defines their relations. Therefore, the domain model can also be used as a taxonomy of the IoT.

The *information model* specifies the data semantics of the domain model. That is, it refers to the knowledge and behavior of the entities considered in the domain model since they are responsible for either keeping track of certain information or performing specific tasks (it describes which type of information the entities are responsible for).

The *communication model*, in turn, addresses the main communication paradigms necessary for connecting entities, ensuring interoperability between heterogeneous networks. The proposed communication model is structured in a seven-layer stack and describes how communication has to be managed by each layer in order to achieve the interoperability features required in the IoT. It also describes the actors (communicating elements) and the channel model for communication in IoT.

The RA of IoT-A mainly consists of "views" and "perspectives," which vary depending on the requirements of each specific application. Figure 1.2 illustrates that the perspectives "evolution and interoperability," "performance and scalability," "trust, security, and privacy," and "availability

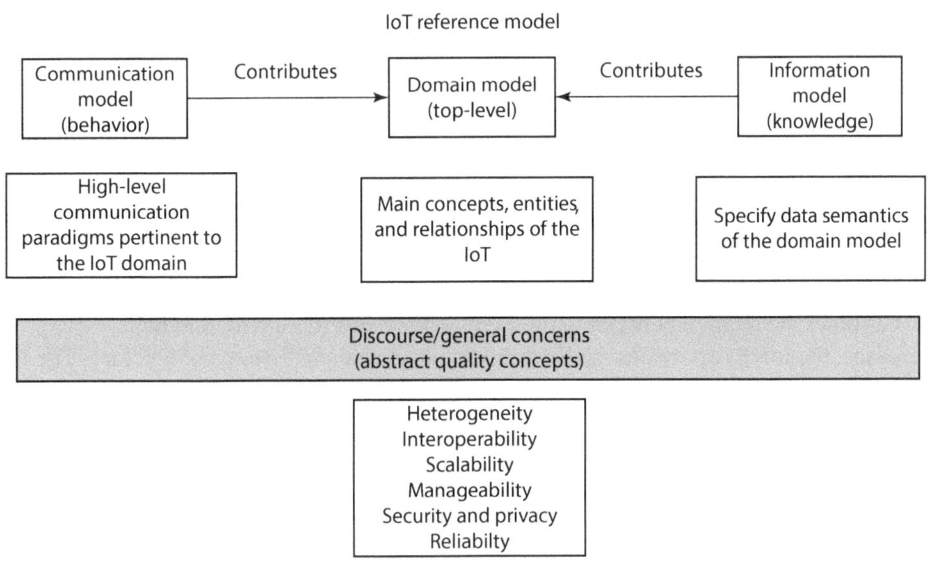

FIGURE 1.1 RM Proposed by IoT-A (Adapted from Bassi et al., 2013)

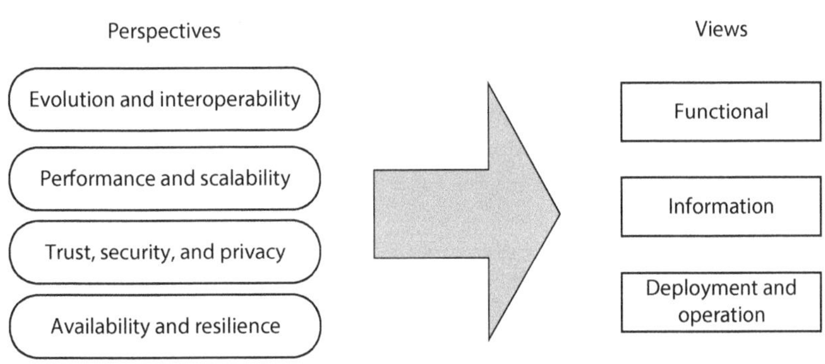

FIGURE 1.2 Perspectives and Views of IoT-A (Adapted from Bassi et al., 2013)

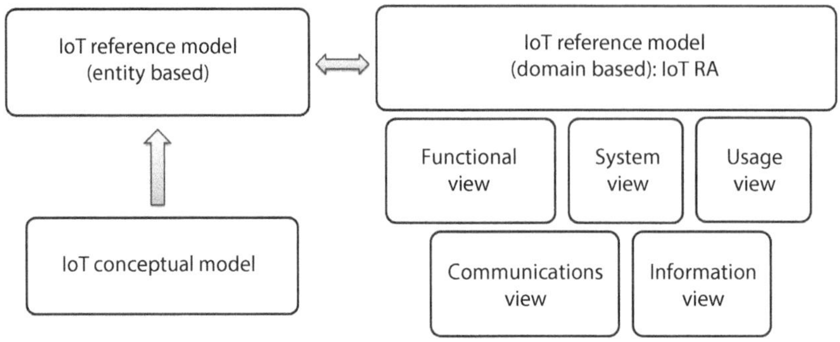

FIGURE 1.3 Relation between CM, RM, and RA (Adapted from ISO/IEC, 2016)

and resilience" are applied to all the views: the "functional" view, the "information" view, and the "deployment and operation" view, respectively.

While applying perspectives to views, not every view is impacted by the perspectives in the same manner or grade. For example, the perspectives have a high impact when applied to the operation view.

1.3.2 IoT RA

IoT RA, created by the ISO/IEC[29] (CD 30141), envisions the construction of an IoT system based on a generic IoT conceptual model (CM) that includes the most important characteristics and domains of IoT. Then, it uses the CM as a basis to create a high-level system-based RM. This reference model is, in turn, structured in five architectural views (functional view, system view, user view, information view, and communication view) from different perspectives, which compose the RA itself. Figure 1.3 shows the relation between these three components (CM, RM, and RA).

In essence, the IoT RA provides the basics to create a concrete system architecture. The IoT RA is considered an application-specific architecture or a "target system architecture" since the RA can adapt to the requirements of a specific system, like an agricultural system, a smart home/building, a smart city, and so forth. The original standard was published in 2018, which was replaced by a new version released in 2024.

1.3.3 IEEE P2413

IEEE P2413 is based on ISO/IEC/IEEE 42010:2011: "Systems and Software Engineering Architecture Description."[30] The goal is not to create a new standard but to address common aspects of different application domains of the IoT. The IEEE working group is collaborating with ISO, ITU-T, and the Industrial Internet Consortium (IIC), among others, with the common goal of achieving better standards for the IoT in all its areas of application. The focus is on achieving interoperability, together with other quality attributes such as protection, privacy, security, and safety. Some of these challenges are further discussed in Section 1.6.

1.3.4 Industrial RAs

The IIRA[31] is a standard-based open architecture for Industrial Internet Systems (IISs), proposed by the IIC Technology Working Group, whose members are companies like AT&T, Cisco, IBM, General Electric, and Intel. The Industrial Internet is considered an IoT system, enabling intelligent industrial operations and focusing on key characteristics for this type of system: safety, security, and resilience. IISs cover energy, healthcare, manufacturing, the public sector, transportation, and related industrial systems.

TABLE 1.1
IIRA Overview: Viewpoints and Their Concerns and Stakeholders

Viewpoints	Concerns	Crosscutting Concerns		Stakeholders
Business	Identification of stakeholders, business vision, values, and objectives of an Industrial IoT system	Safety	Security	Decision makers, product managers, system engineers
Usage	Expected system usage			System engineers, product managers, other users
Functional	Functional components, interfaces and interactions between them			System architects, developers, integrators
Implementation	Technologies needed, communication protocols, life cycle			System architects, developers, integrators, system operators

The Industrial Internet Architecture Framework (IIAF) is based on ISO/IEC/IEEE 42010:2011, and as such, it uses the same constructs and common terms, such as viewpoints, concerns, and stakeholders, as well as views and models.[32] The IIRA is the result of applying IIAF to the Industrial IoT systems. Table 1.1 shows an overview of the IIRA. Each viewpoint influences the viewpoints below it. In turn, lower viewpoints validate and sometimes cause revisions in the higher viewpoints. There are some crosscutting concerns, such as security and safety, which are discussed in other reports from the IIC.

Initially designed for the German industry, RAMI 4.0 is a result of cooperation between: Plattform Industrie 4.0 (Industrie 4.0 [I4.0] is considered a specialization within IoT); some German associations, like BITKOM, VDMA, and ZVEI; and several German companies (Adolphs et al., 2015). Their main goal was to achieve a common understanding of what is necessary to evolve from current industries and make I4.0 a reality. To do that, it was necessary to develop an architecture model to be used as a reference in this migration.

RAMI 4.0 focuses on the optimization of central industrial processes, namely, research and development, production, logistics, and service. It describes the structures and functions of the I4.0 components, based on existing and relevant standards.

Figure 1.4 shows the RAMI 4.0 architecture model, which is a three-dimensional model, where the horizontal axis represents the life cycle of systems or products, distinguishing between "type" (life cycle of a product from the idea to the product, going through design, development, and testing) and "instance" (it represents the manufacturing of a type; its life cycle goes from manufacturing, selling, and delivering to the client to being installed in a particular system). The vertical axis (six layers), in turn, corresponds to the IT perspective of an I4.0 component, meaning that it breaks complex projects into smaller parts, like business processes, functional descriptions, communications behavior, and hardware/assets. Finally, the third axis represents a functional hierarchy, which does not refer to equipment classes or hierarchical levels of the automation pyramid but to grouping functionalities and responsibilities within the factories. It contains core aspects of I4.0, such as field devices, control devices, stations, work centers, and enterprises, but expands the hierarchy levels of the IEC 62264 standard by adding "product" and "connected world."

1.3.5 Other Reference Models and Architectures for IoT

So far, there have been several contributions to create RMs for IoT, most of them based on IoT-A. In fact, up to now, several architectures have been proposed, but as they are designed for a specific IoT application, they cannot be used as a reference since they do not adapt to other applications' requirements (Yin et al., 2015; Pang, 2013; Vlacheas et al., 2013; Domingo, 2012; Yun and Yuxin, 2010).

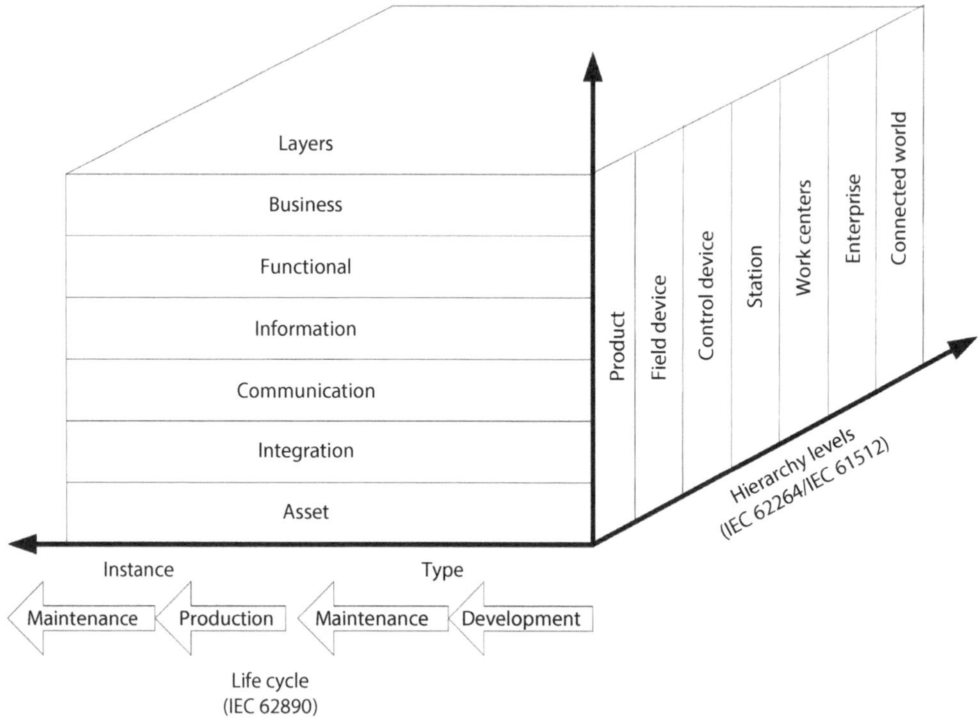

FIGURE 1.4 RAMI 4.0 (Adapted from RAMI 4.0 Reference Architectural Model for Industrie 4.0, https://www.isa.org/intech-home/2019/march-april/features/rami-4-0-reference-architectural-model-for-industr.)

An alternative architectural stack for the "web of things" consists of "levels of functionalities," with each level composed of a set of application protocols and tools (Guinard and Trifa, 2016). The idea is to provide developers with the necessary tools to implement IoT products and applications and to maximize reuse and interoperability. In this section, only some of the most significant proposals that might be considered as a reference are briefly described.

1.3.5.1 Cisco Reference Model

In 2014, Cisco proposed a seven-layer RM (Cisco, 2014), which is represented in Figure 1.5, giving a more practical point of view. The lowest level includes the physical devices and controllers (the *things*); then there is connectivity and, above that, edge (fog) computing, where some initial aggregation, elimination of data duplication, and analysis can be carried out. The lower three levels, in turn, are considered operational technology (OT). The top four levels relate to IT. The lowest level in the IT part of the stack is storage, and this is followed, going toward the top, by data abstraction, applications, collaboration, and (business) processes.

1.3.5.2 Reference IoT Layered Architecture

Every IoT RA must include some essential components, such as interoperability and integration components, context-aware computing techniques, and security guidelines for the whole architecture (Karzel et al., 2016). The resulting proposed architecture is RILA. RILA is a more concrete architecture, intended to be easier to comprehend for customers and industry than the high-level IoT-A. It not only provides guidelines on how to put IoT-A into practice but also demonstrates that this architecture can really be implemented using actual use cases. RILA acts between things, devices, and the user.

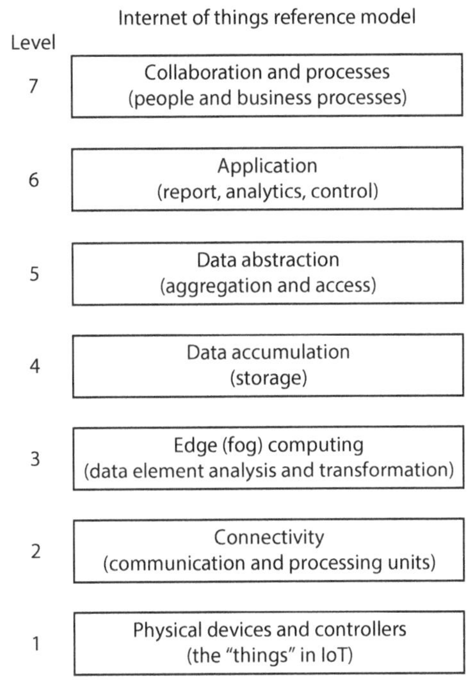

FIGURE 1.5 IoT RM Proposed by Cisco (Adapted from Cisco, The Internet of Things reference model, 2014, https://dl.icdst.org/pdfs/files4/0f1d1327c5195d1922175dd77878b9fb.pdf.)

RILA consists of six layers, as depicted in Figure 1.6. Besides these layers, there are two cross-section layers, "security" and "management," that affect all other layers.

The *device integration layer* includes all the different types of devices, receives their measurements, and communicates actions. This layer can be seen as a translator that speaks many languages (Karzel et al., 2016). The output of the sensors and tags, as well as the input of the actuators, depends on the protocol they implement.

The *device management layer* is responsible for receiving device registrations and sensor measurements from the device integration layer and for communicating status changes for actuators to the device integration layer. Then, the device integration layer checks if the status change (i.e., the action) conforms with the respective actuator and translates the status change to the actuator. The device management layer controls the devices that are connected to the system; every change to a device's registration, as well as new measurement data, should be communicated from the device integration layer to the device management layer, so the information can be updated and stored.

Normally, the *data management layer* is a central database (but it can also be a data warehouse or even a complete data farm, in the case of larger IoT systems) that stores all data of a thing. Thus, the implementation of the data management layer strongly depends on the use case (Karzel et al., 2016).

The *context management layer* defines the central business logic and is responsible for tasks like defining the goals of the thing, consuming and producing the context situations of the things, evaluating the context situation toward the goals, triggering actions that will help to fulfill the goal according to the evaluated rules, and finally, publishing context situations for other things.

The *thing integration layer* is responsible for finding other things to communicate, verifies if communication with the new thing is possible, and is responsible for a registration mechanism.

The *application integration layer* connects the user to the thing, being considered the service layer or even a simple user interface. The concrete implementation of the layer depends on the use case.

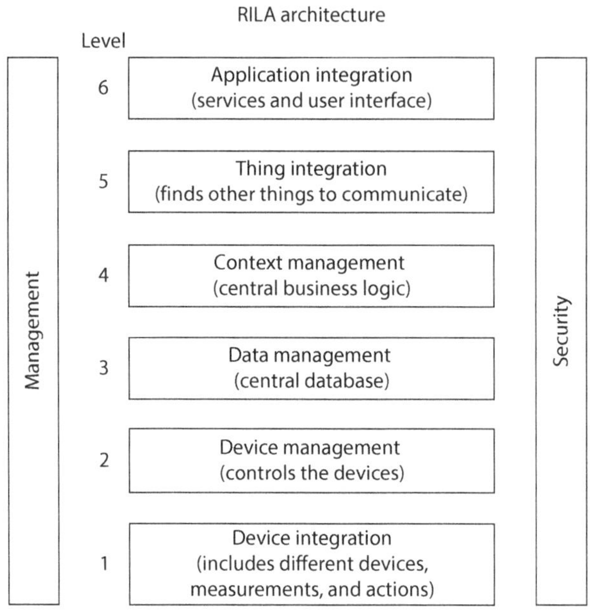

FIGURE 1.6 Reference IoT Layered Architecture (Adapted from Karzel et al., A reference architecture for the Internet of Things, https://www.infoq.com/articles/internet-of-things-reference-architecture/.)

1.4 ENABLING TECHNOLOGIES

In the previous section, a few international organizations that are engaged in defining and developing common standards for the IoT were already mentioned. Many of these are integrated into alliances and consortia that include members from industry, SDOs, manufacturers, network and service providers, academia, and research laboratories. oneM2M, for example, consists of eight important SDOs from all around the world: ARIB (Japan), ATIS (United States), CCSA (China), ETSI (Europe), TIA (United States), TSDSI (India), TTA (Korea), and TTC (Japan), which have come together with six industry partners and consortia (Broadband Forum, Continua Alliance, GlobalPlatform, HGI, Next Generation M2M Consortium, and OMA) and more than 200 member organizations, including companies (like Alcatel, Nokia, Huawei, Deutsche Telekom, and Qualcomm) and universities.

This section presents a short overview of some of the technologies that are being used for connected object identification and discovery, communication, and devices. More detailed discussions on common IoT-enabling technologies and protocols can be found in Alsubaei et al. (2018). In the final part of this section, we will see some of the currently available online platforms for IoT project prototyping, development, and testing.

1.4.1 Identification and Discovery

For the things to be identified within a distributed networked system, a unique identifier is needed for each thing. And here a thing can be anything ranging from a physical to a virtual object, an event, or a person. Data such as time and location – either geographic (coordinates) or within a network (Uniform Resource Identifier [URI] or IP) – can be used for identification purposes. The Electronic Product Code (EPC) is mostly used for counting and tracking goods in the supply chains without human intervention. While RFID-driven EPCs are attached to physical objects, URIs and IPs allow the identification and discovery of an object's presence on the web. HyperCat,[33] for example, is a solution that allows us to expose any number of URIs, together with additional information attached to them, in a Resource Description Framework (RDF)-like format.

IP addresses are also used as identifiers for networked objects, together with name labels, which can be used for human readability and resource discovery using a naming service, such as mDNS.[34] mDNS is an IETF project that aims to adapt the well-established Domain Name System (DNS) programming interfaces and formats to small networks where there are no name servers available. People, in turn, can be identified through the devices they carry along or through more sophisticated techniques, such as biometric data obtained through face or iris recognition, fingerprints, voice, and so on. The ISO IoT technology mind map includes more identification-related technologies and possible solutions.

In 2016, ISO/IEC published the 29161 standard (ISO/IEC, 2016), which aims to ensure full compatibility among different identification forms.

1.4.2 Communication Patterns and Protocols

RFC 7452[35] describes four basic communication patterns for IoT environments: device-to-device communication pattern, device-to-cloud communication pattern, device-to-gateway communication pattern, and back-end data-sharing pattern.

The device-to-device pattern is applied when two devices communicate directly, normally using a wireless network. There are several protocol stacks available to carry out this type of communication. Depending on the usage scenario, the protocol stack may include, for instance, Bluetooth or IEEE 802.15.4, IPv6, User Datagram Protocol, and Constrained Application Protocol.

The device-to-cloud communication pattern is used when data captured by the device from the environment is uploaded to an application service provider. Communication is based on IP, but when the device manufacturer and the application service provider are the same, the integration of other devices may be difficult. For this not to happen, the protocols used to communicate with the server need to be made available.

The device-to-gateway communication pattern may be used when the system contains non-IP devices, when support for legacy devices is needed, or when additional security functionality must be implemented. Gateways can also be mobile, providing only temporary connections to the Internet. Smartphones are an example of this.

The back-end data-sharing pattern is used when there is a need to analyze combined data from several sources. RESTful APIs can be used, although they are not standardized. This pattern may allow users to move their data from one IoT service to another (Rose et al., 2015).

Beyond some well-known communication standards, like Bluetooth, Wi-Fi, or GSM, IoT systems use many more. It is worth pointing out that new communication standards are being developed specifically for some of the IoT scenarios. For example, when the system is composed of devices with constrained resources across wide area networks, low-power wide area network solutions must be applied.

On the one hand, the last two decades or so registered immense growth in the field of mobile telecommunications, with devices that have more and more resources in terms of processors, memory, sensors, and connectivity. As such, 3GPP and the mobile operators have been making a huge effort toward new standards for mobile communications with even larger bandwidths, higher bit rates, and support for multimedia live streaming.

On the other hand, solutions for the scenarios that will not so much include resource-rich devices but mostly resource-constrained devices stayed behind and only appeared recently. In 2016, 3GPP released a new standard for the IoT called NB-IoT, which is supported by leading manufacturers and by the world's 20 largest mobile operators. NB-IoT is a technology that allows us to bidirectionally and securely connect multiple sensors and devices with low bandwidth requirements. It is a technology that requires low energy consumption (more than ten years of autonomy) and allows strong penetration of the radio signal in indoor environments.

In the meantime, many other low-power solutions for wide area networks, such as LoRa, NWave, and Sigfox, have been developed and adopted by many IoT deployments. It is predictable that there will be a convergence among these technologies in the near future, and some of them will eventually prevail, while others will disappear or remain in use in legacy deployments. Table 1.2 summarizes some of the communication protocols and standards currently in use.

TABLE 1.2
Overview of Communication Technologies and Standards for IoT

Name	Frequency	Range	Examples	Standards
BLE	2.4 GHz	1–100 m >100 m	Headsets, wearables, sports and fitness, healthcare, proximity, automotive	IEEE 802.15.1 Bluetooth SIG
EnOcean	315, 868, 902 MHz	300 m outdoor 30 m indoors	Monitoring and control systems, building automation, transportation, logistics	ISO/IEC 14543-3-10
GSM	Europe: 900 MHz and 1.8 GHz		Mobile phones, asset tracking, smart meter, M2M	3GPP
LoRa	Sub-1 GHz ISM band	2–5 km urban 15 km suburban 45 rural	Smart city, long range, M2M	LoRaWAN
NB-IoT	700–900 MHz	10–15 km rural deep indoor penetration	Smart meters, event detectors, smart city, smart home, industrial monitoring	3GPP LTE Release 13
NFC	13.56 MHz	Under 0.2 m	Smart wallets, smart cards, action tags, access control	ISO/IEC 18092 ISO/IEC 14443-2, -3, -4 JIS X6319-4
NWave	Sub-1 GHz ISM band	Up to 10 km	Agriculture, smart city, smart meter, logistics, environmental	Weightless
RFID DASH7	120–150 kHz (LF), 13.56 MHz (HF), 2450–5800 MHz (microwave), 3.1–10 GHz (microwave), 433 MHz (UHF), 865–868 MHz (Europe), 902–928 MHz (North America) (UHF)	10 cm–200 m 0–5 km	Road tolls, building access, inventory, goods tracking, building automation, smart energy, smart city logistics	ISO 18000
Sigfox	900 MHz	3–10 km urban 30–50 km rural	Smart meters, remote monitoring, security	
Weightless	470–790 MHz	Up to 10 km	Smart meters, traffic sensors, industrial monitoring	Weightless
Wi-Fi	2.4 GHz, 3.6 GHz, 4.9/5 GHz	Up to 100 m	Routers, tablets, smartphones, laptops	IEEE 802.11
Z-Wave	ISM band 865–926 MHz	100 m	Monitoring and control for home and light commercial environments	Z-Wave Recommendation ITU G.9959
ZigBee	2.4 GHz; 784 MHz in China, 868 MHz in Europe, and 915 MHz in the United States and Australia	10–20 m	Home and building automation, WSN, industrial control	IEEE 802.15.4

Source: Data from Postscapes, IoT technology guidebook, IoT Technology | An overview to what makes up Internet of Things (IoT) systems and devices. From sensors and communication protocols to APIs and machine learning, 2019. https://www.postscapes.com/internet-of-things-technologies/; Opensensors, How to choose the best connectivity network for your Project, 2015, https://www.linkedin.com/pulse/how-choose-best-iot-connectivity-network-your-project-james-moulding/; ETSI (European Telecommunications Standards Institute), SmartM2M; IoT standards landscape and future evolutions, ETSI TR 103 375 V1.1.1 (2016-10), 2016, http://www.etsi.org/deliver/etsi_tr/103300_103399/103375/01.01.01_60/tr_103375v010101p.pdf.

Note: M2M, machine-to-machine; LF, low frequency; HF, high frequency; UHF, ultra-high frequency.

1.4.3 Devices and Test Beds

A wide range of specialized and multipurpose sensors are available on the market, as well as many SBCs with embedded sensors or with support for several sensors and actuators. Wearable devices are also acquiring more and more enthusiasts, especially in the area of sports and physical activity tracking, sometimes in addition to the already well-equipped smartphones with several embedded sensors.

Sensors enable devices to capture data from their environment. They can be categorized in different ways. Here is a nonexhaustive list of possible categories and some examples:

- *Location*: GPS, GLONASS, Galileo, Wi-Fi, Bluetooth, ultra-wideband (UWB)
- *Biometric*: Fingerprint, iris, face
- *Acoustic*: Microphone
- *Environmental*: Temperature, humidity, pressure
- *Motion*: Accelerometer, gyroscope

Actuators, on the other hand, allow devices to act on their environment and may be of different types, like hydraulic, pneumatic, electric, mechanical, or piezoelectric.

SBCs are becoming more and more popular among both hobbyists and researchers (Samaila et al., 2017; Meruje et al., 2025). They are low cost (mostly less than $100), provide a reasonable amount of processing power and RAM, sometimes have embedded sensors or support the connection of external sensors, and support wireless connectivity (Wi-Fi, Bluetooth Low Energy [BLE], and ZigBee). Mostly, these are Linux machines. Some of the most popular examples are Arduino, Raspberry Pi, BeagleBone Black, and Pine 64.[36] Their characteristics are summarized in Table 1.3.

As an alternative to deploying one's own hardware on site, there are laboratories around the world where the physical devices are located and can be tested remotely, especially for large-scale scenario testing. Examples include the FIT IoT-lab,[37] located in France, and iMinds,[38] in Belgium.

TABLE 1.3
Characteristics of Some of the Currently Available Single-Board Computers for IoT

	Models	CPU	RAM	Operating System	Price ($)	Connectivity	Embedded Sensors
Arduino	20+ models	ATmega, ATSAM, AR9331, etc.	0.5 kB–16 MB	Linux	30	Ethernet, Wi-Fi, extension boards	Extensions
Beagle board	BeagleBone Black	AM335x, 1 GHz; ARM Cortex A8	512 MB–1 GB	Linux	50	Ethernet	Extensions
Intel	Edison, Joule, Galileo	Intel Atom, Intel Quark	256 MB–4 GB	Linux	20–60	Wi-Fi, Bluetooth	
Pine 64	Pine A64, Pine A64+	ARM Cortex A53, 1.2 GHz	512 MB–2 GB	Linux, Android, Windows IoT	15–29	Ethernet	
Raspberry Pi	Zero, RPi 1 A+, RPi 2 B+, RPi 3 B	ARM1176, 1 GHz; ARM1176; ARM Cortex A7; ARMv8, 1.2 GHz	512 MB–1 GB	Linux, Windows IoT	20–40	Wi-Fi, BLE, Ethernet, extensions	SenseHAT, other

All of them are parts of OneLab,[39] which offers several test beds, giving developers a means to quickly implement and test their projects in controlled environments, being able to evaluate them before the actual deployment takes place.

OneLab is not restricted to the IoT area; it also offers test beds for other areas, like cloud computing and software-defined networks. Regarding IoT, there is one test bed specifically targeted at *smart city* applications and another for *connected commerce*, which addresses logistics monitoring (e.g., monitoring the quality of food and its transportation during the whole supply chain).

1.4.4 Cloud, Fog, and Edge Computing

To gain insights from collected sensor data and make the right decisions, it must be first processed and analyzed. Complex data analytics algorithms, including machine learning models, are usually used to find patterns in this data (Byabazaire et al., 2025; Samaraweera et al., 2025). The data also needs to be stored for future use or even further processing. However, in IoT, connected devices often have constrained resources and limited power capacities, making it difficult to accomplish such tasks locally. Thus, offloading collected data to more resourceful processing and storage units is necessary. Cloud, fog, and edge computing are three popular computing paradigms that can meet this requirement (Happ, 2018; Byabazaire et al., 2025).

Cloud computing provides the necessary processing power and storage facilities by offering a pool of shared and virtually unlimited resources and services (Hassan, 2011; Mell and Grance, 2011). Cloud services are typically hosted on the Internet, where users can easily subscribe/unsubscribe, scale up/down used services, and pay only for their actual use of these services. This flexibility allows for scalability where business requirements and workloads dynamically change. In addition to these services, cloud providers provide reliable, nonvolatile, replicated storage, and rapid and configurable networking for data distribution. Popular cloud providers like Amazon, Microsoft, and Google offer various cloud services ranging from virtualized hardware resources (e.g., servers and storage capacities) to complete software systems (e.g., data warehousing and visualization tools). The former form is known as infrastructure as a service (IaaS), whereas the latter is known as software as a service (SaaS). Cloud providers also offer platforms that enable users to manage subscribed services and build custom applications around them, which is known as platform as a service (PaaS).

While cloud computing works well in many scenarios, it is not always the best offloading mechanism. In scenarios where Internet connectivity is unstable, latency is not tolerated, or privacy and data sovereignty are critical, data processing and storage need to take place at or close to its sources instead of sending it to the cloud (i.e., the core network). Fog computing and edge computing are two concepts that bring these services closer to the edges of the network (Bonomi et al., 2012; Shi et al., 2016). The difference between these two models is not very clear in the literature and is sometimes used interchangeably. However, the industry often uses the term edge computing to refer to computing taking place on IoT devices and gateways, smartphones, laptops, and on-premise servers, while it uses the term fog computing to refer to the layer that sits between these edge devices/nodes and the cloud (i.e., nearby datacenters), which often has more powerful resources. Figure 1.7 illustrates the three paradigms.

We should note that neither fog computing nor edge computing intends to replace cloud computing. Instead, they complement it by providing an intermediary layer between the end devices and backend cloud services for initial data preprocessing (e.g., data cleansing and transformation) and/or temporary storage. The field refers to this as the cloud-edge continuum, which combines the central cloud computing model (collocated data centers) and the fog/edge decentralized models (dispersed devices). Processing IoT data at the network edges enables: improved performance for latency-sensitive applications (e.g., autonomous vehicles); better control over data privacy (e.g., sensitive data can be encrypted or filtered out); improved reliability (e.g., in the case of unreliable

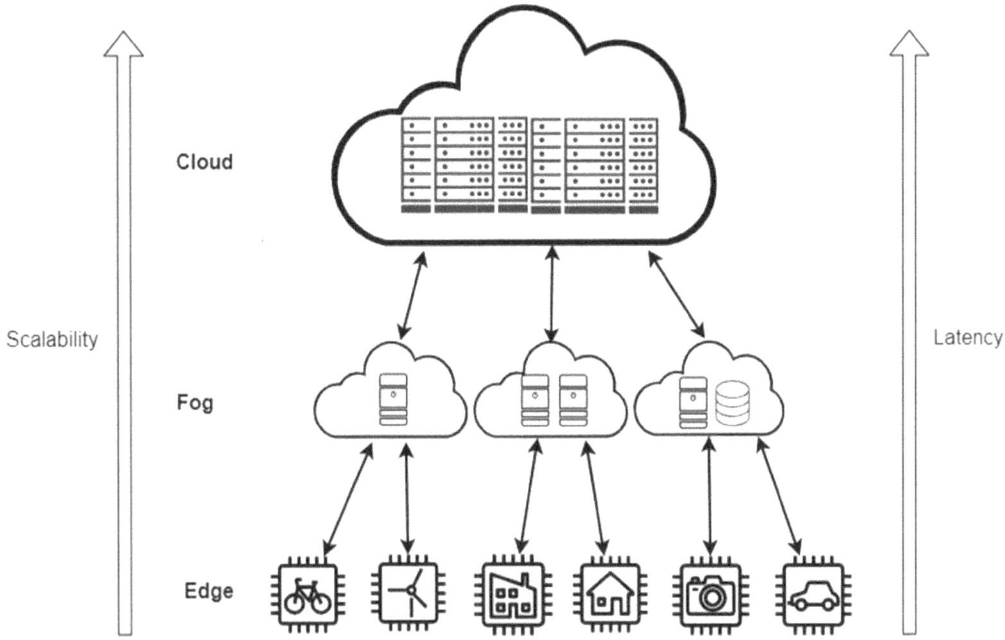

FIGURE 1.7 Cloud, Fog, and Edge Computing

Internet connectivity); and cost savings (i.e., cloud services at a large scale are not cheap and sending data in and out of cloud servers incurs fees).

1.5 APPLICATION AREAS: AN OVERVIEW

A symbiosis between platforms, applications, devices, and services gives the ability to improve citizens' well-being and quality of life. The great potentialities offered by the IoT make the development of a huge number of applications possible, while it also plays a crucial role in the so-called fourth Industrial Revolution (I4.0). The IIC[40] was created with the goal of transforming the industry through intelligent, interconnected objects that may improve performance, lower costs, and increase reliability. This consortium considers that industry involves the areas of energy, healthcare, manufacturing, smart cities, and transportation. The ICC is now part of the Digital Twins Consortium.

Nevertheless, the areas of application cover various sectors of society and are grouped in diverse ways in the literature. There are at least two fairly complete and interesting classification schemes of the application domains (Atzori et al., 2010; AIOTI, 2015).[41] Considering the most relevant research made on applications, but essentially these two sources, an overview of the main application domains is shown in Figure 1.8.

As the figure depicts, the main areas of application are smart cities, healthcare, smart homes and buildings, mobility and transportation, energy, industry, agriculture, and the environment/planet. Note that some applications, like environmental monitoring, can fit into different groups, like smart cities, smart buildings, the environment, and industry. It is worth mentioning that in the context of IoT, it is not possible to develop a "one-size-fits-all" solution. For example, a solution for home environmental monitoring may not be adequate for industrial ambient monitoring due to different types of physical conditions and relevant parameters. An industrial setting requires different levels of accuracy, security, and robustness for the deployed sensors and software, while a home environment usually does not impose such restrictions. Some examples of these applications are briefly introduced below. Many of these applications are discussed in great detail in Hassan (2018).

1.5.1 Smart Cities

Finding ways to use technology to improve the quality of life in a city has become one of the most popular research topics in the area of IoT applications (Yin et al., 2015; Zanella et al., 2014; Vlacheas et al., 2013; Minoli and Occhiogrosso, 2018). Smart city solutions include several areas, ranging from water and waste management (like smart dumpsters) (Hong et al., 2014; Phithakkitnukoon et al., 2013), lighting control (Castro et al., 2013), energy (Kyriazis et al., 2013), transportation, traffic, and parking management, to building efficiency, services, and safety (Perera et al., 2014).

Even though smart buildings, transportation and mobility, and energy are sometimes included under the smart cities umbrella, Figure 1.5 dedicates a separate branch to each, due to the importance that these areas are receiving.

Some examples of real-life implementations occur in the cities of London[42] and Greenwich in the United Kingdom; Santander (Sanchez et al., 2011), Barcelona, and Murcia in Spain; Amsterdam in Holland; Aarhus in Denmark; Oulu (Gil-Castineira et al., 2011) in Finland; several towns in Korea; and Bordeaux in France. AT&T is helping cities deploy integrated smart city solutions and will deliver solutions in some cities in the United States. The CITYkeys[43] project, funded by the European Commission's Horizon 2020 program, aims at developing an evaluation framework to compare smart city solutions across European cities.

1.5.2 Healthcare

IoT technologies may bring significant benefits to the healthcare domain, namely, in two areas: clinical care and remote monitoring (Russo and Eriksson, 2018; Fong et al., 2018; Tang and Winoto, 2018; Sagahyroon et al., 2017; Kumar et al., 2025a). Basically, the use of small, low-cost, and low-power wearable biosensors can enhance the quality of life of people suffering from chronic diseases or even during emergencies, either inside (Dohr et al., 2010; Bui and Zorzi, 2011; X. Li et al., 2011; Domingo, 2012; Doukas and Maglogiannis, 2012; Yang et al., 2014) or outside their homes (Domingo, 2012; Doukas and Maglogiannis, 2012). Elderly tracking or ambient assisted living encompasses technical systems to support elderly people in their daily routine to allow an independent and safe lifestyle as long as possible. This is a special case of healthcare that has gained increased importance due to the problem of population aging (Dohr et al., 2010; Yang et al., 2014). Another interesting example is using IoT in noninvasive glucose-level sensing for diabetes management (Istepanian et al., 2011) or using an IoT platform to improve the lives of people with disabilities (Domingo, 2012). Besides these common use cases, IoT is also being used in tracking and monitoring medical supplies (Ranga et al., 2025). Chapters 8 and 9 provide in detail two examples of IoT applications in healthcare.

1.5.3 Smart Homes and Smart Buildings

Smart home applications can range from elderly monitoring (Yang et al., 2014) and home automation, in either air conditioning control and monitoring or lighting control, to consumption monitoring and energy saving (Wei and Li, 2011; Kelly et al., 2013), or even solar greenhouse production (Wang et al., 2004).

The focal point of IoT applications for "smart buildings" or "intelligent buildings" is essentially cost savings, providing the building with some intelligence, through building automation (Bugeja et al., 2018). These applications mainly focus on air conditioning and lighting monitoring and control and on consumption monitoring and energy saving (Wei and Li, 2011; Moreno et al., 2014; Brad and Murar, 2014; Ji et al., 2014). Security is also an important issue; therefore, fire and intrusion monitoring are also crucial (Li, 2013; Ryu, 2015; Li et al., 2013). Intel, Telit, IBM, and many others have several solutions for smart building implementations (Zhong and Dong, 2011).

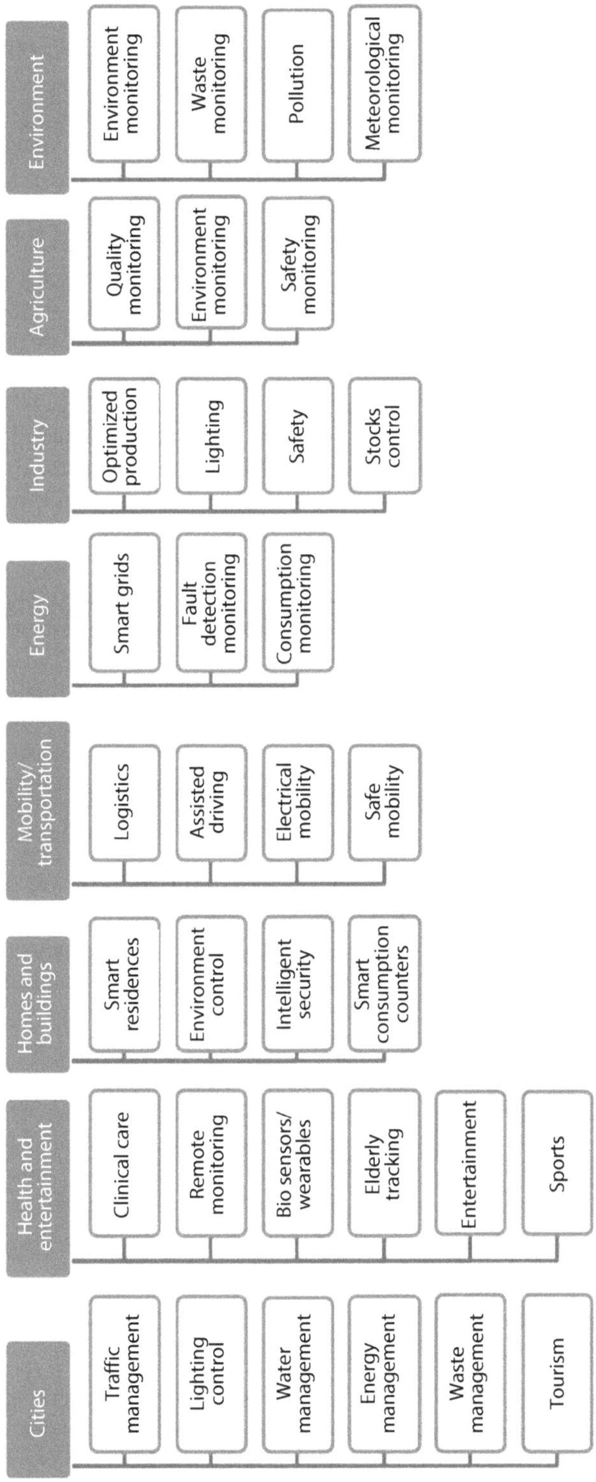

FIGURE 1.8 Applications Grouped in Domains (Based on Atzori et al., 2010; AIOTI, 2015)

1.5.4 Mobility and Transportation

All types of vehicles in a city (cars, trains, buses, and bicycles) are becoming more equipped with sensors and/or actuators, resulting in a network composed of a set of mobile sensors. Both roads and rails, as well as transported goods, are also equipped with tags and sensors that send important information to traffic control sites (Ben-Daya et al., 2019; Zantalis et al., 2019; Singh et al., 2022; Ranga et al., 2025). This not only allows monitoring of the status of the transported goods but also allows the creation of innovative solutions, allowing transportation vehicles to better route the traffic or providing tourists with appropriate transportation information.

Moreover, modern cars are also equipped with several sensors, forming a kind of in-vehicle network that provides kinematics information, automotive diagnostic services, and so forth. Cars can be further equipped with external sensing devices to monitor specific physical parameters, such as pollution, humidity, and temperature. Thus, the concept of "smart vehicles" emerges.

If properly collected and delivered, such data can contribute to making road transport greener, smarter, and safer (Zouganeli and Svinnset, 2009; Campolo et al., 2012). For example, driving recommendations that aim at eco-efficiency for public transportation and reducing fuel consumption and emissions can be provided (Tielert et al., 2010; Kyriazis et al., 2013). Mobile applications, such as Google Traffic or Waze,[44] rely on user-contributed data to monitor traffic conditions. Smart traffic light infrastructures can be used to improve the lives of drivers or make cycling or driving in cities safer and smoother. For example, combining data from smartphones carried by cyclists and traffic data gathered from different kinds of sensors deployed in the traffic light infrastructure of a city may allow for an intelligent traffic light orchestration, letting cyclists drive smoothly without unnecessary stopping at each crossroad (Anagnostopoulos et al., 2016). Another specific area of application is modern logistics, which refers to monitoring the whole process of the physical movement of goods from suppliers to demanders in order to ensure their quality (Zhengxia and Laisheng, 2010; Zhang et al., 2011).

1.5.5 Energy

The smart grid is a recent kind of intelligent power system that can improve energy efficiency, reduce environmental impact, improve the safety and reliability of the electricity supply, and reduce the electricity transmission of the grid. The integration of IoT technology in smart grids can help to implement fault detection and monitoring, as well as consumption monitoring, through the installation of energy sensors (Yun and Yuxin, 2010; L. Li et al., 2011; Liu et al., 2006; Bui et al., 2012; Ou et al., 2012; Yang et al., 2017; Minoli and Occhiogrosso, 2018; Finco and Minoli, 2018; Ahmad and Zhang, 2021).

Other groups of related solutions envision the heat and energy management in homes and buildings to accomplish an energy savings purpose (Kyriazis et al., 2013; Sundramoorthy et al., 2010). Using IoT technology to collect data on energy consumption can also help to improve the energy efficiency and competitiveness of manufacturing companies at the energy production level (Shrouf and Miragliotta, 2015).

1.5.6 Smart Manufacturing

The design and operation of a manufacturing system need numerous types of decision-making at various levels of its activities. Therefore, IoT can be applied to develop modern manufacturing enterprises characterized by dynamic and distributed environments (Da Xu et al., 2014). In these environments, IoT technology can be used to serve a variety of purposes, from environment control, lighting control, and safety to production optimization, error/fault detection and correction, and automatic control of stocks (Fantana et al., 2013; Tao et al., 2014; Bi et al., 2014; Willner, 2018).

1.5.7 Smart Agriculture

Modern agriculture has a different set of requirements than traditional agriculture. It must be high yield, high quality, efficient, safe, and ecological (Shifeng et al., 2011). IoT technology has contributed to agriculture modernization and improvement (Shifeng et al., 2011; Bo and Wang, 2011). For example, sensors have been successfully deployed for irrigation control, fertilization and pest control, and animal and greenhouse monitoring, leading to what is known today as precision farming (Abbasi et al., 2014; Zhang et al., 2018).

1.5.8 Environment/Smart Planet

IoT can be used to allow for the development and management of sustainable cities, covering issues like environmental monitoring (Fang et al., 2014), pollution control (Du et al., 2013; Fang et al., 2014), meteorological monitoring (Du et al., 2013), disaster monitoring (Chaturvedi et al., 2018; Kumar et al., 2025b), or waste management (Hong et al., 2014).

IoT technology can be used to tackle rapid urbanization and related environmental problems, allowing us to study the environment, planning, and construction issues in order to increase our understanding of how to integrate urban development and ecological processes for sustainable city construction.

Chapter 7 offers an excellent application that uses IoT in environmental monitoring and disaster prediction.

1.6 CHALLENGES

Currently, IoT is one of the main accelerators of technological innovation, being one of the areas with greater potential for the transformation of society and the economy (Schoder, 2025). As such, all the involved stakeholders, ranging from technologists to developers, companies, and users, face several challenges that remain to be tackled (Hassan et al., 2017; Vos, 2017; Sliwa, 2018). Experience from areas such as distributed systems, networks, mobile and ubiquitous computing, context awareness, and WSN could be considered a good starting point for seeking appropriate solutions for issues, such as interoperability, openness, security, scalability, and failure handling in the scope of the IoT systems.

Based on a nonexhaustive literature survey, a set of challenges is shortly discussed in the next sections. Research publications were grouped into the five previously mentioned categories, as shown in Table 1.4.

TABLE 1.4
Overview of IoT Challenges

Challenges	References
Heterogeneity and interoperability	Ortiz et al. (2014), Serbanati et al. (2011), Al-Fuqaha et al. (2015), Rose et al. (2015), Borgia (2014), Miorandi et al. (2012), Atzori et al. (2010), Alur et al. (2016)
Openness	Perera et al. (2014), Alur et al. (2016)
Security, privacy, and trust	Perera et al. (2014), Serbanati et al. (2011), Ortiz et al. (2014), Al-Fuqaha et al. (2015), Borgia (2014), Rose et al. (2015), Miorandi et al. (2012), Whitmore et al. (2014), Atzori et al. (2010), Taivalsaari and Mikkonen (2017), Alur et al. (2016)
Scalability	Alur et al. (2016), Al-Fuqaha et al. (2015), Borgia (2014), Miorandi et al. (2012)
Failure handling	Alur et al. (2016), Ortiz et al. (2014), Taivalsaari and Mikkonen (2017), and Miorandi et al. (2012)

1.6.1 Interoperability

Heterogeneity has been a great challenge in distributed systems, as a variety of networks, hardware, different operating systems, and programming languages started to coexist within the same system. IoT systems of the future will be composed of humans, machines, things, and groups of them. To accomplish the functioning of such a network, seamless communication and cooperation among all the components is crucial (Ortiz et al., 2014; Hazra et al., 2021; Noaman et al., 2022; Albouq et al., 2022). Developers and programmers also need to be prepared to cope with multidevice, always-on, highly dynamic, and distributed systems and adapt and update their programming skills to this new paradigm (Taivalsaari and Mikkonen, 2017).

Even though there are other challenges for the IoT, interoperability remains one of the most challenging goals for IoT systems, unless an RA and a set of standards are developed (Atzori et al., 2010; Rose et al., 2015; Serbanati et al., 2011). There has been an effort by the ISO/IEC to create an RA, the IoT RA. This architecture gathers consensus from several organizations (WG 10, 2015).

Furthermore, to interoperate, IoT components must be identifiable and discoverable by other components (Atzori et al., 2010; Borgia, 2014; Miorandi et al., 2012; Ortiz et al., 2014; Serbanati et al., 2011), as discussed in Section 1.1. After the discovery and identification process, components must be able to somehow communicate (Al-Fuqaha et al., 2015; Alur et al., 2016; Borgia, 2014; Miorandi et al., 2012; Ortiz et al., 2014). Section 1.2 gave an overview of the communication protocols and patterns that are currently being used and developed. It is imperative that in the near future, existing and new wireless technologies, such as NB-IoT, LoRaWAN, and Sigfox, be thoroughly tested and further developed to achieve steps in the direction of having standards for connectivity among IoT devices.

1.6.2 Openness

The openness of a system is the degree to which it can be extended and reimplemented in new ways. IoT systems must be prepared to share their data and resources with other systems in sometimes unpredictable ways (Alur et al., 2016; Perera et al., 2014; Ahlgren et al., 2016; Rahman and Hussain, 2020) In order to make sustainable IoT systems, "openness must provide a correct balance between access to functionality, human interaction, and privacy and security" (Alur et al., 2016).

1.6.3 Security, Privacy, and Trust

In addition to the current already complex security and privacy landscape, IoT introduces considerably more security and privacy issues. Often, IoT systems rely on wireless communications that intrinsically pose security problems. Additionally, the large amount of data generated raises new concerns not only about managing, processing, and analyzing such an amount of data but also about how to ensure data confidentiality. IoT systems, especially those that collect sensitive data (e.g., healthcare systems), need to be secured at all layers, from the physical layer all the way up to the application layer (Perera et al., 2014; Rose et al., 2015; Chanal and Kakkasageri, 2020; Nebbione and Calzarossa, 2020; Ogonji et al., 2020; Tournier et al., 2021; Schiller, 2022). Existing IoT devices and systems have been shown to be particularly vulnerable to denial of service attacks and other cybersecurity attacks (Farid, 2025; Zhao et al., 2025). Only with adequate security and data protection mechanisms in place can the IoT systems gain trust from the users (Ortiz et al., 2014; Rose et al., 2015). Security and privacy issues should be considered from the very beginning of the system design (Miorandi et al., 2012). Moreover, security zones and dynamic trust management in such environments are equally important (Lee et al., 2025).

Caution is advised in the way data is processed, particularly ensuring it is kept anonymous until official data protection authorities make formal recommendations (Serbanati et al., 2011). In fact, this is an area of ongoing work, as more research is needed to overcome these issues. Many end-user companies and customers point to security as the main reason not to have embraced IoT yet, while others find that data protection and privacy are the greatest barriers to the development of the IoT. An assessment of the extent to which existing data protection regulations fully address the IoT needs to be carried out in order to establish what actions still need to be taken. Key legal issues that might challenge the use of IoT need to be discussed, as well as ways to overcome them. In regard to personal data processing, the EU has published several regulations and directives, including the General Data Protection Regulation.[45]

Nevertheless, so far, some security frameworks have been proposed to provide confidentiality, integrity, and authentication (Serbanati et al., 2011; Al-Turjman and Alturjman, 2018). However, these frameworks add some communication and processing overhead to achieve their goal. Security requirements for IoT systems may be grouped into four sets (Borgia, 2014): secure authentication and authorization, secure configuration and data transmission, secure data storage, and secure access to data. For all these requirements, it is necessary to keep in mind that many IoT devices are low power and resource constrained (Tiloca and Raza, 2018). As such, some of the conventional techniques may not be adequate and new ones must be developed (e.g., lightweight cryptography; McKay, 2017). The use of emerging technologies like blockchain, machine learning, and radio frequency fingerprinting techniques, which have been gaining traction in recent years, also seems to improve security in IoT applications (Pulkkis et al., 2018; Tahsien et al., 2020; Wang et al., 2025; Li and Cetin, 2025). Many of these issues and resolutions are covered in great detail in Chapters 2–5 of this book.

1.6.4 Scalability

A scalable system continues to work effectively even when the amount of resources and the number of users increase significantly. In regard to the IoT, billions of devices are already connected to the Internet, and many more are expected to be added in the near future. Many of these devices will be mobile and will have low power capacities and unstable connections. As such, the traditional solutions may not be enough to guarantee proper functioning of the networks (Alur et al., 2016). At least two levels of scalability may be considered in the scope of IoT: network scalability and data scalability (Borgia, 2014). As the network of interconnected objects grows, the importance of interoperability will become crucial to successful adoption and implementation. The same goes with data security and privacy. Issues related to energy consumption will also need to be tackled (Miorandi et al., 2012).

The volume and variety of IoT data and the velocity at which it is being generated can be overwhelming (i.e., big data), which will require new and more efficient ways to handle it (Cisco, 2014). Additionally, with so many different types of devices generating data, there are many reasons why this data may be stored in disparate data storage: There might be too much data to put in one place; transferring data into a central database might consume too much processing power, so it is recommended that data is retrieved separately from the data generation process; devices might be geographically dispersed, and it is advisable to process data locally in order to obtain some processing optimization; there might be the need to separate raw data from data that represents an event; and finally, different kinds of data processing might be required (e.g., batch processing and real-time processing). For these reasons, the data abstraction level must process many different things, which include integrating multiple data formats from different sources. Ensuring consistent semantics of data across various sources is of extreme relevance. Finally, confirming that data is complete for the higher-level application is also as important (Cisco, 2014).

1.6.5 Failure Handling

Failure handling includes detecting, masking, and tolerating failures; recovering from failures; and redundancy of resources. In such a complex, dynamic, and heterogeneous environment as the case with IoT, systems are required to be able to self-configure, self-diagnose, and self-repair (Alur et al.,

2016). Gateways, as more resourceful components of the IoT systems, maybe the right place to implement self-management fault, configuration, accounting, performance, and security features (Al-Fuqaha et al., 2015).

Systems that successfully meet failure-handling requirements are going to gain trust more rapidly among their users (Ortiz et al., 2014; Bagchi et al., 2020; Berger et al., 2021; Aboubakar et al., 2022). This, however, makes it hard for developers to find the right balance between application logic and error handling (Taivalsaari and Mikkonen, 2017). Nevertheless, recent advances in artificial intelligence and machine learning are expected to help in addressing these issues (Samaraweera et al., 2025; Byabazaire et al., 2025; Tirana and Chatzopoulos, 2025; Bhattarai et al., 2025). By implementing intelligent mechanisms on IoT devices and gateways, as well as backend servers (whether on the cloud or at the edge, or a mix of both), the efficiency of data processing and failure handling can improve considerably. Such techniques could allow us to generate intelligent mechanisms for dynamic anomaly detection and failure handling (among many other use cases) instead of relying on static, preprogrammed rules.

1.7 CONCLUSION

Statista estimates that there are approximately 18 billion IoT-connected devices today with annual revenue of $390.5 billion worldwide.[46] It also predicts that by 2033, there will be nearly 40 billion IoT devices, with annual revenue exceeding $934 billion worldwide. Despite this foreseen growth, the IoT ecosystem is complex with many layers that need to be understood well. Also, many players are involved in its implementation, including device makers, communication and IT service providers, platform providers, and software vendors. This chapter introduced the main topics of IoT, namely, its definitions, reference models, and RAs, enabling technologies, standards, main application domains, and challenges. As explained previously, international bodies, enterprises, academia, and industry are working together on a common definition and a RA for the IoT, as well as on standards that will enable interoperability among systems. Solutions to tackle other challenges, such as security and privacy, openness, scalability, and failure handling are also being explored. Many are already on the table, and new ones are still emerging.

As a network of devices that communicate autonomously and continuously connected to the Internet, the IoT has several social, individual, economic, and environmental implications. As discussed in Section 1.5, IoT applications are already having a positive impact on many aspects of life. The development of smart cities, for example, in terms of infrastructure, transport, and buildings, has had a significant societal impact by improving the efficiency and sustainability of a whole range of urban services. IoT also plays a crucial role in Industry 4.0 (and that will be extended to Industry 5.0), in which industrial sites are being transformed through intelligent and interconnected objects that improve performance, lower costs, and increase reliability.

5G/6G, artificial intelligence, edge computing, embedded machine learning, augmented reality, virtual reality, and digital twins are some of the currently emerging trends for the IoT. The potential of the IoT associated with these trends points toward a promising future. Over time, the results of the collaborative work done by different groups from industry, academia, and standardization bodies will become evident. It is already happening.

NOTES

1 This chapter was originally published in Hassan, Q.F, Khan, A.R. and Madani, S.A. (eds.), 2017. *Internet of Things: Challenges, Advances, and Applications*. It has been updated and expanded by the editor for inclusion in this book.
2 http://info.cern.ch/.
3 https://datatracker.ietf.org/doc/html/rfc2460.

4 http://www.iot-conference.org/iot2008/.
5 https://www.arduino.cc/.
6 https://www.raspberrypi.org/.
7 https://beagleboard.org/black.
8 https://onelab.eu/.
9 http://eur-lex.europa.eu/LexUriServ/LexUriServ.do?uri=COM:2009:0278:FIN:EN:PDF.
10 http://www.onem2m.org/.
11 https://aioti.eu/
12 https://www.statista.com/statistics/1183457/iot-connected-devices-worldwide/.
13 http://www.internet-of-things-research.eu/about_iot.htm.
14 https://www.iso.org/committee/45020.html.
15 https://www.iso.org/files/live/sites/isoorg/files/developing_standards/docs/en/internet_of_things_report-jtc1.pdf.
16 https://tools.ietf.org/html/rfc7452.
17 http://iot.ieee.org/definition.html.
18 https://standards.ieee.org/wp-content/uploads/import/documents/other/iot.pdf.
19 https://standards.ieee.org/ieee/2413/6226/.
20 http://www.gsma.com/.
21 https://www.gsma.com/solutions-and-impact/technologies/internet-of-things/.
22 https://www.itu.int/en/ITU-T/gsi/iot/Pages/default.aspx.
23 https://www.itu.int/en/ITU-T/studygroups/2017-2020/20/Pages/default.aspx.
24 https://www.iiconsortium.org/IIRA/.
25 https://www.iso.org/obp/ui/#iso%3Astd%3Aiso-iec%3A29182%3A-1%3Aed-1%3Av1%3Aen.
26 https://www.iso.org/standard/88800.html.
27 https://www.w3.org/WoT/IG/wiki/images/9/9a/10N0536_CD_text_of_ISO_IEC_30141.pdf.
28 http://www.gs1.org.
29 https://www.iso.org/standard/88800.html.
30 https://www.iso.org/standard/74393.html.
31 https://www.iiconsortium.org/IIRA/.
32 http://www.iiconsortium.org/IIC_PUB_G1_V1.80_2017-01-31.pdf.
33 https://iot.ieee.org/articles-publications/newsletter/january-2016/hypercat-resource-discovery-on-the-internet-of-things.html.
34 http://multicastdns.org/.
35 https://datatracker.ietf.org/doc/html/rfc7452.
36 https://pine64.org/.
37 https://www.iot-lab.info/.
38 https://www.europeana-space.eu/partners/iminds-vzw-iminds-belgium/.
39 https://onelab.eu/.
40 https://www.iiconsortium.org/.
41 We should note that while these areas represent the most common "commercial" IoT application domains, the technology is also being widely used in other domains, including military applications and defense systems (Douglass et al., 2022; Sliwa, 2025).
42 http://www.organicity.eu.
43 https://cordis.europa.eu/project/id/646440/reporting.
44 https://www.waze.com/live-map/.
45 https://gdpr-info.eu/.
46 https://www.statista.com/statistics/1194709/iot-revenue-worldwide/.

REFERENCES

Abbasi, A.Z., Islam, N. and Shaikh, Z.A., 2014. A review of wireless sensors and networks' applications in agriculture. *Computer Standards & Interfaces*, 36(2), pp.263–270. https://doi.org/10.1016/j.csi.2011.03.004.

Aboubakar, M., Kellil, M. and Roux, P., 2022. A review of IoT network management: Current status and perspectives. *Journal of King Saud University-Computer and Information Sciences*, 34(7), pp.4163–4176. https://doi.org/10.1016/j.jksuci.2021.03.006.

Adolphs, P., Bedenbender, H., Dirzus, D., Ehlich, M., Epple, U., Hankel, M., Heidel, R., Hoffmeister, M., Huhle, H., Kärcher, B. and Koziolek, H., 2015. *Reference Architecture Model Industrie 4.0 (RAMI4. 0)*. ZVEI and VDI, status report.

Ahlgren, B., Hidell, M. and Ngai, E.C.H., 2016. Internet of things for smart cities: Interoperability and open data. *IEEE Internet Computing*, 20(6), pp.52–56. https://doi.org/10.1109/mic.2016.124.

Ahmad, R. and Alsmadi, I., 2021. Machine learning approaches to IoT security: A systematic literature review. *Internet of Things*, 14, p.100365. https://doi.org/10.1016/j.iot.2021.100365.

Ahmad, T. and Zhang, D., 2021. Using the internet of things in smart energy systems and networks. *Sustainable Cities and Society*, 68, p.102783. https://doi.org/10.1016/j.scs.2021.102783.

AIOTI (Alliance for the Internet of Things). 2015. Internet of Things applications. https://aioti.eu/wp-content/uploads/2017/03/AIOTIWG01Report2015-Applications.pdf.

Albouq, S.S., Abi Sen, A.A., Almashf, N., Yamin, M., Alshanqiti, A. and Bahbouh, N.M., 2022. A survey of interoperability challenges and solutions for dealing with them in IoT environment. *IEEE Access*, 10, pp.36416–36428. https://doi.org/10.1109/access.2022.3162219.

Al-Fuqaha, A., Guizani, M., Mohammadi, M., Aledhari, M. and Ayyash, M., 2015. Internet of things: A survey on enabling technologies, protocols, and applications. *IEEE Communications Surveys Tutorials*, 17(4), pp.2347–2376. https://doi.org/10.1109/comst.2015.2444095.

Alsubaei, F., Abuhussein, A. and Shiva, S., 2018. An overview of enabling technologies for the Internet of Things. In Hassan, Q.F. (ed.), *Internet of Things A to Z: Technologies and applications*, pp.77–112. Wiley & Sons. https://doi.org/10.1002/9781119456735.ch3.

Al-Turjman, F. and Alturjman, S., 2018. Confidential smart-sensing framework in the IoT era. *The Journal of Supercomputing*, 74(10), pp.5187–5198. https://doi.org/10.1007/s11227-018-2524-1.

Alur, R., Berger, E., Drobnis, A.W., Fix, L., Fu, K., Hager, G.D., Lopresti, D., Nahrstedt, K., Mynatt, E., Patel, S., Rexford, J., Stankovic, J.A. and Zorn, B., 2016. Systems computing challenges in the Internet of Things. ArXiv Preprint ArXiv:1604.02980. https://arxiv.org/pdf/1604.02980.

Anagnostopoulos, T., Ferreira, D., Samodelkin, A., Ahmed, M. and Kostakos, V., 2016. Cyclist-aware traffic lights through distributed smartphone sensing. *Pervasive and Mobile Computing*, 31(C), pp.22–36. https://doi.org/10.1016/j.pmcj.2016.01.012.

Atzori, L., Iera, A. and Morabito, G., 2010. The Internet of Things: A survey. *Computer Networks*, 54(15), pp.2787–2805. https://doi.org/10.1016/j.comnet.2010.05.010.

Bagchi, S., Abdelzaher, T.F., Govindan, R., Shenoy, P., Atrey, A., Ghosh, P. and Xu, R., 2020. New frontiers in IoT: Networking, systems, reliability, and security challenges. *IEEE Internet of Things Journal*, 7(12), pp.11330–11346. https://doi.org/10.1109/jiot.2020.3007690.

Bassi, A., Bauer, M., Fiedler, M., Kramp, T., Van Kranenburg, R., Lange, S. and Meissner, S., 2013. *Enabling things to talk: Designing IoT solutions with the IoT architectural reference model*, pp.163–211. Springer. https://doi.org/10.1007/978-3-642-40403-0.

Bauer, M., Bui, N., De Loof, J., Magerkurth, C., Nettsträter, A., Stefa, J. and Walewski, J.W., 2013. IoT reference model. In Bassi, A., et al. (eds.), *Enabling things to talk: Designing IoT solutions with the IoT architectural reference model*, pp.113–162. Springer. https://doi.org/10.1007/978-3-642-40 403-0_7.

Ben-Daya, M., Hassini, E. and Bahroun, Z., 2019. Internet of Things and supply chain management: A literature review. *International Journal of Production Research*, 57(15–16), pp.4719–4742. https://doi.org/10.1080/00207543.2017.1402140.

Berger, C., Eichhammer, P., Reiser, H.P., Domaschka, J., Hauck, F.J. and Habiger, G., 2021. A survey on resilience in the iot: Taxonomy, classification, and discussion of resilience mechanisms. *ACM Computing Surveys (CSUR)*, 54(7), pp.1–39. https://doi.org/10.1145/3462513.

Bhattarai, A., Samaraweera, C., Peng, D., Liu, Y. and Sharif, H., 2025. Tackling integration challenges of machine learning in diverse Internet of Things: A spread-out architectural solution. In Hassan, Q.F. (ed.), *Advances in Internet of Things: Challenges, solutions, and emerging technologies*. Chapman and Hall/ CRC. https://doi.org/10.1201/9781003506638.

Bi, Z., Da Xu, L. and Wang, C., 2014. Internet of Things for enterprise systems of modern manufacturing. *IEEE Transactions on Industrial Informatics*, 10(2), pp.1537–1546. https://doi.org/10.1109/tii.2014.2300338.

Bo, Y. and Wang, H., 2011. The application of cloud computing and the Internet of Things in agriculture and forestry. In *2011 International Joint Conference on Service Sciences (IJCSS)*, Taipei, Taiwan, pp.168–172.

Bonomi, F., Milito, R., Zhu, J. and Addepalli, S., 2012. Fog computing and its role in the Internet of Things. In *Proceedings of the First Edition of the MCC Workshop on Mobile Cloud Computing*, Association for Computing Machinery. New York, United States, pp.13–16. https://doi.org/10.1145/2342509.2342513.

Borgia, E., 2014. The Internet of Things vision: Key features, applications and open issues. *Computer Communications*, 54, pp.1–31. https://doi.org/10.1109/ijcss.2011.40.

Brad, S. and Murar, M., 2014. Smart buildings using IoT technologies. *Construction of Unique Buildings and Structures*, 5(20), pp.15–27. https://unistroy.spbstu.ru/userfiles/files/2014/5(20)/2_brad_20.pdf.

Bugeja, J., Jacobsson, A. and Davidsson, P., 2018. Smart connected homes. In Hassan, Q.F. (ed.), *Internet of Things A to Z: Technologies and applications*, pp.359–384. Wiley & Sons. https://doi.org/10.1002/9781119456735.ch13.

Bui, N., Castellani, A.P., Casari, P. and Zorzi, M., 2012. The Internet of energy: A web-enabled smart grid system. *IEEE Network*, 26(4), pp.39–45. https://doi.org/10.1109/mnet.2012.6246751.

Bui, N. and Zorzi, M., 2011. Health care applications: A solution based on the Internet of things. In *Proceedings of the 4th International Symposium on Applied Sciences in Biomedical and Communication Technologies*, Barcelona, p.131. https://doi.org/10.1145/2093698.2093829.

Byabazaire, J., Tirana, J., Chouliaras, A., Koutsos, V., Aslanidis, T., Panagiotidis, I. and Chatzopoulos, D., 2025. Deep learning and the Internet of Things: Applications, challenges and opportunities. In Hassan, Q.F. (ed.), *Internet of Things A to Z: Technologies and applications*, Second Editions. Wiley and Sons, Hobokon, New Jersey. In Press.

Campolo, C., Iera, A., Molinaro, A., Yuri Paratore, S. and Ruggeri, G., 2012. SMeaRTCaR: An integrated smartphone-based platform to support traffic management applications. In *2012 First International Workshop on Vehicular Traffic Management for Smart Cities (VTM)*, Dublin, pp.1–6. https://doi.org/10.1109/vtm.2012.6398700.

Castro, M., Jara, A.J. and Skarmeta, A.F.G., 2013. Smart lighting solutions for smart cities. In 2013 *27th International Conference on Advanced Information Networking and Applications Workshops (WAINA)*, Barcelona, pp.1374–1379. https://doi.org/10.1109/waina.2013.254.

Chanal, P.M. and Kakkasageri, M.S., 2020. Security and privacy in IoT: A survey. *Wireless Personal Communications*, 115(2), pp.1667–1693. https://doi.org/10.1007/s11277-020-07649-9.

Chaturvedi, P., Thakur, K.K., Mali, N., Kala, V.U., Kumar, S., Yadav, S. and Dutt, V., 2018. A low-cost IoT framework for landslide prediction and risk communication. In Hassan, Q.F. (ed.), *Internet of Things A to Z: Technologies and applications*, pp.593–610. Wiley and Sons. https://doi.org/10.1002/9781119456735.ch21.

Cisco, 2014. The Internet of Things reference model. In *Internet of Things World Forum*, San Jose, CA, pp.1–12.

Da Xu, L., He, W. and Li, S., 2014. Internet of Things in industries: A survey. *IEEE Transactions on Industrial Informatics*, 10(4), pp.2233–2243. https://doi.org/10.1109/tii.2014.2300753.

Dohr, A., Modre-Osprian, R., Drobics, M., Hayn, D. and Schreier, G., 2010. The Internet of Things for ambient assisted living. *ITNG*, 10, pp.804–809. https://doi.org/10.1109/itng.2010.104.

Domingo, M.C., 2012. An overview of the Internet of Things for people with disabilities. *Journal of Network and Computer Applications*, 35(2), pp.584–596. https://doi.org/10.1016/j.jnca.2011.10.015.

Douglass, R., Gremban, K., Swami, A. and Gerali, S., 2022. *IoT for Defense and National Security*. Wiley & Sons. https://www. doi.org/10.1002/9781119892199.

Doukas, C. and Maglogiannis, I., 2012. Bringing IoT and cloud computing towards pervasive healthcare. In *2012 Sixth International Conference on Innovative Mobile and Internet Services in Ubiquitous Computing (IMIS)*, Palermo, Italy, pp.922–926. https://doi.org/10.1109/imis.2012.26.

Du, K., Mu, C., Deng, J. and Yuan, F., 2013. Study on atmospheric visibility variations and the impacts of meteorological parameters using high temporal resolution data: An application of environmental Internet of things in China. *International Journal of Sustainable Development & World Ecology*, 20(3), pp.238–247. https://doi.org/10.1080/13504509.2013.783886.

Eteng, A.A., Abdul Rahim, S.K. and Leow, C.Y., 2018. RFID in the Internet of Things. In Hassan, Q.F. (ed.), *Internet of Things A to Z: Technologies and applications*, pp.135–152. Wiley & Sons. https://doi.org/10.1002/9781119456735.ch5.

Fang, S., Da Xu, L., Zhu, Y., Ahati, J., Pei, H., Yan, J. and Liu, Z., 2014. An integrated system for regional environmental monitoring and management based on Internet of things. *IEEE Transactions on Industrial Informatics*, 10(2), pp.1596–1605. https://doi.org/10.1109/tii.2014.2302638.

Fantana, N., Riedel, T., Schlick, J., Ferber, S., Hupp, J., Miles, S., Michahelles, F. and Svensson, S., 2013. *IoT applications – Value creation for industry*, p.153. River Publishers Series in Communications. River Publishers. https://doi.org/10.1201/9781003338659-3.

Farid, F., 2025. A threat analysis framework for cyberattacks in smart cities: Ransomware in focus. In Hassan, Q.F. (ed.), *Advances in Internet of Things: Challenges, solutions, and emerging technologies*. Chapman and Hall/CRC. https://doi.org/10.1201/9781003506638.

Finco, L. and Minoli, D., 2018. Implementing the Internet of Things for renewable energy. In Hassan, Q.F. (ed.), *Internet of Things A to Z: Technologies and applications*, pp.425–446. Wiley & Sons. https://doi.org/10.1002/9781119456735.ch15.

Fong, B., Fong, A.C.M. and Li, C.K., 2018. Internet of Things in smart ambulance and emergency medicine. In Hassan, Q.F. (ed.), *Internet of Things A to Z: Technologies and applications*, pp.475–506. Wiley & Sons. https://doi.org/10.1002/9781119456735.ch17.

Gershenfeld, N., 1999. *When things start to think*. Henry Holt and Co. https://doi.org/10.2307/20049565.

Gil-Castineira, F., Costa-Montenegro, E., Gonzalez-Castano, F.J., López-Bravo, C., Ojala, T. and Bose, R., 2011. Experiences inside the ubiquitous Oulu smart city. *Computer*, 44(6), pp.48–55. https://doi.org/10.1109/mc.2011.132.

GSMA (GSM Association), 2016. GSMA IoT security guidelines. https://www.gsma.com/solutions-and-impact/technologies/internet-of-things/iot-security/iot-security-guidelines/.

Gubbi, J., Buyya, R., Marusic, S. and Palaniswami, M., 2013. Internet of Things (IoT): A vision, architectural elements, and future directions. *Future Generation Computer Systems*, 29(7), pp.1645–1660. https://doi.org/10.1016/j.future.2013.01.010.

Guinard, D. and Trifa, V., 2016. *Building the web of things: With examples in node. js and raspberry pi*. Simon and Schuster.

Happ, D., 2018. Cloud and fog computing in the Internet of Things. In Hassan, Q.F. (ed.), *Internet of Things A to Z: Technologies and applications*, pp.113–134. Wiley & Sons. https://doi.org/10.1002/9781119456735.ch4.

Hassan, Q.F., 2011. Demystifying cloud computing. *Journal of Defense Software Engineering*, 24, pp.16–21. https://www.researchgate.net/publication/215795101_Demystifying_Cloud_Computing.

Hassan, Q.F. ed., 2018. *Internet of Things A to Z: Technologies and applications*. John Wiley & Sons. https://doi.org/10.1002/9781119456735.

Hassan, Q.F., Khan, A.R. and Madani, S.A., eds., 2017. *Internet of Things: Challenges, advances, and applications*. Chapman and Hall/CRC. https://doi.org/10.4324/9781315155005.

Hazra, A., Adhikari, M., Amgoth, T. and Srirama, S.N., 2021. A comprehensive survey on interoperability for IIoT: Taxonomy, standards, and future directions. *ACM Computing Surveys (CSUR)*, 55(1), pp.1–35. https://doi.org/10.1145/3485130.

Hong, I., Park, S., Lee, B., Lee, J., Jeong, D. and Park, S., 2014. IoT-based smart garbage system for efficient food waste management. *Scientific World Journal*, 2014(2014), p.646953. https://doi.org/10.1155/2014/646953.

ISO/IEC (International Organization for Standardization/International Electrotechnical Commission), 2014. Study report on IoT reference architectures/frameworks. https://www.iso.org/files/live/sites/isoorg/files/developing_standards/docs/en/internet_of_things_report-jtc1.pdf.

ISO/IEC (International Organization for Standardization/International Electrotechnical Commission), 2016. Information technology – Data structure – Unique identification for the Internet of things. ISO/IEC 29161:2016. August. https://www.iso.org/standard/45240.html#:~:text=The%20unique%20identification%20is%20a,use%20with%20any%20IoT%20media.

Istepanian, R.S.H., Hu, S., Philip, N.Y. and Sungoor, A., 2011. The potential of Internet of M-health things 'm-IoT' for non-invasive glucose level sensing. In *2011 Annual International Conference of the IEEE Engineering in Medicine and Biology Society*, Boston, MA, pp.5264–5266. https://doi.org/10.1109/iembs.2011.6091302.

ITU-T (International Telecommunications Union Telecommunication Standardization Sector), 2005. *ITU Internet reports, the Internet of Things*. ITU-T.

ITU-T (International Telecommunications Union Telecommunication Standardization Sector), 2012. *Overview of the Internet of Things*. Y.2060. https://www.itu.int/rec/t-rec-y.2060-201206-i.

Jakobs, K., 2018. On Standardizing the Internet of Things and its applications. In Hassan, Q.F. (ed.), *Internet of Things A to Z: Technologies and applications*, pp.191–218. Wiley & Sons. https://doi.org/10.1002/9781119456735.ch7.

Ji, S.W., Yun Teng, H. and Feng Su, J., 2014. The application and development of the Internet of things in intelligent buildings. *Advanced Materials Research*, 834, pp.1854–1857. https://doi.org/10.4028/www.scientific.net/amr.834-836.1854.

Karzel, D., Marginean, H. and Tran, T.-S., 2016. A reference architecture for the Internet of Things. January 29. https://www.infoq.com/articles/internet-of-things-reference-architecture.

Kelly, S.D.T., Suryadevara, K. and Chandra Mukhopadhyay, S., 2013. Towards the implementation of IoT for environmental condition monitoring in homes. *IEEE Sensors Journal*, 13(10), pp.3846–3853. https://doi.org/10.1109/jsen.2013.2263379.

Kumar, P., Husain, A., Priyanka, P., Kumar, A., Sankhayan, S., Uday, K.V. and Dutt, V., 2025b. GeoSense: A smart IoT-Driven framework for monitoring landslides in the himalayas. In Hassan, Q.F. (ed.), *Advances in Internet of Things: Challenges, solutions, and emerging technologies*. Chapman and Hall/CRC. https://doi.org/10.1201/9781003506638.

Kumar, Y., Kapur, R., Sankhyan, S., Premchand, Sharma, S., Bhavsar, A. and Dutt, V., 2025a. IoTMaxHRV: A low-cost IoT device for heart rate variability assessment. In Hassan, Q.F. (ed.), *Advances in Internet of Things: Challenges, solutions, and emerging technologies*. Chapman and Hall/CRC. https://doi.org/10.1201/9781003506638.

Kyriazis, D., Varvarigou, T., White, D., Rossi, A. and Cooper, J., 2013. Sustainable smart city IoT applications: Heat and electricity management & eco-conscious cruise control for public transportation. In *2013 IEEE 14th International Symposium and Workshops on a World of Wireless, Mobile and Multimedia Networks (WoWMoM)*, Madrid, pp.1–5. https://doi.org/10.1109/wowmom.2013.6583500.

Lee, B., Imtiaz, B., McLoughlin, K. and Ye, Y., 2025. On the role of dynamic trust management in the evolving Internet of Things. In Hassan, Q.F. (ed.), *Internet of Things A to Z: Technologies and applications*, Second Edition. Wiley & Sons. In Press.

Li, B. and Cetin, E., 2025. Enhancing IoT security with radio frequency fingerprinting: Traditional and deep learning-based approaches. In Hassan, Q.F. (ed.), *Advances in Internet of Things: Challenges, solutions, and emerging technologies*. Chapman and Hall/CRC. https://doi.org/10.1201/9781003506638.

Li, L., Xiaoguang, H., Ke, C. and Ketai, H., 2011. The applications of Wi-Fi-based wireless sensor network in Internet of Things and smart grid. In *2011 6th IEEE Conference on Industrial Electronics and Applications*, Beijing, pp.789–793. https://doi.org/10.1109/iciea.2011.5975693.

Li, X., 2013. Multi-day and multi-stay travel planning using geo-tagged photos. In *Proceedings of the Second ACM SIGSPATIAL International Workshop on Crowdsourced and Volunteered Geographic Information* (GEOCROWD '13), Orlando, FL, pp.1–8. https://doi.org/10.1145/2534732.2534733.

Li, X., Lu, R., Liang, X., Shen, X., Chen, J. and Lin, X., 2011. Smart community: An Internet of things application. *IEEE Communications Magazine*, 49(11), pp.68–75. https://doi.org/10.1109/mcom.2011.6069711.

Li, Z., Wang, T., Gong, Z. and Li, N., 2013. Forewarning technology and application for monitoring low temperature disaster in solar greenhouses based on Internet of Things. *Transactions of the Chinese Society of Agricultural Engineering*, 29(4), pp.229–236. https://www.ingentaconnect.com/content/tcsae/tcsae/2013/00000029/00000004/art00029?crawler=true&mimetype=application/pdf.

Liu, X., Ceorner, M.D. and Shenoy, P., 2006. Ferret: RFID localization for pervasive multimedia. In *UbiComp 2006: Ubiquitous computing*, Orange County, CA, pp.422–440. https://doi.org/10.1007/11853565_25.

McKay, K.A., Bassham, L., Turan, M.S. and Mouha, N., 2017. *Report on lightweight cryptography*. NIST. https://doi.org/10.6028/nist.ir.8114.

Mell, P. and Grance, T., 2011. *The NIST definition of cloud computing*. National Institute of Standards and Technology. https://doi.org/10.6028/nist.sp.800-145.

Meruje, M., Samaila, M.G., Franqueira, V.N., Freire, M.M. and Inácio, P.R.M., 2018. A tutorial introduction to IoT design and prototyping with examples. In Hassan, Q.F. (ed.), *Internet of things A to Z: Technologies and applications*, First Edition. Wiley & Sons, pp. 153–190. In press.

Minerva, R., Biru, A. and Rotondi, D., 2015. *Towards a definition of the Internet of Things (IoT)*. IEEE Internet of Things. https://iot.ieee.org/images/files/pdf/IEEE_IoT_Towards_Definition_Internet_of_Things_Revision1_27MAY15.pdf.

Minoli, D., 2013. *Building the Internet of Things with IPv6 and MIPv6: The evolving world of M2M communications*. Wiley. https://doi.org/10.1002/9781118647059.

Minoli, D. and Occhiogrosso, B., 2018a. Internet of Things applications for smart cities. In Hassan, Q.F. (ed.), *Internet of Things A to Z: Technologies and applications*, pp.319–358. Wiley & Sons. https://doi.org/10.1002/9781119456735.ch12.

Minoli, D. and Occhiogrosso, B., 2018b. The emerging "Energy Internet of Things". In Hassan, Q.F. (ed.), *Internet of Things A to Z: Technologies and applications*, pp.385–424. Wiley & Sons. https://doi.org/10.1002/9781119456735.ch14.

Miorandi, D., Sicari, S., De Pellegrini, F. and Chlamtac, I., 2012. Internet of Things: Vision, applications and research challenges. *Ad Hoc Networks*, 10(7), pp.1497–1516. https://doi.org/10.1016/j.adhoc.2012.02.016.

Moreno, M.V., Zamora, M.A. and Skarmeta, A.F., 2014. User-centric smart buildings for energy sustainable smart cities. *Transactions on Emerging Telecommunications Technologies*, 25(1), pp.41–55. https://doi.org/10.1002/ett.2771.

Nebbione, G. and Calzarossa, M.C., 2020. Security of IoT application layer protocols: Challenges and findings. *Future Internet*, 12(3), p.55. https://doi.org/10.3390/fi12030055.

Noaman, M., Khan, M.S., Abrar, M.F., Ali, S., Alvi, A. and Saleem, M.A., 2022. Challenges in integration of heterogeneous internet of things. *Scientific Programming*, 2022(1), p.8626882. https://doi.org/10.1155/2022/8626882.

Ogonji, M.M., Okeyo, G. and Wafula, J.M., 2020. A survey on privacy and security of Internet of Things. *Computer Science Review*, 38, p.100312. https://doi.org/10.1016/j.cosrev.2020.100312.

Ortiz, A.M., Hussein, D., Park, S., Han, S.N. and Crespi, N., 2014. The cluster between Internet of Things and social networks: Review and research challenges. *IEEE Internet of Things Journal*, 1(3), pp.206–215. https://doi.org/10.1109/jiot.2014.2318835.

Ou, Q., Zhen, Y., Li, X., Zhang, Y. and Zeng, L., 2012. Application of Internet of Things in smart grid power transmission. In *2012 Third FTRA International Conference on Mobile, Ubiquitous, and Intelligent Computing (MUSIC)*, Vancouver, BC, pp.96–100. https://doi.org/10.1109/music.2012.24.

Pang, Z., 2013. Technologies and architectures of the Internet-of-Things (IoT) for health and well-being. Doctoral thesis, KTH, School of Information and Communication Technology (ICT), Electronic Systems, Kista, Sweden. https://www.diva-portal.org/smash/record.jsf?pid=diva2:621384.

Perera, C., Zaslavsky, A., Christen, P. and Georgakopoulos, D., 2014. Context aware computing for the Internet of Things: A survey. *IEEE Communications Surveys and Tutorials*, 16(1), pp.414–454. https://doi.org/10.1109/surv.2013.042313.00197.

Phithakkitnukoon, S., Wolf, M.I., Offenhuber, D., Lee, D., Biderman, A. and Ratti, C., 2013. Tracking trash. *IEEE Pervasive Computing*, 12(2), pp.38–48. https://doi.org/10.1109/mprv.2013.37.

Pulkkis, G., Karlsson, J. and Westerlund, M., 2018. Blockchain-based security solutions for IoT systems. In Hassan, Q.F. (ed.), *Internet of Things A to Z: Technologies and applications*, pp.255–274. Wiley & Sons. https://doi.org/10.1002/9781119456735.ch9.

Rahman, H. and Hussain, M.I., 2020. A comprehensive survey on semantic interoperability for Internet of Things: State-of-the-art and research challenges. *Transactions on Emerging Telecommunications Technologies*, 31(12), p.e3902. https://doi.org/10.1002/ett.3902.

Ranga, S., Vasudev, T., Kumar, P., Sankhyan, S., Kumar, Y., Uday, K.V. and Dutt, V., 2025. IoT-MediGuard: An advanced monitoring and predictive analytics system for transportation of medical supplies. In Hassan, Q.F. (ed.), *Advances in Internet of Things: Challenges, solutions, and emerging technologies*. Chapman and Hall/CRC. https://doi.org/10.1201/9781003506638.

Rose, K., Eldridge, S. and Chapin, L., 2015. The internet of things: An overview. *The Internet Society (ISOC)*, 80(15), pp.1–53. https://www.internetsociety.org/wp-content/uploads/2021/01/ISOC-IoT-Overview-20151014_0.pdf.

Russo, N.L. and Eriksson, J., 2018. The Internet of Things and people in health care. In Hassan, Q.F. (ed.), *Internet of Things A to Z: Technologies and applications*, pp.447–474. Wiley & Sons. https://doi.org/10.1002/9781119456735.ch16.

Ryu, C-S., 2015. IoT-based intelligent for fire emergency response systems. *International Journal of Smart Home*, 9(3), pp.161–168. https://doi.org/10.14257/ijsh.2015.9.3.15.

Sagahyroon, A., Aburukba, R. and Aloul, F., 2017. The Internet of things and e-Health: Remote patients monitoring. In Hassan, Q.F., Khan, A.R. and Madani, S.A. (eds.), *Internet of Things: Challenges, advances, and applications*, pp.303–319. Chapman and Hall/CRC.

Samaila, M.G., Sequeiros, J.B., Correia, A.F., Freire, M.M. and Inácio, P.R., 2017. IoT hardware development platforms: Past, present, and future. In Hassan, Q.F., Khan, A.R. and Madani, S.A. (eds.), *Internet of Things: Challenges, advances, and applications*, pp.107–139. Chapman and Hall/CRC.

Samaraweera, C., Bhattarai, A., Peng, D. and Sharif, H., 2025. Artificial Intelligence and the Internet of Things: Algorithms, Applications, and Challenges. In Hassan, Q.F. (ed.), *Internet of Things A to Z: Technologies and applications*, Second Editions. Wiley and Sons. In Press.

Sanchez, L., Antonio Galache, J., Gutierrez, V., Manuel Hernandez, J., Bernat, J., Gluhak, A. andGarcia, T., 2011. Smartsantander: The meeting point between future Internet research and experimentation and the smart cities. In *Future Network and Mobile Summit (FutureNetw)*, pp.1–8. Warsaw. https://www.smart-santander.eu/downloads/Presentations/futurenetworksummit2011.pdf.

Schiller, E., Aidoo, A., Fuhrer, J., Stahl, J., Ziörjen, M. and Stiller, B., 2022. Landscape of IoT security. *Computer Science Review*, 44, p.100467. https://doi.org/10.1016/j.cosrev.2022.100467.

Schoder, D., 2025. Introduction to the Internet of Things. In Hassan, Q.F. (ed.), *Internet of Things A to Z: Technologies and applications*, Second Edition. Wiley & Sons. In Press.

Serbanati, A., Medaglia, C.M., Ceipidor, U.B. and Turcu, C., 2011. Building blocks of the internet of things: State of the art and beyond. *Deploying RFID-Challenges, Solutions, and Open Issues*, 395, pp.116–124. https://www.intechopen.com/chapters/17872.

Shi, W., Cao, J., Zhang, Q., Li, Y. and Xu, L., 2016. Edge computing: Vision and challenges. *IEEE Internet of Things Journal*, 3(5), pp.637–646. https://doi.org/10.1109/jiot.2016.2579198.

Shifeng, Y., Chungui, F., Yuanyuan, H. andShipping, Z., 2011. Application of IOT in agriculture. *Journal of Agricultural Mechanization Research*, 7, pp.190–193.

Shrouf, F. and Miragliotta, G., 2015. Energy management based on Internet of things: Practices and framework for adoption in production management. *Journal of Cleaner Production*, 100, pp.235–246. https://doi.org/10.1016/j.jclepro.2015.03.055.

Singh, P., Elmi, Z., Meriga, V.K., Pasha, J. and Dulebenets, M.A., 2022. Internet of Things for sustainable railway transportation: Past, present, and future. *Cleaner Logistics and Supply Chain*, 4, p.100065. https://doi.org/10.1016/j.clscn.2022.100065.

Sliwa, J., 2018. Environment, people, and time as factors in the Internet of Things technical revolution. In *Internet of Things A to Z: Technologies and applications*, pp.51–76. Wiley and Sons. https://doi.org/10.1002/9781119456735.ch2.

Sliwa, J., 2025. Internet of military things and weaponized AI: Technology in the age of conflict. In Hassan, Q.F. (ed.), *Internet of Things A to Z: Technologies and applications*, Second Edition. Wiley & Sons. In Press.

Sundramoorthy, V., Liu, Q., Cooper, G., Linge, N. and Cooper, J., 2010. DEHEMS: A user-driven domestic energy monitoring seystem. In *Internet of Things (IOT)*, Tokyo, pp.1–8. https://doi.org/10.1109/iot.2010.5678451.

Tahsien, S.M., Karimipour, H. and Spachos, P., 2020. Machine learning based solutions for security of Internet of Things (IoT): A survey. *Journal of Network and Computer Applications*, 161, p.102630. https://doi.org/10.1016/j.jnca.2020.102630.

Taivalsaari, A. and Mikkonen, T., 2017. A roadmap to the programmable world: Software challenges in the IoT era. *IEEE Software*, 34(1), pp.72–80. https://doi.org/10.1109/ms.2017.26.

Tang, T. and Winoto, P., 2018. An Internet of Things approach to "Read" the emotion of children with autism spectrum disorder. In Hassan, Q.F. (ed.), *Internet of Things A to Z: Technologies and applications*. Wiley & Sons. https://doi.org/10.1002/9781119456735.ch20.

Tao, F., Zuo, Y., Da Xu, L. and Zhang, L., 2014. IoT-based intelligent perception and access of manufacturing resource toward cloud manufacturing. *IEEE Transactions on Industrial Informatics*, 10(2), pp.1547– 1557. https://doi.org/10.1109/TII.2014.2306397.

Tawalbeh, L.A., Muheidat, F., Tawalbeh, M. and Quwaider, M., 2020. IoT Privacy and security: Challenges and solutions. *Applied Sciences*, 10(12), p.4102. https://doi.org/10.3390/app10124102.

Tielert, T., Killat, M., Hartenstein, H., Luz, R., Hausberger, S. and Benz, T., 2010. The impact of traffic-light- to-vehicle communication on fuel consumption and emissions. In *2010 Interenet of Things (IOT)*, Tokyo, pp.1–8. https://doi.org/10.1109/iot.2010.5678454.

Tiloca, M. and Raza, S., 2018. Security mechanisms and technologies for constrained IoT devices. In Hassan, Q.F. (ed.), *Internet of Things A to Z: Technologies and applications*, pp.219–254. Wiley & Sons. https://doi.org/10.1002/9781119456735.ch8.

Tirana, J. and Chatzopoulos, D., 2025. Split learning and synergetic inference: When IoT collaborates with the cloud-edge continuum. In Hassan, Q.F. (ed.), *Advances in Internet of Things: Challenges, solutions, and emerging technologies*. Chapman and Hall/CRC. https://doi.org/10.1201/9781003506638.

Tournier, J., Lesueur, F., Le Mouël, F., Guyon, L. and Ben-Hassine, H., 2021. A survey of IoT protocols and their security issues through the lens of a generic IoT stack. *Internet of Things*, 16, p.100264. https://doi.org/10.1016/j.iot.2020.100264.

Vlacheas, P., Giaffreda, R., Stavroulaki, V., Kelaidonis, D., Foteinos, V., Poulios, G., Demestichas, P., Somov, A., Rahim Biswas, A. and Moessner, K., 2013. Enabling smart cities through a cognitive management framework for the Internet of Things. *IEEE Communications Magazine*, 51(6), pp.102–111. https://doi.org/10.1109/mcom.2013.6525602.

Vos, M., 2017. Organizational implementation and management challenges in the Internet of Things. In Hassan, Q.F., Khan, A.R., and Madani, S.A. (eds.), *Internet of Things: Challenges, advances, and applications*, pp. 33–48. Chapman and Hall/CRC.

Wang, X., Song Dong, J., Chin, C., Ravipriya Hettiarachchi, S. and Zhang, D., 2004. Semantic space: An infrastructure for smart spaces. *IEEE Pervasive Computing*, 3(3), pp.32–39. https://doi.org/10.1109/MPRV.2004.1321026.

Wang, Y., Zhao, J., Qiang, Y., Zhao, J. and Li, K., 2025. Enhancing IoT systems through blockchain integration: Consensus mechanisms and security enhancements. In Hassan, Q.F. (ed.), *Advances in Internet of Things: Challenges, solutions, and emerging technologies*. Chapman and Hall/CRC. https://doi.org/10.1201/9781003506638.

Wei, C. and Li, Y., 2011. Design of energy consumption monitoring and energy-saving management system of intelligent building based on the Internet of things. In *2011 International Conference on Electronics, Communications and Control (ICECC)*, Ningbo, China, pp.3650–3652. https://doi.org/10.1109/icecc.2011.6066758.

Weiser, M., 1991. The computer for the 21st century. *Scientific American*, 265(3), pp.94–104. https://doi.org/10.1016/b978-0-08-051574-8.50097-2.

Weyrich, M. and Ebert, C., 2016. Reference architectures for the Internet of Things. *IEEE Software*, 33(1), pp.112–16. https://doi.org/10.1109/ms.2016.20.

WG 10, 2015. ISO/IEC JTC1/WG 10 Working Group on Internet of Things. https://jtc1info.org/sd-2-history/jtc1-working-groups/wg-10/.

Whitmore, A., Agarwal, A. and Da Xu, L., 2014. The Internet of Things – A survey of topics and trends. *Information Systems Frontiers*, 17(2), pp.261–274. https://doi.org/10.1007/s10796-014-9489-2.

Willner, A., 2018. The industrial Internet of Things. In Hassan, Q.F. (ed.), *Internet of Things A to Z: Technologies and applications*, pp.293–318. Wiley & Sons. https://doi.org/10.1002/9781119456735.ch11.

Yang, G., Xie, L., Mäntysalo, M., Zhou, X., Pang, Z., Da Xu, L., Kao-Walter, S., Chen, Q. and Zheng, L.-R., 2014. A health-IoT platform based on the integration of intelligent packaging, unobtrusive bio-sensor, and intelligent medicine box. *IEEE Transactions on Industrial Informatics*, 10(4), pp.2180–2191. https://doi.org/10.1109/tii.2014.2307795.

Yang, Q., Ehsan, A., Jiang, L., Zhao, H. and Cheng, M., 2017. The Internet of Things in electric distribution networks. In Hassan, Q.F., Khan, A.R. and Madani, S.A. (eds.), *Internet of Things: Challenges, advances, and applications*. Chapman and Hall/CRC, pp. 231–252.

Yao, X., Farha, F., Li, R., Psychoula, I., Chen, L. and Ning, H., 2021. Security and privacy issues of physical objects in the IoT: Challenges and opportunities. *Digital Communications and Networks*, 7(3), pp.373–384. https://doi.org/10.1016/j.dcan.2020.09.001.

Yin, C.T., Xiong, Z., Chen, H., Yuan Wang, J., Cooper, D. and David, B., 2015. A literature survey on smart cities. *Science China Information Sciences*, 58(10), pp.1–18. https://www.researchgate.net/profile/Jingyuan-Wang-14/publication/281670019_A_literature_survey_on_smart_cities/links/5c0e2f0092851c39ebe1e619/A-literature-survey-on-smart-cities.pdf.

Yun, M. and Yuxin, B., 2010. Research on the architecture and key technology of Internet of Things (IoT) applied on smart grid. In *2010 International Conference on Advances in Energy Engineering (ICAEE)*, Beijing, pp.69–72. https://doi.org/10.1109/ICAEE.2010.5557611.

Zakon, R.H., 2016. Hobbes' Internet Timeline 25. The Definitive ARPAnet & Internet History. https://www.zakon.org/robert/internet/timeline/#Growth.

Zanella, A., Bui, N., Castellani, A., Vangelista, L. and Zorzi, M., 2014. Internet of Things for smart cities. *IEEE Internet of Things Journal*, 1(1), pp.22–32. https://doi.org/10.1109/jiot.2014.2306328.

Zantalis, F., Koulouras, G., Karabetsos, S. and Kandris, D., 2019. A review of machine learning and IoT in smart transportation. *Future Internet*, 11(4), p.94. https://doi.org/10.3390/fi11040094.

Zhang, L., Ibibia, K., Dabipi, W. and Brown, L., 2018. Internet of Things applications in agriculture. In Hassan, Q.F. (ed.), *Internet of Things A to Z*. John Wiley and Sons Inc. https://doi.org/10.1002/9781119456735.ch18.

Zhang, Y., Chen, B. and Lu, X., 2011. Intelligent monitoring system on refrigerator trucks based on the Internet of Things. In *International Conference on Wireless Communications and Applications*, Sanya, China, pp.201–206. https://doi.org/10.1007/978-3-642-29157-9_19.

Zhao, D., Li, Q., Lv, J., Li, R., Xiao, J., Wu, K., Zou, Q. and Li, K., 2025. Security and privacy in smart homes: Challenges and latest developments. In Hassan, Q.F. (ed.), *Advances in Internet of Things: Challenges, solutions, and emerging technologies*. Chapman and Hall/CRC. https://doi.org/10.1201/9781003506638.
Zhengxia, W. and Laisheng, X., 2010. Modern logistics monitoring platform based on the Internet of Things. In *2010 International Conference on Intelligent Computation Technology and Automation (ICICTA)*, Changsha, China, pp.726–731. https://doi.org/10.1109/icicta.2010.650.
Zhong, Y. and Dong, Y.-T., 2011. Application of the Internet of Things to security automation system for intelligent buildings. *Internet of Things Technologies*, 4, p.041.
Zouganeli, E. and Svinnset, I.E., 2009. Connected objects and the Internet of Things – A paradigm shift. In *2009 International Conference on Photonics in Switching*, Pisa, pp.1–4. https://doi.org/10.1109/ps.2009.5307747.

2 Security and Privacy in Smart Homes
Challenges and Latest Developments

Dan Zhao, Qing Li, Qingsong Zou, Jingyu Xiao, Kaidong Wu, Ruoyu Li, Jianhui Lv, Yong Jiang and Keqin Li

2.1 INTRODUCTION

Over the past decade, smart homes have become more and more popular among consumers who desire intelligent and convenient living spaces. Smart homes integrate a variety of Internet-enabled Internet of Things (IoT) devices, such as smart meters, smart locks, and surveillance cameras, which are controlled and monitored remotely through smartphones or tablets (Bugeja et al., 2018). Smart homes are closely intertwined with consumers' personal privacy and safety: many smart devices are equipped with sensors that monitor consumers' surroundings and even their health metrics; some smart devices actively engage with our living areas to carry out their intended tasks. Protecting users against privacy and safety threats in smart homes has become an important issue.

The first line of defense for keeping IoT traffic safe from privacy threats is to encrypt the data. Encryption ensures that eavesdroppers cannot steal sensitive information by simply checking the plaintext content of data packets. Despite this effort, methods like fingerprinting can still manage to obtain sensitive data and infer information like device types, as well as the state of their physical environments, just by analyzing the traffic patterns. To establish defenses against fingerprinting, a range of protective strategies, e.g., virtual private network (VPN) tunneling and traffic morphing, have been developed. A VPN tunnel establishes encrypted connections between devices and servers so that attackers cannot gain access to the traffic. Traffic morphing reshapes traffic by padding packets or fiddling with their timing. By changing the patterns of traffic's footprint, the morphed data confuses fingerprinting models trained on the original data, leading them to make wrong inference decisions.

Security threats in IoT systems can stem from multiple sources, e.g., interactions between devices and the cloud and cyberattacks. In recent years, trigger-action platforms (TAPs) such as IFTTT, Apple Home, and Samsung SmartThings have been used by increasingly more users. TAPs allow users to configure rules that define the behaviors of multiple smart devices according to their wills. However, IoT devices can sometimes perform unwanted or even hazardous actions following the rules configured by users (Chen et al., 2022). Such risks may be caused by benign logic errors made by users, particularly when configuring the combined behaviors of more than one device. For example, the rules "If the user leaves home, then close the window" and "If the temperature is above 25 degrees, open the window" could conflict, resulting in the window being opened when the user is not at home but the temperature rises above 25°. They can also be the result of malicious attacks that obtain excessively high privileges to access the devices. IoT devices are also often victims of cyberattacks like brute force attacks, vulnerability exploits, and IoT-specific malware.

 DOI: 10.1201/9781003506638-3

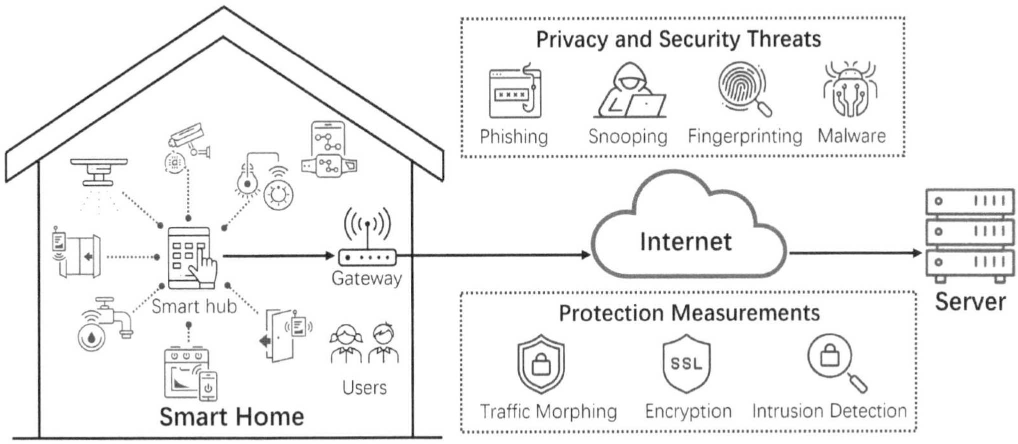

FIGURE 2.1 Privacy and Security Considerations in Smart Homes

Conventional network intrusion detection systems (NIDS) designed for general networks are often not suitable for IoT systems. This is because the NIDS for IoT systems must overcome a few unique challenges, including the heterogeneity of IoT protocols, diverse communication modes, varying traffic patterns, and scalable deployments. In recent years, some NIDS specifically designed for IoT have been proposed (Barradas et al., 2021; Fu et al., 2023; Tang et al., 2020; Dong et al., 2023; Qu et al., 2023). The most foundational and widely used IoT NIDS is signature-based attack detection. More advanced NIDS for IoT, e.g., statistical model-based and learning-based NIDS, are also developed for better intrusion detection performance. Additionally, user behavior prediction, which uses methods like Markov models and advanced AI technologies such as long short-term memory (LSTM), graph neural networks (GNNs), and transformers, also helps identify abnormal behaviors in smart homes.

Figure 2.1 briefly illustrates privacy and security considerations in smart homes. When IoT devices exchange data with remote servers through the Internet, they are susceptible to various forms of privacy and security threats.

In this chapter, we discuss the security and privacy threats faced by IoT systems and where we are in protecting IoT systems against these threats. We also review some of the state-of-the-art methods for protecting IoT systems from privacy and security risks. While we particularly focus on smart home scenarios in this chapter, many of the techniques discussed here, including fingerprinting, traffic morphing, and NIDS, are also applicable to a broader context of IoT networks, e.g., smart cities and industrial IoT systems.

2.2 CHALLENGES AND CURRENT STATUS IN IOT SECURITY AND PRIVACY

2.2.1 Why Do IoT Security and Privacy Issues Matter?

According to statistics published by Transforma Insights[1] (2022), the active number of IoT devices worldwide is expected to surge to 29.4 billion in 2030. Many IoT devices use users' sensitive information and can directly impact their physical environments.

IoT medical devices, e.g., remote patient monitoring (RPM) systems, glucose monitors, and heart-rate sensors, collect health metrics such as heart rate, body temperature, glucose levels, and blood pressure (Sagahyroon et al., 2017). This data is often sent to healthcare providers through the Internet. These devices can also perform simple medical tasks, such as automatically administering insulin. Malicious attackers could tamper with these devices to put the health and safety of users in danger. For example, an attacker could deactivate a glucose monitoring device to stop insulin supply to a user, which would pose a serious threat to the user's health.

Smart homes integrate various IoT devices to create automated and smart homes (Bugeja et al., 2018). Popular smart home devices include smart lighting systems, smart home appliances, security cameras, and smart door locks. Smart homes provide considerable convenience and peace of mind to users. However, these IoT devices can also fall victim to attacks that put users' physical safety in danger. For instance, criminals may hack security cameras to spy on users' activities at home. Attackers may gain control of smart locks to unlock exterior doors to facilitate physical intrusion.

Nowadays, users are becoming more aware of the potential security vulnerabilities that come with IoT devices. As a result, many of them are now willing to increase their budget for more secure IoT devices to avoid privacy and security risks (Emami-Naeini et al., 2023).

2.2.2 What Makes Current IoT Devices Vulnerable to Privacy and Security Risks?

Previously, numerous IoT devices were released with critical security and privacy flaws, raising alarm among a wide range of groups, from end-users to manufacturers, and even Internet Service Providers (ISPs). There are a few root causes behind these vulnerabilities.

First, companies making IoT devices might be more focused on profit than on making sure their products are secure. They might not feel the need to follow the best security and privacy standards. Smaller companies might not have the right skills to implement strong security features. These companies might rush to put their products on the market. They are often eager to grab a piece of the market quickly as long as their product can work and connect to the Internet without investing time to attend to safety considerations.

Second, IoT devices usually come with very limited storage and computational resources compared to general-purpose computing devices like personal computers and smartphones. Though comprehensive security measures, such as authentication and encryption, have been designed for conventional computer networks, IoT devices often struggle to implement these measures.

Third, many IoT devices persistently operate with initially shipped firmware that becomes outdated over time. Ideally, firmware should be updated regularly to patch newly identified security vulnerabilities and defend against new attacks. However, many IoT devices lack automated update mechanisms. Vendors also frequently lack motivation to maintain health update cycles out of profit considerations.

Fourth, users may also lack technical expertise and security awareness. Very often, users stick with the default or use easily guessable passwords. They also tend to lack interest in manually updating firmware, especially when they have a wide array of IoT devices deployed across their properties. As reported by the Dawes Centre for Future Crime at UCL[2] in 2020, merely 52% of consumers regularly update software for their PCs and smartphones, and only 32% adhere to government password advice. The situation is likely even worse with IoT devices.

2.3 CURRENT IOT PRIVACY THREATS AND DEFENSE MEASURES

Many IoT devices, especially those in smart homes, continuously monitor and send data related to users' private activities to cloud servers via the Internet, thereby exposing user privacy to potential risks. Smart home network traffic is susceptible to interception by passive eavesdroppers, such as ISPs and Wi-Fi eavesdroppers. As a basic attempt to mitigate privacy breaches, most IoT devices use encryption to safeguard data traffic. Popular encryption algorithms include symmetric encryption algorithms, e.g., Advanced Encryption Standard (AES), and asymmetric encryption algorithms, such as Elliptic-Curve Cryptography (ECC), Digital Signature Algorithm (DSA) and Rivest–Shamir–Adleman (RSA). Research into lightweight cryptography tailored for IoT devices with limited resources and minimal processing capabilities is also intensively underway (Rana et al., 2022).

Though encrypting traffic can deter passive observers from accessing sensitive information directly through deep packet inspection, techniques like fingerprinting can still unveil sensitive data from encrypted streams. Fingerprinting exploits the limited-purpose nature of IoT devices by analyzing encrypted traffic patterns, such as send/receive rates, packet sizes, or Wi-Fi signal patterns, to infer device types or activities. Mainstream defense mechanisms to counter the threats posed by fingerprinting include VPN tunneling and traffic morphing. VPN tunneling enhances privacy by masking packet headers, making device identification more challenging. Traffic morphing obfuscates or shapes IoT traffic to conceal traffic patterns that might disclose actual device status and user activities. Through methods like packet padding and timing adjustments, traffic morphing misleads the fingerprinting model trained on original traffic data into making incorrect classification results. Ideally, traffic morphing solutions should not negatively affect IoT device performance, must not exceed data limits, and should not necessitate alterations to proprietary device software. Moreover, since NIDS rely on original traffic patterns, traffic morphing should be designed carefully so as not to impair these systems' effectiveness.

In this section, we discuss how attackers infer sensitive information from IoT traffic patterns and review some of the latest IoT traffic fingerprinting methods. We then introduce countermeasures against IoT traffic fingerprinting, especially recent traffic morphing approaches.

2.3.1 Privacy Breaches via IoT Traffic Fingerprinting

The invasion of privacy has become a significant concern among users of electronic devices for many years. Researchers have found ways to locate and track general mobile computing devices in in-building scenarios through radio frequency (RF) signals (Bahl and Padmanabhan, 2000) over the past two decades. Jafari et al. (2018) achieved the identification of ZigBee IoT devices by applying deep learning (DL) algorithms to RF signals from these devices.

In recent years, researchers have investigated the traffic traces from a variety of IoT devices. Their findings demonstrate that, though encrypted traffic conceals the plaintext within the IoT traffic, the traffic still often displays distinctive characteristics in its patterns and the metadata (Dong et al., 2020; Li et al., 2024). These features could be exploited to extract information beyond mere locating and tracking devices. The most frequently utilized attributes in IoT device fingerprinting techniques include the 5-tuples (source/destination IP addresses/ports) and traffic statistics (rates and flow durations).

An example of research studies done in this area include Apthorpe et al. (2017), which presented a simple example that infers user activities through TCP/IP headers and traffic rates of IoT devices. IoT traffic analysis was conducted on four smart home devices: a smart home hub, a sleep monitor, a security camera, and a smart switch, all communicating with cloud servers using the TLS/SSL protocol. By passively monitoring the traffic, an adversary could identify flows and label device types by simply analyzing the metadata, e.g., TCP/IP packet headers. Since an IoT device typically has limited purposes, the adversary could even successfully infer user interactions from the variation in traffic rate. For example, traffic rate spikes from a sleep monitor can indicate when the user gets into and out of bed.

Before we dive into discussing the latest IoT fingerprinting methods, we would like to provide a more intuitive and illustrative portrayal of IoT device traffic patterns. This can help in understanding the core concept that traffic features, such as packet length, directions, and intervals, vary across different IoT devices and may be analyzed using techniques like ML models for device fingerprinting purposes. We conducted two sets of case studies on several IoT devices in real-life scenarios and presented their traffic patterns in the form of diagrams and tables for illustration purposes.

In the first case study, we tracked the traffic of a smart desk light (Mijia 1S[3]), a standing fan (Mijia Standing Fan 1X[4]), and a heater (Smartmi Intelligent Convector Heater 1s[5]) over a two-month period and plotted scatter diagrams of their packet lengths in a 3.5-hour time window as examples in Figures 2.2–2.4. These three devices were all connected to an automation platform, i.e., the MIJIA app. The desk light and heater used the TCP protocol, while the standing fan used UDP. In these plots, the segment to the left of the vertical dashed line represents the period when the automation platform is not being utilized, while the segment to the right shows when it is in use. When the automation platform is in use, the shaded periods indicate the device is on, and the unshaded periods indicate the device is off. From these diagrams, we can observe with the naked eye that the traffic patterns, such as packet lengths, directions, and intervals, vary among these devices. Consistent with several existing studies, e.g., Ma et al. (2020), we also note that IoT devices' traffic typically falls into two categories: active traffic, and idle traffic. Active traffic, highlighted by stars in Figures 2.2–2.4, is generated when users interact with IoT devices, such as when turning the heater on or off or when a scheduled task is executed. It is usually abrupt and presents task-dependent characteristics, such as variations in packet lengths and frequencies of occurrence. Idle traffic, marked by dots in Figures 2.2–2.4, is likely generated by routine data exchanges between IoT devices and cloud servers. Idle traffic normally occurs in the background for purposes like device state synchronization and sensor data updates. This kind of traffic, which can persist throughout the lifespan of devices, often displays periodicity. Particularly, the idle traffic from sensorless devices, e.g., the desk light in Figure 2.3, may follow fixed cycles. For idle traffic from devices with sensors, the cycle is related to the frequency of changes of the monitored metric, e.g., variations in room temperature for a smart heater.

In the second case study, we tracked the traffic from a Xiaomi smart plug over ten days. As summarized in Table 2.1, there exist sets of packets with specific lengths that appear periodically. Note that the order of the packets can change due to possible retransmission and reordering. For example, a set of three packets with lengths of 175, 447, and 191 appear every 5 minutes.

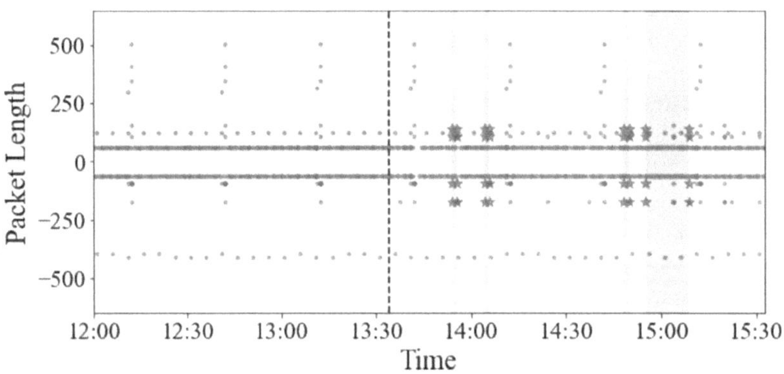

FIGURE 2.2 Scatter Diagram of Packet Length from a Standing Fan

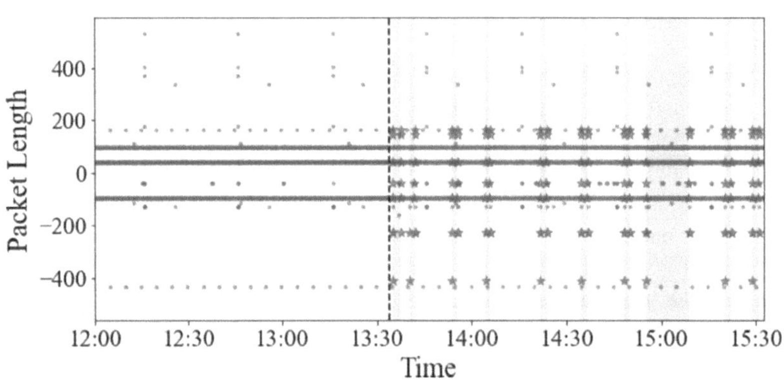

FIGURE 2.3 Scatter Diagram of Packet Length from a Desk Light

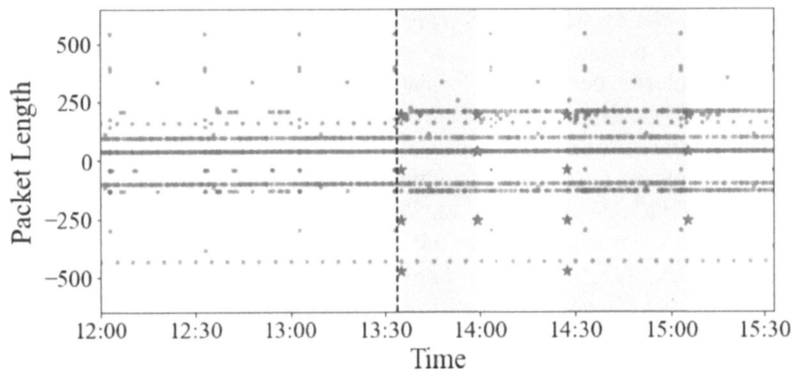

FIGURE 2.4 Scatter Diagram of Packet Length from a Heater

TABLE 2.1
A Xiaomi Plug Demonstrates Periodic Bursts of Key Packets

Lengths of Periodically Appearing Set of Packets	Period	Protocol
74, −74	30 s	UDP
111, −111, 60	40 s	TCP
175, −447, 191	5 min	TCP
351, −143	15 min	TCP
543, −143, 431, −399	30 min	TCP

Note: The minus sign indicates the direction from WAN to LAN, and vice versa.

From these case studies, one can observe that the traffic on IoT devices can exhibit spatial and temporal patterns that vary across different devices. Many existing IoT device fingerprinting methods exploit traffic patterns to identify IoT devices.

Next, we introduce some representative existing IoT device fingerprinting methods. We will provide some insights into the trends of fingerprinting methods and how they evolve as networks become more sophisticated, more privacy protections are in place, and more advanced models, e.g., ML and DL algorithms, become available. We discuss the choices of traffic features (i.e., signatures) and the models utilized by the latest fingerprinting methods. The goal is to raise awareness among vendors and users about the essence of the threats they face and give them a sense of what the future may hold so that they can be better prepared. Existing IoT device fingerprinting methods can be classified into two categories: signature-based IoT device fingerprinting and machine learning (ML)-based methods.

Signature-based IoT device fingerprinting methods use basic signatures like MAC addresses and DHCP messages. This type of method allows real-time inference of the device type. However, these signatures are now quite often ineffective for fingerprinting purposes with modern ISPs. Some newer research has tried using destination IP addresses and domains from DNS queries as IoT signatures instead (Saidi et al., 2020). However, devices from the same or different vendors can share the same public cloud services, leading to ambiguity in the signatures. As a result, the effectiveness of these signatures is also questionable. Recently, Trimananda et al. (2020) observed that smart home device traffic typically consists of pairs of packets: a request packet in the uplink direction from an IoT device to a cloud server, and a reply packet in the downlink direction, with distinguishable lengths and directions across different devices and events. Based on this observation, packet-level signatures comprising only the lengths and directions were identified. However, this signature is only valid with TCP traffic, while a significant portion of IoT traffic uses UDP.

ML-based methods identify IoT devices by feeding carefully constructed traffic feature vectors into ML algorithms. These feature vectors consist of a variety of features across various granularities. At the finest granularity, per packet information, e.g., port, timestamp, TCP flag, and protocol, is frequently employed (Meidan et al., 2020). From a more coarse-grained level, flow statistics, such as mean/variance/maximum/minimum of packet length/count/inter-arrival times, are also seen in many works (Thangavelu et al., 2019). Some of the existing works have adopted supervised learning algorithms, such as LightGBM (Meidan et al., 2020) and Random Forest (Miettinen et al., 2017). While ML-based methods provide more flexible solutions than signature-based methods, their use in practical ISP networks faces significant hurdles. The widespread presence of middleboxes in modern networks, e.g., VPN gateways and Network Address Translation (NAT) gateways, tends to conceal traffic features that these methods depend on. This is because traffic passing through these devices may be subject to processes like traffic fusion, encryption, and encapsulation. For instance, when traffic flows through an SSL/TLS VPN, e.g., OpenVPN, its original Layer 4 header is no longer visible due to encapsulation with a new TCP/UDP header.

DL-based IoT fingerprinting methods mark recent advancements in the research of IoT fingerprinting techniques. DL algorithms are often more powerful than traditional ML algorithms since they can automatically learn implicit and complex patterns, especially in the spatial-temporal dimension, where manually defining such features could be inefficient or impractical. In the scope of IoT fingerprinting, DL algorithms can be exploited to mine the sequential patterns embedded across consecutive packets in IoT traffic. Dong et al. (2020) designed a sequence-learning algorithm (i.e., bidirectional LSTM)-based approach that effectively mines contextual information from adjacent packets to accurately classify packets. The model demonstrated efficiency in NAT scenarios, yet showed limitations when applied to VPN scenarios. Ma et al. (2020) proposed a convolutional neural network (CNN)-based method to accurately identify IoT devices in open network environments (e.g., VPNs) where IoT traffic and other background traffic are mixed. According to observations similar to those in Case Study 1, the proposed method extracts important sequence profiles from mixed traffic and utilizes their matching patterns in mixed traffic to form spatio-temporal fingerprints. Finally, a classical CNN model is employed to achieve device classification.

Up to this point, DL-based IoT device fingerprinting techniques have achieved some satisfying classification performance. However, they have some clear shortcomings too. First, DL models consume substantial amounts of storage and computation resources. They are unable to make real-time inferences when processing high-speed network traffic, even with the help of GPU acceleration. Second, these models require caching packets over a period of time to gather enough spatial/temporal relationships, which further drives up inference time.

Some of the latest efforts have explored leveraging programmable data planes to achieve line-speed IoT device fingerprinting under high-speed traffic. Unlike conventional switches that only support limited hardwired processing of traffic, programmable switches support customization of processing on packets to realize desired functionalities with TB-level throughput and only a microsecond-level processing delay. Li et al. (2024) proposed an IoT device fingerprinting framework, DeviceRadar, that leverages a programmable data plane for real-time inference with 40 Gbps throughput and state-of-the-art accuracy. Following the observations from Case Study 2, this work defines the "key packets"- packets that exhibit sufficient frequency and periodic patterns and size variations distinguishable across devices. DeviceRadar employs a DL model to create a packet embedding table to map packet sizes to a high-dimensional space, which can reveal spatial correlations among packet sizes. The embedding table is eventually transformed into a packet neighboring probability matrix, illustrating the likelihood that packet sizes observed over a short period originate from the same device. Feature vectors derived from the key packets and the matrix are then used for per-device classification to identify the device type.

The IoT fingerprinting methods discussed above are compromising the privacy of users in today's IoT systems. As we look to the future, these techniques may further evolve to resolve challenges that they are currently facing, e.g., overlapping patterns that are introduced by the surging number of IoT devices and the complexities that arise from the presence of middleboxes. They may also progress to infer even more detailed information about IoT devices, such as their exact models and software versions. Vendors of IoT devices and researchers should continuously monitor the evolution of IoT fingerprinting methods to stay informed about the latest threats.

2.3.2 Strategies for Defending against IoT Traffic Fingerprinting

As the first line of defense against privacy breaches, encryption algorithms encrypt the plaintext payload in data packets to prevent eavesdroppers from directly obtaining sensitive information from the traffic. However, encryption alone is far from enough to protect users' privacy in the face of today's more advanced privacy attack techniques. IoT fingerprinting, which was discussed in the previous section, is a representative of advanced attacks that can penetrate encrypted IoT traffic. Since each IoT device typically performs a limited number of specialized functions, their traffic patterns are mostly simple and exhibit obvious differences across different devices. IoT fingerprinting exploits this property to infer sensitive information, such as the type of device, from analyzing the encrypted traffic. Therefore, new defense strategies are necessary to further protect IoT traffic from fingerprinting attacks.

Traffic morphing is an effective defense method against fingerprinting. Traffic morphing has been proven effective in defending against various forms of fingerprinting, including website fingerprinting, video fingerprinting, and IoT fingerprinting. Traffic morphing modifies the original traffic by using methods such as packet padding, packet delaying, and inserting dummy packets. Traffic morphing changes the traffic patterns so that fingerprinting models, which are trained on original traffic data, produce incorrect classification results.

Random padding is one of the most straightforward and efficient traffic-morphing strategies. It is now integrated into IoT smart platforms, like the Samsung SmartThings platform. Dyer et al. (2012) tested the random padding method on the network maximum transmission unit (MTU) and found it can reduce the accuracy of some traffic analysis attacks. However, while random padding methods provide simple, effective, and easy-to-deploy countermeasures against fingerprinting, they fail against the latest attacks. In addition, these methods lack effective overhead control mechanisms and are likely to introduce unexpected overhead.

To overcome these shortcomings, several heuristic traffic-morphing schemes have been proposed. Heuristic traffic morphing schemes aim to avoid inference from fingerprinting models with little overhead through mathematical reasoning. Xiong et al. (2018) proposed a locally differentially private packet obfuscation mechanism to defend against packet-size side-channel attacks in IoT networks. This work formulates the traffic morphing problem as the optimization of the conditional probability distribution between the original packet sizes and obfuscated packet sizes. The authors solve the optimization problem to obtain the optimal obfuscation mechanism for minimizing bandwidth overhead under privacy constraints. Pinheiro et al. (2021) proposed an adaptive packet pad ding method based on Software Defined Networking (SDN), which utilizes a Representational State Transfer (REST) interface to indicate the padding mechanism. This mechanism ensures the modification of the length of packets generated by connected devices.

Heuristic traffic morphing schemes have achieved better robustness against complex attacks and allow for adjustable overhead to balance defensive performance and cost. However, they are currently challenged by the latest ML-based fingerprinting models, especially the DL-based ones. ML/DL-based fingerprinting models designed for IoT traffic exhibit excellent fingerprinting performance and can break through the defense line provided by conventional defense mechanisms.

Recently, adversarial learning techniques targeting ML and DL models have been utilized to defend against ML/DL-based fingerprinting models. Adversarial ML, or adversarial attacks, generates adversarial samples to evade ML models, DL models, or other intelligent systems. Typically, adversarial ML aims to make the model produce incorrect inferences by applying minimal perturbations. In the field of computer vision, smaller perturbations are desirable since they make attacks less noticeable. In defending against IoT fingerprinting, the aim of adversarial learning is to deceive fingerprinting models into making incorrect inferences by applying minimal perturbations to the traffic.

However, adversarial learning techniques originally invented for the computer vision field cannot be directly applied to the traffic morphing domain. In the computer vision domain, adversarial learning algorithms can usually alter the feature vectors with minimal constraints. For example, in (Brendel et al., 2018), only upper and lower bound constraints on pixel values are imposed on the adversarial samples. But in the network domain, defenders normally face more strict constraints when applying perturbations to traffic. Certain packet information, such as protocol type, flags, and payload length, commonly used as fingerprinting features by attackers, cannot be arbitrarily changed to maintain communication integrity. Therefore, careful considerations are required when applying adversarial methods in network scenarios.

Nasr et al. (2021) proposed a blind adversarial algorithm that morphs network traffic by generating adversarial perturbations. This generative adversarial network (GAN)-based scheme uses the generated adversarial perturbations to guide modifications in the distribution of packet lengths, transmission delays, and the insertion of dummy packets within the data streams. It can be employed for defense against most traffic analysis tasks and is capable of generating perturbed traffic in real time.

In Tan et al. (2023), a reinforcement learning-based approach for generating adversarial traffic samples was presented. Powered by a timely reward model, the learning agent generates an optimal strategy to perturb the target fingerprint recognition model while retaining the original traffic function. This approach requires minimal knowledge of the fingerprint model and supports real-time deformation. However, the cost of running deep models in real time to generate deformation schemes remains high.

It is not difficult to notice that both of the above methods employ black-box or gray-box settings. The black-box setting in adversarial learning indicates a complete lack of knowledge about the target model's details, e.g., the adversaries' fingerprinting models. In contrast, the white-box setting means full access to the details of the target model is available, including model architecture, gradients, and weights of individual neurons, which is impractical in most network scenarios. The gray-box setting lies in between, where the trainer can obtain limited knowledge of the target model, such as the probabilities of various labels in the model's output. For example, the method proposed in Nasr et al. (2021) requires knowledge of the target model's loss function. Clearly, black-box or gray-box settings are more viable and, therefore, preferred over the white-box setting in real-world network environments.

2.3.3 Harmonizing IoT Traffic Sniffing Defense Mechanisms with NIDS

The traffic morphing techniques discussed in the previous section have demonstrated considerable protective capabilities against IoT device fingerprinting in experimental settings. However, to deploy them in actual complex network environments, some problems still need to be solved.

One of the challenges is the fact that modifications to traffic made by traffic morphing techniques can disrupt the normal operation of other concurrent applications within the networks. NIDS are examples of such applications whose normal functioning can be influenced by traffic morphing. NIDS distinguishes malicious traffic from benign traffic by learning their patterns. Current traffic morphing strategies are mostly implemented without considering NIDS. As a result, the morphed traffic may fall into the intrusion detection system's malicious confidence interval, and be misclassified as malicious traffic. The cost of these false positive judgments is high since they can invoke expert review or cause disruptions to normal network functionality.

Figure 2.5 is an illustration of the conflicts between fingerprinting countermeasures and NIDS. In this figure, traffic samples from IoT devices i, j, and k, which are of different types, are represented

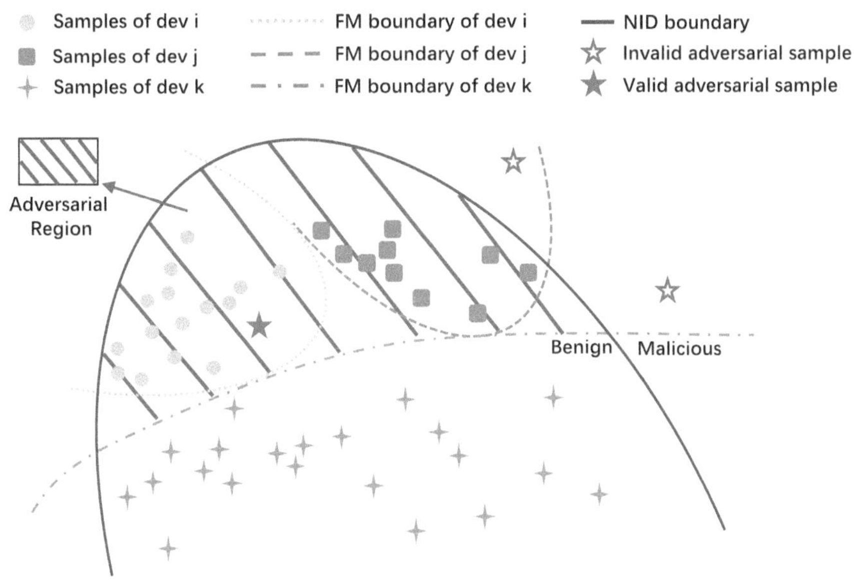

FIGURE 2.5 A Brief Illustration of the Dilemma between Fingerprinting Countermeasures and NIDS(FM is the Abbreviation for the Fingerprinting Model)

by dots, squares, and cross stars, respectively. The traffic samples from these three devices are spatially partitioned by their respective fingerprinting models, whose boundaries are represented by dotted, dashed, and dash-dot curves, respectively. The solid curve represents the boundary of a NIDS model. Apparently, all samples from the three devices are classified as benign by this NIDS model.

An ideal model should accurately fit the data distribution of the traffic samples and form precise boundaries. In such a perfect case, no misclassifications should occur. Unfortunately, perfect models are not realistic. In practice, the NIDS's decision boundary might not always match up exactly with that of the fingerprinting models.

The three five-pointed stars (two shallow ones and a solid one) in Figure 2.5 represent adversarial samples generated by an adversarial learning model for device k's traffic without considering the NIDS. All these adversarial samples effectively serve the purpose of traffic morphing because they are misclassified by device k's fingerprinting model as not originating from device k. However, due to the imperfect boundaries produced by the models, two of the adversarial samples (shallow five-pointed stars) trigger false alarms in the NIDS. Only the adversarial sample located in the adversarial region (the solid five-pointed star) is effective for traffic morphing without causing a false alarm from the NIDS, i.e., it is correctly classified as benign by the NIDS while avoiding being recognized as device k's traffic by the fingerprinting model.

In fact, adversarial samples should be generated to mostly fall within the adversarial region. Only this region could fulfill traffic morphing purposes without disturbing the operation of the NIDS. The main challenge here is to characterize the decision boundaries of the fingerprinting model and NID model and accurately depict the adversarial region. The problem can be transformed into a general adversarial learning task once the adversarial region is correctly identified. However, current adversarial learning methods require considerable modifications to adapt to the specific fingerprinting and NID models. This motivates future research work on traffic morphing solutions that consider the latest IoT fingerprinting attack algorithms and provide effective morphing defenses without compromising the effectiveness of NIDS. These solutions may exploit collaboration between the cloud and user sides so that NIDS can receive accurate information needed for identifying adversarial regions and negotiating more customized shaping methods.

2.4 IOT SECURITY CHALLENGES AND RECENT SOLUTIONS

Since numerous IoT devices can interact directly with the physical environments surrounding users, the security of IoT systems is of notable importance. Security threats in IoT systems can come from multiple sources, including interactions between devices and their environments and cyberattacks.

Over the past years, TAPs have become the breeding ground for security and privacy threats. TAPs allow users to configure trigger-action rules that define the behaviors of smart devices in their homes. However, with user-defined rules, devices may not always behave as expected. Sometimes, these unexpected behaviors may even cause hazards. The unexpected device behaviors may simply originate from benign logic errors in multiple rules. For example, a user may define a rule that opens a water valve for the fire sprinklers when a fire alarm goes off, but also another rule that shuts it off if there is a suspected water leak. TAPs also frequently suffer from the overprivilege issue, i.e., granting excessively high authorization levels to applications. Attackers may take advantage of this incorrect authorization level and conduct malicious activities on smart devices.

Another major source of IoT security threats is cyberattacks. Though NIDS for conventional computer networks have been around for decades, they must undergo significant re-design to fit the IoT ecosystems. Particularly, NIDS must address several unique challenges of IoT ecosystems. These challenges include the diversity of IoT protocols, the various modes of communication, and ever-changing traffic patterns. These systems should also be scalable to accommodate the continuously growing number of smart devices.

The most foundational and popular NIDS rely on signature-based attack detection. More recent works design NIDS based on statistical models and cutting-edge ML/DL models, which have been proven to enhance the performance of IoT NIDS for better infrastructure protection. Also, recent DL technologies such as LSTM, GNN, and transformer models have made learning user behaviors possible. Accurate prediction of user behaviors can assist users and vendors in identifying abnormal behaviors.

This section discusses methods for securing TAPs, NIDS designed for IoT systems, and how user behavior learning helps identify abnormal behaviors in smart homes.

2.4.1 IoT TAP Safety and Integrity Strategies

TAPs allow users to define customized "if-then" rules that enable interoperability between different smart devices and web servers. Popular TAPs include IFTTT, Apple Home, Xiaomi Mijia, and Samsung SmartThings. TAPs have made it much easier for users to build automation systems that better fit their preferences. However, they may also expose users to unexpected security risks. The security risks of TAPs mainly fall into two types.

The first type of TAP security risk is caused by the overprivilege issue with the TAP platform. Such problems have been identified on various TAP platforms. For instance, Fernandes et al. (2017) report that 75% of IFTTT's channels were granted overprivileged OAuth tokens, which are at risk of being misused by attackers to manipulate devices and data for malicious purposes.

Several research projects have attempted to resolve the overprivileged security risk. Fernandes et al. (2018) designed the Decentralized Trigger-Action Platform (DTAP), an extension to the OAuth 2.0 protocol used by mainstream TAP platforms. DTAP introduces two OAuth tokens of different granularity. The bearer first requests a Transfer Token (XToken), the coarse-grained token, from the trigger and action channels. Then, the XTokens are used by the bearer to further negotiate a Rule-Specific Token, i.e., the fine-grained token, from an online service that permits the user to execute a single function with the necessary parameters. A more recent study (Chen et al., 2022) designed minTAP to minimize data access in TAPs and enforce the least-privilege principle. MinTAP utilizes language-based data minimization to ensure only the necessary part of user data is released to TAPs and blocks unrelated API access.

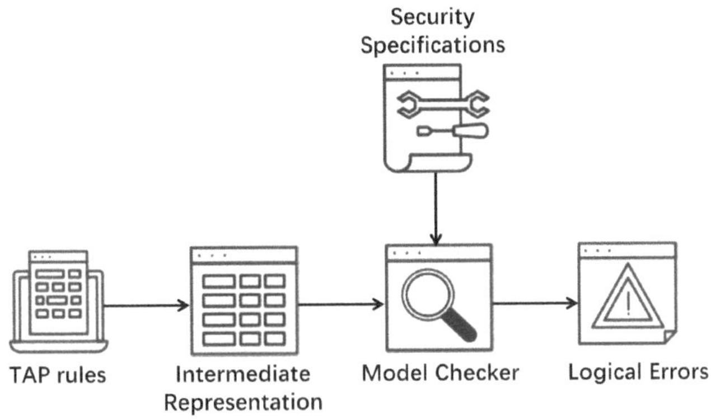

FIGURE 2.6 Model-Checking-Based TAP Verification Systems

The second TAP security risk is caused by logical errors, or "TAP Bugs," which can arise when multiple co-existing rules/apps lead to unintended interactions among devices, ultimately posing a threat to the physical security of users. Common examples of such errors include multiple rules that generate conflicting actions and rules that block the triggers of other rules. These errors may cause security hazards in the physical world. For instance, consider two rules: "If the user leaves home, then close the window" and "If the temperature is above 25 degrees, open the window." If the user leaves home and the temperature rises above 25°, the window may be left open, though the user may think it is closed. Logical errors can arise from unintentional mistakes made by users themselves or could be deliberately introduced by malicious adversaries into TAP rules to intentionally harm users.

Many TAP logical error detection methods are built upon model checking, a verification method for systematically checking whether a system meets specific specifications by exploring all possible states. Figure 2.6 illustrates a common framework for model-checking-based TAP verification systems. In this model, TAP rules are first converted into intermediate representations that are suitable for model checkers. Then, the model checkers verify the intermediate representations against the specified security specifications and output vulnerabilities. An example of a security property could be "The water valve must not be turned off when a fire alarm water sprinkler is on." In this case, the model checker verifies whether executing TAP rules, which involve actions such as turning off the water valve and activating the water sprinkler, breaches this security property.

Researchers have designed many implementations of model checking-based TAP verification systems, which differ in specific details, such as intermediate representation formats, checking methods, and applicable TAP platforms. Wang et al. (2019) proposed a model checking-based TAP verification system called iRuler, which is compatible with IFTTT. iRuler first uses a rule parser to interpret rules from IFTTT into rule representations in a predefined format. A model builder then generates intermediate representations of the IoT deployment from the rule representations, deployment configurations, and device/service metadata. Specifically, trigger-action rules and state transitions are modeled as rewrite rules in the intermediate representations. Last, a model checker, i.e., rewriting modulo SMT, checks over intermediate representations to detect logical errors. Other examples of model checking-based TAP verification systems include TAPInspector (Yu and Liu, 2022) and IoTCOM (Alhanahnah et al., 2020), which use graph structures, specifically the Inter-procedural Control Flow Graph (ICFG), as intermediate representations. TAPInspector uses a symbolic model checker for model checking, while IoTCOM uses a bounded model checker. TAPInspector and IoTCOM are both compatible with SmartThings and IFTTT platforms.

There are also several non-model checking-based logical error detection systems. HOMEGUARD (Chi et al., 2020), for instance, implements a satisfiability modulo theories (SMT)-based method for the SmartThings platform. Unlike model-checking-based ones, this approach does not require safety specification input from users. HOMEGUARD uses a Rule Extractor and a Configuration Collector to model the automation within a smart home. The symbolic executor-based Rule Extractor carries out path-level analyses on the Abstract Syntax Tree representations of SmartApps. This process extracts rule semantics in a structured rule representation format that encloses the trigger, condition, and action of a rule. The Configuration Collector collects configuration information about a smart home through instrumenting SmartApps, e.g., devices bound to the app and different values for variables. HOMEGUARD defines a set of cross-app interference threat types and encodes each type into an SMT model. It checks the rules and configurations against the SMT models to detect potential risks.

2.4.2 Intrusion Detection Strategies for Safeguarding IoT Infrastructure from Cyber Threats

IoT systems have suffered from many cyberattacks in the past decades. The Mirai botnet is among the most notorious attacks in the history of IoT, along with BASHLITE and Carna. In 2016, the Mirai botnet infected 600k IoT devices with weak or default passwords through brute-forcing and launched a massive 620 Gbps Distributed Denial of Service attack toward targets such as Krebs on Security, OVH, and Dyn. Even now, the Mirai botnet and its constantly evolving variants are still posing significant threats to IoT systems.

Though basic security measures like encryption, authentication, and access control have been put in place, IoT networks are still prone to cyberattacks. Various NIDS have been designed to ensure the cybersecurity of conventional computer networks since the '80s. However, these NIDS must be redesigned to meet the specific requirements of IoT systems. First, they should be able to process encrypted traffic, since a major portion of today's Internet traffic is encrypted. Second, they should process traffic with negligible latency so as not to impact the quality of service. Third, they should be lightweight enough to deploy on IoT devices with limited hardware specifications.

Traditionally, many IoT NIDS have been deployed using a rule-based approach, which protects legitimate device traffic from network attacks by filtering network traffic through pre-configured rules. These rule-based approaches are generally divided into two categories based on the rule configuration strategy: signature-based NIDS and specification-based NIDS. In signature-based NIDS, incoming network traffic is compared against known attack signatures stored in the database. If a match is found, the system triggers an alarm. Specification-based NIDS work by setting up a bunch of rules according to the behavioral profiles of network devices, including details on protocols and routing tables. Traffic that disobeys these predefined rules would trigger an alarm. Rule-based approaches enjoy the benefits of low deployment costs and high throughput. However, they have several limitations, including high rates of false positives and/or false negatives, the inability to detect zero-day attacks, and vulnerability to evasion tactics.

Nowadays, anomaly detection-based NIDS are receiving increasing attention due to their potential to solve these problems. Anomaly detection-based NIDS profiles the normal behavioral traffic of a device through features extracted from network traffic. If the difference between the detected network traffic and the profile exceeds a set threshold, the traffic is considered malicious, and an alert is generated. To profile the complex normal network traffic, one promising avenue is to explore artificial intelligence (AI) and ML, especially DL algorithms, to improve system security. CICFlowMeter[6] and Barradas et al. (2021) used ML algorithms to perform anomaly detection. Tang et al. (2020) used LSTM to model the normal plaintext in the HTTP payload. Mirsky et al. (2018) used ensemble MLP autoencoders to detect malicious traffic.

Feature extraction, which processes raw input and prepares input features for learning models, also requires considerable prior knowledge. From the perspective of feature extraction granularity, anomaly detection-based NIDS can be categorized into three types: packet-level approaches, flow-level approaches, and multilevel approaches.

Packet-level approaches utilize the attribute fields in the headers of a packet (such as the TCP header and the IP header), including the packet size, the direction of the packet, the packet payload, and statistics of the numerical features. Cho et al. (2009) proposed a method that extracts features by computing the sum of TCP header fields, packet length, and number of connected links for each node. Mirsky et al. (2018) proposed Kitsune, which extracts features by considering the packet size as the original information and calculating statistics that are aggregated along with the channel, socket, and source IP, among others. In this way, the features essentially contain multi-granularity features for each packet. Tang et al. (2020) proposed ZeroWall, which models normal HTTP plaintext payloads using LSTM. Marin et al. (2019) used encrypted packet payloads directly as features. Even for encrypted traffic, information above the Layer 4 header can be exploited for classification (Rezaei and Liu, 2019).

Typically, a flow is defined as a bi-directional packet sequence uniquely identified by the 5-tuple (SrcIP, SrcPort, DstIP, DstPort, L4 Protocol). Flow-level features include time-series elements such as packet-size sequences and packet-interval sequences, along with statistics computed from these sequences, such as the mean and variance. CICFlowMeter provides a generalized and comprehensive flow-level feature extraction tool that is widely used in NIDS. For each flow, it extracts more than 80 statistical network traffic features, including Duration, Number of Packets, Number of Bytes, and Length of Packets. Li et al. (2022) proposed IoTEnsemble, which aggregates flows using fuzzy 5-tuple to reduce overhead, and then uses the aggregated packet-size sequence and packet-interval sequences as features. Fu et al. (2021) proposed Whisper, which performs Discrete Fourier Transformation on packet-size sequences and packet-interval sequences to gain frequency features, as a robust mechanism for revision attacks.

Packet-level features and flow-level features each have their own strengths. Packet-level features provide the fine-grained details of an individual packet and are therefore suitable for detecting clearly defined anomalies. Flow-level features provide a big picture of how traffic changes, which is helpful for identifying anomalies that only form their shapes over time.

Some of the latest works exploit a combination of packet-level and flow-level features for detecting anomalies, i.e., multilevel approaches. Dong et al. (2023) proposed HorusEye, a two-stage anomaly detection scheme. It uses flow-level features for initial screening of traffic in the first stage, and packet-level features for deeper investigation of suspicious flows in the second stage. Qu et al. (2023) proposed an input-agnostic hierarchical framework for multilevel (packet, flow, and trace) feature extraction of network traffic to exploit hierarchical information in the network.

While existing NIDS have achieved considerable success, there is still much room for improvement. First, the false positive rate (FPR) of anomaly detection-based NIDS could be further reduced. Since all alerts must be manually inspected, reducing the number of false positives can significantly save labor costs. Second, there is a gap between the alert from NIDS and the depth of information ideally desired for manual verification. The vast majority of learning-based NIDS offer little interpretation of their inference results, yet manual verification demands human-readable explanations. This calls for NIDS with better interpretability. Third, current NIDS are generally designed for specific network environments. Adapting them to new scenarios is challenging due to changes in the normal traffic profiles. There has been some noteworthy work on addressing these issues. Fu et al. (2023) proposed pVoxel, which reduces the FPR by analyzing the density of the alert distribution and removing false positives from low-density regions. Wei et al. (2023) proposed xNIDS, which achieves state-of-the-art performance in interpreting NIDS by capturing feature dependencies in structured data. Yang et al. (2021) proposed CADE to improve the transferability of the NIDS against concept drift by detecting drifting samples that deviate from existing classes and providing explanations of the detected drift's reasons.

2.4.3 Leveraging User Behavior Learning for Abnormal Behavior Detection

In the process of people's daily use of smart home devices, user-device interaction (UDI) data is generated. However, abnormal behavior can also appear in the sequence of normal behaviors, which may pose substantial security risks within smart homes. These abnormal behaviors can originate from two primary sources: First, improper operations by users can cause abnormal behaviors, such as inadvertently activating the air conditioner's cooling mode during the winter or forgetting to close a water valve. Second, malicious attackers can exploit vulnerabilities within IoT devices and platforms, taking unauthorized control of these devices. For example, hackers can compromise IoT platforms, allowing them to turn off security cameras and manipulate home automation systems, creating opportunities for burglary.

Prior research has revealed that adversaries can leak personal information and gain physical access to the home (Jia et al., 2017). In Fernandes et al. (2016), spoof attacks are employed to exploit automation rules and trigger unexpected device actions. Chi et al. (2022) applied delay-based attacks to disrupt cross-platform IoT information exchanges, resulting in unexpected interactions that render IoT devices and smart homes insecure. These attacks are designed to cause smart home devices to perform actions that deviate from users' normal behaviors, posing significant security threats. These security concerns emphasize the urgency of robust behavioral modeling methods and enhanced security measures to safeguard smart home environments.

A growing body of research aims to propose algorithms that exploit the interaction data between smart home users and devices to enhance the security of smart homes. Such user behavior learning methods mainly leverage advanced DL methods (e.g., transformer and graph neural networks) to model user behavior and mine user intentions, behavioral habits, and operating context. As shown in Figure 2.7, user behavior sequence data is collected for training user behavior learning models; once trained, the model can be deployed to detect anomalies. From a methodological perspective, user behavior learning methods can be divided into two categories: supervised behavior learning and unsupervised behavior learning.

Supervised behavior learning methods aim to predict the subsequent behavior of the user for anomaly behavior identification. IoTBeholder (Zou et al., 2023) utilizes attention-based LSTM to predict user behavior from history sequences. SmartSense (Jeon et al., 2022) leverages a query-based transformer to model contextual information of user behavior sequences. DeepUDI (Xiao, Zou, Li, Zhao, Li, Tang, et al., 2023) and SmartUDI (Xiao, Zou, Li, Zhao, Li, Weng, et al., 2023) use relational gated graph neural networks, capsule neural networks, and contrastive learning to model users' routines, intents, and multi-level periodicities. These methods consider that behavior that violates model predictions is highly likely abnormal.

Unsupervised behavior learning methods aim to learn behavior sequence context from benign behavior data to achieve anomaly detection. 6thSense (Sikder et al., 2017) utilizes Naive Bayes to detect malicious behavior associated with sensors in smart homes. Aegis (Sikder et al., 2019) utilizes a Markov Chain to detect malicious behavior sequences. Both methods use probability as a metric, and behavioral sequences with lower probability are more likely to be considered anomalous. ARGUS (Rieger et al., 2023) uses an Autoencoder based on Gated Recurrent Units to calculate

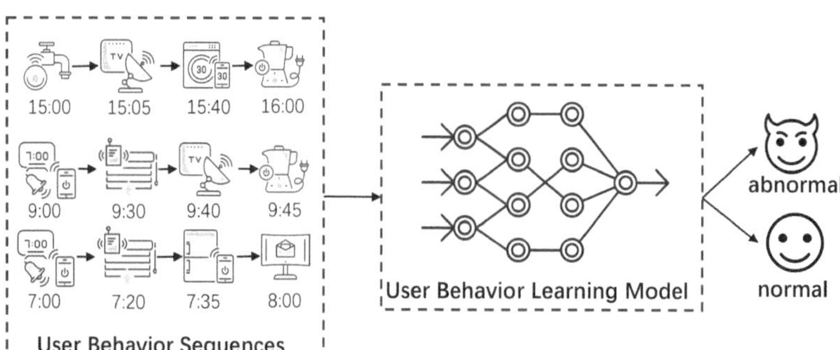

FIGURE 2.7 User Behavior Learning for Anomaly Detection

the anomaly score of user behavior sequences. ARGUS is trained on benign data sets to detect infiltration attacks. The anomaly score is the reconstruction error of the behavioral sequence. The larger the reconstruction error, the higher the possibility of abnormal behavior.

There is room for improvement in today's behavior learning models. They mainly depend on behavior patterns to spot anomalies, but they do not take into account things like temperature, humidity, and lighting in the home, which can be important. Because of this, they might not catch anomalies that are actually related to the environment. Also, in a smart home with several residents, their behaviors can mix and affect each other. Current models do not consider this, which can lead them to label normal behaviors as anomalies by mistake. Moreover, IoT environments are always changing as user behaviors and interactions with devices shift. Existing solutions cannot keep up with these changes. Future work could tackle these problems by including more context from the environment, understanding how different residents' behaviors relate to each other, and using incremental training methods.

2.5 CONCLUSION

In this chapter, we discussed some privacy and security challenges in current IoT systems and briefly explored the latest technologies for protecting IoT systems from safety issues. As smart devices become an indispensable part of our daily lives, safety and security flaws could damage the user experience and threaten personal safety. IoT systems face various privacy and security challenges because of the unique constraints of the hardware and the huge number of devices involved.

Despite ongoing efforts to safeguard IoT ecosystems, the future still presents significant challenges. In addition to the future research directions for each specific topic that we have discussed at the end of each section, there are common challenges that all research on the privacy and security of IoT systems must confront. First, as the functionality and interoperability of smart devices continue to increase, the attack surface of IoT systems may also expand. Vulnerabilities in any part of the ecosystem – be it apps, smartphones, cloud platforms, automation platforms, or third-party services like content delivery networks (CDNs) – can lead to security and privacy issues. Second, many IoT devices are now adopting AI technologies for intelligent functionalities. For instance, surveillance cameras and smart door locks can integrate image recognition. Voice assistants from various vendors, such as Amazon Alexa, Google Assistant, and Apple Siri, are enabled by many smart home devices to facilitate the hands-free operation of devices. However, AI technologies themselves are facing security concerns. Additionally, the transition from IPv4 to IPv6 is in progress to solve the problem of exhausting IPv4 address space. IPv6 allows IoT devices to directly connect to the Internet without going through smart hubs and includes built-in support for IPsec, which secures and encrypts IoT traffic. However, after the transition to IPv6, the protection layer provided by gateways will no longer exist. The development of new defense strategies is therefore required. Researchers, vendors, and regulators should join forces to keep IoT spaces secure from emerging threats.

ACKNOWLEDGMENTS

This work is supported by the Major Key Project of PCL under grant No. PCL2023A06-4.

NOTES

1 Transforma Insights (2022), 'Global IoT Forecast Report, 2022–2032'. https://transformainsights.com/research/reports/global-iot-forecast-report-2032.
2 "How secure is consumer IoT?" The Dawes Centre for Future Crime at UCL, https://www.ucl.ac.uk/future-crime/sites/future_crime/files/ucl_dawes_centre_consumer_iot_briefing_august_2020.pdf.
3 www.mi.com/global/product/mi-led-desk-lamp-1s/.

4 www.mi.com/airfan1x.
5 www.smartmiglobal.com/pages/convector-heater-1s.
6 https://www.unb.ca/cic/research/applications.html.

REFERENCES

Alhanahnah, M., Stevens, C. and Bagheri, H. (2020), Scalable analysis of interaction threats in IoT systems, in *Proceedings of the 29th ACM SIGSOFT International Symposium on Software Testing and Analysis*, ISSTA 2020, ACM, New York, July 18–22, pp. 272–285. https://doi.org/10.1145/3395363.3397347

Apthorpe, N. J., Reisman, D. and Feamster, N. (2017), A smart home is no castle: Privacy vulnerabilities of encrypted IoT traffic [online]. [Preprint]. [Accessed 28 April 2024]. https://arxiv.org/abs/1705.06805

Bahl, P. and Padmanabhan, V. N. (2000), RADAR: An in-building RF-based user location and tracking system, in *Proceedings IEEE INFOCOM 2000*, Tel Aviv, IEEE Computer Society, Israel, March 26–30, pp. 775–784. https://doi.org/10.1109/INFCOM.2000.832252

Barradas, D., Santos, N., Rodrigues, L., Signorello, S., Ramos, F. M. V. and Madeira, A. (2021), FlowLens: Enabling efficient flow classification for ML-based network security applications, in *Proceedings 2021 Network and Distributed System Security Symposium*, Virtual Event, The Internet Society, San Diego, CA, February 21, pp. 775–784–25. https://doi.org/10.14722/ndss.2021.24067

Brendel, W., Rauber, J. and Bethge, M. (2018), Decision-based adversarial attacks: Reliable attacks against black-box machine learning models, in *6th International Conference on Learning Representations*, ICLR 2018, Vancouver, BC, Canada, April 30–May 3. https://openreview.net/forum?id=SyZI0GWCZ

Bugeja, J., Jacobsson, A. and Davidsson, P. (2018), Smart connected homes, in Hassan, Q. (ed.), *Internet of Things A to Z: Technologies and Applications*. Wiley-IEEE Press, pp. 359–384. https://doi.org/10.1002/9781119456735.ch13

Chen, X., Zhang, X., Elliot, M., Wang, X. and Wang, F. (2022), Fix the leaking tap: A survey of Trigger-Action Programming (TAP) security issues, detection techniques and solutions, *Computers & Security* 120, 102812. https://doi.org/10.1016/j.cose.2022.102812

Chen, Y., Alhanahnah, M., Sabelfeld, A., Chatterjee, R. and Fernandes, E. (2022), Practical data access minimization in Trigger-Action platforms, in *31st USENIX Security Symposium (USENIX Security 22)*, USENIX Association, Boston, MA, August 10–12, pp. 2929–2945. https://www.usenix.org/conference/usenixsecurity22/presentation/chen-yunang-practical

Chi, H., Fu, C., Zeng, Q. and Du, X. (2022), Delay wreaks havoc on your smart home: Delay-based automation interference attacks, in *Proceedings of IEEE Symposium on Security and Privacy (SP)*, IEEE Computer Society, San Francisco, CA, May 22–26, pp. 285–302. https://doi.org/10.1109/SP46214.2022.9833620

Chi, H., Zeng, Q., Du, X. and Yu, J. (2020), Cross-app interference threats in smart homes: Categorization, detection and handling, in *2020 50th Annual IEEE/IFIP International Conference on Dependable Systems and Networks (DSN)*, IEEE, Valencia, Spain, June 29–July 2, pp. 411–423. https://doi.org/10.1109/DSN48063.2020.00056

Cho, E.J., Kim, J.H. and Hong, C.S. (2009). Attack model and detection scheme for Botnet on 6LoWPAN, in *Management Enabling the Future Internet for Changing Business and New Computing Services*. APNOMS 2009. Lecture Notes in Computer Science, vol 5787. Springer, Berlin, Heidelberg. https://doi.org/10.1007/978-3-642-04492-2_66

Dong, S., Li, Z., Tang, D., Chen, J., Sun, M. and Zhang, K. (2020), Your smart home can't keep a secret: Towards automated fingerprinting of IoT traffic, in *Proceedings of the 15th ACM Asia Conference on Computer and Communications Security*, ACM, Taipei, China, October 5–9, pp. 47–59. https://doi.org/10.1145/3320269.3384732

Dong, Y., Li, Q., Wu, K., Li, R., Zhao, D., Tyson, G., Peng, J., Jiang, Y., Xia, S. and Xu, M. (2023), HorusEye: A realtime IoT malicious traffic detection framework using programmable switches, in *32nd USENIX Security Symposium (USENIX Security 23)*, USENIX Association, Anaheim, CA, August 9–11, pp. 571–588. https://www.usenix.org/conference/usenixsecurity23/presentation/dong-yutao

Dyer, K. P., Coull, S. E., Ristenpart, T. and Shrimpton, T. (2012), Peek-a-Boo, I still see you: Why efficient traffic analysis countermeasures fail, in *Proceedings of IEEE Symposium on Security and Privacy (SP)*, IEEE Computer Society, San Francisco, CA, May 20–23, pp. 332–346. https://doi.org/10.1109/SP.2012.28

Emami-Naeini, P., Dheenadhayalan, J., Agarwal, Y. and Cranor, F. C. (2023), Are consumers willing to pay for security and privacy of IoT devices? in *32nd USENIX Security Symposium (USENIX Security 23)*, USENIX Association, Anaheim, CA, August 9–11, pp. 1505–1522. https://www.usenix.org/conference/usenixsecurity23/presentation/emami-naeini

Fernandes, E., Jung, J. and Prakash, A. (2016), Security analysis of emerging smart home applications, in *Proceedings of IEEE Symposium on Security and Privacy (SP)*, IEEE Computer Society, San Jose, CA, May 22–26, pp. 636–654. https://doi.org/10.1109/SP.2016.44

Fernandes, E., Rahmati, A., Jung, J. and Prakash, A. (2017), Decoupled-IFTTT: Constraining privilege in trigger-action platforms for the internet of things [online]. [Preprint]. [Accessed 28 April 2024]. https://doi.org/10.48550/arXiv.1707.00405

Fernandes, E., Rahmati, A., Jung, J. and Prakash, A. (2018), Decentralized action integrity for Trigger-Action IoT platforms, in *Proceedings 2018 Network and Distributed System Security Symposium*, The Internet Society, San Diego, CA, February 18–21. https://doi.org/10.14722/ndss.2018.23121

Fu, C., Li, Q., Shen, M. and Xu, K. (2021), Realtime robust malicious traffic detection via frequency domain analysis, in *Proceedings of the 2021 ACM SIGSAC Conference on Computer and Communications Security (CCS)*, Virtual Event, ACM, Republic of Korea, November 15–19, pp. 3431–3446. https://doi.org/10.1145/3460120.3484585

Fu, C., Li, Q., Xu, K. and Wu, J. (2023), Point cloud analysis for ML-based malicious traffic detection: Reducing majorities of false positive alarms, in *Proceedings of the 2023 ACM SIGSAC Conference on Computer and Communications Security (CCS)*, ACM, Copenhagen, Denmark, November 26–30, pp. 1005–1019. https://doi.org/10.1145/3576915.3616631

Jafari, H., Omotere, O., Adesina, D., Wu, H.-H. and Qian, L. (2018), IoT devices fingerprinting using deep learning, in *IEEE Military Communications Conference (MILCOM)*, Los Angeles, CA, October 29–31, pp. 1–9. https://doi.org/10.1109/MILCOM.2018.8599826

Jeon, H., Kim, J., Yoon, H., Lee, J. and Kang, U. (2022), Accurate action recommendation for smart home via two-level encoders and commonsense knowledge, in *Proceedings of the 31th ACM International Conference on Information & Knowledge Management (CIKM)*, ACM, Atlanta, GA, October 17–21, pp. 1–10. https://doi.org/10.1145/3511808.3557226

Jia, Y. J., Chen, Q. A., Wang, S., Rahmati, A., Fernandes, E., Mao, Z. M. and Prakash, A. (2017), ContexIoT: Towards providing contextual integrity to appified IoT platforms, in *Proceedings 2017 Network and Distributed System Security Symposium*, The Internet Society, San Diego, CA, February 26–March 1, pp. 2–2. https://doi.org/10.14722/ndss.2017.23051

Li, R., Li, Q., Huang, Y., Zhang, W., Zhu, P. and Jiang, Y. (2022), IoTEnsemble: Detection of botnet attacks on Internet of Things, in *ESORICS 2022: 27th European Symposium on Research in Computer Security*, Copenhagen, Denmark, September 26–30, pp. 569–588. https://doi.org/10.1007/978-3-031-17146-8_28

Li, R., Li, Q., Lin, T., Zou, Q., Zhao, D., Huang, Y., Tyson, G., Xie, G. and Jiang, Y. (2024), DeviceRadar: Online IoT device fingerprinting in ISPs using programmable switches [online]. [Preprint]. [Accessed 28 April 2024]. https://doi.org/10.48550/arXiv.2404.12738

Ma, X., Qu, J., Li, J., Lui, J. C. S., Li, Z. and Guan, X. (2020), Pinpointing hidden IoT devices via spatial-temporal traffic fingerprinting, in *Proceedings IEEE INFOCOM 2020*, 06–09 July, Toronto, ON, Canada, pp. 894–903. https://doi.org/10.1109/INFOCOM41043.2020.9155346

Marin, G., Casas, P. and Capdehourat, G. (2019), Deep in the dark-deep learning-based malware traffic detection without expert knowledge, in *2019 IEEE Security and Privacy Workshops (SPW)*, IEEE Computer Society, 19–23 May, Los Alamitos, CA, pp. 36–42. https://doi.org/10.1109/SPW.2019.00019

Meidan, Y., Sachidananda, V., Peng, H., Sagron, R., Elovici, Y. and Shabtai, A. (2020), A novel approach for detecting vulnerable IoT devices connected behind a home NAT, *Computers & Security* 97, 101968. https://doi.org/10.1016/j.cose.2020.101968

Miettinen, M., Marchal, S., Hafeez, I., Asokan, N., Sadeghi, A. and Tarkoma, S. (2017), IoT SENTINEL: Automated device-type identification for security enforcement in IoT, in *37th IEEE International Conference on Distributed Computing Systems (ICDCS)*, IEEE Computer Society, Atlanta, GA, June 5–8, pp. 2177–2184. https://doi.org/10.1109/ICDCS.2017.283

Mirsky, Y., Doitshman, T., Elovici, Y. and Shabtai, A. (2018), Kitsune: An ensemble of autoencoders for online network intrusion detection, in *Proceedings 2018 Network and Distributed System Security Symposium*, The Internet Society, San Diego, CA, February 18–21. https://doi.org/10.14722/ndss.2018.23204

Nasr, M., Bahramali, A. and Houmansadr, A. (2021), Defeating DNN-based traffic analysis systems in real-time with blind adversarial perturbations, in *30th USENIX Security Symposium (USENIX Security 21)*, USENIX Association, Virtual, August 11–13, pp. 2705–2722. https://www.usenix.org/conference/usenixsecurity21/presentation/nasr

Pinheiro, A. J., de Araujo-Filho, P. F., de Bezerra, J. M. and Campelo, D. R. (2021), Adaptive packet padding approach for smart home networks: A tradeoff between privacy and performance, *IEEE Internet of Things Journal* 8(5), 3930–3938. https://doi.org/10.1109/JIOT.2020.3025988

Qu, J., Ma, X., Li, J., Luo, X., Xue, L., Zhang, J., Li, Z., Feng, L. and Guan, X. (2023), An input-agnostic hierarchical deep learning framework for traffic fingerprinting, in *32nd USENIX Security Symposium (USENIX Security 23)*, USENIX Association, Anaheim, CA, August 9–11, pp. 589–606. https://www.usenix.org/conference/usenixsecurity23/presentation/qu

Rana, M., Mamun, Q., and Islam, R. (2022). Lightweight cryptography in IoT networks: A survey, *Future Generation Computer Systems* 129, 77–89. https://doi.org/10.1016/j.future.2021.11.011

Rezaei, S. and Liu, X. (2019), Deep learning for encrypted traffic classification: An overview, *IEEE Communications Magazine* 57(5), 76–81. https://doi.org/10.1109/MCOM.2019.1800819

Rieger, P., Chilese, M., Mohamed, R., Miettinen, M., Fereidooni, H. and Sadeghi, A.-R. (2023), ARGUS: Context-based detection of stealthy IoT infiltration attacks, in *32nd USENIX Security Symposium (USENIX Security 23)*, USENIX Association, Anaheim, CA, August 9–11, pp. 4301–4318. https://www.usenix.org/conference/usenixsecurity23/presentation/rieger

Sagahyroon, A., Aburukba, R. and Aloul, F. (2017), The Internet of Things and e-Health: Remote patients monitoring, in Hassan, Q., Khan, A. R., and Madani, S. A. (eds.), *Internet of Things: Challenges, Advances, and Applications*. Chapman and Hall/CRC, pp. 303–319.

Saidi, S. J., Mandalari, A. M., Kolcun, R., Haddadi, H., Dubois, D. J., Choffnes, D., Smaragdakis, G. and Feldmann, A. (2020), A haystack full of needles: Scalable detection of IoT devices in the wild, in *Proceedings of the ACM Internet Measurement Conference (IMC '20)*. Association for Computing Machinery, New York, October 27–29, pp. 87–100. https://doi.org/10.1145/3419394.3423650

Sikder, A. K., Aksu, H. and Uluagac, A. S. (2017), 6thSense: A context-aware sensor-based attack detector for smart devices, in *26nd USENIX Security Symposium (USENIX Security 17)*, USENIX Association, Vancouver, BC, August 16–18, pp. 397–414. https://www.usenix.org/conference/usenixsecurity17/technical-sessions/presentation/sikder

Sikder, A. K., Babun, L., Aksu, H. and Uluagac, A. S. (2019), Aegis: A context-aware security framework for smart home systems, in *Proceedings of the 35th Annual Computer Security Applications Conference (ACSAC)*, ACM, San Juan, Puerto Rico, December 9–13, pp. 28–41. https://doi.org/10.1145/3359789.3359840

Tan, S., Yu, S., Liu, W., He, D. and Chan, S. (2023), You can glimpse but you cannot identify: Protect IoT devices from being fingerprinted, *IEEE Transactions on Dependable and Secure Computing* 21(3), 1–14. https://doi.org/10.1109/TDSC.2023.3275850

Tang, R., Yang, Z., Li, Z., Meng, W., Wang, H., Li, Q., Sun, Y., Pei, D., Wei, T., Xu, Y. and Liu, Y. (2020), ZeroWall: Detecting zero-day web attacks through encoder-decoder recurrent neural networks, in *Proceedings IEEE INFOCOM 2020*, 06–09 July, Toronto, ON, Canada, pp. 2479–2488. https://doi.org/10.1109/INFOCOM41043.2020.9155278

Thangavelu, V., Divakaran, D. M., Sairam, R., Bhunia, S. S. and Gurusamy, M. (2019), DEFT: A distributed IoT fingerprinting technique, *IEEE Internet of Things Journal* 6(1), 940–952. https://doi.org/10.1109/JIOT.2018.2865604

Trimananda, R., Varmarken, J., Markopoulou, A. and Demsky, B. (2020), Packet-level signatures for smart home devices, in *Proceedings 2020 Network and Distributed System Security Symposium*, The Internet Society, San Diego, CA, February 23–26. https://doi.org/10.14722/ndss.2020.24097

Wang, Q., Datta, P., Yang, W., Liu, S., Bates, A. and Gunter, C. A. (2019), Charting the attack surface of trigger-action IoT platforms, in *Proceedings of the 2019 ACM SIGSAC Conference on Computer and Communications Security (CCS)*, ACM, London, November 11–15, pp. 1439–1453. https://doi.org/10.1145/3319535.3345662

Wei, F., Li, H., Zhao, Z. and Hu, H. (2023), xNIDS: Explaining deep learning-based network intrusion detection systems for active intrusion responses, in *32nd USENIX Security Symposium (USENIX Security 23)*, USENIX Association, Anaheim, CA, August 9–11, pp. 4337–4354. https://www.usenix.org/conference/usenixsecurity23/presentation/wei-feng

Xiao, J., Zou, Q., Li, Q., Zhao, D., Li, K., Tang, W., Zhou, R. and Jiang, Y. (2023), User device interaction prediction via relational gated graph attention network and intent-aware encoder, in *Proceedings of the 2023 International Conference on Autonomous Agents and Multiagent Systems (AAMAS)*, AAMAS, London, May 29–June 2, pp. 1634–1642. dl.acm.org/doi/abs/10.5555/3545946.3598820

Xiao, J., Zou, Q., Li, Q., Zhao, D., Li, K., Weng, Z., Li, R. and Jiang, Y. (2023), I know your intent: Graph-enhanced intent-aware user device interaction prediction via contrastive learning, *Proceedings of the ACM on Interactive, Mobile, Wearable and Ubiquitous Technologies (IMUWT/UbiComp)* 7(3), 1–28. https://doi.org/10.1145/3610906

Xiong, S., Sarwate, A. D. and Mandayam, N. B. (2018), Defending against packet-size side-channel attacks in IoT networks, in *2018 IEEE International Conference on Acoustics, Speech and Signal Processing (ICASSP)*, IEEE, Calgary, AB, Canada, April 15–20, pp. 2027–2031. https://doi.org/10.1109/ICASSP.2018.8461330

Yang, L., Guo, W., Hao, Q., Ciptadi, A., Ahmadzadeh, A., Xing, X. and Wang, G. (2021), CADE: Detecting and explaining concept drift samples for security applications, in *30th USENIX Security Symposium (USENIX Security 21)*, USENIX Association, Virtual, August 11–13, pp. 2327–2344. https://www.usenix.org/conference/usenixsecurity21/presentation/yang-limin

Yu, Y. and Liu, J. (2022), TAPInspector: Safety and liveness verification of concurrent trigger-action IoT systems, *IEEE Transactions on Information Forensics and Security* 17, 3773–3788. https://doi.org/10.1109/TIFS.2022.3214084

Zou, Q., Li, Q., Li, R., Huang, Y., Tyson, G., Xiao, J. and Jiang, Y. (2023), IoTBeholder: A privacy snooping attack on user habitual behaviors from smart home Wi-Fi traffic, *Proceedings of the ACM on Interactive, Mobile, Wearable and Ubiquitous Technologies (IMWUT/UbiComp)* 7(1), 1–26. https://doi.org/10.1145/3580890

3 Enhancing IoT Security with Radio Frequency Fingerprinting

Traditional and Deep Learning-Based Approaches

Bo Li and Ediz Cetin

3.1 INTRODUCTION

The *Internet of Things* (IoT) is one of the most promising global infrastructures of an interconnected network of heterogeneous wireless devices and sensors (Hassan, 2018). The rapid expansion and increasing pervasiveness of the IoT devices and the nascent application scenarios enabled by rapid developments and advances in wireless connectivity and device miniaturization mean that IoT devices can now be connected in wider and denser ecosystems serving a variety of applications. While the opportunity is considerable, there are also risks and challenges, such as security vulnerabilities as wireless applications where IoT devices are used become more ubiquitous (Zhang et al., 2017; Butun, Österberg & Song, 2019), leading to growing concerns and hence increasing emphasis on security and in particular device-level security (Neshenko et al., 2019; Hassija et al., 2019).

Identifying and authenticating IoT devices to ensure trustworthy communications between them is one of the key approaches to enhancing security. Given their limited computational resources and power, traditional high-level security solutions, which are complex and resource intensive, are not suitable for use in these IoT devices (Surendran, Nassef & Beheshti, 2018; Tiloca & Raza, 2018; Mousavi et al., 2021). Hence, there is a need to develop novel approaches to enhance wireless communications security specifically for IoT devices where energy is always highly restricted and low-complexity approaches for security are of great importance (Xu et al., 2015; Xie et al., 2024).

Different from the traditional cryptographic-based methods, *Radio Frequency Fingerprinting* (RFF), the most widely explored wireless device fingerprinting technology of generating device signatures at the physical layer, enables the receiver to identify transmitter(s) based on physical layer processing. By extracting RF fingerprints of transmitter(s), the characteristics imparted on the transmitted signals by the analog front-end components can be used to distinguish them. RFF techniques have drawn considerable attention as they can provide a layer of protection for trustworthy communications and thus enhance the security of wireless networks (Ureten & Serinken, 2007; Brik et al., 2008). RFF techniques make use of the intrinsic characteristics of devices. By extracting these characteristics from the *Radio Frequency* (RF) signals, wireless devices can be distinguished accordingly. Operating at the *Physical Layer* (PHY), RFF techniques for IoT security work on the received signal and can be easily integrated into wireless networks without changing the communications standards used, such as Wi-Fi (Jagannath, Jagannath & Kumar, 2022a), ZigBee (Zhang et al., 2022), and LoRa (Shen et al., 2021).

 DOI: 10.1201/9781003506638-4

This chapter focuses on the RFF techniques for enhancing IoT security. It introduces and provides background on wireless device fingerprinting. Various types of RFF techniques, ranging from traditional approaches to state-of-the-art deep learning-based approaches, are described. This chapter concludes with a discussion of the challenges and future trends.

3.2 A GLIMPSE INTO RFF TECHNIQUES

The premise of RFF is to extract unique features imparted by the analog components of the RF transceivers, such as oscillators, mixers, filters, amplifiers, and antennas – which, like our unique fingerprints, are hard to impersonate – on the transmitted signals to identify them. These RF features come from imperfections caused during hardware manufacturing and are unique to each device, even if they are of the same model and made by the same manufacturer. While these RF features among mass-produced wireless devices can be minute, they are still significant enough to be used as distinguishing factors.

Conventional RFF approaches utilize a variety of handcrafted features on the physical layer for device identification or authentication. In contrast, deep learning-based approaches rely on neural networks to automatically extract features from a certain format or representation of received signals. In general, RFF techniques can be classified into two broad areas: conventional and deep-learning-based, as shown in Figure 3.1.

Conventional approaches for RFF have focused on identifying transmitters by obtaining certain characteristics such as carrier frequency offset (Brik et al., 2008) and amplifier nonlinearity (Huang et al., 2016) through different algorithms such as *Support Vector Machine* (SVM) (Robyns et al., 2017), Random Forest (RndF) (Patel, Temple & Baldwin, 2014), and *Multiple Discriminant Analysis* (MDA) (Bihl, Bauer & Temple, 2016). The conventional approaches can be subdivided into two further sub-categories depending on the signals used for feature extraction, i.e., *transient-based* or *steady-state signal-based* approaches, which are discussed in Section 3.3.

With the deep learning technology showing significant impact on a variety of applications in recent years, including computer vision (He et al., 2016), natural language processing (Young et al., 2018), and speech recognition (Deng, Hinton & Kingsbury, 2013), it has also been introduced to RFF techniques for its principal advantage of extracting features automatically. When applied to solving RFF problems, deep learning techniques have been shown to outperform the conventional approaches (Jagannath, Kane & Jagannath, 2022b), and this further motivates researchers to explore and reapply the conventional techniques combined with deep learning in innovative ways. The following sub-sections provide details of these conventional and deep learning-based RFF approaches.

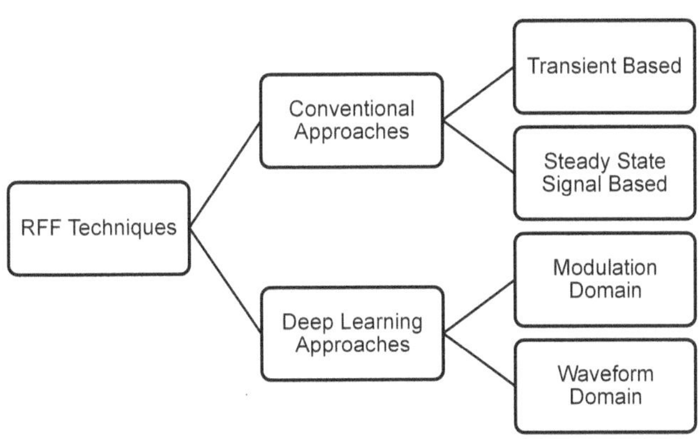

FIGURE 3.1 Radio Frequency Fingerprinting Approach Classification

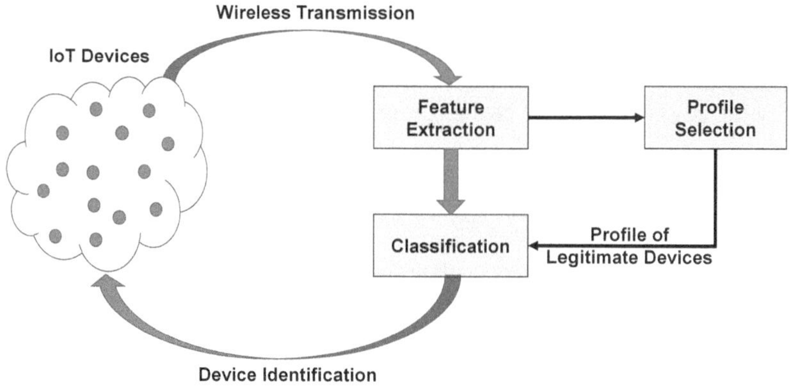

FIGURE 3.2 A Conventional RFF Framework for Wireless Device Identification

3.3 CONVENTIONAL RFF TECHNIQUES

Conventional RFF techniques operate by manually identifying the effective features. A variety of physical layer features have been studied and utilized in conventional RFF approaches, ranging from known preambles within a data packet to transient portions of signals. A conventional RF fingerprinting framework for wireless device identification is shown in Figure 3.2.

First, the signals transmitted from wireless devices are collected and features are extracted from them to create profiles of these devices. The classifier is then used to calculate the probability of a match between the feature extracted from the received signal and each of the created profiles of legitimate devices. It is important to note that, when such approaches are applied to a large scale of wireless devices, the identification performance declines rapidly as the features become less distinguishable among the large number of wireless devices.

Depending on the type of signal used for feature extraction, conventional RFF techniques can be divided into *transient* signal and *steady-state* signal techniques (Shaw & Kinsner, 1997). These are discussed in the following subsections.

3.3.1 Transient-Based Approaches

Transient-based approaches rely on the assumption that signals generated during the short transient phase when a radio transmitter is activated (or deactivated) can be detected and they have characteristics that can be unique, which can be analyzed to identify the radio transmitter (Ureten & Serinken, 1999). These approaches capture signal segments that transition from channel noise to signal amplitude or vice versa. These transitions typically occur during *power-up* or *power-down* phases of a transmitter, allowing for RF fingerprint extraction. Since these *transients* are very short in time, typically lasting for a few microseconds, it is very challenging to detect and capture them. Hence, solutions based on this approach require extremely high sampling rates.

The transient-based approaches have four distinct phases of operation (Hall, Barbeau & Kranakis, 2003):

1. *Pre-processing:* The objective of preprocessing is to extract the signal features that will be used to detect the start of the transient. To achieve this, signal information from the time (amplitude) or frequency domains (frequency or phase) is typically utilized.
2. *Transient Start Detection:* There is always an abrupt change between the channel noise and the start point of the original signal transmission. The transient start detection phase attempts to locate this start point based on the extracted features from the pre-processing phase. There are some classic methods available for transient start detection, including

Bayesian Step Change Detection (BSCD), *Bayesian Ramp Change Detection* (BRCD), *Variance Fractal Dimension Threshold Detection* (VFDTD), and *Phase Detection* (PD), to name a few.
3. *Feature Extraction:* When the transient segment of the signal is located, feature extraction is then performed on the transient to create a fingerprint. The method of extraction based on the concept of fractal dimension is the most popular, followed by wavelet-based extraction.
4. *Classification:* When fingerprints are available, various techniques, such as the parallel analog network (Boser et al., 1991) or *Probabilistic Neural Network* (PNN) (Specht, 1990), can be used for classification.

The following details the methods used for transient start detection, as this is the most challenging phase of operation.

3.3.1.1 Bayesian Step Change Detection

A transient signal is non-stationary, resulting in varying statistical properties over time. To detect the transient, a sliding window can be used to divide the signal into successive segments, BSCD utilizes the fractal dimensions of these successive segments of the signal, which are calculated by the Higuchi method (Higuchi, 1988), to find the most likely segment as the start of the transient. Several techniques reported in the literature utilize the Bayesian step change detector for transients' detection (Ureten & Serinken, 1999; Klein et al., 2009). Ureten and Serinken (1999) analyzed radio transmissions from 30 different radio transmitters, and detection was accomplished using the fractal dimension measure followed by a Bayesian step-change detector. The results show that this approach allows the detection of the transient in less than ten samples on average. In Klein and Temple (2009), a Fractal-BSCD approach was proposed that achieves a detection rate of above 90% with three different 802.11a devices. The fractal derivation can be found in Bassey et al. (2019) and can be calculated using the Higuchi method (Higuchi, 1988).

3.3.1.2 Bayesian Ramp Change Detection

BRCD is a modification of the BSCD method with the capability to detect gradual changes or "ramps," in addition to abrupt changes in the time series data. This method incorporates parameters that describe both the mean and the trend (slope) of the data over time when the power of the signal gently increases. Ureten and Serinken (2005) designed a Bayesian ramp change detector to detect the turn-on transients of Wi-Fi radios. It was observed that the power levels of Wi-Fi devices increase gradually to avoid the spread of radio signals to adjacent frequency channels and that the power-up ramp at the turn-on point is a unique characteristic inherent to each device. Hence, the Wi-Fi turn-on transients were collected for feature extraction. Compared to other approaches that use abrupt change detectors, the approach provides better results as the power-up ramp lasts longer and contains more information for identification.

3.3.1.3 Variance Fractal Dimension Threshold Detection

Several approaches that utilize multifractal analysis for transient extraction are also reported in the literature (Shaw & Kinsner, 1997; Kinsner et al., 2003). In Shaw and Kinsner (1997), the transient signal, which is a non-stationary signal with time-varying fractality, is characterized by a model consisting of several local fractal dimensions of the signal. As a result, the profile of the transient is more accurate and contains more information for identification through multifractal analysis. This approach achieved a 92.5% detection rate with eight different transmitters, including three Kenwood models, three Force models, and two Yaesu models.

3.3.1.4 Phase Detection

PD methods utilize phase characteristics to extract transient signals rather than the amplitude characteristics used in the aforementioned methods. Proposed by Hall, Barbeau, and Kranakis (2003), this approach is a more robust solution as the phase of a signal is less susceptible to noise and interference, and the slope of the phase becomes and remains linear from the start of the transient portion of the signal. The detection rate of this approach was approximately 85%–95% with ten Bluetooth devices. The feature combination method has been used in the literature to improve the performance of PD. In Hall, Barbeau, and Kranakis (2004), a feature vector is derived from three signal components, namely amplitude, phase, and frequency. This feature vector is then used to represent its corresponding transmitter as its device profile. The probability of a match is calculated by the classifier using a Kalman filter to decide if the extracted feature vector from the observed signal matches the feature vector from one of the legitimate devices. Fourteen 802.11b transceivers were used, and the approach achieved a detection rate of 94%–100%. In Barbeau, Hall and Kranakis (2006), a frequency domain approach is proposed where the amplitude envelope, instantaneous phase, and frequency characteristics calculated by the *Discrete Wavelet Transform* were utilized as a feature vector to enhance the characterization of the transient signal. The preliminary results presented showed a detection rate of 93% and an average false alarm rate of 5% with ten Bluetooth devices.

3.3.2 Steady-State Signal-Based Approaches

Though transient-based approaches have attracted major attention in the early years of RFF research, the difficulty of capturing transients motivated researchers to consider alternative approaches based on extracting features from steady-state signals instead. To this end, researchers have explored a variety of feature extraction methods, and these approaches gradually became dominant before deep learning-based techniques were introduced into this field of research. The extracted features used can be categorized into *location-independent* and *location-dependent* ones.

3.3.2.1 Location-Independent Features

Location-independent features are relevant to the hardware implementation of individual devices. The imperfection caused during the manufacturing, though subtle, results in unique fingerprints and allows for identification. The most commonly used location-independent features are as follows:

- *Preamble Feature:* There are approaches reported in the literature that utilized *preambles* for feature extraction. Gerdes et al. (2006) explored the possibility of distinguishing the same model Ethernet cards manufactured by the same company using steady-state signals transmitted by wireline transmitters. During their experiments, the features were extracted from the IEEE Ethernet 802.3 frame preamble. They designed a classifier where a matched filter was applied to render a final decision regarding the relationship between the observed transmitter and each of the legitimate devices. The approach achieved less than 1% *False-Reject Rate* (FRR) for 16 devices of 3 different models. The approach in Kennedy, Scanlon, and Buddhikot (2008) utilized the baseband preamble signal and used the *Fast Fourier Transform* to preprocess the captured signal for feature vector generation. The results of laboratory experiments showed an identification accuracy of 85% at *Signal-to-Noise Ratio* (SNR) of 15 dB for 20 phones, of which half were the same model.
- *Power Amplifier Feature:* Dolatshahi, Polak, and Goeckel (2010) focused on the *Input/Output* (I/O) characteristics of power amplifiers in transmitters. The imperfections caused by the production process were studied as the feature to identify transmitters. To simplify the problem, the authors demonstrated their approach under a scenario where only two users were trying to connect to one wireless access point. Two MAXIM MAX2242 power amplifiers were simulated using the *Advanced Design System* (ADS) software, with a transmit center frequency of 2.45 GHz. The feature vector they utilized was the coefficients of a memoryless fourth-order Volterra series representation. The probability of detection error was low, below 1% with an SNR of 10 dB. One of the drawbacks of this

solution is its dependence on the power level of a transmitter. This is a major limitation as the power level varies as the transmitter switches to different operating modes over time, and hence the characteristics change accordingly. To address the issue that characteristics extracted from power amplifiers are influenced by varying temperature levels, in Polak, Dolatshahi, and Goeckel (2011), a new approach utilizing the I/O characteristics of the *Digital-to-Analog Converter* and the power amplifier of transmitters to discover illegal transmissions was proposed. The feature vectors were derived from the models of the non-linear transmitter components and four different testing algorithms were applied for evaluating the performance. During the experiment of 50,000 simulation trials with eight IEEE 802.11b devices from the same manufacturer, 100% identification accuracy was observed.

- *RF Oscillator Feature:* In Polak and Goeckel (2015), the characteristics of RF oscillators implemented as *Phased-Locked-Loops* (PLLs) have been studied. For those oscillators, the PLL's phase noise is extracted as the device's RF fingerprint. These extracted fingerprints are unique because they depend on the variations of the PLL circuit's components, and the variations are present even with identical circuit designs. During the experiments, eight Analog Devices ADF4360-1 oscillators were utilized, operating at 2.4 GHz. The results show stable performance even with varying SNR levels.
- *Multi-Feature:* There are approaches reported in the literature that utilize multiple location-independent features. A system named *PAssive RAdiometric Device Identification System* (PARADIS) was developed by Brik et al. (2008), which utilizes features including magnitude and phase errors, *In-phase* (I) and *Quadrature* (Q) signal origin offset and SYNC correlation of the IEEE 802.11 frame to identify the source *Network Interface Card* (NIC) through passive radio-frequency analysis. This system was able to effectively differentiate more than 130 identical IEEE 802.11 NICs, achieving 99% identification accuracy regardless of the ambient noise and the fluctuations of the wireless channel. Tu et al. (2019) exploited the combination of four different signal feature sequences derived from the time domain, wavelet domain, short-time Fourier transform (STFT), and Wigner-Ville distribution, respectively, to identify six Zigbee devices, achieving an identification accuracy above 96%.

3.3.2.2 Location-Dependent Features

Location-dependent features are related to the channel and propagation where the wireless IoT devices operate. The most common feature is the *Received Signal Strength* (RSS). The RSS fingerprints have been extensively studied and are primarily utilized in wireless device identification, intrusion detection in *Radio Frequency Identification* (RFID) applications, and related areas (Danev et al., 2012; Han et al., 2015). Another feature that is used is the *Channel State Information* (CSI). The unique wireless *Channel State Information at the Receiver* (CSIR) of a legitimate user, such as channel frequency response and channel impulse response, can be used to authenticate subsequent transmissions from this user. Using this approach, spoofers with significantly different CSIs from that of the legitimate user of a different spatial location can be identified and their attempt to access can be denied (Faria & Cheriton, 2006; Li et al., 2006; Xiao et al., 2007). The two commonly used features are as follows:

- *CSIR Features:* The channel frequency response was leveraged to enhance the security of physical layer authentication. Xiao et al. (2008) proposed a method for authenticating mobile wireless transmitters in a building environment using channel frequency responses to differentiate between legitimate users and potential intruders. The authentication process consists of two parts, namely the inter-burst and intra-burst authentications to address the terminal mobility problem. Simulation results using the ray-tracing software WiSE showed that the detection rate is around 96%, with a false alarm rate of 6%, when the transmitter moved at a speed of 1.43 m/s. An alternative to channel frequency response is the channel impulse response. Tugnait and Kim (2010) explored a single-carrier time-domain approach through residual testing. By using the CSI acquired from a previous transmission, they generated the linear innovations for the current transmission and tested whether the two transmissions shared the same underlying channel. Their hypothesis was that if the two transmissions shared the same underlying channel, the residuals would exhibit whiteness. The simulation result shows that intruder/spoofer can be correctly identified with an accuracy of 93%.

- *RSS Feature:* Li et al. (2017) have studied the unique identification problem, i.e., detecting whether an individual is carrying more than one active/passive wireless device, using received RF signal strength fingerprints, which are collected and modeled as time series. To achieve unique identification, the similarity of these time series is calculated. Inspired by word-matching research in speech recognition, *Dynamic Time Warping* (DTW) was adopted instead of the Euclidean distance to measure the similarity of two time series of RSS data that may vary in time or speed. During each simulation, 40 nodes were randomly selected to move together, and the perfect result, i.e., no false positive or incorrect results, could be obtained with an appropriate DTW threshold.

3.4 DEEP LEARNING APPROACHES FOR RFF

Recently, deep learning techniques have become more and more popular. The principal advantage of deep learning technology is that it can extract features automatically, which has brought *artificial intelligence* (AI) to far-reaching applications such as self-driving cars (Ramos et al., 2017), drug discovery (Chen et al., 2018), fraud detection in credit card transactions (Roy et al., 2018), and automatic game playing (Justesen et al., 2019), to name a few. It has been considered a revolutionary technology that would be predominant in the next few decades, and its performance is progressing rapidly.

The framework of wireless device identification using RFF and deep learning is shown in Figure 3.3. As it can be seen from the figure, the first step is raw data collection from wireless IoT devices, followed by signal preprocessing to generate the appropriate inputs to the deep learning model. After the deep learning model is trained and tested, it can be used for device identification.

Different types of deep learning models have been designed for RFF techniques. In Robyns et al. (2017), two types of deep learning models, *Multilayer Perceptron* (MLP) and *Convolutional Neural Network* (CNN), have been used for device classification tasks and they outperformed an SVM-based model. During the experiments, 22 LoRa devices were utilized, and the approach achieved 99% identification accuracy for identical chipsets. In Wu et al. (2018), a *Long Short-Term*

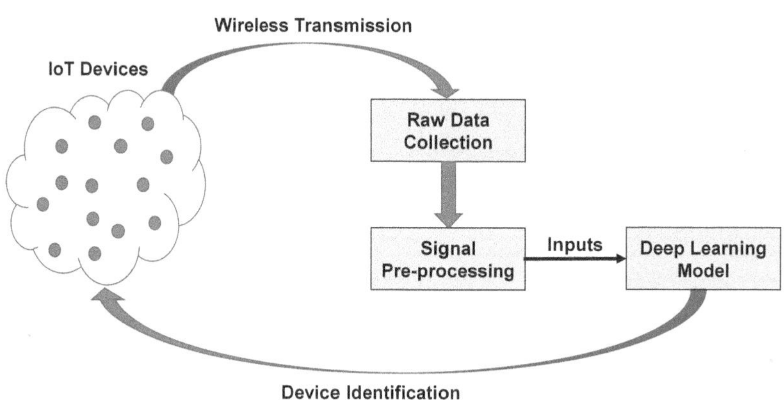

FIGURE 3.3 A Deep Learning RFF Framework for Wireless Device Identification

Memory (LSTM)-based *Recurrent Neural Network* (RNN) was designed to identify identical wireless devices when they are working at an SNR of 12 dB. With this approach, the raw samples at the output of the ADC were collected and divided into sequences for training and testing purposes. This approach achieved more than 95% identification accuracy with six *Universal Software Radio Peripheral* (USRP)-2900 transmitters at an SNR of 12 dB. Roy et al. (2019) proposed an adversarial learning approach using *Generative Adversarial Nets* (GANs). The GAN exploited the *I/Q* imbalance as fingerprints for identifying and classifying the transmitters, achieving 99.9% accuracy in classifying trusted transmitters from fake ones with eight USRP devices.

Deep learning provides a powerful framework for RF fingerprinting whereby features are automatically extracted, which provides researchers with more flexibility. Compared with machine learning and other algorithm-based classical techniques, deep learning has shown its potential when applied to the same task (Robyns et al., 2017). However, when introducing deep learning to deal with RF signals, the major challenge is how to best present the signal information of the collected raw data to the neural networks so that they can extract and learn the features effectively. To address this challenge, researchers have explored a variety of signal preprocessing methods to transfer raw data into effective representations for training and testing of deep learning models.

Depending on where the raw data is collected at the receiver, the learning-based RFF approaches can be classified into *modulation* and *waveform* domains, as shown in Figure 3.4. The modulation domain approaches focus on the intermediate data from the demodulator. In contrast, the waveform domain approaches utilize the samples of the received signals that are not processed by the blocks in the receiver chain and hence retain the most intact characteristics. Typically, the raw data samples are collected at the output of the ADC. Waveform domain approaches have attracted more attention from researchers given the scope to explore a variety of preprocessing methods on the baseband signals. During the first few years of introducing deep learning to RFF, some studies in the waveform domain used raw sample sequences as inputs and achieved high identification accuracies (Riyaz et al., 2018). In recent years, various preprocessing methods have been proposed to improve the identification accuracy performance further (Xie et al., 2024). The waveform domain approaches can be further categorized into *raw sample*-based and *feature*-based approaches.

This section briefly introduces some supervised deep learning models that have been widely used for RF fingerprinting. It also provides detailed discussions about approaches of different categories, i.e., raw sample-based waveform domain approaches, feature-based waveform domain approaches, and modulation domain approaches.

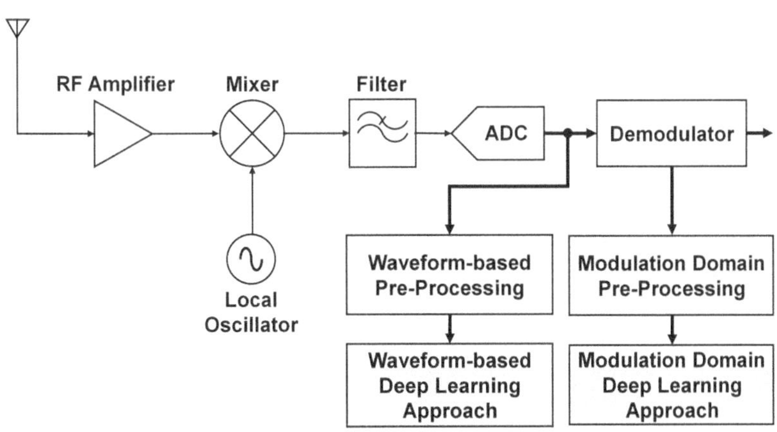

FIGURE 3.4 The Data Collection Points for Waveform and Modulation Domain Deep Learning Approaches on a Traditional Receiver Structure

3.4.1 Overview of Deep Learning Architectures

The concept of AI was introduced by computer scientists in the late 1950s (Buchanan, 2005). However, it has seen obstacles and advanced slowly, mainly due to the lack of computational power. In the early days of AI, researchers paid attention to machine learning as a main approach to achieve the goal of *general AI*, which has the ability to mimic human intelligence. Machine learning relies on algorithms such as decision tree learning and inductive logic programming to train the machine on data to make a determination or prediction. However, it did not show the capability or potential to achieve the ultimate goal of general AI (Carbonell, Michalski & Mitchell, 1983).

Neural networks, specifically called artificial neural networks, were inspired by the neuron connections (biological neural networks) in our brains and make up the backbone of deep learning algorithms. As the state-of-the-art technique for machine learning, deep neural networks have shown their excellent performance to implement practical applications of machine learning and achieve the ultimate goal of general AI. Deep learning is a specific subfield of machine learning, and the *deep* in deep learning refers to successive layers of representations. In contrast, machine learning, sometimes called *shallow* learning, often involves one or two layers of representations of the data. It is common to see deep learning models that have tens or even hundreds of successive layers.

Different types of neural networks have been introduced into RFF approaches, including MLPs, CNNs, RNNs, LSTMs (popular variants of RNNs), and GANs. MLPs are a type of feedforward neural network without cycles or loops in the architecture and are commonly used for classification, regression, and pattern recognition. They generally consist of an input layer, one or more hidden layers, and an output layer, with each neuron in one layer connecting to every neuron in the next layer. CNNs have been widely used for image classification and speech recognition. Their architecture includes three main types of layers, namely *the convolutional* layer, *pooling* layer, and *fully connected* (FC) layer. Different from MLPs and CNNs, which assume inputs and outputs are independent of each other, RANs take information from prior inputs to influence the current input and output, making them suitable to solve language translation, natural language processing, speech recognition, and image captioning problems. RNNs have three layers, which are *input, recurrent hidden*, and *output* layers. LSTMs are a type of RNN designed to overcome the problem of the long-term dependencies of traditional RNNs. The memory cells of LSTMs can maintain information over long periods, which allows them to capture long-term dependencies in sequential data effectively. GANs, on the other hand, consist of two neural networks, a *generator* and a *discriminator*. The generator generates new data as much as possible from input data, while the discriminator evaluates the authenticity of the generated data. They are used in applications, such as image synthesis, data augmentation, text-to-image generation, image-to-image generation, etc. These approaches enable researchers to focus on exploring robust and specific deep learning architectures tailored for RF fingerprinting rather than spending significant time and resources on feature engineering.

3.4.2 Raw Sample-Based Waveform Domain Approaches

With support from a deep learning framework, the performance of RFF techniques is improving rapidly, and approaches with varying levels of complexity covering various application domains have been proposed in recent years.

There have been proposals that do not specify which features they target and simply utilize raw sample sequences collected at the output of ADC as signal representation, so the feature extraction is totally dependent on the neural networks they have designed. Motivated by modulation classification studied for recognizing the modulation type of a received radio signal as a key task in *Dynamic Spectrum Access* research (O'Shea, Corgan & Clancy, 2016; Riyaz et al., 2018), they proposed an approach whereby millions of sub-sequences were generated by manipulating a sliding window over a large number of captured *I/Q* samples with the sliding window moving by one sample at a time. They then used the sub-sequences to train a neural network and were successful in identifying five USRP devices with above 91% identification accuracy.

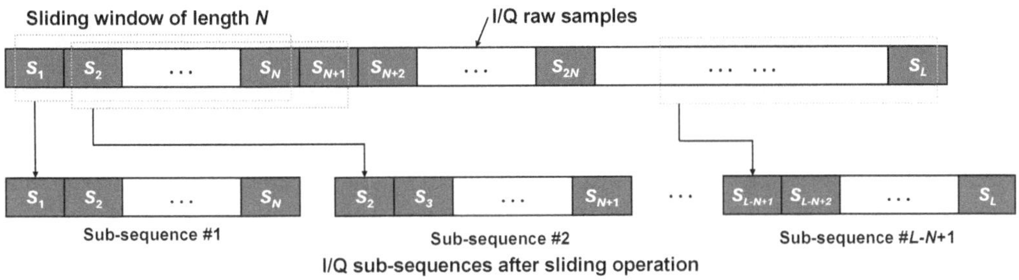

FIGURE 3.5 An Illustration of Sliding Operation over *I*/*Q* Sequences to Generate Sub-sequences

Raw sample-based waveform domain approaches leverage the power of deep learning models by directly feeding raw sample sequences to deep learning models for training and validation purposes, expecting the deep learning models to extract and learn underlying features or patterns from the raw data. In these approaches, the captured *I*/*Q* samples are divided into sub-sequences, which are then fed into a deep learning model for training and validation purposes. Figure 3.5 shows a sliding window operation over *I*/*Q* sequences $s_1 \ldots s_L$ to generate sub-sequences as inputs to a deep learning model (Riyaz et al., 2018).

The key advantage of raw sample-based approaches is the simplicity of the preprocessing. This makes them suitable for use in scenarios where computational resources are limited. However, a notable drawback of these techniques is the sheer volume of inputs, which can significantly prolong the training and validation phases, and may require substantial storage capacity. Another drawback is time and location dependence. When classifiers are trained on raw *I*/*Q* samples under a given channel and then tested on *I*/*Q* samples obtained under different channels, or when transmitters experience very similar channel conditions, their performance degrades severely (Sankhe et al., 2019).

In Riyaz et al. (2018), a CNN framework was proposed utilizing raw sample sequences as input. One USRP B210 was used to capture signals over the air from five USRP B210 devices at a 1.92 MS/s sampling rate at a center frequency of 2.45 GHz. They collected approximately 20 million samples per device. The raw sample sub-sequences of length 128 were generated by manipulating a sliding window on the captured *I*/*Q* samples with the sliding window moving by one sample at a time. Each complex value is represented as a two-dimensional (2-D) real value, which results in the dimension of the input data growing to 2 × 128. The CNN architecture, consisting of four layers, two convolutional layers and two FC or dense layers and the *Rectified Linear Unit* (`ReLU`), was used as the activation functions. The approach was implemented in Keras[1] running on top of TensorFlow[2] on an NVIDIA Cuda-enabled Tesla K40m GPU.[3] The deep learning approach achieved 98% accuracy for five devices, outperforming SVM and logistic regression algorithms, which achieved ~55% accuracy with two devices and ~33% accuracy with five devices. The impact of distance on RF fingerprinting has been analyzed through experiments collecting data over a distance ranging between 2 and 50 ft (0.6–15 m) with steps of 4 ft (1.2 m). The results show that the proposed approach achieved a classification accuracy greater than 95% up to the distance of 34 ft (~10 m).

The performance of different deep learning models has been evaluated in the scheme proposed by Jafari et al. (2018). The ability of three different deep learning models has been compared, including MLP, CNN, and RNN, using samples from ZigBee devices. The MLP model contained four dense layers of size 256, 256, and 128, respectively, and six-classes neurons at the output, with the first three layers using `ReLU` activation functions except for a `Softmax` activation on the output layer. The CNN model was a four-layer network consisting of two convolutional layers and two dense FC layers with `ReLU` activation functions utilized except for the `Softmax` activation on the output layer. 2-D zero padding was used after the convolutional layers to preserve the spatial size of the input volume, so the input and output width and height are the same. The adopted LSTM model consisted of three stacked LSTM layers, containing 32 memory blocks in each of the three layers and a dense layer, enabling learning and remembering over long sequences. Tanh activation functions were used on the LSTM layer and Softmax activation on the output layer. The samples were captured at varying SNR levels to demonstrate the robustness of the proposed models to various wireless channel conditions. Six ZigBee devices were configured to transmit at different SNR levels of 0, –1, –5, –10, and –15 dB and one NI USRP 293x was used to capture the *I/Q* samples. The size of the captured data for each device at every SNR was approximately 10 GB for 5 minutes. As a result, 300 GB of data were collected for six devices in total. The impact of input length was analyzed by generating inputs of different lengths varying between 16, 32, 64, and 128. The experiments demonstrated that increasing the input length size did not have any major impact on the performance. With 50 epochs, the MLP model achieved an identification accuracy of 96.3%, whereas the CNN model achieved 94.7%, and the LSTM model achieved 76.0% accuracy, respectively.

Sankhe et al. (2019) proposed a CNN framework for RF fingerprinting called ORACLE – *Optimized Radio classification through CNNs*, where a CNN architecture requiring only raw *I/Q* samples for an idealized static channel environment without channel estimation or prior knowledge of the communications protocol was used. The CNN architecture consisted of four layers, with two convolution layers and two FC (or dense) layers. Each convolution layer was followed by a `ReLU` activation, which performed a predetermined non-linear transformation on each element of the convolved output. The work demonstrated the performance of ORACLE to classify COTS Wi-Fi devices using an external database, which contained labeled *I/Q* samples collected from 140 devices (phones, tablets, laptops, and drones) from 122 manufacturers. The raw *I/Q* samples were pre-processed using a sliding window (Riyaz et al., 2018), resulting in input of length 128 formatted into a 2-D array of real values of size 2 × 128. The CNN architecture was implemented in Keras running on top of TensorFlow on a system with eight NVIDIA Cuda-enabled Tesla K80m GPUs. ORACLE achieved up to 99% classification accuracy for up to 100 different devices, with the performance dropping slightly to 96% for 140 devices. To further evaluate ORACLE's performance with identical devices, 16 high-end X310 USRP *Software Defined Radios* (SDRs) were used as transmitters for data collection. With this setup, the accuracy was close to 98.6%. However, the performance degraded severely due to the limitations of raw *I/Q* samples in dynamic channels where multipath reflection and fading distort the samples to the extent that the classifier can no longer correctly identify devices. To address the time and location dependence of the raw sample-based approach, the work also demonstrated the performance for dynamic channels with the same CNN model using data in the modulation domain, which is discussed in Section 4.4.

3.4.3 Feature-Based Waveform Domain Approaches

Unlike the raw sample-based approaches, there are a variety of feature engineering methods proposed along with different deep learning models for RF fingerprinting. The main premise is to generate a better format or representation of received signals for the deep learning models to extract patterns. General flow for feature-based waveform domain approaches, as depicted in Figure 3.6. As can be observed from the figure, raw samples of received signals – either captured over-the-air or synthetically generated – are first transformed into another representation before being used as inputs to train the deep learning models. The trained model is then used for new signal classification.

3.4.3.1 Channel State Information

Apart from features from hardware imperfections, RF fingerprints extracted from CSI have also been used for authentication within *Industrial Wireless Sensor Networks*. In Liao et al. (2019a), RF fingerprinting was used for detecting spoofing attacks in *Multiple-Input Multiple-Output Orthogonal Frequency Division Multiplexing* systems. A CNN model was trained using the CSIs of different transmitters. Afterwards, the trained CNN was used to authenticate legitimate transmitters or detect spoofers, with the approach achieving 90% authentication accuracy.In Liao et al.

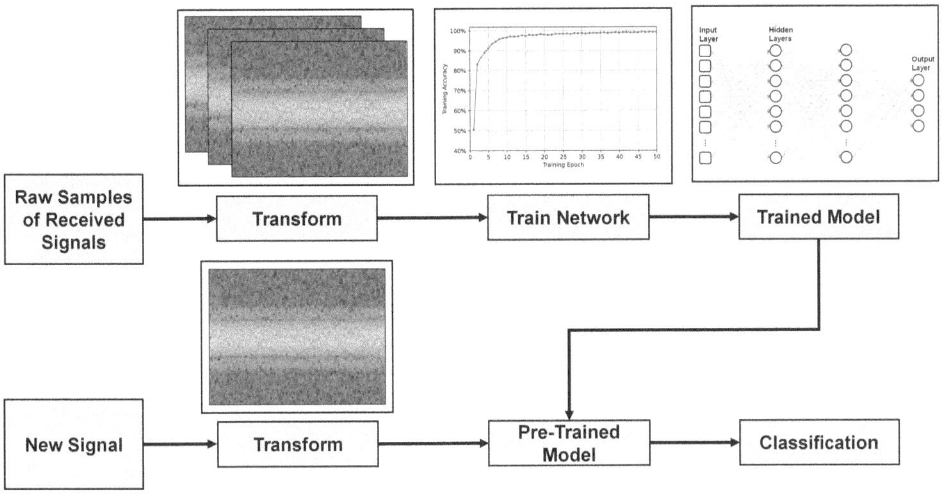

FIGURE 3.6 Framework of Feature-Based Waveform Domain Approaches

(2019b), three schemes based on MLP, CNN, and *Convolution Pre-processing Neural Network* (CPNN) were demonstrated. The MLP model achieved 95.5% and 77.25% identification accuracy with four sensor nodes and eight sensor nodes, respectively, while the CNN model achieved 86.25% and 67.87% identification accuracy, and the CPNN model achieved 85.25% and 66.75% identification accuracy, respectively.

3.4.3.2 Differential Constellation Trace Figure

Differential Constellation Trace Figure (DCTF)-based approaches have been proposed in a series of studies by Peng et al. (2019); Wang et al. (2020); Fu et al. (2023). 2D DCTF representation was utilized for wireless device identification. The DCTFs were generated from the samples of received signals without requiring any synchronization or further processing at the receiver. In Peng et al. (2019), the proposed approach achieved 99.1% and 93.8% identification accuracy with a CNN model at the SNR levels of 30 dB and 15 dB, respectively, when classifying 54 target ZigBee devices. In Wang et al. (2020), the experimental results for six mobile phones could reach classification accuracies of 92.97% and 99.77% using a CNN model with SNR of 25 dB and 50 dB, respectively. In Fu et al. (2023), the classification accuracy of the DCTF scheme was 89.59% using a CNN model for six 5G mobile phones, where the SNR of the system was 25 dB.

3.4.3.3 Heat Constellation Trace Figure

Heat Constellation Trace Figure (HCTF) is proposed to handle the synchronization of an RF signal and represents the heat of its trace. Peng et al. (2021) proposed a novel RFF identification method based on HCTF, and a *Slice Integration Cooperation* method was employed to enhance the identification performance. RF signals from seven VSG25A vector signal generators were captured for classification in a noise-free environment. Noise was added before the HCTF generation, and a CNN model was utilized. The identification accuracy of the proposed scheme was 91.07% at the SNR level of 0 dB and higher than 99.64% when the SNR was greater than 5 dB.

3.4.3.4 Signal Waveform

In Li and Cetin (2021b), a graphical waveform domain-based deep learning approach for RF fingerprinting for device identification was proposed and validated using 12 TI CC2538 Zigbee devices. Raw samples were captured over-the-air using a USRP B210 device and preprocessed into grayscale (gray) and black and white (binary) waveform images, as shown in Figure 3.7,

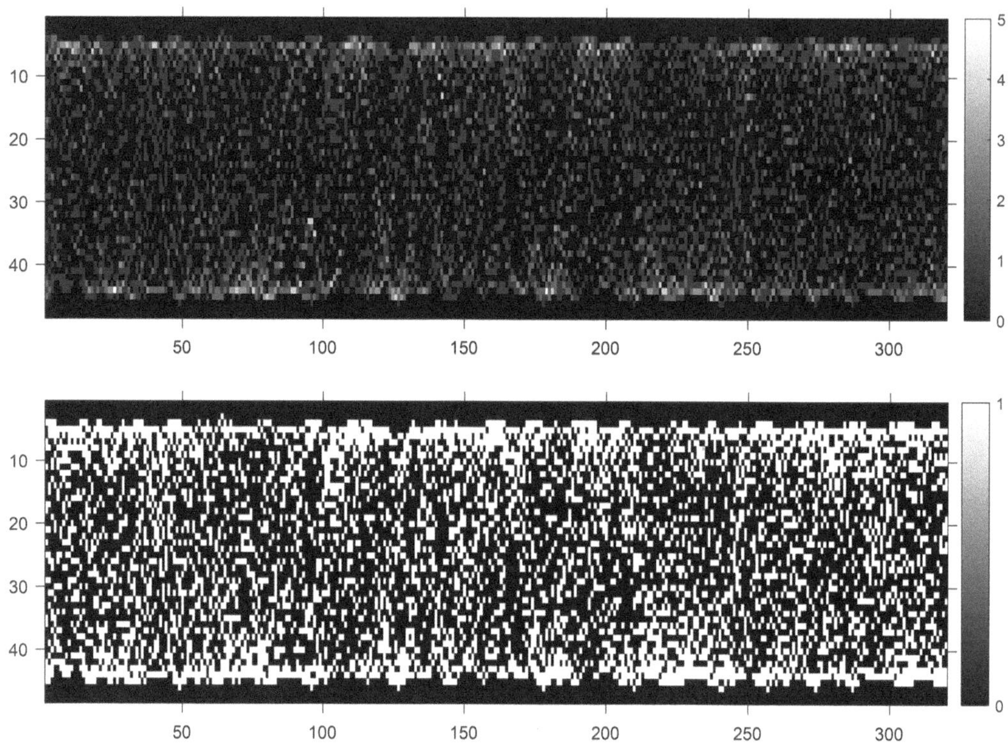

FIGURE 3.7 Generated (a) Gray and (b) Binary Waveform Images (Li & Cetin, 2021b)

which are then used as inputs to a MLP model for training and testing. The performance of the proposed approach considering varying-size gray and binary images was evaluated. Gray images maintain both waveform patterns and additional intensity information, whereas binary images only provide the former. The approach achieved near 99% identification accuracy when a gray waveform image of size 48 × 320 was used and 97% identification accuracy for the same size binary image. The proposed approach is modulation-agnostic and can be applied to any device or platform that utilizes RF signals. It enables longer datasets to be captured in an image without impacting the structure and the complexity of the subsequent deep learning processing. It also provides flexibility as the parameters of the pre-processing method can be potentially optimized to suit a desired identification performance given a specific application and/or device with certain computational resources.

Further research was carried out that generates *dual-layer images,* which are a stacked combination of images from the *I/Q* samples, as shown in Figure 3.8, to preserve the link between the two constituent parts of a complex signal (Chavez, Li & Cetin, 2022). This link might contain features that can increase the device identification accuracy. *Additive White Gaussian Noise* was added to the captured signals in Li and Cetin (2021b) to change their SNR to compare the performance of the dual-layer image approach with that of the single-layer approach under various SNR levels. MLP and CNN models were trained on the *single-* and *dual-*layer images and achieved identification accuracy above 99%. However, with varying SNR, the identification accuracy of the CNN model trained from the *single-layer images* reduced to 74%, whereas the reduction when the dual-layer images were used was less at 89%. It is noticeable that the dual-layer approach saw less performance reduction, highlighting its robustness in challenging environments. The experimental results also showed that the link between the *I/Q* samples can be preserved for better and more stable performance with the trade-off of a nearly doubled number of deep learning parameters.

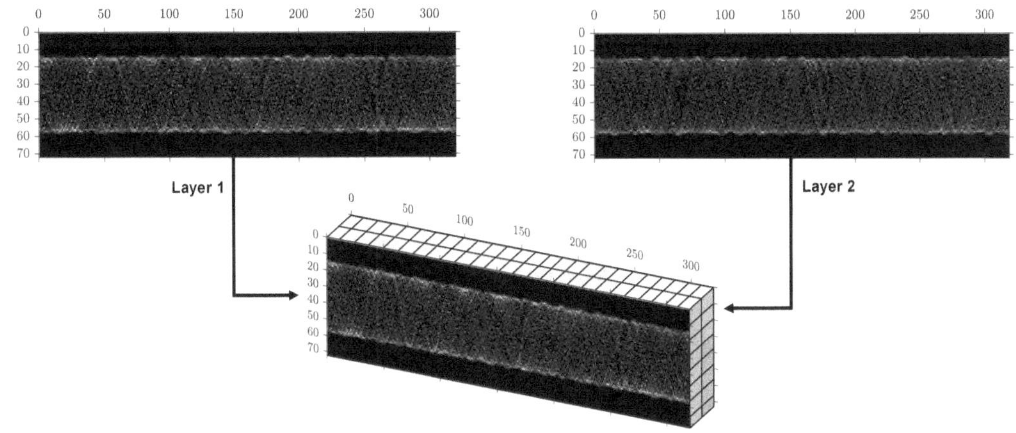

FIGURE 3.8 Generated Dual-Layer Waveform Image of Size 72 × 320 with Layer 1 (*I* samples) and Layer 2 (*Q* samples) (Chavez, Li & Cetin, 2022)

3.4.3.5 Short-Time Fourier Transform

Transformation methods, such as the one based on STFT, have been studied for RFF techniques. Li and Cetin (2024) presented a graphical deep learning approach in the *Time-Frequency* (TF) domain based on STFT, where the intensity information, e.g., the magnitude of the STFT, was used to generate 2-D image inputs for training and testing of the deep learning models for RF fingerprinting and identification. The TF representation provides the time-localized frequency information, enabling the observation of the signal's frequency content variations as time progresses. It is this information that the proposed approach exploited for RF fingerprinting and identification. The performance of the proposed approach was evaluated using two different over-the-air captured datasets from 12 Zigbee devices. Two-stage experimentation was conducted, such that in the first stage the signals were captured in a makeshift RF chamber to guarantee that the impairments rather than the interference from other signals were the target for classification. In the second stage, the signals collected over the air under real-life conditions were used to evaluate the performance. The experimental results show that the proposed approach using a CNN model achieves nearly 100% identification accuracy based on data captured in a makeshift RF chamber and an accuracy of 98% on the dataset, which was captured under real-life conditions, demonstrating the robustness of the proposed approach.

3.4.3.6 Density Trace Plot

Density Trace Plot (DTP) has been explored to leverage RF fingerprints originating from symbol transition trajectories for transmitter authentication. Focusing on *I*/*Q* imbalances (Cetin, Kale & Morling, 2007) as the source impairment for RF fingerprints, Huang et al. (2024) investigated the feasibility of three types of DTPs based on constellation, eye, and phase traces. They utilized three deep learning architectures for performance evaluation, namely 2D-CNN, 2D-CNN+bi-directional long short-term memory (biLSTM), and 3D-CNN. Overall, five ADALM-PLUTO SDRs were used as transmitters for data collection and performance evaluation in both wired and wireless conditions with modulation schemes of *Binary Phase Shift Keying* and *Quadrature Amplitude Modulation* (QAM), including 4-QAM, 16-QAM, and 64-QAM, achieving identification accuracy of 96.7% across signals of varying modulation complexities.

3.4.4 Modulation Domain Approaches

It is not common to see modulation domain approaches because there are fewer characteristics available when compared to the raw data when the received signal has been processed through some of the receiver blocks, such as frequency offset estimation and carrier estimation.

Sankhe et al. (2019) proposed a CNN framework, as previously mentioned in Section 4.2, for RF fingerprinting called ORACLE, which works with demodulated symbols instead of raw *I/Q* samples to classify transmitters under dynamic channels. The performance of the trained CNN classifier was evaluated with 16 X310 USRP SDRs. With samples collected from these SDRs through a cable, ORACLE easily distinguished them, achieving a classification accuracy of 99.76%. This indicates that the pre-trained classifier was able to identify the USRP SDRs accurately. Next, data were collected over the wireless channel for performance evaluation by conducting the experiments in two different locations: (1) in a lab, which represents a typical indoor environment and (2) in a more open recreational area, which has fewer reflections. ORACLE achieved higher than 99.5% identification accuracy in both environments, which proves that the unique patterns created by the impairments can still be detected, even in the presence of channel noise and after being processed by the receiver blocks.

Li and Cetin (2021a) have proposed a graphical deep learning approach for RF fingerprinting and demonstrated that the constellation graph can indeed be used to achieve high identification accuracy. The demodulated symbols, i.e., constellation points, were preprocessed to generate the corresponding image inputs, as shown in Figure 3.9, before tuning the deep learning models. With the approach, two types of inputs were generated, namely gray (grayscale) and binary (black and white) image inputs, for performance evaluation and comparison. To generate these input images, the sequence of demodulated symbols was first divided into sub-sequences, and these sub-sequences were then pre-processed to generate the image inputs for subsequent deep learning processing, leading to the proposed graphical deep learning approach to RFF.

The approach in Li and Cetin (2021a) can be paired with either CNN or MLP models for device identification. The use of images, as opposed to demodulated symbols, provides a compact way

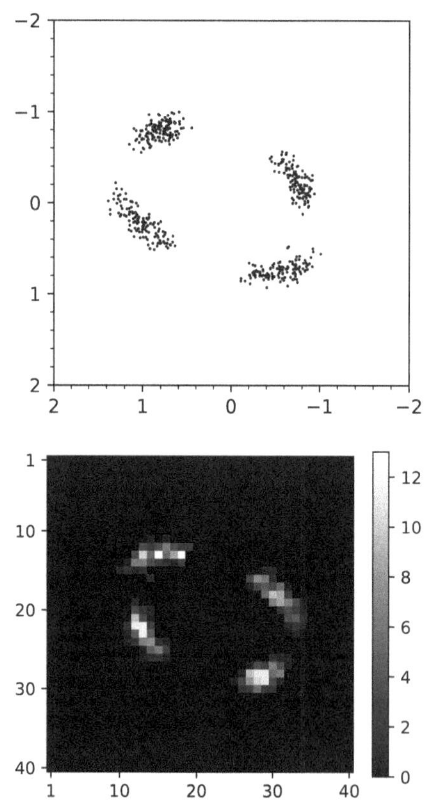

FIGURE 3.9 Pre-processing: (a) Before and (b) After Quantization Process (Li & Cetin, 2021a)

to capture signal characteristics in a fixed-size image irrespective of the number of demodulated symbols. This enables longer datasets to be captured without impacting the structure and the complexity of the subsequent deep learning processing. The performance of the graphical approach was evaluated using a two-stage process consisting of coarse- and fine-grained evaluation, carried out using synthetic and over-the-air captured datasets, respectively, with the approach achieving near 100% identification accuracy. Considering the performance based on the over-the-air captured data, the approach achieved near 100% identification accuracy when grayscale images of size 40 × 40 were used with CNN or MLP models, whereas in the case of binary images of the same size, the identification accuracy was near 98% for the MLP and 90% for the CNN models, demonstrating the efficacy of the proposed approach paired with different deep learning models.

3.5 CHALLENGES AND FUTURE TRENDS

As detailed in Sections 3 and 4, various RFF approaches have been developed and applied as a novel protective layer for enhancing IoT security. As discussed in Sections 3 and 4, each RFF approach employs different techniques leveraging the unique RF characteristics of devices. A comparison of the main RFF approaches is presented in Table 3.1.

There is no doubt that deep learning will play a dominant role in the future of RFF techniques. Recently, researchers started to put more emphasis on applying RFF techniques to solve real-world problems. Recently, RFF techniques with deep learning have been studied and applied to identify friendly/authorized *Unmanned Aerial Vehicles* (UAVs) and prevent them from being spoofed by malicious users (Reus-Muns & Chowdhury, 2021). In Soltani et al. (2020), the authors proposed a deep learning approach using raw *I/Q* samples of transmitters to address the issue of trust via a combination of techniques such as a multi-classifier with a score-based two-level aggregation method and data augmentation to make the framework more robust. The experiments were conducted by flying seven identical DJI M100 UAVs inside an RF anechoic chamber, with the proposed approach achieving an overall accuracy of 91% with the multi-classifier and up to 95% accuracy when data augmentation was added to the scheme. Mohanti et al. (2020) designed an MLP model to identify friendly/authorized UAVs. The proposed framework, called AirID, works on the physical layer

TABLE 3.1
Comparison of Main RFF Approaches

Approach		Features/Methods	Advantages	Disadvantages
Conventional	Transient signal-based	Transient signal	Unique characteristics	Very difficult to detect and capture the transient
	Steady-state signal-based	Location-independent features: power amplifier, RF oscillator, preamble	Imperfections of devices are inherited in many components	Difficult to distinguish trivial feature variations
		Location-dependent features: RSS, CSIR	Low complexity	Inefficient when devices move, performance degrades in dynamic environments
Deep learning based	Waveform domain	Raw sample	Low complexity of data pre-processing	Sheer volume of inputs, performance is location-dependent
	Feature-based	Feature engineering	High accuracy with proper feature engineering	Requires complex computation
	Modulation domain	Demodulated symbol	Low complexity of data pre-processing	Fewer characteristics compared to raw samples

and was implemented by mounting ETTUS B200mini SDRs to the DJI Matrice M100 UAVs. By intentionally inserting distortions in the transmitted *I/Q* samples from each UAV for the receiver to identify device identity, the work achieved 98% identification accuracy.

In Reus-Muns et al. (2020), the RFF technique was used to address the issue of trust in future 5G networks when authenticating *Base Stations* (BSs). The proposed approach was performed on a software-defined platform, namely the *Platform for Open Wireless Data-driven Experimental Research* (POWDER; Breen et al., 2020). POWDER was designed as a facility of wireless sensor networks for researchers to implement and test their future wireless technologies. Data collected from the POWDER testbed include Wi-Fi, LTE, and 5G *New Radio* (NR) signals. Experiments were conducted based on POWDER data using neural networks with triplet-loss functions (Koch, Zemel & Salakhutdinov, 2015; Schroff, Kalenichenko & Philbin, 2015). The approach achieved a BS detection accuracy of above 99% when the model was trained and tested on the same day and 92.97% accuracy when it was trained and tested on different days.

IoT networks can have many nodes, which poses a challenge for RFF solutions. RFF with deep learning has been applied to identify a large number of wireless devices. In Jian et al. (2020), a deep CNN architecture was used to identify 10,000 devices consisting of 5,117 Wi-Fi and 5,000 *Automatic Dependent Surveillance-Broadcast* (ADS-B) devices, based on the *I/Q* samples. The preprocessing method used was similar to that of Riyaz et al. (2018). In Robinson et al. (2020), the authors designed an *Augmented Dilated Causal Convolution* (ADCC) network that combines *Dilated Causal Convolution* (DCC) layers with traditional convolutional layers. The approach is inspired by the success of WaveNet (Van Den Oord et al., 2016), which was a deep learning model for audio generation operated directly at the waveform level. The experiments were conducted on a portion of the original dataset, which consisted of signals from over 10,000 ADS-B devices and more than 53,000 Wi-Fi devices using either the IEEE 802.11a (5 GHz) or IEEE 802.11g (2.4 GHz) standard. The model achieved above 60% identification accuracy on the largest device population size of 10,000. The performance, however, was above 90% when only considering no more than 500 devices.

There are, however, several critical challenges and open problems that must be addressed for the application of RFF in IoT networks. First, the dynamic environment impacts the RFF features. Variables, such as weather, temperature, and other environmental factors, have an impact on the inherent characteristics of the components, such as oscillators, resulting in performance degradation. Second, IoT network devices are always energy and resource limited. However, there is a notable lack of comprehensive analysis regarding the complexities of RFF techniques. Lastly, the integration of RFF techniques into IoT networks requires attention to system-level design. While RFF is a compelling alternative that could potentially replace existing security solutions, it demands careful consideration at the system level.

3.6 CONCLUSIONS

This chapter provided an overview of the RFF techniques available for adoption in enhancing the security of the IoT devices and networks. The key idea of RFF is to extract characteristics from transmitted signals to generate unique signatures, which can then be used to distinguish the transmitters. This provides a generic solution for authenticating or identifying IoT wireless devices. Various types of RFF techniques, ranging from the traditional approaches to the state-of-the-art deep learning-based approaches, were presented and discussed. The former focuses on specific features inherited in signals. However, deep learning-based approaches rely on the capability of deep learning models to automatically learn the features of the data. As the IoT ecosystem continues to evolve and expand, more deep learning-based RFF techniques are expected to be carried out in two directions. First, deep learning models will be designed specifically for RFF techniques in embedded systems. Currently, the deep learning models are mainly for computer vision and speech recognition research, which always require substantial computational resources. There is a need for deep learning models optimized for implementation on embedded systems with limited computational and power resources while maintaining the effectiveness and performances of the RFF approaches. Another direction is innovative feature engineering methods. Future research work is needed to continue exploring better signal representations as inputs to the deep learning models for improved and robust identification performance. The process of transforming raw data, namely feature engineering, plays a crucial role in the performance of deep learning-based RFF approaches as it provides the representation to the model to learn, resulting in better performance. Future feature engineering methods, i.e., pre-processing methods, will be tailored for the lightweight deep learning models designed specifically for RFF in embedded systems, aiming to find better signal representations that can enhance the accuracy, robustness, and efficiency of RFF approaches.

NOTES

1 https://keras.io/.
2 https://www.tensorflow.org/.
3 https://www.nvidia.com/.

REFERENCES

Barbeau, M., Hall, J. and Kranakis, E., 2006, October. Detection of rogue devices in bluetooth networks using radio frequency fingerprinting. In *Proceedings of the 3rd IASTED International Conference on Communications and Computer Networks,* Lima, Peru, *CCN* (pp. 4–6). https://doi.org/10.22215/etd/2006-06353.

Bassey, J., Adesina, D., Li, X., Qian, L., Aved, A. and Kroecker, T., 2019, June. Intrusion detection for IoT devices based on RF fingerprinting using deep learning. In *2019 Fourth International Conference on Fog and Mobile Edge Computing (FMEC),* Rome, Italy, (pp. 98–104). https://doi.org/10.1109/FMEC.2019.8795319.

Bihl, T.J., Bauer, K.W. and Temple, M.A., 2016. Feature selection for RF fingerprinting with multiple discriminant analysis and using ZigBee device emissions. *IEEE Transactions on Information Forensics and Security, 11*(8), pp. 1862–1874. https://doi.org/10.1109/TIFS.2016.2561902.

Boser, B.E., Sackinger, E., Bromley, J., Le Cun, Y. and Jackel, L.D., 1991. An analog neural network processor with programmable topology. *IEEE Journal of Solid-State* Circuits, 26(12), pp. 2017–2025. https://doi.org/10.1109/4.104196.

Breen, J., Buffmire, A., Duerig, J., Dutt, K., Eide, E., Hibler, M., Johnson, D., Kasera, S.K., Lewis, E., Maas, D. and Orange, A., 2020, September. POWDER: Platform for open wireless data-driven experimental research. In *Proceedings of the 14th International Workshop on Wireless Network Testbeds, Experimental Evaluation & Characterization,* London, UK, (pp. 17–24). https://doi.org/10.1145/3411276.3412204.

Brik, V., Banerjee, S., Gruteser, M. and Oh, S., 2008, September. Wireless device identification with radiometric signatures. In *Proceedings of the 14th ACM International Conference on Mobile Computing and Networking,* San Francisco, CA, (pp. 116–127). https://doi.org/10.1145/1409944.1409959.

Buchanan, B.G., 2005. A (very) brief history of artificial intelligence. *Ai Magazine, 26*(4), pp. 53–53.

Butun, I., Österberg, P. and Song, H., 2019. Security of the Internet of Things: Vulnerabilities, attacks, and countermeasures. *IEEE Communications Surveys & Tutorials, 22*(1), pp. 616–644. https://doi.org/10.1109/COMST.2019.2953364.

Carbonell, J.G., Michalski, R.S. and Mitchell, T.M., 1983. An overview of machine learning. *Machine Learning, 1*, pp. 3–23. https://doi.org/10.1016/B978-0-08-051054-5.50005-4.

Cetin, E., Kale, I. and Morling, R.C.S., 2007. Living and dealing with RF impairments in communication transceivers. In *IEEE International Symposium on Circuits and Systems (ISCAS),* New Orleans, LA, (pp. 21–24). https://doi.org/10.1109/ISCAS.2007.378172.

Chavez, F., Li, B. and Cetin, E., 2022, August. Dual-layer waveform domain deep learning approach for RF fingerprinting. In *2022 IEEE 65th International Midwest Symposium on Circuits and Systems (MWSCAS),* Fukuoka, Japan, (pp. 1–5). https://doi.org/10.1109/MWSCAS54063.2022.9859498.

Chen, H., Engkvist, O., Wang, Y., Olivecrona, M. and Blaschke, T., 2018. The rise of deep learning in drug discovery. *Drug Discovery Today, 23*(6), pp. 1241–1250. https://doi.org/10.1016/j.drudis.2018.01.039.

Danev, B., Capkun, S., Jayaram Masti, R. and Benjamin, T.S., 2012. Towards practical identification of HF RFID devices. *ACM Transactions on Information and System Security (TISSEC), 15*(2), pp. 1–24. https://doi.org/10.1145/2240276.2240278.

Deng, L., Hinton, G. and Kingsbury, B., 2013, May. New types of deep neural network learning for speech recognition and related applications: An overview. In *2013 IEEE International Conference on Acoustics, Speech and Signal Processing,* Vancouver, BC, (pp. 8599–8603). https://doi.org/10.1109/ICASSP.2013.6639344.

Dolatshahi, S., Polak, A. and Goeckel, D.L., 2010, November. Identification of wireless users via power amplifier imperfections. In *2010 Conference Record of the Forty Fourth Asilomar Conference on Signals, Systems and Computers,* Pacific Grove, CA, (pp. 1553–1557). https://doi.org/10.1109/ACSSC.2010.5757798.

Faria, D.B. and Cheriton, D.R., 2006, September. Detecting identity-based attacks in wireless networks using signalprints. In *Proceedings of the 5th ACM Workshop on Wireless Security (WiSe '06),* Los Angeles, California, (pp. 43–52). https://doi.org/10.1145/1161289.1161298.

Fu, H., Dong, H., Yin, J. and Peng, L., 2023. Radio frequency fingerprint identification for 5G mobile devices using DCTF and deep learning. *Entropy, 26*(1), p. 38. https://doi.org/10.3390/e26010038.

Gerdes, R., Daniels, T., Mina, M. and Russell, S., 2006. Device identification via analog signal fingerprinting: A matched filter approach. In *144 Proceedings of the Network and Distributed System Security Symposium (NDSS)* San Diego, California, (p. 78). https://www.ndss-symposium.org/wp-content/uploads/2017/09/Device-Identification-via-Analog-Signal-Fingerprinting-A-Matched-Filter-Approach-Ryan-M.-Thomas.pdf.

Hall, J., Barbeau, M. and Kranakis, E., 2003. Detection of transient in radio frequency fingerprinting using signal phase. *Wireless and Optical Communications, 9*, p. 13. https://people.scs.carleton.ca/~barbeau/Publications/2003/hall03.pdf.

Hall, J., Barbeau, M. and Kranakis, E., 2004. Enhancing intrusion detection in wireless networks using radio frequency fingerprinting. *Communications, Internet, and Information Technology, 1*. https://people.scs.carleton.ca/~kranakis/Papers/IDSRFFv4-4.pdf.

Han, J., Qian, C., Yang, P., Ma, D., Jiang, Z., Xi, W. and Zhao, J., 2015. GenePrint: Generic and accurate physical-layer identification for UHF RFID tags. *IEEE/ACM Transactions on Networking, 24*(2), pp. 846–858. https://doi.org/10.1109/TNET.2015.2391300.

Hassan, Q.F. ed., 2018. *Internet of Things A to Z: Technologies and Applications*. John Wiley & Sons. https://doi.org/10.1002/9781119456735.

Hassija, V., Chamola, V., Saxena, V., Jain, D., Goyal, P. and Sikdar, B., 2019. A survey on IoT security: Application areas, security threats, and solution architectures. *IEEE Access, 7*, pp. 82721–82743. https://doi.org/10.1109/ACCESS.2019.2924045.

He, K., Zhang, X., Ren, S. and Sun, J., 2016, June. Deep residual learning for image recognition. In *2016 IEEE Conference on Computer Vision and Pattern Recognition (CVPR),* Las Vegas, NV, (pp. 770–778). https://doi.org/10.48550/arXiv.1512.03385.

Higuchi, T., 1988. Approach to an irregular time series on the basis of the fractal theory. *Physica D: Nonlinear Phenomena, 31*(2), pp. 277–283. https://doi.org/10.1016/0167-2789(88)90081-4.

Huang, D., Al-Hourani, A., Sithamparanathan, K. and Rowe, W.S., 2024. Deep learning methods for IoT device authentication using symbols density trace plot. *IEEE Internet of Things Journal, 11*(10), pp. 18167–18179. https://doi.org/10.1109/JIOT.2024.3361892.

Huang, G., Yuan, Y., Wang, X. and Huang, Z., 2016. Specific emitter identification based on nonlinear dynamical characteristics. *Canadian Journal of Electrical and Computer Engineering, 39*(1), pp. 34–41. https://doi.org/10.1109/CJECE.2015.2496143.

Jafari, H., Omotere, O., Adesina, D., Wu, H.H. and Qian, L., 2018, October. IoT devices fingerprinting using deep learning. In *MILCOM 2018-2018 IEEE Military Communications Conference (MILCOM),* Los Angeles, CA, (pp. 1–9). https://doi.org/10.1109/MILCOM.2018.8599826.

Jagannath, A., Jagannath, J. and Kumar, P.S.P.V., 2022a. A comprehensive survey on radio frequency (RF) fingerprinting: Traditional approaches, deep learning, and open challenges. *Computer Networks, 219*, p. 109455. https://doi.org/10.1016/j.comnet.2022.109455.

Jagannath, A., Kane, Z. and Jagannath, J., 2022b, December. RF fingerprinting needs attention: Multi-task approach for real-world WiFi and bluetooth. In *GLOBECOM 2022-2022 IEEE Global Communications Conference,* Rio de Janeiro, Brazil, (pp. 4607–4612). https://doi.org/10.1109/GLOBECOM48099.2022.10001572.

Jian, T., Rendon, B.C., Ojuba, E., Soltani, N., Wang, Z., Sankhe, K., Gritsenko, A., Dy, J., Chowdhury, K. and Ioannidis, S., 2020. Deep learning for RF fingerprinting: A massive experimental study. *IEEE Internet of Things Magazine, 3*(1), pp. 50–57. https://doi.org/10.1109/IOTM.0001.1900065.

Justesen, N., Bontrager, P., Togelius, J. and Risi, S., 2019. Deep learning for video game playing. *IEEE Transactions on Games*, *12*(1), pp. 1–20. https://doi.org/10.1109/TG.2019.2896986.

Kennedy, I.O., Scanlon, P. and Buddhikot, M.M., 2008, October. Passive steady state rf fingerprinting: A cognitive technique for scalable deployment of co-channel femto cell underlays. In *2008 3rd IEEE Symposium on New Frontiers in Dynamic Spectrum Access Networks,* Chicago, IL, (pp. 1–12). https://doi.org/10.1109/DYSPAN.2008.46.

Kinsner, W., Cheung, V., Cannons, K., Pear, J. and Martin, T., 2003, September. Signal classification through multifractal analysis and complex domain neural networks. In *Proceedings of IEEE Conference on Cognitive Informatics,* London, (pp. 41–46). https://doi.org/10.1109/COGINF.2003.1225951.

Klein, R., Temple, M.A., Mendenhall, M.J. and Reising, D.R., 2009, June. Sensitivity analysis of burst detection and RF fingerprinting classification performance. In *2009 IEEE International Conference on Communications,* Dresden, Germany, (pp. 1–5). https://doi.org/10.1109/ICC.2009.5199451.

Koch, G., Zemel, R. and Salakhutdinov, R., 2015, July. Siamese neural networks for one-shot image recognition. In *ICML Deep Learning Workshop* (Vol. 2, No. 1). http://www.cs.cmu.edu/~rsalakhu/papers/oneshot1.pdf.

Li, B. and Cetin, E., 2021a. Design and evaluation of a graphical deep learning approach for RF fingerprinting. *IEEE Sensors Journal, 21*(17), pp. 19462–19468. https://doi.org/10.1109/JSEN.2021.3088137.

Li, B. and Cetin, E., 2021b, May. Waveform domain deep learning approach for RF fingerprinting. In *2021 IEEE International Symposium on Circuits and Systems (ISCAS),* Daegu, Korea, (pp. 1–5). https://doi.org/10.1109/ISCAS51556.2021.9401486.

Li, B. and Cetin, E., 2024. A graphical deep learning approach to RF fingerprinting in the time-frequency domain. *IEEE Sensors Journal, 24*(10), pp. 16984–16990. https://doi.org/10.1109/JSEN.2024.3383437.

Li, Q., Fan, H., Sun, W., Li, J., Chen, L. and Liu, Z., 2017. Fingerprints in the air: Unique identification of wireless devices using RF RSS fingerprints. *IEEE Sensors Journal, 17*(11), pp. 3568–3579. https://doi.org/10.1109/JSEN.2017.2685564.

Li, Z., Xu, W., Miller, R. and Trappe, W., 2006, September. Securing wireless systems via lower layer enforcements. In *Proceedings of the 5th ACM Workshop on Wireless Security,* Los Angeles California, (pp. 33–42). https://doi.org/10.1145/1161289.1161297.

Liao, R., Wen, H., Pan, F., Song, H., Xu, A. and Jiang, Y., 2019a, March. A novel physical layer authentication method with convolutional neural network. In *2019 IEEE International Conference on Artificial Intelligence and Computer Applications (ICAICA),* Dalian, China, (pp. 231–235). https://doi.org/10.1109/ICAICA.2019.8873460.

Liao, R.F., Wen, H., Wu, J., Pan, F., Xu, A., Jiang, Y., Xie, F. and Cao, M., 2019b. Deep-learning-based physical layer authentication for industrial wireless sensor networks. *Sensors*, *19*(11), p. 2440. https://doi.org/10.3390/s19112440.

Mohanti, S., Soltani, N., Sankhe, K., Jaisinghani, D., Di Felice, M. and Chowdhury, K., 2020, December. AirID: Injecting a custom RF fingerprint for enhanced UAV identification using deep learning. In *GLOBECOM 2020-2020 IEEE Global Communications Conference,* Taipei, Taiwan, (pp. 1–6). https://doi.org/10.1109/GLOBECOM42002.2020.9322561.

Mousavi, S.K., Ghaffari, A., Besharat, S. and Afshari, H., 2021. Security of internet of things based on cryptographic algorithms: A survey. *Wireless Networks*, *27*(2), pp. 1515–1555. https://doi.org/10.1007/s11276-020-02535-5.

Neshenko, N., Bou-Harb, E., Crichigno, J., Kaddoum, G. and Ghani, N., 2019. Demystifying IoT security: An exhaustive survey on IoT vulnerabilities and a first empirical look on Internet-scale IoT exploitations. *IEEE Communications Surveys & Tutorials, 21*(3), pp. 2702–2733. https://doi.org/10.1109/COMST.2019.2910750.

O'Shea, T.J., Corgan, J. and Clancy, T.C., 2016. Convolutional radio modulation recognition networks. In *Engineering Applications of Neural Networks: 17th International Conference, EANN 2016*, Aberdeen, September 2–5, 2016, Proceedings 17 (pp. 213–226). https://doi.org/10.48550/arXiv.1602.04105.

Patel, H.J., Temple, M.A. and Baldwin, R.O., 2014. Improving ZigBee device network authentication using ensemble decision tree classifiers with radio frequency distinct native attribute fingerprinting. *IEEE Transactions on Reliability, 64*(1), pp. 221–233. https://doi.org/10.1109/TR.2014.2372432.

Peng, L., Zhang, J., Liu, M. and Hu, A., 2019. Deep learning based RF fingerprint identification using differential constellation trace figure. *IEEE Transactions on Vehicular Technology*, *69*(1), pp. 1091–1095. https://doi.org/10.1109/TVT.2019.2950670.

Peng, Y., Liu, P., Wang, Y., Gui, G., Adebisi, B. and Gacanin, H., 2021. Radio frequency fingerprint identification based on slice integration cooperation and heat constellation trace figure. *IEEE Wireless Communications Letters, 11*(3), pp. 543–547. https://doi.org/10.1109/LWC.2021.3135932.

Polak, A.C., Dolatshahi, S. and Goeckel, D.L., 2011. Identifying wireless users via transmitter imperfections. *IEEE Journal on Selected Areas in Communications, 29*(7), pp. 1469–1479. https://doi.org/10.1109/JSAC.2011.110812.

Polak, A.C. and Goeckel, D.L., 2015. Wireless device identification based on RF oscillator imperfections. *IEEE Transactions on Information Forensics and Security, 10*(12), pp. 2492–2501. https://doi.org/10.1109/TIFS.2015.2464778.

Ramos, S., Gehrig, S., Pinggera, P., Franke, U. and Rother, C., 2017, June. Detecting unexpected obstacles for self-driving cars: Fusing deep learning and geometric modeling. In *2017 IEEE Intelligent Vehicles Symposium (IV)*, Los Angeles, CA, (pp. 1025–1032). https://doi.org/10.1109/IVS.2017.7995849.

Reus-Muns, G. and Chowdhury, K.R., 2021. Classifying UAVs with proprietary waveforms via preamble feature extraction and federated learning. *IEEE Transactions on Vehicular Technology, 70*(7), pp. 6279–6290. https://doi.org/10.1109/TVT.2021.3081049.

Reus-Muns, G., Jaisinghani, D., Sankhe, K. and Chowdhury, K.R., 2020, December. Trust in 5G open RANs through machine learning: RF fingerprinting on the POWDER PAWR platform. In *GLOBECOM 2020-2020 IEEE Global Communications Conference,* Taipei, Taiwan, (pp. 1–6). https://doi.org/10.1109/GLOBECOM42002.2020.9348261.

Riyaz, S., Sankhe, K., Ioannidis, S. and Chowdhury, K., 2018. Deep learning convolutional neural networks for radio identification. *IEEE Communications Magazine, 56*(9), pp. 146–152. https://doi.org/10.1109/MCOM.2018.1800153.

Robinson, J., Kuzdeba, S., Stankowicz, J. and Carmack, J.M., 2020, January. Dilated causal convolutional model for RF fingerprinting. In *2020 10th Annual Computing and Communication Workshop and Conference (CCWC)*, Las Vegas, NV, (pp. 0157–0162). https://doi.org/10.1109/CCWC47524.2020.9031257.

Robyns, P., Marin, E., Lamotte, W., Quax, P., Singelée, D. and Preneel, B., 2017, July. Physical-layer fingerprinting of LoRa devices using supervised and zero-shot learning. In *Proceedings of the 10th ACM Conference on Security and Privacy in Wireless and Mobile Networks,* Boston, Massachusetts, (pp. 58–63). https://doi.org/10.1145/3098243.3098267.

Roy, A., Sun, J., Mahoney, R., Alonzi, L., Adams, S. and Beling, P., 2018, April. Deep learning detecting fraud in credit card transactions. In *2018 systems and information engineering design symposium (SIEDS),* Charlottesville, VA, (pp. 129–134). https://doi.org/10.1109/SIEDS.2018.8374722.

Roy, D., Mukherjee, T., Chatterjee, M. and Pasiliao, E., 2019, April. Detection of rogue RF transmitters using generative adversarial nets. In *2019 IEEE Wireless Communications and Networking Conference (WCNC),* Marrakesh, Morocco, (pp. 1–7). https://doi.org/10.1109/WCNC.2019.8885548.

Sankhe, K., Belgiovine, M., Zhou, F., Riyaz, S., Ioannidis, S. and Chowdhury, K., 2019, April. ORACLE: Optimized radio classification through convolutional neural networks. In *IEEE INFOCOM 2019-IEEE Conference on Computer Communications,* Paris, France, (pp. 370–378). https://doi.org/10.48550/arXiv.1812.01124.

Schroff, F., Kalenichenko, D. and Philbin, J., 2015. Facenet: A unified embedding for face recognition and clustering. In *Proceedings of the IEEE Conference on Computer Vision and Pattern Recognition (CVPR)*, Boston, MA, (pp. 815–823). https://doi.org/10.48550/arXiv.1503.03832.

Shaw, D. and Kinsner, W., 1997, May. Multifractal modelling of radio transmitter transients for classification. In *IEEE WESCANEX 97 Communications, Power and Computing. Conference Proceedings,* Winnipeg, MB, (pp. 306–312). https://doi.org/10.1109/WESCAN.1997.627159.

Shen, G., Zhang, J., Marshall, A., Peng, L. and Wang, X., 2021. Radio frequency fingerprint identification for LoRa using deep learning. *IEEE Journal on Selected Areas in Communications, 39*(8), pp. 2604–2616. https://doi.org/10.1109/JSAC.2021.3087250.

Soltani, N., Reus-Muns, G., Salehi, B., Dy, J., Ioannidis, S. and Chowdhury, K., 2020. RF fingerprinting unmanned aerial vehicles with non-standard transmitter waveforms. *IEEE Transactions on Vehicular Technology, 69*(12), pp. 15518–15531. https://doi.org/10.1109/TVT.2020.3042128.

Specht, D.F., 1990. Probabilistic neural networks. *Neural Networks*, 3(1), pp. 109–118. https://doi.org/10.1016/0893-6080(90)90049-Q.

Surendran, S., Nassef, A. and Beheshti, B.D., 2018, May. A survey of cryptographic algorithms for IoT devices. In *2018 IEEE Long Island Systems, Applications and Technology Conference (LISAT),* Farmingdale, NY, (pp. 1–8). https://doi.org/10.1109/LISAT.2018.8378034.

Tiloca, M. and Raza, S., 2018. Security mechanisms and technologies for constrained IoT devices, in Hassan, Q. (ed.) *Internet of Things A to Z: Technologies and Applications.* Wiley-IEEE Press, pp. 255–274. https://doi.org/10.1002/9781119456735.ch8.

Tu, Y., Zhang, Z., Li, Y., Wang, C. and Xiao, Y., 2019. Research on the Internet of Things device recognition based on RF-fingerprinting. *IEEE Access*, *7*, pp. 37426–37431. https://doi.org/10.1109/ACCESS.2019.2904657.

Tugnait, J.K. and Kim, H., 2010, January. A channel-based hypothesis testing approach to enhance user authentication in wireless networks. In *2010 Second International Conference on COMmunication Systems and NETworks (COMSNETS 2010)*, Bangalore, (pp. 1–9). https://doi.org/10.1109/COMSNETS.2010.5432018.

Ureten, O. and Serinken, N., 1999, June. Bayesian detection of radio transmitter turn-on transients. In *NSIp* (pp. 830–834). https://doi.org/10.1049/el:19991369.

Ureten, O. and Serinken, N., 2005. Bayesian detection of Wi-Fi transmitter RF fingerprints. *Electronics Letters*, *41*(6), pp. 373–374. https://doi.org/10.1049/el:20057769.

Ureten, O. and Serinken, N., 2007. Wireless security through RF fingerprinting. *Canadian Journal of Electrical and Computer Engineering*, *32*(1), pp. 27–33. https://doi.org/10.1109/CJECE.2007.364330.

Van Den Oord, A., Dieleman, S., Zen, H., Simonyan, K., Vinyals, O., Graves, A., Kalchbrenner, N., Senior, A. and Kavukcuoglu, K., 2016. Wavenet: A generative model for raw audio. arXiv preprint arXiv:1609.03499. https://doi.org/10.48550/arXiv.1609.03499.

Wang, S., Peng, L., Fu, H., Hu, A. and Zhou, X., 2020, July. A convolutional neural network-based RF fingerprinting identification scheme for mobile phones. In *IEEE INFOCOM 2020-IEEE Conference on Computer Communications Workshops (INFOCOM WKSHPS),* Toronto, ON, (pp. 115–120). https://doi.org/10.1109/INFOCOMWKSHPS50562.2020.9163058.

Wu, Q., Feres, C., Kuzmenko, D., Zhi, D., Yu, Z., Liu, X. and 'Leo'Liu, X., 2018. Deep learning based RF fingerprinting for device identification and wireless security. *Electronics Letters*, *54*(24), pp. 1405–1407. https://doi.org/10.1049/el.2018.6404.

Xiao, L., Greenstein, L., Mandayam, N. and Trappe, W., 2007, June. Fingerprints in the ether: Using the physical layer for wireless authentication. In *2007 IEEE International Conference on Communications ,* Glasgow, (pp. 4646–4651). https://doi.org/10.1109/ICC.2007.767.

Xiao, L., Greenstein, L., Mandayam, N. and Trappe, W., 2008, May. A physical-layer technique to enhance authentication for mobile terminals. In *2008 IEEE International Conference on Communications ,* Beijing, (pp. 1520–1524). https://doi.org/10.1109/ICC.2008.294.

Xie, L., Peng, L., Zhang, J. and Hu, A., 2024. Radio frequency fingerprint identification for Internet of Things: A survey. *Security and Safety*, *3*, p. 2023022. https://doi.org/10.1051/sands/2023022.

Xu, Q., Zheng, R., Saad, W. and Han, Z., 2015. Device fingerprinting in wireless networks: Challenges and opportunities. *IEEE Communications Surveys & Tutorials, 18*(1), pp. 94–104. https://doi.org/10.1109/COMST.2015.2476338.

Young, T., Hazarika, D., Poria, S. and Cambria, E., 2018. Recent trends in deep learning based natural language processing. *IEEE Computational Intelligence Magazine, 13*(3), pp. 55–75. https://doi.org/10.48550/arXiv.1708.02709.

Zhang, J., Duong, T.Q., Woods, R. and Marshall, A., 2017. Securing wireless communications of the internet of things from the physical layer, an overview. *Entropy*, *19*(8), p. 420. https://doi.org/10.3390/e19080420.

Zhang, Z., Hu, A., Xu, W., Yu, J. and Yang, Y., 2022. An artificial radio frequency fingerprint embedding scheme for device identification. *IEEE Communications Letters, 26*(5), pp. 974–978. https://doi.org/10.1109/LCOMM.2022.3148037.

4 Strengthening IoT Systems with Blockchain

Consensus Mechanisms and Security Enhancements

Yingsen Wang, Jun Zhao, Yan Qiang, Juanjuan Zhao and Keqin Li

4.1 INTRODUCTION

A wide range of businesses are being affected by the Internet of Things (IoT) in today's quickly changing digital world (Hassan, 2018; Ogonji, Okeyo, and Wafula, 2020). Its revolutionary effect is drastically altering our lifestyles by upending long-standing conventions about labor, travel, and energy conservation (Castillo-Effer et al., 2004). Because it automates operations and allows for rapid information transmission, the IoT is improving efficiency across a wide range of enterprises. But while the IoT grows, so does the complicated network of worries it raises about sophisticated cyber threats and data privacy violations (Siegfried, Rosenthal, and Benlian, 2020; Stoyanova et al., 2020). Addressing these crucial security challenges requires our full attention and commitment since they directly affect our financial security and safety (Medaglia and Serbanati, 2010; Michelin et al., 2018).

Because blockchain is a distributed, tamper-proof ledger system, it offers new possibilities for addressing IoT security issues (Gao, Hatcher, and Yu, 2018; Zheng et al., 2018). Information inside the IoT may be safeguarded against theft and manipulation by putting it on the blockchain, which offers transparency, security, and the immutability of data (Wu et al., 2019; Xie et al., 2020). The decentralized structure of the blockchain enhances the overall security and dependability of the IoT ecosystem by guarding against attacks focused on data and single points of failure (Xie et al., 2019). Because of this, combining blockchain technology with the IoT can contribute to safer and more dependable smart living and working environments for humans (Iuon-Chang and Tzu-Chun, 2017; X. Wang et al., 2019).

While blockchain technology has a lot of potential for IoT development, it currently struggles to meet the latency, energy consumption, and scalability requirements of IoT scenarios (Miglani et al., 2020). In order to overcome these performance difficulties, consensus mechanisms are essential technologies and determinants (W. Wang et al., 2019). Consensus mechanisms are the cornerstone of blockchain technology and are crucial for preserving data integrity and network security. They verify and authorize all of the network's nodes' transactions and blocks (Xiao et al., 2020). Blockchain uses consensus processes to decide whether new data should be consistent or not. A completely autonomous IoT network requires dispersed communication between its many sensors and devices, or nodes (Novo, 2018). Consensus mechanisms are algorithms or rules that synchronize disconnected nodes to achieve agreement, allowing multiple stakeholders to agree on the authenticity and sequencing of transactions in the absence of a central governing authority (Mingxiao et al., 2017). Consensus platforms attempt to guarantee the consistency and dependability of transactions

DOI: 10.1201/9781003506638-5

throughout IoT nodes, constraining fraud and data tampering. Different consensus techniques employ different algorithms and regulations, leading to various degrees of security, efficiency, and energy consumption (Bach, Mihaljevic, and Zagar, 2018; Baliga, 2020).

In this chapter, we present several blockchain consensus techniques that can operate with resource-constrained IoT devices and networks. While there are several consensus methods, we will concentrate on proof-of-X (PoX), byzantine fault tolerance (BFT), directed acyclic graph (DAG), and block alliance consensus (BAC) since they take different approaches to energy efficiency, scalability, and fault tolerance. This focus enables us to give a thorough assessment of their appropriateness for IoT contexts, distinguishing our work by addressing the specific problems and performance needs of resource-constrained IoT systems. IoT networks commonly use devices with little processing and communication power, which makes them unable to sustain high bandwidth usage or do complex computations. We explore methods to lower computation requirements and accelerate the convergence of possible consensus systems in order to tackle these issues. Additionally, we look into the possible applications of currently in-use consensus algorithms for building blockchain-powered IoT networks. In order to help readers better understand the potential applications of these common consensus methods in the context of the IoT, we shall showcase their performance towards the conclusion of this chapter.

4.2 OVERVIEW OF IOT BLOCKCHAIN

This section explores the merging of blockchain technology and the IoT. The intention is to provide a thorough analysis of the ways in which blockchain technology might enhance scalability, security, and reliability of IoT infrastructures.

4.2.1 Blockchain Technology

The underlying technology of encrypted digital currencies like Bitcoin is called blockchain. By integrating distributed data storage, encryption, smart contracts, peer-to-peer networks, and distributed consensus processes, blockchain creates a ledger that is kept in a tree structure or chain-like DAG (Zheng et al., 2018). In this ledger, data encryption and verification are the main applications of cryptography. The resulting data is subsequently usable in the market after going through a number of logical processing stages using different computer technologies (Monrat, Schelen, and Andersson, 2019). Presently, blockchain is categorized into three modes based on application scenarios, requirements, and environments: public blockchain, consortium blockchain, and private blockchain (Belotti et al., 2019).

4.2.1.1 The Data Structure of Blockchain

Decentralization, coherence, and openness, three essential characteristics of blockchain construction that are essential to its general validity and operational effectiveness, are supported by the technology. The blockchain is a distributed database that is made up of a series of blocks, each of which contains a set of events (Ahram et al., 2017). Each block in the blockchain structure contains three main components: a header, a list of transactions, and a cryptographic hash of the previous block. Figure 4.1 illustrates the typical data structure of a blockchain. Hash values (Hash 1, Hash 2, etc.) are used to uniquely identify transactions. Transactions (Tx1, Tx2, etc.) are the actual data entries within each block. The Merkle root allows for efficient and secure verification of the data within the block.

The block header contains essential data like the block date, a hash link to the previous block, and a nonce value that the process of mining requires to carry out the blockchain's consensus mechanism. This header acts as an individual identification for each block in the chain, ensuring its chronology and cryptography connection (Xu, Chen, and Kou, 2019). The blockchain's data payload consists of a set of transactions throughout each block. Participants in the network start these

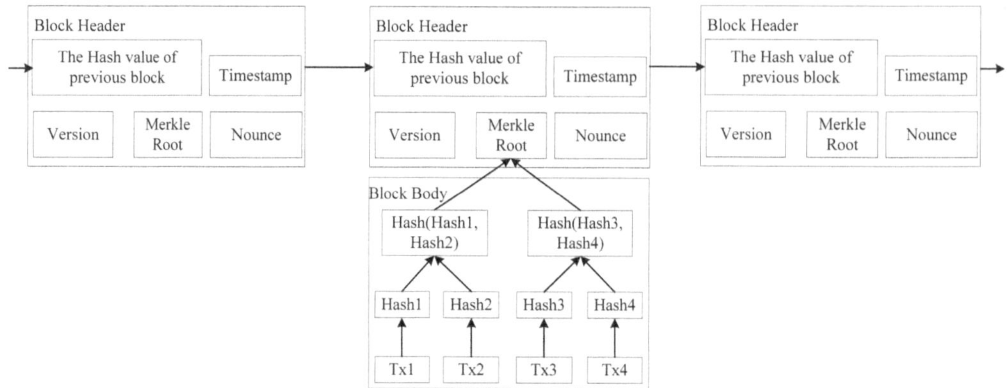

FIGURE 4.1 The Data Structure of a Classic Blockchain

transactions. Typically, such transactions are recorded in a standard format to maintain consistency and interchange across several blockchain platforms (Guo and Yu, 2022). Furthermore, the blockchain's tamper-resistance and immutability depend on the "previous hash," or cryptographic hash, of the previous block (Gorkhali, Li, and Shrestha, 2020). Any effort to tamper would be practically impossible to pull off and would be detected instantly. Because the hash of one block is included in the header of every other block, changing the contents of one block requires recalculating the hashes of every other block.

4.2.1.2 Cryptographic Algorithms in Blockchain

We talk about asymmetric encryption techniques and hashing in this section. Because they offer data confidentiality, secrecy, and immutability, hash algorithms are crucial to blockchain technology. Hashing algorithms, also called cryptographic hash functions, take any length of input and output a digest, which is a fixed-size result (Efanov and Roschin, 2018). It is nearly impossible to infer the true input from the hash value since this output serves as a unique and irreversible representation of the inputs. Among the many vital functions performed by the hash algorithms employed in blockchain technology are *data integrity* and *Merkle trees.* Transactions involving hashing algorithms are rather random and unpredictable. Every block generates a unique hash value. Because of this, a distributed blockchain's data integrity may be audited. A new hash value would result from altering any data within a block, warning other network users of any tampering efforts. Hash functions are employed by Merkle trees in blockchain data structures to efficiently arrange and verify the integrity of large amounts of information. Merkle trees employ a method of iteratively hashing pairs of data blocks until a sole root hash is reached, enabling swift verification of vast datasets while minimizing computational effort.

SHA-256 (Secure Hash Algorithm 256-bit) and SHA-3 (Secure Hash Algorithm 3), which are well-known for their cryptographic strength and collision resistance, are often used hash algorithms in blockchain technology (Wang, Duan, and Zhu, 2018). To guarantee the safety and reliability of blockchain networks, the above-described algorithms undergo a thorough cryptographic examination to ensure their resilience against brute-force, collision, and pre-image attacks.

Assertive encryption, sometimes referred to as cryptography with public keys, is a fundamental component of blockchain technology that permits digital signatures, secure communication, and cryptographic key management (Chen et al., 2023). Two mathematically related keys – a public key and a private key – are used in asymmetric encryption, as opposed to symmetric encryption, which employs a single key for both encryption and decryption. Figure 4.2 shows an example of asymmetric encryption.

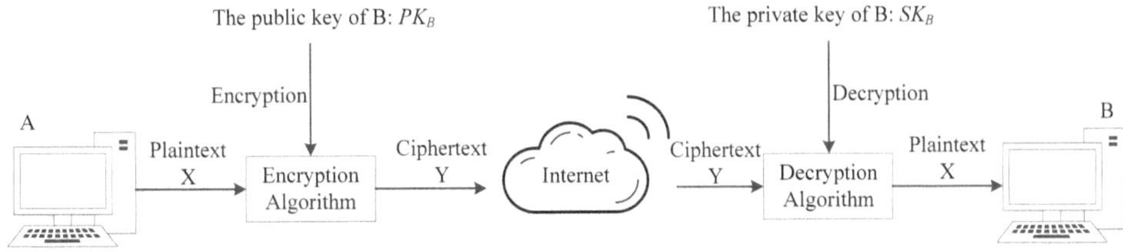

FIGURE 4.2 The Asymmetric Encryption

Asymmetric encryption makes it possible to establish and validate digital signatures, which act as cryptographic proof of dependability and validity for exchanges and occurrences on the blockchain (Zhai et al., 2019). Users can sign data with their private key to prove ownership and authenticity, and other users can verify the signature's validity using the corresponding public key (Ugochukwu et al., 2022). Elliptic curve cryptography and Rivest–Shamir–Adleman are two widely used asymmetric encryption algorithms that are very computationally efficient and have good safety features. To ensure that the algorithms are resilient to many forms of attacks, including brute-force attacks and key recovery, they undergo a comprehensive cryptographic examination procedure.

4.2.1.3 Smart Contract

Smart contract implementation, which is transforming the way contracts are executed and maintained in decentralized systems, is one of the most important developments in blockchain technology (Mohanta, Panda, and Jena, 2018). Smart contracts essentially function as automated agreements with terms embedded directly in the code. Their autonomous actions in enforcing and executing pre-established protocols eliminate the need for intermediaries. This allows us to streamline and automate difficult business procedures.

Within the blockchain architecture, on Ethereum-like systems, smart contracts are written and executed in Turing-complete languages like Solidity. The blockchain ensures that smart contracts are implemented in an immutable, transparent, and decentralized manner (Khan et al., 2021). When certain conditions are met, smart contracts can self-start predefined actions, including moving digital assets, updating data, or triggering other smart contracts.

There are drawbacks with regard to security, scalability, and legal enforceability, despite the fact that smart contracts have many advantages. Malicious parties may take advantage of or suffer financial losses due to vulnerabilities in the smart contract code, such as coding flaws or security holes (Macrinici, Cartofeanu, and Gao, 2018). Since conventional legal frameworks might not adequately account for the particularities of blockchain-based contracts, there is also ongoing discussion over the enforceability of smart contracts in court.

4.2.2 Blockchain-Based IoT Networks

The IoT enables bidirectional movement between data analytics and physical domains and also allows for the integration of digital information systems with real-world objects by establishing a broadly distributed network. The IoT technology autonomously collects data from its environment and wirelessly sends it to the edge or cloud for processing and/or storage (Ogonji, Okeyo, and Wafula, 2020). The considerable advances in data security and transparency that have come from the integration of blockchain technology with IoT have made the development of more secure and efficient intelligent IoT systems feasible.

4.2.2.1 The Superiority of Blockchain Technology for IoT Applications

Because IoT applications frequently entail the collection and transfer of sensitive data, blockchain technology is a great fit because of its innate capacity to improve data security and user privacy.

Distinct from alternative technologies, blockchain utilizes cryptographic methods to safeguard information throughout the transaction process, guaranteeing that information stays private and is shielded from unwanted access. Because IoT devices frequently have limited resources and might not be able to independently deploy strong security measures, this is especially crucial. Additionally, because blockchain is decentralized, it is less vulnerable to distributed denial-of-service assaults, which are frequent in centralized systems. Because blockchain distributes data and control among several nodes, it ensures that IoT services remain reliable and accessible even in the case of a network failure. The demand for safe, scalable data management is growing as the number of IoT devices increases. Blockchain's robust solution to these issues boosts confidence in and acceptance of IoT technologies.

4.2.2.2 IoT Hierarchical Architecture

Three hierarchical levels comprise the architectural design of the IoT: perception, network, and application. Two essential elements of this architecture are modularity and scalability (Jabraeil Jamali et al., 2020; Ai and Cui, 2022). The IoT architecture is based on the perception layer, which is primarily in charge of data collection. It is composed of several sensors and devices for monitoring and tracking physical aspects of the environment, such as temperature, humidity, and position. The network layer manages information transfer, establishing secure and efficient message transmission between the perception layer and other higher-level layers. Among the communication technologies in this tier are Bluetooth and wireless networks. The end-user experience is within the purview of the application layer, which is the highest layer in the architecture. This layer transforms data and services into concrete applications, such as smart city management, industrial automation, intelligent transportation systems, and healthcare solutions (Pan and Yang, 2018).

The data processing layer, service layer, and security layer are important layers in addition to the fundamental three. Access control, authentication, encryption, and security protocols are all part of the security layer, which is responsible for ensuring the security and integrity of the data during collection, transmission, and processing. In order to process and analyze large amounts of data in real time and extract useful information, the data processing layer makes use of edge computing and cloud computing technologies. The data is subsequently presented to users through interfaces provided by the service layer, which also provides automatic monitoring and remote diagnostics, among other features that make the data immediately usable for users.

4.2.2.3 Blockchain-Enhanced IoT System Architecture

As the number of IoT devices increases quickly, it is becoming increasingly important to ensure their administration and operation are sustainable. Sustaining the long-term reliability of IoT systems requires optimizing the distribution of dispersed computing resources and consistently raising energy efficiency (Pan and Yang, 2018; Makhdoom et al., 2019). In the meantime, to solve the heterogeneity and interoperability challenges of devices, industry alliances and international standardization groups are working to build uniform communication protocols and interface standards. The IoT technology still confronts obstacles including data privacy protection, network security, system dependability, and economies of scale, despite its enormous promise in a variety of industries like smart cities, industrial automation, and healthcare (Liu et al., 2020). A ground-breaking development in IoT design, the combination of blockchain and IoT addresses important issues including device autonomy, data security, and privacy protection. Because of its intrinsic immutability and decentralization, blockchain technology – a distributed ledger – offers a reliable way to guarantee the validity and integrity of data produced by IoT devices. Every device in an IoT context has the potential to function as a network node, producing and maintaining data as well as contributing to data generation (Wang et al., 2022a). Blockchain and IoT integration improve the network's resistance to assaults and fault tolerance. Figure 4.3 depicts the interaction and contrast between blockchain and IoT architectures.

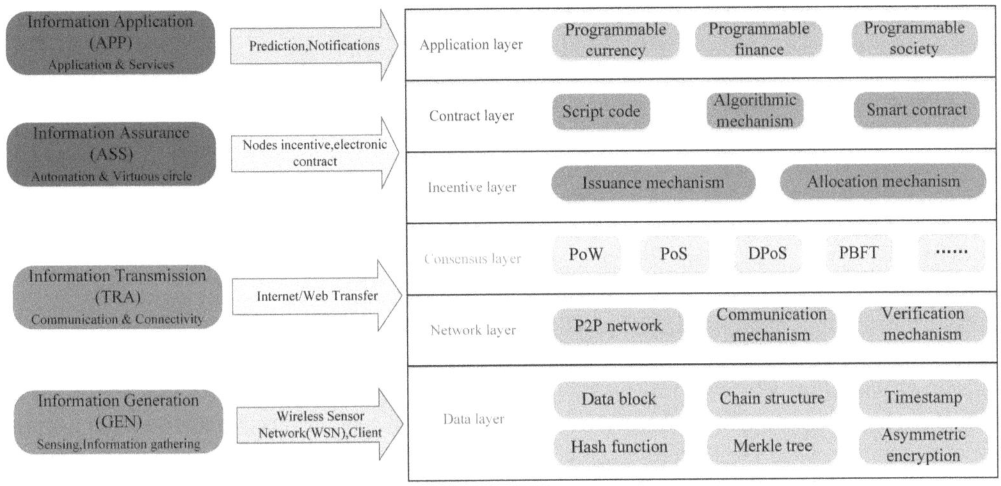

FIGURE 4.3 The Connection between IoT and Blockchain

Included in the process of creating information is the data source, which can be information gathered from wireless sensors. This data source is similar to the data layer of the blockchain and ensures data integrity and anonymity. P2P technology makes it possible for a direct node-to-node link and data transfer in blockchain networks. Data transmission across networks is necessary following data generation. Like the network layer of a blockchain, this also has a consensus layer that makes sure all nodes agree that the data is accurate and true. Blockchain uses node incentives and smart contracts, which correspond to the incentive layer and contract layer in blockchain, to ensure automation and positive loops that guarantee information security.

The IoT blockchain architecture has three primary versions: consortium chains, private chains, and public chains. Regarding participant scope, goal, and authorization management, each offers special features. The first blockchain, referred to as the public chain, provided a fully decentralized, anonymous, and open ecosystem. Every node may join the public chain network, broadcast transactions, validate transactions, and receive data. A single company or entity makes up the highly centralized private chain. It is mostly used for internal administration, which includes processing and storing data within a company, and is restricted to certain individuals or group members (Maftei et al., 2023). Private chains can be customized to fulfill specific requirements and expedite transactions. Combining elements of public and private chains, consortium networks are weakly decentralized blockchains under the authority of several organizations. Banks, government agencies, and large corporations are examples of respectable establishments. Public chains lack the privacy protection and transaction efficiency of consortium chains. Consortium chains reduce the dangers associated with a single enterprise as opposed to private chains, therefore improving security and confidence overall.

Because of its access control, identity verification, and job allocation, consortium chains, one of the three architectural variants of blockchain, have emerged as the go-to platform for business-to-business cooperation. Consortium chains, in particular, are considered the best way to promote security and trust in large-scale IoT systems because of their relatively high controllability and security, their efficient administration, and their operational capabilities. Access control and identity verification in consortium chains offer an environment that is verifiable and impervious to tampering, which is necessary for the high-frequency interaction and data exchange that IoT devices often demand in automated systems. Consortium chains must preserve the decentralized structure of distributed networks while guaranteeing dependable communication among members of the network.

4.2.3 IoT-Blockchain Applications

The integration of IoT and blockchain is transforming the industry by increasing market transparency, security, and efficiency (Pulkkis, Karlsson, and Westerlund, 2018). This technology pairing not only revolutionizes operational protocols but also opens up new avenues for innovation across various sectors. In this section, we discuss two examples of using IoT integrated with blockchain in practice: in the energy and e-commerce industries, and identify how they differ and how they accentuate the benefits of the integration of IoT with blockchain into modern structures and systems.

LO3 Energy (Khalili, Disfani, and Ahmadi, 2021) has designed a breakthrough system called TransActive Grid, which creates an atmosphere for the distributed sale of solar power. This system is incorporated into a micro-grid configuration where local solar power generators can directly sell excess energy to their immediate customers as well as manage the supply-demand balance within a localized grid location. This innovative approach also helps reduce transmission losses while enhancing energy communitarianism at the same time. As a result of their reliance on blockchain, the plans developed through LO3's system guarantee the consistency, clarity, and efficiency of the involved transactions without requiring the existence of a central surveillance body. This setup comprises proof-of-work (PoW) consensus to provide security to the energy transactions in the microgrid, which in turn is essential for enhancing the reliability of the network. Employment of such a technology is consistent with overall energy management initiatives; for instance, New York State's "Reforming the Energy Vision" has the vision of enhancing the grid's effectiveness, reliability, and competency, among others. LO3 depicts the application of blockchain in the energy industry.

JD.COM (Tang, Jia, and Yang, 2022) is an e-commerce company that developed its technological facet with the "JD Blockchain Open Platform." This is specifically aimed at lending integration with the help of blockchain and is more inclined toward letting the business entities acquire, improve, and deploy this application quickly. It can also include services involving the blockchain gateway, node, and consensus network, where the consensus technique can be categorized as the BFT type to ensure security and stability. Based on this architecture, three main roles are specified in the UDP/IP network. Other nodes can be divided into consensus peers, gateway peers, and IoT devices. To be precise, you can call them full nodes; these are the ones that can or have to maintain the blockchain and have the ability to process transactions. This layer comprises several gateway peers responsible for interfacing with both the blockchain network and the IoT devices for input or output requirements. The proper functioning of this decentralized system has positive effects on the improvement of the transaction process and provides a mechanism for secure information exchange in the network.

4.3 BLOCKCHAIN CONSENSUS METHODS IN IOT

The ability of consensus mechanisms to allow several nodes to concur on whether to add a new block makes them crucial to the functioning of blockchain networks. These strategies enable the decentralized character of blockchain networks. Bitcoin uses PoW, which is the most popular consensus method. In the past several years, a wide range of consensus-building techniques have been created and used in diverse settings. This section will introduce, compare, and examine these typical consensus methods, as well as examine if they are suitable for adoption in blockchain-integrated IoT networks.

4.3.1 Common Consensus Mechanisms

4.3.1.1 Proof-of-X

A key element of blockchain consensus mechanisms, the PoX method functions on the basis of proving ownership or authority over particular resources or traits, represented by the variable "X." In order for users of this method to participate in the process of building blocks or approving transactions inside the blockchain network, they must furnish cryptographic proof of their ownership or control of these resources. These users are referred to as validators or miners. The particular resource or property denoted by "X" may differ based on how the blockchain protocol is constructed.

PoW: A fundamental component of blockchain consensus algorithms that is widely utilized in Bitcoin and other cryptocurrencies. PoW participants, sometimes called miners, compete to employ processing power to figure out challenging cryptographic riddles (Salimitari et al., 2017; Salimitari and Chatterjee, 2018). Their main goal is to find a nonce value that, when added to the block contents, results in a hash that meets predefined requirements; this is often a string of leading zeros. The first miner to find the correct nonce and produce a hash that is accepted can add a new block to the blockchain that contains a collection of verified transactions. The use of PoW in the context of IoT faces several challenges. PoW requires a lot of resources, including high computational power and energy consumption, and therefore, resource-constrained IoT devices with limited processing power and energy supply are not fit for it (Eyal and Sirer, 2018). Further, while real-time transaction processing is frequently crucial in IoT applications, the delay caused by PoW's competitive mining process may hinder it.

Despite these challenges, PoW may still be helpful in some IoT scenarios, particularly in environments where security and decentralization are high concerns. PoW-based blockchains, for instance, can be a powerful option for industrial IoT applications where unchangeable and trustless data recording is crucial. On the other hand, alternative consensus methods can provide IoT deployments with more scalable and efficient solutions, better satisfying the latency and resource requirements of IoT networks. A couple of them are practical Byzantine fault tolerance (PBFT) and proof-of-stake (PoS).

- *PoS:* Depending on their level of network investment, individuals commonly referred to as "forgers" or "bonded validators" are selected to forge new blocks and validate transactions in a PoS system. PoS is a well-known consensus method in blockchain technology that replaces the energy-consuming PoW algorithm. PoS selects validators based on the quantity of cryptocurrency tokens they own and agree to "stake" as security, in contrast to PoW, which depends on computing effort and energy consumption for transaction validation and network security (Salimitari, Chatterjee, and Fallah, 2020). The percentage of Bitcoin that a person stakes against the total amount in circulation directly affects their chances of getting chosen as a validator (King and Nadal, 2012). Since they stand to lose their staked tokens if they attempt to approve fraudulent transactions, validators are motivated to act honorably and validate transactions accurately.

  When it comes to the IoT, in particular, the PoS algorithm offers several advantages over PoW. Primarily, PoS is more suitable for IoT devices with limited processing power and energy supply since it consumes significantly less energy and computational power than PoW. By eliminating the need for expensive mining gear, PoS also reduces the barrier to entry for network participation. To make matters better, PoS can help blockchain networks become more scalable and efficient, which will allow for faster transaction processing and lower latency. For the needs of a wide range of IoT applications, these advancements are essential. Better environmental sustainability is also displayed by PoS-based blockchains, as they do not depend on energy-intensive mining processes.

Delegated Proof-of-Stake (DPoS): A consensus technique that is often employed in place of more traditional PoW and PoS algorithms. DPoS achieves both decentralization and scalability by assigning the duty of transaction validation to a predetermined number of chosen delegates or nodes, rather than allowing direct involvement from all network users in the consensus process (Skh Saad and Raja Mohd Radzi, 2020). Through voting, network participants select a small number of delegates who will be in charge of validating transactions and adding new blocks using the DPoS technique. These delegates are often selected on the basis of their reputation, level of technical expertise, and level of network involvement. Because there are a lot fewer delegates than there are participants overall, reaching a consensus is made easier and requires fewer processing resources.

- *Proof-of-Activity (PoA):* A consensus mechanism for blockchain networks that blends aspects of PoW and PoS. In order to generate fresh blocks and validate transactions, miners first participate in a PoW phase of PoA (Skh Saad and Raja Mohd Radzi, 2020). In contrast to conventional PoW algorithms, PoA demands that miners pledge a certain quantity of Bitcoin as security during the PoW stage. After a block has been successfully mined during the PoW phase, the network moves onto a PoS phase where validators – selected according to their network stake – confirm the block's authenticity and suggest the next block in the chain. Since validators stand to lose their staked tokens if they approve fraudulent transactions, they are motivated to perform honorably and validate transactions accurately.

  Numerous advantages exist for IoT applications using the PoA technique. The first way that PoA lowers the energy usage of traditional PoW approaches is by combining PoW with PoS. With this hybrid approach, the computational effort happens in the PoW stage and agreement is achieved by staking in the PoS phase. PoA becomes more suitable for IoT devices with limited processing and energy capacity and is more ecologically friendly with this strategy. For real-time IoT applications where timely data processing is necessary, PoA also provides reduced latency and quicker transaction processing times than PoW-based algorithms. Furthermore, by lowering the processing power required for block production and validation, PoA enhances the scalability of blockchain networks. In light of the growing number of IoT devices, this makes network expansion easier.

4.3.1.2 BFT-Based Consensus

In blockchain networks, the BFT consensus is a key technique meant to establish agreement across dispersed nodes even when there are bad or malicious individuals present (Driscoll et al., 2003). BFT consensus, which has its roots in the Byzantine Generals Problem, guarantees the consistency and integrity of transactions throughout the network by accepting Byzantine faults, such as dishonest nodes or contradicting information being sent by different nodes. In Figure 4.4, the issue scenario is depicted. A key problem in fault tolerance and distributed computing is the Byzantine Generals' Problem, which involves three Byzantine generals, one of whom is a traitor. One of the main issues is that a treacherous general may transmit contradictory signals, making it difficult for the devoted generals to come to an agreement. In order to decide on a coordinated course of action, like invading or withdrawing, the loyal generals must confer with one another and make sure they can get beyond the ambiguity created by the treacherous general's perhaps misleading signals. The idea is to come up with a system or formula that, even with the traitor present, enables the obedient generals to come to a consensus and prevent disastrous missteps or poor choices. Consensus is reached in a BFT-based blockchain network by means of a sequence of communication rounds in which participating nodes propose and exchange messages, including cryptographic signatures and transaction data. Nodes jointly decide on the legitimacy of transactions and the arrangement of blocks in the blockchain by voting and validating messages (Castro and Liskov, 2002). Blockchain networks can benefit from the durability of BFT consensus algorithms against malicious assaults, as well as their high throughput and low latency. Because of these features, BFT is especially well-suited for applications like real-time data processing and financial transactions that call for high transaction volume and quick confirmation times.

- *PBFT:* A consensus technique used in distributed systems, such as blockchain networks, to provide fault tolerance. Because it can handle high throughput and low latency applications very well, PBFT is a preferred option for permissioned blockchain networks. Figure 4.5 shows how the PBFT consensus method works. In PBFT, a subset of nodes, referred to as replicas, votes many rounds in order to reach a consensus. In order to reach consensus, the

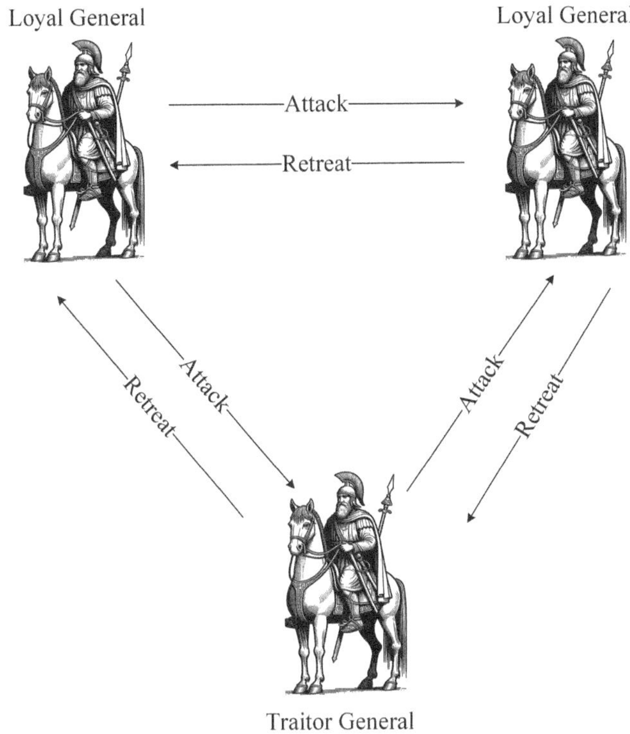

FIGURE 4.4 Byzantine Problem with Three Generals

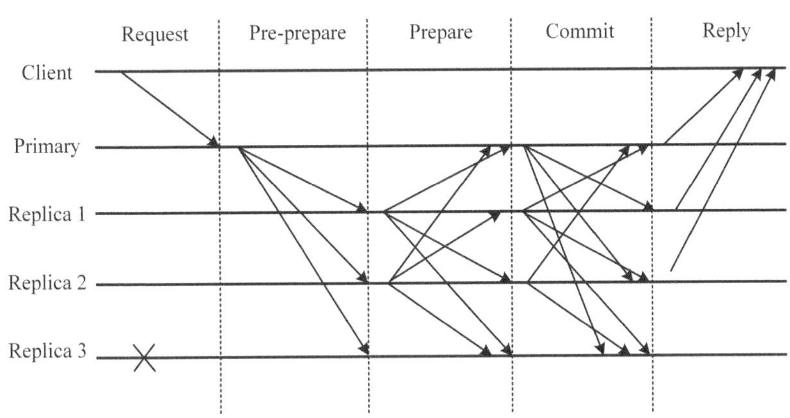

FIGURE 4.5 PBFT Consensus Process

method requires that at least two-thirds of the copies agree on the order and validity of transactions. As long as there are less than one-third of the total number of erroneous copies, PBFT may withstand Byzantine faults like malicious or malfunctioning nodes. Replica 3 is either delayed or not participating in the expected manner. The primary replica might initiate a view change (the election of a new primary) if it detects that a replica is faulty or unresponsive. The consensus process will continue as long as the number of faulty replicas does not exceed the tolerance limit of the protocol.

PBFT's capacity to reach an agreement quickly – often in a few message exchange rounds – is one of its primary features. Because they need to handle transactions in real time, apps like financial services and IoT platforms benefit greatly from PBFT's efficiency. Strong liveness and security guarantees are also provided by PBFT in order to guarantee that only valid transactions are added to the blockchain and that the network continues to function even in the event of Byzantine failures. For blockchain-based systems to remain valid and unbroken, such dependability is necessary, particularly for crucial IoT applications where data quality and consistency are crucial. To further enhance its scalability, PBFT provides concurrent transaction processing and asynchronous communication between copies.

- *Zyzzyva:* A speculative BFT algorithm. Unlike PBFT, which ensures consistency regardless of the primary node's behavior through reliable propagation and requires an algorithmic complexity of $O(N^2)$ messages to achieve consensus across the network, Zyzzyva considers two different system states. One state accounts for the primary node behaving maliciously or encountering excessive network delay, while the other assumes it operates normally under favorable network conditions. In the former case, where the primary node is deemed trustworthy, Zyzzyva employs rapid message broadcasting with a low message complexity of $O(N)$. In the latter scenario, Zyzzyva regresses to a more conventional BFT method to preserve system consistency if additional nodes notice suspicious activity coming from the primary node. Zyzzyva, however, runs the danger of engaging in speculative activity. Algorithm performance can be hampered by untrustworthy primary nodes or bad network circumstances; these factors have a major impact on the algorithm's efficiency. There is an increase in latency and complexity when switching from the fast BFT method to the regular BFT method. Zyzzyva finds performance limitations similar to those in PBFT in the second scenario, although it operates with reduced message complexity under typical conditions in the context of large-scale networks.
- *Delegated Byzantine Fault Tolerance (DBFT):* To achieve consensus across distributed nodes in blockchain networks, even when faced with Byzantine faults, a consensus technique called DBFT is often used. Other network users assign delegates or validators, a group of reliable nodes, to handle the construction and validation of blocks in DBFT. A vote procedure or nomination process is frequently used to choose these delegates, who are then chosen according to predetermined standards, including reputation, stakeholder interest, and other factors. Blocks are submitted and approved by the chosen delegates over the course of several rounds of the DBFT algorithm. As the other delegates assess and decide if the suggestions are genuine, each delegate offers a block in turn throughout each round. When a supermajority of delegates, typically two-thirds or more, agree that a proposed block is acceptable, consensus is obtained. The procedure continues until an agreement is obtained or a timeout occurs if consensus is not attained in a given round.

  Assuming that most delegates are reliable and behave in the network's best interest, DBFT protects network security and integrity. Compared to other consensus techniques, block validation by a small, trusted set of nodes reduces the possibility of malicious activity and speeds up transaction processing. However, as the choice of delegates may be affected by elements like reputation or income, DBFT also poses a risk to centralization. NEO, a cryptocurrency, makes use of this consensus process. This method and PBFT have many of the same desirable features. However, its average block generation delay of 15 seconds is unsatisfactory for the IoT.
- *Stellar Consensus Protocol (SCP):* A decentralized consensus mechanism that the Stellar blockchain network uses to bring dispersed nodes to agreement. The goals of SCP are to maintain the decentralization, security, and consistency of the Stellar network (Mazières, 2017). SCP functions fundamentally as a federated Byzantine agreement protocol, in which quorums are formed from network nodes. Every quorum is a group of nodes that concur on the order of blocks and the legitimacy of transactions as a whole. Quorum slices,

or collections of nodes that cross to create a quorum, are the foundation of SCP. SCP guarantees that nodes can agree even with few direct connections to one another by employing quorum slices. The nomination phase and the vote phase are the two stages that make up the SCP procedure. The nomination protocol, which starts the process, is where new values – also referred to as candidate values – are put out for approval. Every node in the quorum receives these suggested values, and each node votes for one of the candidate values. At the end of this phase, a single value for the slot is decided upon. The voting process then starts, with everyone casting their votes to approve or disapprove of the value that was decided upon during the nomination phase (Dinh *et al.*, 2018). Votes related to the present slot are ultimately counted, and votes designated for abortion are disqualified. When there is not enough agreement to commit or abort a value, there may be a vote to start a higher-value nomination, which works similarly to the execution of a new voting procedure.

Through a series of rounds in which nodes converse and exchange messages to propose, validate, and approve transactions, SCP reaches consensus (Mazières, 2017). The protocol makes use of the federated voting idea, in which every node casts a vote within its quorum to determine if proposed transactions and blocks are legitimate. When a supermajority of nodes in all quorums concur on the legitimacy and sequence of transactions, consensus is obtained. One noteworthy characteristic of SCP is its resilience to certain Byzantine faults, including node outages and network partitions, without compromising consensus. Because of its fault tolerance, SCP may be used in decentralized networks such as Stellar, where nodes might be hostile or unreliable.

- *Ripple:* Employs the Ripple Protocol Consensus Algorithm (RPCA), a decentralized consensus technique. Unlike other blockchain-based consensus algorithms like PoW or PoS, the RPCA consensus algorithm does not rely on mining or a particular set of validators (Chase and MacBrough, 2018). The Unique Node List (UNL) is a dynamic, unique list of dependable participants maintained by each Ripple server in the RPCA. Every server has faith in the assessments of other servers about its UNL. These servers independently suggest and decide whether or not a transaction is legitimate. Consensus is obtained when a supermajority of UNL servers agree that a transaction is valid. Rapid transaction confirmation speeds (between 3 and 5 seconds) and scalability against double-spending attacks are the fundamental tenets of the Ripple consensus algorithm (Ometov *et al.*, 2020). A transaction is said to have attained consensus when 80% of the nodes inside a UNL concur on it. Ripple enables these goals and eliminates the need for energy-intensive mining processes by allowing servers to autonomously propose and approve transactions. Reliability and security of the network are further enhanced by the decentralized structure of RPCA, which prevents any one party from controlling it.
- *ByzCoin Consensus Algorithm (BCA):* Decentralized blockchain technology ByzCoin, which aims to provide secure transaction execution and distributed ledger maintenance, is mostly based on the BCA (Salimitari, Chatterjee, and Fallah, 2020). With potential bad actors or network failures in mind, BCA aims to enable a network of nodes to reach a consensus on the validity and order of transactions. It is inspired by the ideas of classical BFT. Network users may perform cryptographic activities safely and effectively because of threshold cryptography, a key component of BCA. Transactions are safely accepted and validated via threshold signatures, even in the event of Byzantine errors.

  ByzCoin accelerates the scalability of PBFT and enables fast commitment of Bitcoin transactions in a matter of seconds by utilizing a community signing procedure called CoSi in combination with a Byzantine consensus method (Syta *et al.*, 2016). It also modifies PBFT to provide a public blockchain, allowing for dynamic membership that complies with PoW, inspired by the architecture of Bitcoin. This method makes use of a tree-structured communication protocol to greatly minimize the delay characteristic of the Bitcoin blockchain (Kokoris-Kogias *et al.*, 2016). It remains susceptible to denial-of-service attacks, nevertheless. Even though the confirmation durations of 15–20 seconds are above the PayPal network's capability, this delay is still beyond the acceptable limits for IoT networks. Furthermore, the PoW technique used is identical to that of bitcoin, making it inappropriate for IoT devices.

4.3.1.3 DAG-Based Consensus

Consensus algorithms based on DAGs have become more well-known in the blockchain community because they have the capacity to get around some of the scalability and efficiency issues that come with conventional blockchain topologies. DAG-based consensus techniques, in contrast to linear blockchain topologies, make use of a graph-like structure in which each transaction is connected to many prior transactions instead of being arranged in blocks. The ability of DAG-based consensus algorithms to execute transactions simultaneously is one of its most notable features. This makes it possible to validate and confirm several transactions at once. Comparing such simultaneous processing to the sequential nature of typical blockchain topologies significantly increases transaction throughput and scalability.

Let us examine the Greedy Heaviest-Observed Subtree (GHOST) protocol first before introducing the DAG method. Ethereum, which embodies blockchain 2.0, adopts an enhanced GHOST protocol as its consensus mechanism. With an average block generation period of 10 minutes for Bitcoin, "temporary forks" can occur when two miners concurrently get the authority to record transactions (becoming block-producing nodes) and publish blocks. Temporary forks become commonplace in the blockchain due to Ethereum's reduction of the average block generation time to 13 seconds; the number of forks is significantly higher than that of Bitcoin. In order to manage temporary splits, Bitcoin uses the "longest chain" concept, which states that nodes can only mine on the longest valid chain and that block rewards can only be granted to nodes who are granted the ability to record transactions on the longest chain. Block rewards are not granted to nodes in forked chains that publish blocks. Because of the "longest chain" theory behind Bitcoin, the network produces orphan blocks, which wastes computing resources. Ethereum, in contrast to Bitcoin, adheres to the "heaviest chain" principle, which takes into account both the chain's length and the nodes' workload. The blockchain also contains uncle blocks that are not part of the main chain. Figure 4.6 shows uncle blocks, the "longest chain," and the "heaviest chain" principles, where blocks E_3 and C_3 are uncle blocks in Ethereum. In addition to chain length, the GHOST protocol places emphasis on the chain's "weight," or the quantity of blocks it contains. Nodes that publish uncle blocks E_3 and C_3, as well as nodes that include uncle blocks C_2, will receive a certain proportion of block rewards.

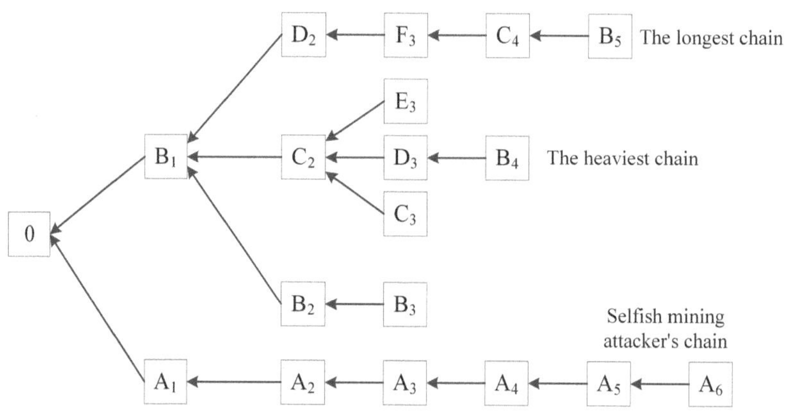

FIGURE 4.6 Ethereum's GHOST Protocol

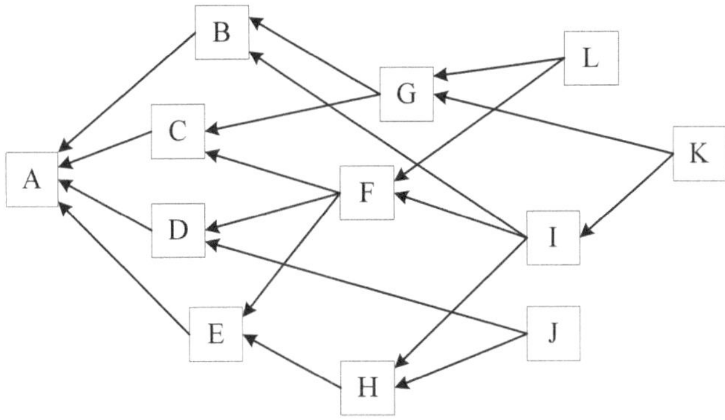

FIGURE 4.7 Directed Acyclic Graph

To further reduce computational waste, the DAG-based consensus mechanism incorporates all blocks into the network, as shown in Figure 4.7. The DAG consists of a set of vertices V and a set of edges E. Each element in the vertex set corresponds to a transaction or block in the DAG ledger, and each element (u, v) in the edge set is a tuple representing the partial order relationship between nodes u and v, indicating that node u indirectly confirms the block of node v. Each new block must reference (point to) several previous blocks, referencing a block $Block(i)$ indicates confirmation and voting for $Block(i)$.

In a DAG structure, no node's processing power is wasted when it mines a block successfully. Nodes do not have to choose which chain to mine on; all they have to do is connect newly mined blocks to the nearest blocks that came before. The distributed ledger technology, which is predicated on a DAG, overcomes the limitation of serial processing in traditional blockchain systems. Regardless of the basic building blocks, blocks, or transactions, a DAG ledger may manage concurrent operations. This ensures a high throughput of transactions, as several transactions may be confirmed in a timely manner without waiting on each other. As Bitcoin has shown, a distributed ledger built on a DAG may use the processing power of all honest nodes to allow the system to absorb bad power up to 50%. Some of the intrinsic problems with DAG that persist are double-spending and problems with transaction sequencing. Double-spending raises blockchain transmission and storage costs while decreasing the system's effective throughput. In order to avoid double-spending and preserve network stability, transaction order is essential. Nodes that are unable to resolve conflicting transactions and validate them consecutively will be unable to uphold the system's integrity.

4.3.1.4 Block Alliance Consensus

BAC is the consensus mechanism we proposed; the detailed process can be found in reference (Wang, Li, et al., 2023; Wang, Yuan, et al., 2023). The BAC consensus mechanism considers the chain idea from blockchain technology and combines it with BFT. It first tries to explain why the communication complexity of BFT in the traditional model is high, to the extent that all nodes may not be able to be synchronized to have the same data. To solve this issue, BAC adapts the chain, with data consistency checks occurring in the following round of blocks. This results in the continuous accumulation of the information, which leads to developing a progressive consensus on the data. With the use of BAC, the scalability, security, and stability of the creation of practical consortium blockchains can be realized, especially for industrial purposes. It uses a policy of decentralization in leadership where the packaging of the block is done by the leader node while other nodes in the network confirm it, thus minimizing cases of competition and, therefore, the formation of blockchains. The guarantee of block consistency is another major aspect of the consensus mechanism, as the candidate nodes are responsible for validating the blocks that are packaged by the leader.

Reliably, an election and credit incentive system raise the likelihood of trustworthy behavior among nodes, with credit figures determining roles and incentives for nodes.

4.3.2 Comparison between Consensus Mechanisms and Their Applicability in IoT

PoX techniques, such as PoW and PoS, depend on computational issues or stake ownership to reach a consensus. PoW employs computationally demanding procedures to safeguard networks; nevertheless, because of its substantial energy usage and expandability, it is not as suitable for IoT networks with restricted resources. On the other hand, PoS offers a more energy-efficient approach by allowing validators to interact based on their stake; nonetheless, centralization issues might arise if significant stakeholder control over the network takes place.

As with PBFT, BFT consensus algorithms prioritize fault tolerance and fast transaction finality, which is ideal for IoT applications that require low latency and quick responses. That being said, BFT protocols frequently experience a quadratic rise in communication cost with an increase in node count. Further, because BFT algorithms need a fixed number of reliable nodes, they pose challenges in decentralized IoT contexts with dynamic membership.

DAG-based consensus approaches allow for consensus to be reached without using traditional blocks and chains. Because DAG designs allow for asynchronous, parallel transactions, they are inherently scalable and efficient for IoT applications with high transaction volumes and concurrency needs. Moreover, the lack of validators and miners in DAG systems eliminates the need for significant processing power and allows lightweight IoT devices to participate in consensus.

A number of factors, including decentralization, energy efficiency, scalability, and latency, must be carefully taken into account when choosing a consensus technique for IoT networks. Table 4.1 shows a comparison between some typical consensus methods. By advancing consensus research and innovation that is especially tailored to IoT demands, we may fully realize the potential of blockchain technology in supporting secure, scalable, and efficient IoT ecosystems.

4.4 EXPERIMENTAL RESULTS AND ANALYSIS

This section presents performance comparison results of some typical consensus mechanisms, addressing key parameters, such as throughput, latency, and scalability. The experimental outcomes are derived from our previous research work (Wang et al., 2022; Wang, Li, et al., 2023; Wang, Yuan, et al., 2023).

TABLE 4.1
A Comparison between Popular IoT Consensus Methods

Consensus	Access	Decentralization	Scalability	Latency	Throughput	IoT suitability
PoW	Public	High	Medium	High	Low	Weak
PoS	All	High	Medium	Medium	Medium	Weak
DPoS	Public	Medium	Medium	Medium	Medium	Weak
PoA	Public	High	High	Medium	Low	Weak
PBFT	Consortium or private	Medium	Low	High	Medium	Medium
ByzCoin	Public	High	High	Medium	High	Medium
DAG	Consortium or private	Medium	High	Low	High	Strong
BAC	Consortium or private	Medium	High	Low	High	Strong

TABLE 4.2
The Meaning of Four Metrics

Blockchain length	Total number of blocks in blockchain at the conclusion of simulation
Average block time	Average time to produce a block in seconds
Total number of stale blocks	A stale block is connected to the blockchain, but it does not belong to the main branch
Last block received by	The number of nodes successfully receiving the last block indicates the overall efficiency of the blockchain system

We utilize the VIBES blockchain simulator in conjunction with Python to assess the performance of classical blockchain algorithms. Apart from VIBES, no existing tool offers the capability to analyze the performance and security attributes of a generic blockchain network (Stoykov, Zhang, and Jacobsen, 2017). VIBES sets itself apart by its capability to simulate blockchain systems beyond Bitcoin and its support for large-scale simulations involving thousands of nodes. All computations are conducted on a Lenovo computer with Windows Ultimate 64-bit, featuring an Intel i7–8550 CPU @ 1.80 GHz and 8.0 GB 2133 MHz LPDDR3 RAM. The software stack comprises Java JDK Version 11.0.10, Scala Version 2.13.5, and Akka Version 2.6.14. VIBES utilizes unsolicited block push, where if a node is aware that none of its peers possess a block, it can directly push it to all of them for block broadcasting. The Master Actor is pivotal in VIBES, orchestrating the execution of events in the blockchain network. Additionally, VIBES offers detailed classifications, functions, network configurations, and metrics for current nodes, which enable the demonstration of the performance of three mechanisms as outlined below.

Four hours is the predetermined duration of the simulation. The simulation experiment for each approach has been performed 50 times using the same number of nodes. The average of these 50 data points is then calculated for each approach. Our analysis focuses on comparing and examining the "Blockchain length," "average block time," "total number of stale blocks," and "last block received by" metrics for PoW, PBFT, and the BAC.

Figure 4.8 presents the comprehensive depiction of nodes and blocks at the conclusion of the simulation. It is evident from Figure 4.8 that "d695ce1f" serves as the primary node, while "6c91d1d9," "40588cb0," "3d96e708," and "035d2044" operate as the remaining ordinary nodes. They numerically depict the geographical arrays of nodes used in the simulation. The blue curved line that connects Europe with Latin America represents a block transfer from one node to another. Especially, it illustrates the process of real-data transfer between nodes residing in different regions in the case of simulation. The Summary section on the right gives important parameters such as the length of the blockchain, average time taken to generate a block, number of nodes, etc., whereas the Events section gives the precise actions and transactions that transpire in the simulation. The "Summary" section in Figure 4.8 encapsulates four key metrics concerning the blocks, as detailed in Table 4.2.

4.4.1 Blockchain Length

The blockchain length comparison of PoW, PBFT, BAC-Tr (BAC-Transaction scenario), and BAC-S (BAC-Sensor scenario) is illustrated in Figure 4.9.

- *PoW:* A near-linear increase in data and block length on the chain is the result of the block production time in PoW blockchain systems, like Bitcoin. These systems are designed with this in mind. The blockchain becomes longer as the number of nodes increases. The results of the experiment show how long the longest blockchain can be, with blocks that are not part of the main branch being referred to as orphan blocks. A difficulty adjustment method is used in PoW to guarantee that the block creation frequency stays constant across predetermined time periods. Insufficient simulated block generation time in VIBES may

FIGURE 4.8 Overall Picture of Nodes and Blocks

prevent the difficulty adjustment mechanism from activating, which would allow network miners to quickly mine new blocks by solving cryptographic challenges within blocks with relative ease. This research only considers the length of the main chain when assessing the blockchain; blocks that are not part of the main branch, such as orphan blocks, are not included in the length of the blockchain.

- *PBFT:* When using PBFT, the main node is in charge of gathering transactions and sending them to other nodes in order to reach a consensus. Because of this method, PBFT may process transactions and reach consensus faster than PoW, especially when there are fewer nodes. However, if the network grows, for example, to a size of more than 100 nodes, the cost of network-wide communication increases since PBFT requires communication across all nodes in each consensus phase. The length and growth rate of the blockchain are impacted by this increased processing delay caused by the network-wide communication cost. This cumulative impact of delay intensifies with increasing node counts, causing the length of the blockchain to generally shrink. This decline in performance indicates that PBFT is better suited for networks with fewer nodes, as scalability in large-scale distributed systems is still a problem.
- *BAC-Tr and BAC-S:* The consensus mechanism is the BAC model that is developed on the mine consortium blockchain network and is aimed at providing accurate data and network security. In fulfilling its task, it works in a certain way where there is a node known as the leader node that has the task of packing the blocks, while other nodes are assigned the task of validating and signing these blocks. The proposed approach is computationally less intensive, thereby making it adaptable to environments with limited resources. Less computational power is required, so it is ideal for organizations with scarce computing resources. Because BAC depends on the validation and confirmation of the majority of nodes in the network to generate new blocks rather than complicated computational problems, its "blockchain length" is longer than that of the other two consensus processes. Even though a new block must be confirmed by at least two blocks of data, the architecture with $O(N)$ communication difficulty greatly lessens the strain of communication

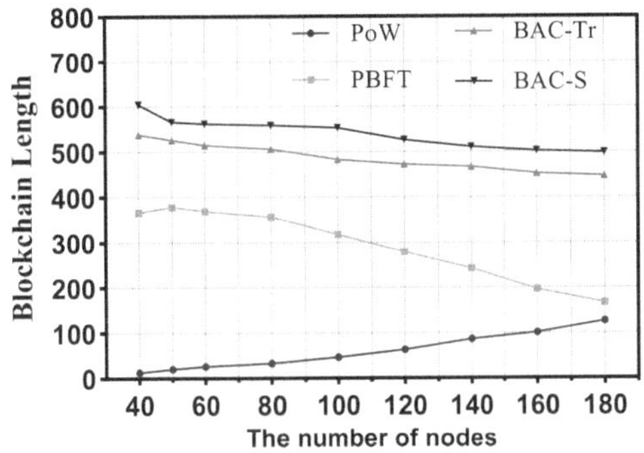

FIGURE 4.9 The Comparison of Blockchain Length

between nodes, particularly in large-scale networks. Even with an increase in node count, the growth of the blockchain remains unaffected because of the BAC's great communication efficiency. BAC-S accomplishes a longer blockchain than BAC-Tr with an equivalent number of nodes by skipping the second round of consensus.

4.4.2 Average Block Time

The comparison of average block time is illustrated in Figure 4.10.

- *PoW:* Under the same block difficulty, a larger node count leads to a shorter block time. This research modifies the block duration to 100 seconds, which makes it easier to specify parameters for other algorithms and ensures the comparability of simulation results. VIBES modifies the block difficulty automatically. The average block duration of a PoW blockchain network, like Bitcoin, ought to be constant, theoretically. This is achieved by adjusting the mining difficulty to ensure that the block time remains consistent, regardless of changes in the network's overall processing power. Nevertheless, the average block time varies significantly in real simulations, as Figure 4.10 illustrates. Numerous variables contribute to this fluctuation: miners' network connection quality, network propagation latency, block propagation algorithms' efficiency, and simulation tools' performance constraints. These variables have an impact on the confirmation time and block propagation speed, which in turn affect the total block duration.
- *PBFT:* In contrast to PoW, PBFT lacks a set block duration and does not produce new blocks through mining. The speed at which each node processes transactions determines how quickly transactions or data blocks are validated in a PBFT network. A reduced total block time is achieved when there are fewer nodes in the network since each node can react and process transactions more rapidly. The time needed to process each block, however, increases dramatically as the network grows, especially around 100 nodes, since each node must synchronize and interact with additional nodes. Because of the increased processing and communication costs in networks with more nodes, the average block time in PBFT networks rises noticeably.
- *BAC-Tr:* BAC guarantees the consistency of the network and the validity of blocks by validating new blocks through a distributed validation procedure that consists of two rounds of consensus confirmation. By limiting the quantity of information sent between nodes, BAC eases the load on network communication in comparison to PBFT. In circumstances where there are more nodes, the communication cost of BAC increases linearly with the number of nodes, whereas the communication overhead of PBFT increases significantly with the

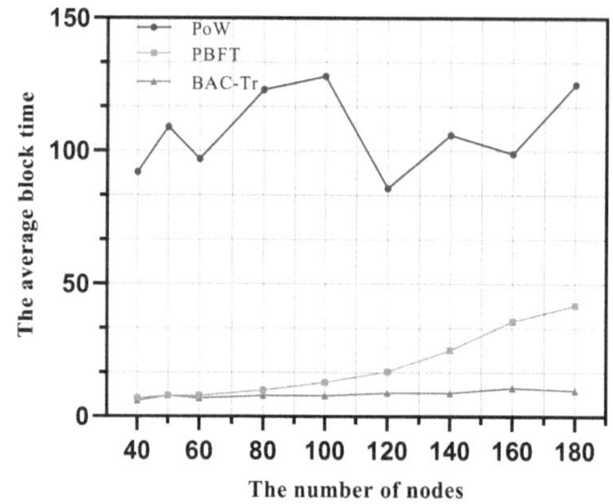

FIGURE 4.10 The Comparison of Average Block Time

square of the node count. Consequently, BAC outperforms PBFT in large-scale networks by achieving shorter average block durations.

4.4.3 Total Number of Stale Blocks

The comparison of the total number of stale blocks is shown in Figure 4.11.

- *PoW:* The overall number of zombie blocks increases as the number of nodes increases, mostly because forks occur often in simulations of PoW blockchains. In PoW, two legitimate chains are momentarily formed if two nodes uncover solutions that nearly simultaneously satisfy the difficulty criterion; however, eventually, only one chain emerges as the primary chain, and the blocks on the other chain turn into zombie blocks. Even though these zombie blocks could be legitimate at one point in time, the network will not accept them as part of the main chain, which will cause the quantity of zombie blocks in the PoW system to increase.
- *PBFT:* The presence of malicious or malfunctioning nodes (also known as Byzantine nodes) in the network can cause inconsistent operations or incorrect message transmission, which can disrupt regular network operations and make reaching consensus more difficult in the PBFT consensus mechanism. The election of a new main node is required to restart consensus operations when a primary node fails, increasing uncertainty in the consensus process. This procedure might result in the reissuing of older blocks or the formation of forks, which would multiply the number of zombie blocks in the network that are unable to join the main chain. A spike in the quantity of zombie blocks indicates that system stability is seriously jeopardized as the network grows, particularly beyond 100 nodes. This problem becomes most noticeable when there are 120 or 160 nodes, highlighting the difficulties PBFT has in preserving efficiency and dependability in big networks.
- *BAC-Tr and BAC-S:* In BAC, the checkpoint node CP is in charge of performing the first validation of newly generated blocks, while the main node P is in charge of creating and broadcasting new blocks. The BAC system seldom creates zombie blocks during regular P operation because of its pre-validation function, which aids in the early detection and elimination of invalid or unlawful blocks. The phenomenon of zombie blocks is commonly linked to network latency, which includes the sluggish transmission of transaction or block data around the network, leading to blocks not being promptly validated by other components of the network. Certain system simulation capabilities, including network

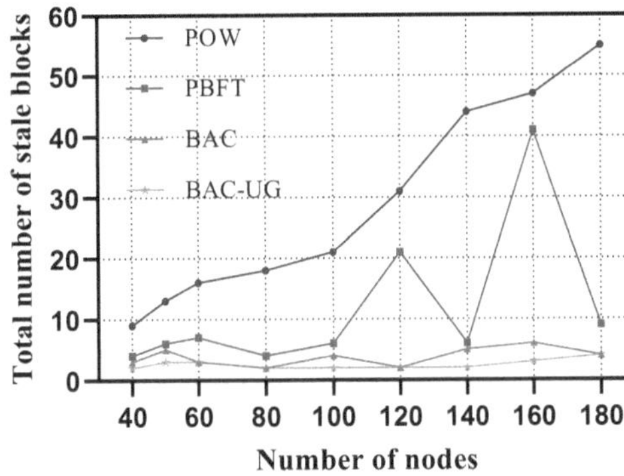

FIGURE 4.11 The Comparison of the Total Number of Stale Blocks

latency simulation, may cause extra delays in the VIBES simulation environment. This might result in a few zombie blocks in the BAC. Even if there are not many of these zombie blocks, their existence shows that in building large-scale distributed ledger systems, network latency must be optimized.

4.5 CONCLUSION

In this chapter, we have explored the integration of blockchain technology with the IoT and examined various blockchain consensus mechanisms applicable to IoT networks. Because they can guarantee security, scalability, and energy efficiency, blockchain consensus techniques are essential for IoT applications. Decentralized and tamper-proof ledgers shield the integrity of sensitive data produced by IoT devices from unwanted access and manipulation. More effective techniques like PBFT and DAG are more appropriate for resource-constrained IoT contexts, in contrast to energy-intensive PoW scenarios. Along with improving network robustness and transparency, these approaches help remove single points of failure and build stakeholder confidence. The main goals of combining blockchain with IoT will soon be resource efficiency, improved interoperability, real-time data processing, and sophisticated security mechanisms. Scalability will be a major focus of future research in order to provide more efficient algorithms for handling the increasing number of IoT devices. These developments are necessary to employ IoT applications across numerous sectors and to address new problems.

REFERENCES

Ahram, T., Sargolzaei, A., Sargolzaei, S., Daniels, J. and Amaba, B. (2017) 'Blockchain Technology Innovations', in *2017 IEEE Technology & Engineering Management Conference (TEMSCON). 2017 IEEE Technology & Engineering Management Conference (TEMSCON)*, Santa Clara, CA: IEEE, pp. 137–141. https://doi.org/10.1109/TEMSCON.2017.7998367.

Ai, Z. and Cui, W. (2022) 'A Proof-of-Transactions Blockchain Consensus Protocol for Large-Scale IoT', *IEEE Internet of Things Journal*, 9(11), pp. 7931–7943. https://doi.org/10.1109/JIOT.2021.3108627.

Bach, L.M., Mihaljevic, B. and Zagar, M. (2018) 'Comparative Analysis of Blockchain Consensus Algorithms', in *2018 41st International Convention on Information and Communication Technology, Electronics and Microelectronics (MIPRO). 2018 41st International Convention on Information and Communication Technology, Electronics and Microelectronics (MIPRO)*, Opatija: IEEE, pp. 1545–1550. https://doi.org/10.23919/MIPRO.2018.8400278.

Baliga, D.A. (2020) 'Understanding Blockchain Consensus Models', *Persistent*, 4(1), p. 14.

Belotti, M., Božić, N., Pujolle, G. and Secci, S. (2019) 'A Vademecum on Blockchain Technologies: When, Which, and How', *IEEE Communications Surveys & Tutorials*, 21(4), pp. 3796–3838. https://doi.org/10.1109/COMST.2019.2928178.

Castillo-Effer, M., Quintela, D.H., Moreno, W., Jordan, R. and Westhoff, W. (2004) 'Wireless Sensor Networks for Flash-flood Alerting', in *Proceedings of the Fifth IEEE International Caracas Conference on Devices, Circuits and Systems, 2004*, Punta Cana, Dominican Republic: IEEE, pp. 142–146. https://doi.org/10.1109/ICCDCS.2004.1393370.

Castro, M. and Liskov, B. (2002) 'Practical Byzantine Fault Tolerance and Proactive Recovery', *ACM Transactions on Computer Systems*, 20(4), pp. 398–461. https://doi.org/10.1145/571637.571640.

Chase, B. and MacBrough, E. (2018) 'Analysis of the XRP Ledger Consensus Protocol'. arXiv. Available at: https://arxiv.org/abs/1802.07242 (Accessed: 19 June 2024).

Chen, C.L., Lim, Z.Y., Xue, X. and Chen, C.H. (2023) 'Special Issue: Symmetric and Asymmetric Encryption in Blockchain', *Symmetry*, 15(2), p. 458. https://doi.org/10.3390/sym15020458.

Dinh, T.T.A., Liu, R., Zhang, M., Chen, G., Ooi, B.C. and Wang, J. (2018) 'Untangling Blockchain: A Data Processing View of Blockchain Systems', *IEEE Transactions on Knowledge and Data Engineering*, 30(7), pp. 1366–1385. https://doi.org/10.1109/TKDE.2017.2781227.

Driscoll, K., Hall, B., Sivencrona, H. and Zumsteg, P. (2003) 'Byzantine Fault Tolerance, from Theory to Reality', in S. Anderson, M. Felici, and B. Littlewood (eds.) *Computer Safety, Reliability, and Security*, Berlin, Heidelberg: Springer (Lecture Notes in Computer Science), pp. 235–248. https://doi.org/10.1007/978-3-540-39878-3_19.

Efanov, D. and Roschin, P. (2018) 'The All-Pervasiveness of the Blockchain Technology', *Procedia Computer Science*, 123, pp. 116–121. https://doi.org/10.1016/j.procs.2018.01.019.

Eyal, I. and Sirer, E.G. (2018) 'Majority Is Not Enough: Bitcoin Mining Is Vulnerable', *Communications of the ACM*, 61(7), pp. 95–102. https://doi.org/10.1145/3212998.

Gao, W., Hatcher, W.G. and Yu, W. (2018) 'A Survey of Blockchain: Techniques, Applications, and Challenges', in *2018 27th International Conference on Computer Communication and Networks (ICCCN)*, Hangzhou: IEEE, pp. 1–11. https://doi.org/10.1109/ICCCN.2018.8487348.

Gorkhali, A., Li, L. and Shrestha, A. (2020) 'Blockchain: A Literature Review', *Journal of Management Analytics*, 7(3), pp. 321–343. https://doi.org/10.1080/23270012.2020.1801529.

Guo, H. and Yu, X. (2022) 'A Survey on Blockchain Technology and Its Security', *Blockchain: Research and Applications*, 3(2), p. 100067. https://doi.org/10.1016/j.bcra.2022.100067.

Hassan, Q.F. (2018) *Internet of Things A to Z: Technologies and Applications*, Hoboken, NJ: John Wiley.

Iuon-Chang, L. and Tzu-Chun, L. (2017) 'A Survey of Blockchain Security Issues and Challenges', *International Journal of Network Security*, 19(5). https://doi.org/10.6633/IJNS.201709.19(5).01.

Jabraeil Jamali, M.A., Bahrami, B., Heidari, A., Allahverdizadeh, P., Norouzi, F., Jabraeil Jamali, M.A., Bahrami, B., Heidari, A., Allahverdizadeh, P. and Norouzi, F. (2020) *Towards the Internet of Things: Architectures, Security, and Applications*, Cham: Springer International Publishing (EAI/Springer Innovations in Communication and Computing). https://doi.org/10.1007/978-3-030-18468-1.

Khalili, S., Disfani, V. and Ahmadi, M. (2021) 'Impact of Blockchain Technology on Electric Power Grids -- A Case Study in LO3 Energy'. arXiv. https://doi.org/10.48550/ARXIV.2106.05395.

Khan, S.N., Loukil, F., Ghedira-Guegan, C., Benkhelifa, E. and Bani-Hani, A. (2021) 'Blockchain Smart Contracts: Applications, Challenges, and Future Trends', *Peer-to-Peer Networking and Applications*, 14(5), pp. 2901–2925. https://doi.org/10.1007/s12083-021-01127-0.

King, S. and Nadal, S. (2012) 'PPCoin: Peer-to-Peer Crypto-Currency with Proof-of-Stake', *self-published paper,* August, 19(1).

Kokoris-Kogias, E., Jovanovic, P., Gailly, N., Khoffi, I., Gasser, L. and Ford, B. (2016) 'Enhancing Bitcoin Security and Performance with Strong Consistency via Collective Signing', *25th usenix security symposium (usenix security 16)*, pp. 279–296.

Liu, T., Wang, Y., Li, Y., Tong, X., Qi, L. and Jiang, N. (2020) 'Privacy Protection Based on Stream Cipher for Spatiotemporal Data in IoT', *IEEE Internet of Things Journal*, 7(9), pp. 7928–7940. https://doi.org/10.1109/JIOT.2020.2990428.

Macrinici, D., Cartofeanu, C. and Gao, S. (2018) 'Smart Contract Applications within Blockchain Technology: A Systematic Mapping Study', *Telematics and Informatics*, 35(8), pp. 2337–2354. https://doi.org/10.1016/j.tele.2018.10.004.

Maftei, A.A., Lavric, A., Petrariu, A.I. and Popa, V. (2023) 'Blockchain for Internet of Things: A Consensus Mechanism Analysis', in *2023 13th International Symposium on Advanced Topics in Electrical Engineering (ATEE)*, Bucharest, Romania: IEEE, pp. 1–5. https://doi.org/10.1109/ATEE58038.2023.10108211.

Makhdoom, I., Abolhasan, M., Abbas, H. and Ni, W. (2019) 'Blockchain's Adoption in IoT: The Challenges, and a Way Forward', *Journal of Network and Computer Applications*, 125, pp. 251–279. https://doi.org/10.1016/j.jnca.2018.10.019.

Mazières, D. (2017) 'The Stellar Consensus Protocol', *A Federated Model for Internet-level Consensus. Version July, 14.*

Medaglia, C.M. and Serbanati, A. (2010) 'An Overview of Privacy and Security Issues in the Internet of Things', in D. Giusto, A. Iera, G. Morabito, and L. Atzori (eds.) *The Internet of Things*, New York, NY: Springer, pp. 389–395. https://doi.org/10.1007/978-1-4419-1674-7_38.

Michelin, R.A., Dorri, A., Steger, M., Lunardi, R.C., Kanhere, S.S., Jurdak, R. and Zorzo, A.F. (2018) 'SpeedyChain: A Framework for Decoupling Data from Blockchain for Smart Cities', in *Proceedings of the 15th EAI International Conference on Mobile and Ubiquitous Systems: Computing, Networking and Services*, New York, NY: Association for Computing Machinery (MobiQuitous '18), pp. 145–154. https://doi.org/10.1145/3286978.3287019.

Miglani, A., Kumar, N., Chamola, V. and Zeadally, S. (2020) 'Blockchain for Internet of Energy Management: Review, Solutions, and Challenges', *Computer Communications*, 151, pp. 395–418. https://doi.org/10.1016/j.comcom.2020.01.014.

Mingxiao, D., Xiaofeng, M., Zhe, Z., Xiangwei, W. and Qijun, C. (2017) 'A Review on Consensus Algorithm of Blockchain', in *2017 IEEE International Conference on Systems, Man, and Cybernetics (SMC)*, Banff, AB: IEEE, pp. 2567–2572. https://doi.org/10.1109/SMC.2017.8123011.

Mohanta, B.K., Panda, S.S. and Jena, D. (2018) 'An Overview of Smart Contract and Use Cases in Blockchain Technology', in *2018 9th International Conference on Computing, Communication and Networking Technologies (ICCCNT)*, Bangalore: IEEE, pp. 1–4. https://doi.org/10.1109/ICCCNT.2018.8494045.

Monrat, A.A., Schelen, O. and Andersson, K. (2019) 'A Survey of Blockchain from the Perspectives of Applications, Challenges, and Opportunities', *IEEE Access*, 7, pp. 117134–117151. https://doi.org/10.1109/ACCESS.2019.2936094.

Novo, O. (2018) 'Blockchain Meets IoT: An Architecture for Scalable Access Management in IoT', *IEEE Internet of Things Journal*, 5(2), pp. 1184–1195. https://doi.org/10.1109/JIOT.2018.2812239.

Ogonji, M.M., Okeyo, G. and Wafula, J.M. (2020) 'A Survey on Privacy and Security of Internet of Things', *Computer Science Review*, 38, p. 100312. https://doi.org/10.1016/j.cosrev.2020.100312.

Ometov, A., Bardinova, Y., Afanasyeva, A., Masek, P., Zhidanov, K., Vanurin, S., Sayfullin, M., Shubina, V., Komarov, M. and Bezzateev, S. (2020) 'An Overview on Blockchain for Smartphones: State-of-the-Art, Consensus, Implementation, Challenges and Future Trends', *IEEE Access*, 8, pp. 103994–104015. https://doi.org/10.1109/ACCESS.2020.2998951.

Pan, J. and Yang, Z. (2018) 'Cybersecurity Challenges and Opportunities in the New "Edge Computing + IoT" World', in *Proceedings of the 2018 ACM International Workshop on Security in Software Defined Networks & Network Function Virtualization*, Tempe, AZ: Association for Computing Machinery (SDN-NFV Sec'18), pp. 29–32. https://doi.org/10.1145/3180465.3180470.

Pulkkis, G., Karlsson, J. and Westerlund, M. (2018) 'Blockchain-Based Security Solutions for IoT Systems', in Q.F. Hassan (ed.) *Internet of Things A to Z*, 1st edn, Hoboken, NJ: Wiley, pp. 255–274. https://doi.org/10.1002/9781119456735.ch9.

Salimitari, M. and Chatterjee, M. (2018) 'A Survey on Consensus Protocols in Blockchain for IoT Networks'. arXiv. https://doi.org/10.48550/ARXIV.1809.05613.

Salimitari, M., Chatterjee, M. and Fallah, Y.P. (2020) 'A Survey on Consensus Methods in Blockchain for Resource-constrained IoT Networks', *Internet of Things*, 11, p. 100212. https://doi.org/10.1016/j.iot.2020.100212.

Salimitari, M., Chatterjee, M., Yuksel, M. and Pasiliao, E. (2017) 'Profit Maximization for Bitcoin Pool Mining: A Prospect Theoretic Approach', in *2017 IEEE 3rd International Conference on Collaboration and Internet Computing (CIC)*, San Jose, CA: IEEE, pp. 267–274. https://doi.org/10.1109/CIC.2017.00043.

Siegfried, N., Rosenthal, T. and Benlian, A. (2020) 'Blockchain and the Industrial Internet of Things: A Requirement Taxonomy and Systematic Fit Analysis', *Journal of Enterprise Information Management*, 35(6), pp. 1454–1476. https://doi.org/10.1108/JEIM-06-2018-0140.

Skh Saad, S.M. and Raja Mohd Radzi, R.Z. (2020) 'Comparative Review of the Blockchain Consensus Algorithm between Proof of Stake (POS) and Delegated Proof of Stake (DPOS)', *International Journal of Innovative Computing*, 10(2). https://doi.org/10.11113/ijic.v10n2.272.

Stoyanova, M., Nikoloudakis, Y., Panagiotakis, S., Pallis, E. and Markakis, E.K. (2020) 'A Survey on the Internet of Things (IoT) Forensics: Challenges, Approaches, and Open Issues', *IEEE Communications Surveys & Tutorials*, 22(2), pp. 1191–1221. https://doi.org/10.1109/COMST.2019.2962586.

Stoykov, L., Zhang, K. and Jacobsen, H.-A. (2017) 'VIBES: Fast Blockchain Simulations for Large-scale Peer-to-peer Networks: Demo', in *Proceedings of the 18th ACM/IFIP/USENIX Middleware Conference: Posters and Demos. Middleware '17: 18th International Middleware Conference*, Las Vegas, Nevada: ACM, pp. 19–20. https://doi.org/10.1145/3155016.3155020.

Syta, E., Tamas, I., Visher, D., Wolinsky, D.I., Jovanovic, P., Gasser, L., Gailly, N., Khoffi, I. and Ford, B. (2016) 'Keeping Authorities "Honest or Bust" with Decentralized Witness Cosigning', in *2016 IEEE Symposium on Security and Privacy (SP)*, San Jose, CA: IEEE, pp. 526–545. https://doi.org/10.1109/SP.2016.38.

Tang, X., Jia, Z. and Yang, W. (2022) 'Blockchain Application Status and Ecology', in X. Tang, X. Deng, and R. Bie (eds.) *Blockchain Application Guide*, Singapore: Springer Nature, pp. 35–48. https://doi.org/10.1007/978-981-19-5260-9_3.

Ugochukwu, N.A., Goyal, S.B., Rajawat, A.S., Islam, S.M., He, J. and Aslam, M. (2022) 'An Innovative Blockchain-Based Secured Logistics Management Architecture: Utilizing an RSA Asymmetric Encryption Method', *Mathematics*, 10(24), p. 4670. https://doi.org/10.3390/math10244670.

Wang, M., Duan, M. and Zhu, J. (2018) 'Research on the Security Criteria of Hash Functions in the Blockchain', in *Proceedings of the 2nd ACM Workshop on Blockchains, Cryptocurrencies, and Contracts. ASIA CCS '18: ACM Asia Conference on Computer and Communications Security*, Incheon Republic of Korea: ACM, pp. 47–55. https://doi.org/10.1145/3205230.3205238.

Wang, W., Hoang, D.T., Hu, P., Xiong, Z., Niyato, D., Wang, P., Wen, Y. and Kim, D.I. (2019) 'A Survey on Consensus Mechanisms and Mining Strategy Management in Blockchain Networks', *IEEE Access*, 7, pp. 2 2328–22370. https://doi.org/10.1109/ACCESS.2019.2896108.

Wang, X., Zha, X., Ni, W., Liu, R.P., Guo, Y.J., Niu, X. and Zheng, K. (2019) 'Survey on blockchain for Internet of Things', *Computer Communications*, 136, pp. 10–29. https://doi.org/10.1016/j.comcom.2019.01.006.

Wang, Y., Li, Y., Jiao, W., Wang, G., Zhao, J., Qiang, Y. and Li, K. (2023) 'An Efficient, Secured, and Infinitely Scalable Consensus Mechanism for Peer-to-Peer Energy Trading Blockchain', *IEEE Transactions on Industry Applications*, 59(5), pp. 5215–5229. https://doi.org/10.1109/TIA.2023.3275551.

Wang, Y., Ma, Y., Qiang, Y., Zhao, J., Li, Y. and Li, K. (2022) 'BAC: A Block Alliance Consensus Mechanism for the Mine Consortium Blockchain', *Concurrency and Computation: Practice and Experience*, 34(27), p. e7344. https://doi.org/10.1002/cpe.7344.

Wang, Y., Yuan, L., Jiao, W., Qiang, Y., Zhao, J., Yang, Q. and Li, K. (2023) 'A Fast and Secured Vehicle-to-Vehicle Energy Trading Based on Blockchain Consensus in the Internet of Electric Vehicles', *IEEE Transactions on Vehicular Technology*, 72(6), pp. 7827–7843. https://doi.org/10.1109/TVT.2023.3239990.

Wu, M., Wang, K., Cai, X., Guo, S., Guo, M. and Rong, C. (2019) 'A Comprehensive Survey of Blockchain: From Theory to IoT Applications and Beyond', *IEEE Internet of Things Journal*, 6(5), pp. 8114–8154. https://doi.org/10.1109/JIOT.2019.2922538.

Xiao, Y., Zhang, N., Lou, W. and Hou, Y.T. (2020) 'A Survey of Distributed Consensus Protocols for Blockchain Networks', *IEEE Communications Surveys & Tutorials*, 22(2), pp. 1432–1465. https://doi.org/10.1109/COMST.2020.2969706.

Xie, J., Tang, H., Huang, T., Yu, F.R., Xie, R., Liu, J. and Liu, Y. (2019) 'A Survey of Blockchain Technology Applied to Smart Cities: Research Issues and Challenges', *IEEE Communications Surveys & Tutorials*, 21(3), pp. 2794–2830. https://doi.org/10.1109/COMST.2019.2899617.

Xie, S., Zheng, Z., Chen, W., Wu, J., Dai, H.N. and Imran, M. (2020) 'Blockchain for Cloud Exchange: A Survey', *Computers & Electrical Engineering*, 81, p. 106526. https://doi.org/10.1016/j.compeleceng.2019.106526.

Xu, M., Chen, X. and Kou, G. (2019) 'A Systematic Review of Blockchain', *Financial Innovation*, 5(1), pp. 1–14. https://doi.org/10.1186/s40854-019-0147-z.

Zhai, S., Yang, Y., Li, J., Qiu, C. and Zhao, J. (2019) 'Research on the Application of Cryptography on the Blockchain', *Journal of Physics: Conference Series*, 1168, p. 032077. https://doi.org/10.1088/1742-659 6/1168/3/032077.

Zheng, Z., Xie, S., Dai, H.N., Chen, X. and Wang, H. (2018) 'Blockchain Challenges and Opportunities: A Survey', *International Journal of Web and Grid Services*, 14(4), pp. 352–375. https://doi.org/10.1504/IJWGS.2018.095647.

5 A Threat Analysis Framework for Cyberattacks in Smart Cities

Ransomware in Focus

Farnaz Farid

5.1 INTRODUCTION

The European Union (EU) first introduced the term Industry 5.0. The main attributes of this new industry standard are sustainability, human-centricity, and resilience (Breque et al., 2021). Since its evolution, relevant research has moved in many directions, and some key directions include human-centric enabling technologies, human-machine collaboration, human-factor integration, and overall sustainability (Gladysz et al., 2023). Human-centric technologies, as the name suggests, investigate the usability and other aspects of wearable sensors, augmented reality, artificial intelligence (AI), cyber-physical systems, and the Internet of Things (IoT) in terms of human usage. The human-machine collaboration focuses on the issues and challenges of creating a more favorable work environment where humans and machines can coexist and work harmoniously. Research on human factor integration addresses potential threats to health and safety arising from Industry 5.0-grade technologies and how to mitigate such challenges. Sustainability research focuses on social and environmental challenges in a smart world, particularly in the context of intelligent manufacturers.

Smart cities are defined as cities built on interconnected systems of smart devices, sensors, and networks, enabling data-driven decision-making where a considerable amount of human-machine collaboration will occur (Minoli and Occhiogrosso, 2018). Industry 5.0, in this context, is an essential standard where various human-centric technologies and services are at its core. Undoubtedly, such shoulder-to-shoulder work among humans and machines in the context of Industry 5.0 will bring the cybersecurity issue into the spotlight (Adel, 2023). Some enablers of Industry 5.0 are IoT, virtual reality (VR) and robotics. In the case of IoT, vulnerabilities and attacks were not new even before the advent of Industry 5.0. Perhaps one of the most devastating attacks in IoT history is the Mirai bot attack, where malware turned networked IoT devices into botnets to perform massive distributed denial-of-service (DDoS) attacks (Raiane, Gomes and Fernandez, 2021). IoT-based surveillance cameras are in demand these days. The pictures and videos captured by these cameras are usually stored in the cloud. Hacking into these cameras and cloud data storage using default/weak passwords and relevant software vulnerabilities is a rising concern for consumers. In 2021, there was this Verdaka security camera hacking incident where at least 150,000 cameras were intruded.[1] The intruders also published the inner settings of hospitals, clinics, mental health institutions, banks, police departments, prisons, schools, and companies like Tesla and Cloudflare. The other IoT attacks that can significantly affect smart cities in the future are similar to the building heating system attacks in Finland and the nuclear facility attacks in Iran in 2010 (Husar, 2022). With the building heating system attack, the cybercriminals successfully turned off the system, followed by a DDoS attack to force the heating controller to reboot the system repeatedly. Such attacks are a real threat to smart cities since interconnected systems can bring down the heating system of the whole city. These tactics can be sophisticatedly used as weapons in cyber

DOI: 10.1201/9781003506638-6

warfare among nation-states. The nation-states can rage a cyberwar alongside the real war, bringing down the critical infrastructure of the opponent country. The nuclear facility attack is another crucial example of how physical military assault can be taken to the cyber world and wage cyber warfare. The risk with such cyberattacks is, in most cases, that it is hard to haunt the real criminals behind such attacks if they are not in the same territory.

Apart from DDoS and bot malware attacks, ransomware is a growing concern for smart cities adapting to Industry 5.0 standards. Ransomware attacks start with malware that encrypts the data of a victim. Then, the attackers demand money in exchange for the decryption key, thus making the system unusable for authorized users for an unlimited time (Hamid et al., 2019). These attacks can disrupt critical infrastructure and services and cause the loss of sensitive data to a great extent. According to the Cloudwards report (Aleksandar, 2020), an organization is exposed to ransomware attacks every 14 seconds. The report published in (Lai, n.d.) predicts that ransomware will cost the global economy $265 billion by 2031. Thus, it is essential to understand the ransomware threat in the context of smart cities and develop effective strategies to prevent and mitigate such attacks.

Moreover, the recent advancement of the ransomware-as-a-service (RaaS) model has intensified the threats, as anyone can launch a ransomware attack with little technical skill and system knowledge. RaaS models adopt a workflow similar to the software-as-a-service (SaaS) model. In the process, already-developed ransomware is rented out to interested parties via the dark web. The interested parties then launch targeted ransomware attacks using the rented toolkit to achieve their objectives. The attackers, in this case, do not need any technical knowledge, as it was previously required. As a result, any novice hacker can launch highly sophisticated ransomware attacks. Smart cities are easy targets for ransomware attacks because of the control and technologies they use. Figure 5.1 depicts a ransomware scenario on smart buildings via a novice hacker.

Despite ransomware attacks being a substantial threat to smart cities, little work has been done in this context (Mello Jr., 2022). To this end, this chapter designs a ransomware threat analysis framework for smart cities to shed some light on vulnerabilities and risks. This framework considers various facets, including attack surfaces, threat actors, attack vectors, comprehensive attack analysis, and mitigation strategies. This chapter discusses the security threats in smart cities and how Industry 5.0 can intensify those threats. This chapter also reviews some related work in the context of mitigating the impacts of ransomware attacks in smart cities. Lastly, it presents an extensive threat modeling framework to help identify and mitigate ransomware attacks.

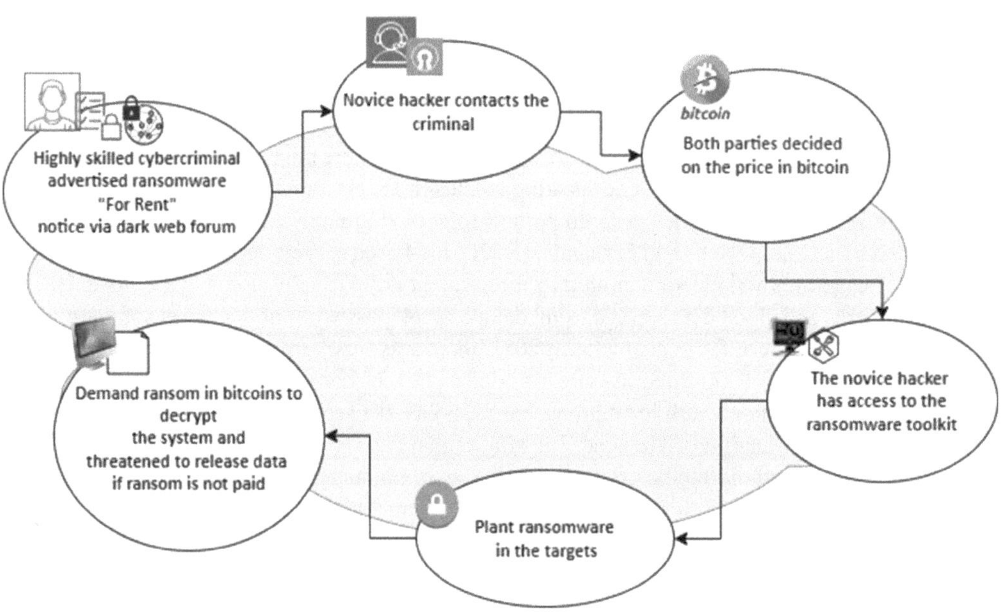

FIGURE 5.1 Ransomware as a Service Model

5.2 SECURITY THREATS IN SMART CITIES

Although there are several definitions of smart cities, they all comprise people, processes, and technologies (Albino, Berardi and Dangelico, 2015). From a security perspective, all these entities can pose different security threats and lead to varying cyberattacks on smart cities. Smart cities have different domains that might be more prone to some attacks than others. The most prominent domains are governance, people, economy, mobility, and environment (Ismagilova et al., 2020). Smart cities rely on complex and comprehensive technological advancements. They use interoperable systems where information is exchanged and control and command run frequently. These interoperable systems can have numerous vulnerabilities if not properly secured (Brooks, 2021). Some examples of such vulnerabilities are access control mechanisms in networked devices, communication protocols, and sensors for monitoring and managing city operations (Kitchin and Dodge, 2017).

Moreover, smart city components, such as sensors, actuators, and control systems, are physically distributed throughout the urban environment. Intelligent systems integrated into smart cities use industrial control systems (ICSs) and supervisory control and data acquisition (SCADA) systems, which are focal points of cyberattacks. Malicious actors can target physical assets associated with these systems, disrupting essential services and compromising public safety (Kitchin and Dodge, 2017). Attacks on relevant critical infrastructures built on these assets may include DDoS attacks, ransomware attacks, data breaches, and unauthorized access to control systems. Like any other system, smart city systems and devices are susceptible to malware that can infect networks and compromise their integrity and functionality. Malicious software can exploit vulnerabilities in the system, compromise data integrity, and disrupt services (Hamid et al., 2019). Insiders with authorized access to intelligent systems pose security risks similar to those of any other organization or system. This includes employees, contractors, or service providers who may misuse their privileges or engage in malicious activities that compromise the integrity and security of the city's infrastructure (Hamid et al., 2019). All these attacks can be directed at critical infrastructures, which are the core services of smart cities. For example, a cyberattack in January 2021 on a water treatment facility in Oldsmar, Florida, allowed the hackers to alter the chemical level of the water supply via a remote-controlled computer (Robles and Perlroth, 2021). In 2018, the city life of Atlanta came to a halt due to a ransomware attack on the IT infrastructure of the city, which affected many smart city domains such as bill payment and policing (McDonald, 2022). Some other recent attacks on critical infrastructure in smart cities are listed in Table 5.1.

The smart city consists of different layers (Tekinerdogan, Köksal, and Çelik, 2023). Various research studies have described smart city layers slightly differently. This work adapts the architecture stated in (Kerner, 2022) to analyze security threats in each layer. Table 5.2 depicts some prominent threats in each layer.

It is apparent that each layer has some common attacks; for example, each layer can be exposed to malware and ransomware attacks. Ransomware attacks are malware attacks that can affect different domains of smart cities. The technology layer is one of the critical building blocks of smart cities since interconnected IoT sensor devices are present in this layer. These sensors can be considered the heart of smart cities. The application layer has the core responsibility for data storage and processing. It is a significant target for cybercriminals since they are looking for data on various entities related to smart cities (Brooks, 2021). However, the attackers can aim for inter-layer attacks. For example, attacks on networking technologies that are part of the technology layer can flow to the application layer through data manipulation, unauthorized surveillance, or even sabotage of critical infrastructure (Kitchin and Dodge, 2017). The bulk of the vast amount of collected data involves personal data. Improper handling of personal information, inadequate anonymization techniques, or unauthorized access to citizen data can erode public trust and lead to privacy violations (Ismagilova et al., 2020). IoT and 5G technologies are critical players in the technology layer. According to the 2020 National Security Agency (NSA) report, 5G technologies will open up new attack vectors for critical infrastructure in smart cities.[2]

Similarly, the Five Eyes security alliance has warned about the security risks posed by the widespread use of 5G technology in smart cities.[3] As a result, the US Cybersecurity and Infrastructure

TABLE 5.1
Cyberattacks in Smart Cities: Some Real-World Incidents

Smart City Domain	Attack Details
Smart governance	• In 2014, the city of Ferguson, Missouri, was hit with a cyberattack that brought down its police website and email network (Buntz, 2018). • Baltimore, May 2019: Some 10,000 local government computers were inaccessible for weeks due to a ransomware attack (McDonald, 2022). • Atlanta, 2018: ransomware attacks caused it to shut down its municipal courts (McDonald, 2022).
Smart grid	• The attack on the Ukrainian electrical grid in December 2015 led to power outages for about 225,000 consumers (McDonald, 2022). • The US Computer Emergency Readiness Team reported a range of cyberattacks targeting energy networks in 2018 (Mohammadpourfard et al., 2021).
Smart water	• January 2021 on a water treatment facility in Oldsmar, Florida (Robles and Perlroth, 2021). • December 2023. Ireland: The hackers interrupted a water utility and cut off the water supply for two days (Kovacs, 2023). • US, 2023: Hackers targeted around ten water facilities in different parts of the US (Lyngaas, 2023).
Smart health	• Ransomware attack on Brno University Hospital (Porter, 2020). • Interpol's warning to health agencies worldwide about a significant rise in targeted ransomware attacks (He et al., 2021).
Smart home	• In November 2021, cybercriminals turned off the heating in two buildings and then launched a DDoS attack against the heating system (Vähäkainu et al., 2022).
Smart industry	• May 2021, Colonial Pipeline attack in the US – attackers got into the system through an exposed password for a VPN account (Kerner, 2022).

Security Agency (CISA) has developed a new set of guidelines around 5G technology usage in smart cities.[4] The UK's National Cyber Security Centre has also raised similar concerns about 5G technology usage in smart cities. For instance, they showed concern about attacks on autonomous vehicle systems and unauthorized and targeted use of data to disadvantage individuals.[5] The Australian Signals Directorate (ASD) has also expressed concerns similar to those of other centers (Budde, 2023). According to statistics published by Statista, the number of IoT devices worldwide will almost double from 15.1 billion in 2020 to more than 29 billion IoT devices in 2030 (Vailshery, 2023). One primary concern for these IoT devices is the nature of the web, which interconnects many devices and can lead to cascading attacks. So, for example, if a hacker hacks into a smart meter, it can laterally move to the whole system and disrupt the water management service. It can flow to other emergency services, such as energy systems.

5.3 RANSOMWARE THREATS IN SMART CITIES

Ransomware is malware that locks up or encrypts files/devices of users, making them inaccessible for legitimate use (Brewer, 2016). After locking up the files or devices, the cybercriminal asks the user or organization to pay a certain amount of money. They also threaten to release critical or sensitive information if the demanded ransom is not paid. Thus, ransomware can be coined as cyber blackmail. The flow of ransomware is depicted in Figure 5.2.

There are different types of ransomware attacks. Table 5.3 describes these types, their potential starting points, and entry layers.

Ransomware attacks can cause severe financial and mental havoc for individuals and organizations (Beaman et al., 2021). These attacks are an imperative concern, especially for smart city dwellers, where critical infrastructure and services rely heavily on interconnected systems and technology (Lamers et al., 2023). Despite implementing cybersecurity measures, ransomware remains a significant threat (Maigida et al., 2019). There are many notable ransomware attacks on critical

TABLE 5.2
Security Threats in Different Layers of Smart Cities

Smart City Layer	Layer Description	Security Threats and Attacks	Threat Description	Impacts
Technology layer	The technology layer comprises sensors, smartphones and all network and communication technologies.	1. Distributed denial-of-service attack (DDoS) 2. Malware injections, including ransomware 3. Device hijacking and spoofing 4. Man in the middle attack 5. Network sniffing 6. Replay attacks 7. Backdoors and malicious firmware 8. Credential-harvesting attacks 9. Physical attacks on devices and networks	The technology layer bears the burden of threats from all types of cyber and physical attacks that those technologies are prone to. For example, IoT devices are prone to spoofing attacks, which will cascade to smart city technology layers as the cities heavily depend on IoT sensors. 5G technologies are another crucial player in the technology layer, and they will open doors for many attacks.	Attacks on technology layers can result in intelligent system service disruptions (availability). Threat actors can temper data and steal credentials using man-in-the-middle and backdoor attacks, thus impacting integrity and confidentiality.
Application Layer	The application layer comprises applications built on data collected by sensors and smartphones. The applications turn that data into alerts, insights and valuable information for users to take action.	1. Data theft 2. Distributed denial-of-service attack (DDoS) 3. Phishing attacks 4. Malware attacks, including ransomware 5. Code injections	Attacks in the application layer can come in various ways. The threat actors can launch phishing, malware, and code injection attacks.	Attacks on the application layer will severely impact confidentiality via data theft, as data can include personally identifiable data (PII), which can lead to many consequences.
Service Layer	The service or usage layer is where the citizens and all stakeholders come together and use those applications to get different smart services, such as healthcare, transport and education.	1. Service disruption 2. Social engineering attacks 3. Malware attacks, including ransomware 4. AI-related deepfake attacks	The service layers can face different attacks as they deal with usage. Users are heavily involved in this layer, and users can pose different threats by falling for social engineering tactics. AI-related deepfakes can be a severe threat to this layer. Malicious threat actors can pose as authenticated persons and jeopardize services.	Attacks in this layer can lead to service disruption (availability), confidentiality, and integrity.

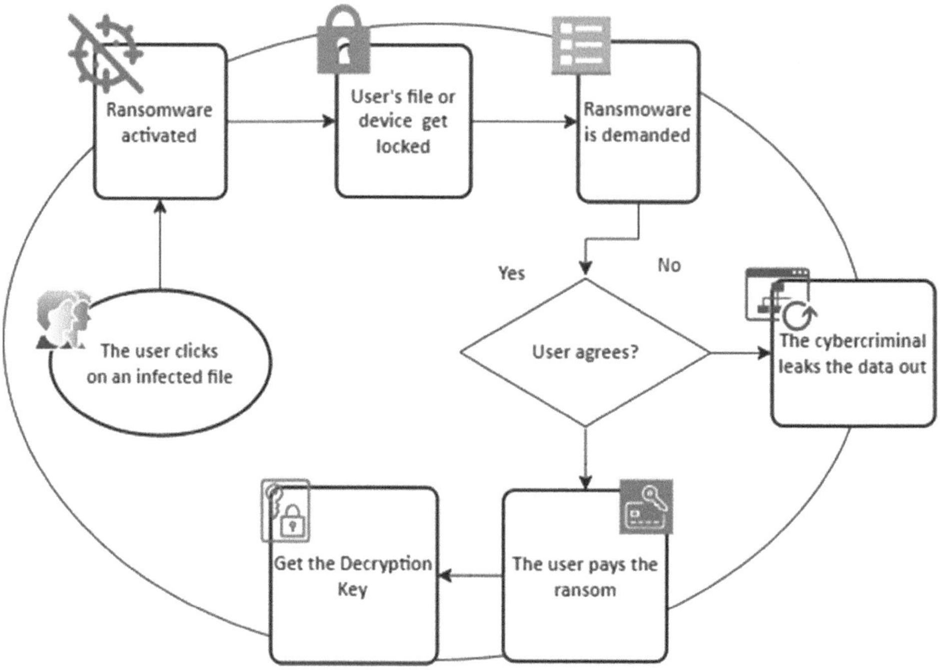

FIGURE 5.2 Ransomware Flow

infrastructure and corporate networks. One of the famous ransomware attacks that has affected the world is Wannacry (Algarni, 2020). The hackers took advantage of a leaked Microsoft Windows hack, EthernBlue, developed by the National Security Agency (NSA). The hacker group "Shadow Brokers" published the code. It affected around 230,000 computers in 150 countries. The attack spread through backdoors. As manifested in the Wannacry attack, the consequences of these disruptions can be significant. The cascading effects can cause severe economic and public safety issues. Moreover, the well-being of individuals who rely on critical infrastructure services can be impacted. For instance, the Colonial Pipeline attack in 2021 resulted in a $4 million ransom payment, and the broader impact on the US East Coast was significant (Plumb, 2023). Other notable ransomware attacks are attacks on the Costa Rican Government in April 2022, the New Mexico Government in January 2022, and the Maryland Department of Health in December 2021. Interested readers can find more in Brooks (2021). Some prominent ransomware attacks have taken place via supply chains. For example, supply chain attacks in 2023 that ended in ransomware attacks are the Norton supply chain attack, the Airbus supply chain attack, the University of California San Francisco (UCSF) Supply Chain Attack, the Microsoft supply chain attack, and the Colonial pipeline attack (Sidhu 2023;Fernando, 2024).

In the case of the UCSF attack, the attackers made the electronic health record system of the hospitals unavailable for several days. This led to surgery cancellation or postponement as clinicians could access patient records. The attack took place by exploiting third-party software, Codecov, a platform for testing software vulnerabilities. Another example is the attack on Airbus, which resulted in a data breach. The attackers accessed personal information associated with over 3,000 Airbus vendors, such as Rockwell Collins and Thales Group. The attack happened through a compromised employee account at Turkish Airlines. As of Norton's attack, the attackers entered Norton's network and stole the personal information of employees. The attackers threatened to release the stolen data if Norton did not pay a ransom. It happened due to a zero-day vulnerability in third-party software MOVEit, which is responsible for file transfer and management. In the case of the Colonial Pipeline attack, the attackers entered its network and encrypted its systems. As a result, the pipeline operation stopped. The incident happened due to a third-party software vulnerability, remote code

TABLE 5.3
Different Types of Ransomware, Starting Points, and Entry Layers

Ransomware Attack	Description	Starting Point	Entry Layer
Crypto-ransomware attack	The ransomware encrypts files and data from any system and needs a decryption key to access files.	The attack can start with phishing and other social engineering attack vectors. For example, a user downloads an attachment or clicks a malicious URL received via emails, social media apps, or SMS. The attack can start with any hacked organization or user account. The attackers intrude on the system with hacked accounts and encrypt some files. Privilege misuse: Insider employees can plant the ransomware in the machines using privileged accounts Backdoors: Some malware is implanted via firmware or software, which helps to implant this type of ransomware.	Technology, application, and service layer
Lockers attack	The lockers lock the user screen and show a ransomware message. The whole system becomes inaccessible.	1. The ransomware can be planted via social engineering attacks. 2. Compromised accounts. 3. Privilege misuse 4. Backdoors	Technology, application, and service layer
Scareware attacks	This type of ransomware tries to generate fear in users by saying their computer/system has been infected with viruses, and they direct the users to pay money to solve the issue. Sometimes, the system becomes unstable due to loads of popups.	1. Drive-by download: The user might download some freebies from the web containing malware. 2. Social engineering 3. Compromised accounts	Technology, application, and service Layer
Doxware attacks	This type of ransomware is a type of blackmailing ransomware. It asks users to pay a certain amount of money, or otherwise, they are threatened with leaking their information.	1. Social engineering 2. Compromised accounts 3. Privilege misuse 4. Drive-by downloads 5. DNS tunneling attack	Technology, application, and service layer
Ransomware-as-a-service (RaaS) attacks	Ransomware-as-a-service is a cybercrime business model where it sells or rents ransomware to interested parties.	1. Social engineering 2. Privilege misuse	Application and service layer

execution, in PulseConnect Secure, a VPN software program. In the case of the Microsoft attack, the attackers accessed Jfrog Artifactory using its exploited vulnerability. They injected malicious code into some of Microsoft's software components. They accessed Microsoft's networks via that software and stole source code and other confidential information.

As smart cities transition from Industry 4.0 to Industry 5.0 standards, there is potential for new cyberattacks as there will be more integration among people, systems, and machines (Czeczot et al., 2023). The leading technologies behind Industry 5.0 are AI, machine learning (ML), robotics, blockchain, big data analytics, IoT, and VR. As a result, the vulnerability and challenges of these technologies will impact cybersecurity in smart cities, along with human threats. For example, attacks on interconnected supply chains or service providers can lead to disruptions across multiple sectors, expanding the impacts (Lamers et al., 2023). Since ransomware attacks are responsible for encrypting the systems illegally, they can have domino effects on various sectors, such as energy, manufacturing, and utilities, affecting the delivery of essential services. Here, a few potential ransomware scenarios are presented to open avenues for more analysis and to understand the intensity of such cyberattacks and their impacts:

- *Ransomware Attacks via AI:* Cyberattacks using AI technologies are slowly gaining momentum. Some of the more common attacks using AI these days are deepfake audio and videos. In recent attacks, using deep voice cloning technologies, the attacker targeted a bank in Hong Kong and an energy company in the UK (Coker, 2022). Ransomware attacks via AI technologies such as generative AI are still in the research stage. For example, the work of Teichmann (2023) showed a step-by-step process where criminals could use generative AI to plan and implement ransomware attacks. AI and ML technologies can be leveraged to generate more sophisticated phishing emails and gather social media profile data more profoundly and swiftly to launch spear phishing attacks. Deepfake-generated videos and voice calls can play a significant role in this context. Someone can pose as an IT support or accounts person and trick people into giving away certificates. Then, using those credentials, they can access the system, steal important data, and plot ransomware. Figure 5.3 depicts such an attack in the context of a healthcare system that adopted the Industry 5.0 standard.

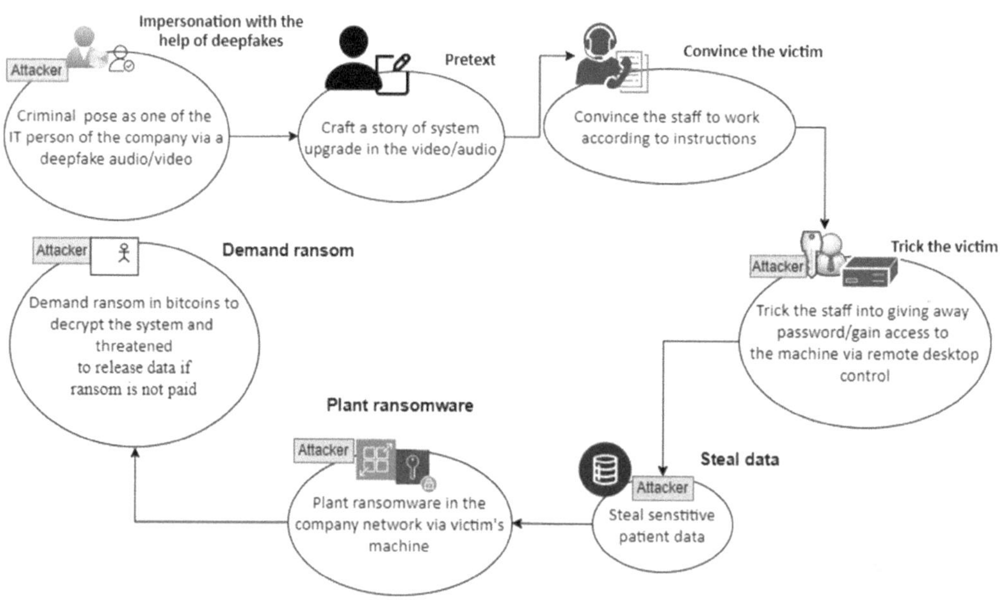

FIGURE 5.3 AI Attacks in an Organization

- *Ransomware Attacks via IoT:* IoT devices are the heart of smart cities. However, in the Industry 5.0 scenario, it would be more integrated into data analytics and deliver personalized experiences to city dwellers. As a result, spoofing one device can have a devastating effect on the entire city. Apart from being directly attacked by ransomware, IoT devices can be catalysts for ransomware attacks in several other ways. They can work as entry points to let malware enter the whole network and system. A spoofed or hacked IoT device can also let the hackers enter the system for lateral movement.
- *Ransomware Attacks via Virtual Reality:* Ransomware attacks via VR technologies are still in the proof-of-concept stage (Farid et al., 2023). This type of attack in VR is possible by tricking users into giving away personal information and by threat actors sabotaging the platform before asking for ransomware.
- *Ransomware Attacks via Robotics:* Since Industry 5.0 will open myriad avenues of human-robot collaboration, researchers are paying great attention to ransomware attacks via robots (Yaacoub et al., 2021). Some attack vectors in robotics could be fake applications developed by third-party vendors, insecure networks, and data storage facilities.

5.4 RELATED WORK

In recent days, ransomware attacks in smart cities have gained significant attention from researchers in industry and academia due to increased attacks in such settings. As a result, we are witnessing a surge in published survey papers stating the risks and impacts of such attacks. For example, the authors Beaman et al. (2021) analyzed the rise in ransomware attacks that occurred during the COVID-19 pandemic. After examining the current prevention and detection methods, they concluded that there is a need for a better understanding of ransomware attack vectors. This study also highlighted the lack of tailored research to address the unique challenges of these attacks in particular sectors such as healthcare, finance, and government. Tatam et al. (2021) explored the limitations, strengths, and potential improvements of current threat modeling methodologies, particularly concerning Advanced Persistent Threats (APTs). Their conclusion suggested developing comprehensive frameworks to recognize and mitigate potential threat scenarios. Some of their valuable insights have inspired the direction of this chapter. Along this line, some researchers recommended a honeypot triangle model to reduce the risk of future ransomware attacks (Waizel, 2022). Bajpai and Enbody (2020) pointed out the constraints of ransomware in smart cities and presented a ransomware risk equation. Many researchers have also proposed ML-based solutions to mitigate cybersecurity threats in smart cities, including ransomware attacks (Lai et al., 2024; Ullah et al., 2020). Deep learning-based solutions have also been proposed to mitigate such attacks (Chen et al., 2021). Although many projects have been proposed in the context of smart cities and cyberattacks, ransomware remains a significant threat. Hence, more strategies and system analyses are required to counteract such attacks (Rehman et al., 2023).

5.5 RANSOMWARE THREAT MODELING

Threat modeling is essential to mitigate ransomware attacks in Industry 5.0, as it has more avenues for human-machine and system interactions. This section designs a comprehensive threat modeling framework for ransomware attacks, using a smart water management system (SWMS) as an example. The steps of the proposed threat modeling framework are detailed in Figure 5.4.

5.5.1 Define Scope

A smart city ecosystem consists of various domains. The most prominent domains discussed in the literature are healthcare, governance, environment, transportation, energy, security and safety, critical infrastructure, and education. Figure 5.5 identifies the primary domains of this ecosystem.

These domains primarily rely on machine-to-machine (M2M) communication and data exchange between sensors and IoT devices. Hence, they have similar critical assets in terms of infrastructure, network, and data-sharing perspectives:

- *Operational Technology (OT) Systems:* These include ICSs, SCADA systems, and other OT devices used in critical infrastructure sectors, such as energy, water, transportation, and manufacturing.
- *Information Technology (IT) Systems:* IT infrastructure, including servers, databases, and network devices, supports critical operations.
- *Communication Networks:* IoT, sensor devices, and M2M communications are core services of any smart city that are possible through communication networks. Communication networks are the key to running smart transportation systems, public safety, and emergency services. Smart utilities connect data and command centers, collecting, processing, and analyzing data to make informed decisions and manage smart city operations. The communication networks in smart cities involve cellular networks (e.g., 4G, 5G), Wi-Fi, Bluetooth, Fiber Optic Networks, etc.
- *Control Systems and Sensors:* Smart cities heavily rely on sensors and control systems embedded in infrastructure components, such as smart grids, traffic management systems, environmental monitoring systems, smart healthcare systems, and building automation systems, which exchange information via communication networks.
- *Data and Information Assets:* Critical infrastructure organizations collect and store large amounts of sensitive data, including customer information, operational data, and intellectual property.
- *Physical Infrastructure:* Physical infrastructure components in a smart city include power generation facilities, transportation systems, hospitals, and water treatment plants.

The threat modeling framework is designed using a SWMS as an example. The following sections expand on each step described in Figure 5.4.

5.5.2 Identify Assets

An asset is a resource with a value attached to it. Assets can be in tangible and intangible forms. For this threat model, both forms of assets are first identified to single out entry points for attackers. SWMS comprises physical infrastructure, applications and services, and network and communication technologies. The critical components (Bello et al., 2023; Addeen et al., 2021) of such systems are as follows:

- *Physical Infrastructure:* The physical infrastructure primarily consists of pipelines, pumps, valves, and reservoirs. Treatment plants, which facilitate the supply and distribution of water throughout the city, are also part of this infrastructure. Water storage facilities

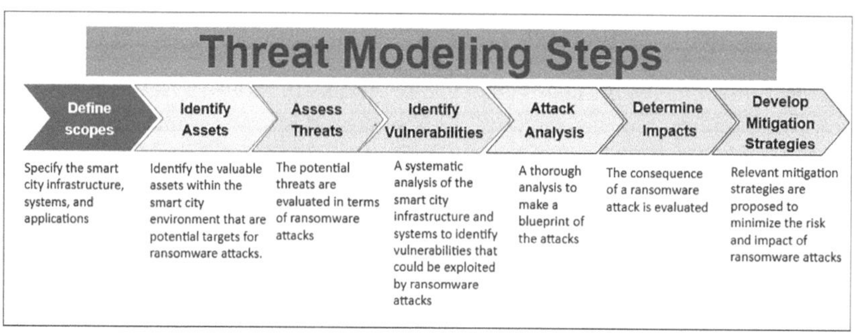

FIGURE 5.4 Adapted Threat Modeling Steps

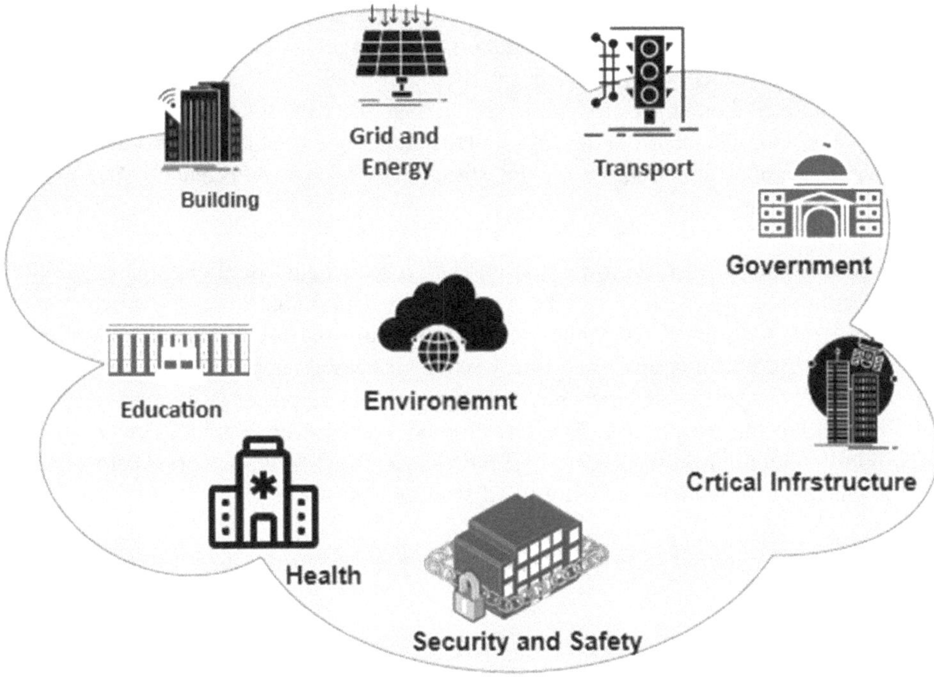

FIGURE 5.5 The Smart City Domains

have water towers or tanks to maintain supply and manage peak demand. The sensor networks are distributed throughout the physical infrastructure to collect real-time water flow, pressure, quality, and leakage data. This sensor network includes smart water meters, flow and pressure sensors, and water quality sensors. The SCADA system comprises remote telemetry units (RTUs) and programmable logic controllers (PLCs). RTUs are deployed throughout the water distribution network to collect sensor data, monitor pressure flow rates, and detect abnormalities or leaks in the system from sensors deployed throughout the water distribution systems. This information is then transmitted to the central SCADA software system for analysis and control.

- *Applications and Services:* SWMSs use different applications and services to collect, process, and analyze data from sensors to monitor the performance, efficiency, and quality of the water distribution network. There are different applications, such as water consumption tracking applications, leak detection and management applications, customer service portals, and integration services. Leak detection applications use advanced algorithms to detect leak locations. The customer portal usually allows customers to access information on various water usages and sometimes gives them a payment portal. Since all systems in a smart city are interconnected, the integration points with other intelligent systems constitute a significant point while designing a threat modeling framework. Water billing and customer information systems can be a primary target for ransomware attacks since these systems contain sensitive information.
- *Networks:* SWMSs use wired, wireless, cellular, and dedicated communication infrastructure. These networks facilitate data transfer and communication among devices, sensors, applications, and the central system. All these network points are entry points for cyberattacks and ransomware attacks.
- *Data Flows:* Data flows are a crucial part of threat modeling. Data flow operations include data collection, storage, processing, and sharing. Data is a significant target of ransomware attacks. Data collection with SWMSs involves gathering information from communication networks, sensors, and monitoring devices. Data storage involves databases or warehouses

that store water consumption, operation logs, sensor readings, customer information, and various other real-time and historical data. Data storage can be a primary target of ransomware attacks since it may contain the personally identifiable data (PII) of customers, financial data, and other sensitive information. Data processing uses advanced algorithms to derive various insights from the data, and the data might be shared with other intelligent systems via application programming interfaces (APIs). These APIs can bring vulnerabilities to the systems as well.

Figure 5.6 shows a SWMS's Dataflow Diagram (DFD) model and highlights the potential threat vectors and vulnerable assets. It recognizes that the threat actors can use several entry points to attack the systems, such as SCADA and intelligent pumps, and use human vulnerabilities. The threat actors can use phishing attacks, a threat vector discussed later, to exploit customers, water suppliers, regulatory agencies or other intelligent system service providers to give away their credentials by clicking the wrong link. The threat actors can also exploit SACDA or smart pump vulnerabilities to compromise the system credentials and launch attacks. Threat actors can launch attacks via sensors, communication, and water distribution networks.

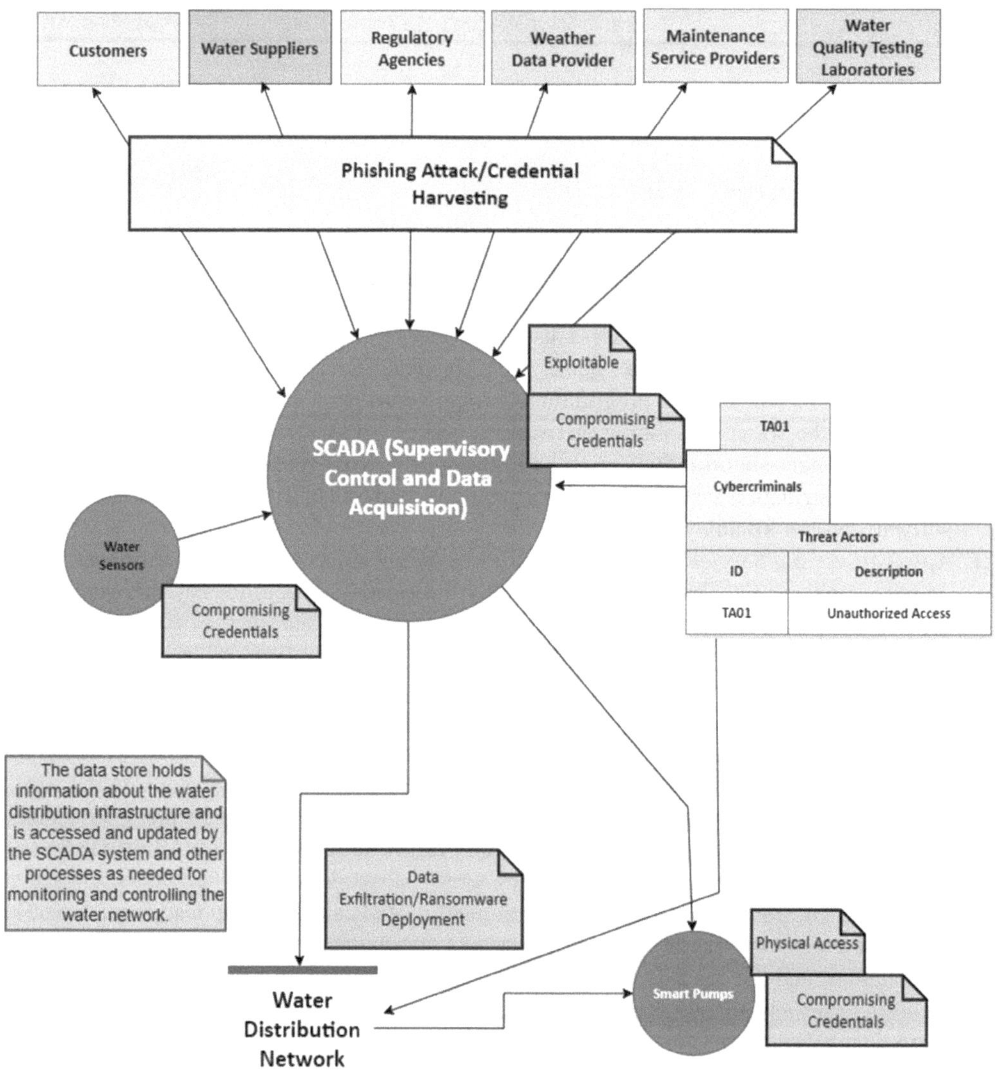

FIGURE 5.6 DFD for Smart Water Management System

5.5.3 Identify Stakeholders

Stakeholders in every system need access to crucial data and system elements, and they are lucrative targets for cyberattacks. Identifying, informing, and safeguarding these stakeholders from becoming such targets is essential. Citizens as service receptors, third-party vendors, government organizations, and service providers are stakeholders in any intelligent system domain. Each of these parties has the potential to launch and be the target of ransomware attacks. For instance, a ransomware attack disrupting services and jeopardizing the integrity of infrastructure components like pumps, valves, or treatment facilities may have severe repercussions for service providers handling the operation of the water management system, necessitating expensive repairs or replacements. A ransomware attack on the water management system could affect the regular environmental protection, public health, and water resource management operations of government organizations. A ransomware attack on the water management system immediately affects customers and the relevant general public, including residential and commercial water users. Water scarcity or quality problems and health and safety hazards, including the spread of waterborne illnesses or pollution accidents, could cause disruptions. The provision of hardware, software, or support services to the SWMS by third-party vendors may result in product or service delivery delays, contract violations, monetary losses, and reputational harm that could negatively impact the relationships of vendors with other smart city stakeholders.

5.5.4 Identify Potential Threat Actors and Attack Vectors

The next step is to identify potential threat actors and attack vectors.

5.5.4.1 Identify Threat Actors

A threat actor is an entity that has the potential to impact the security of an organization (Chapple et al., 2018). Threat actors exploit computer systems, networks, and software vulnerabilities to launch various cyberattacks. Identifying threat actors is an essential step of threat modeling since it helps the relevant stakeholders calculate risks. Threat actors in SWMS include cybercriminals, hacktivists, insiders, and nation-state actors. Their motivations can be indifferent based on context. These motivations may include financial and ideological gain, emotions, political upper hand, winning cyber warfare, showing off technical capabilities, and having fun. These actors might use varying tools and techniques based on their environmental, technical, and financial capabilities.

Cybercriminals in this context might be solely motivated by monetary gains. If it is an organized criminal group, it might use advanced capabilities to infiltrate the water management system and launch the final ransomware attack. If it is a lone wolf or small group, they might use off-the-shelf ransomware tools available on the dark web to launch such attacks. Regardless of their group size or capabilities, these criminals can exploit flaws in the network of water management systems, endpoints, or software to obtain unauthorized access. The other major players can be hacktivists motivated by their ideology and activism. Hacktivists might not like any activities by governments or any party. Thus, it can raise awareness via ransomware attacks on SWMS. However, they might end up not asking for ransom but instead asking for the end of any aggression, injustice, or activity they might disagree with.

On the other hand, insider threats can arise for both malicious and non-malicious undertakings. Insider actors include current, former, and third-party contractors. They pose a significant threat as they know more about the systems and vulnerabilities of any organization. In non-malicious insider threats, employees can be the target of phishing or social engineering attacks, and they can give their credentials. The real threat agents will then use these credentials to explore the water management system and launch a ransomware attack. Malicious insiders may target their own organization for personal grievances, financial incentives, or ideological reasons. Nation-state actors are another major player in cases of such attacks. The nations at war with each other may use critical infrastructure as a new warfront. The nation-state actors might have different motivations, such as winning the cyberwar, gaining geopolitical advantage, and gaining an upper hand in obtaining sensitive information. They can launch such attacks by employing APT groups and other advanced capabilities.

5.5.4.2 Identify Attack Vectors

An attack vector is a path or means by which an attacker or hacker can gain access to a computer or network to deliver a malicious code or outcome (Shacklett, 2021). Since ransomware is a type of malware, it can be an attack vector by itself. However, the threat agents can use other attack vectors to launch final ransomware attacks. To have a clear understanding of ransomware attack paths, attack vectors are divided into two types: first-degree and second-degree attack vectors. First-degree attack vectors are the ones that are launched to gain initial unauthorized access to the system. The second-degree attack vectors are launched to access systems and lateral movement further. Then, the ransomware attack is the final objective of these attack vectors. Figure 5.7 describes these attack vectors. According to Unit 42 Palo Alto's report,[6] the five most common first-degree ransomware attack vectors include social engineering tactics, exploitable software vulnerabilities, brute-force credential attacks, previously compromised credentials, and abuse of trust. Along with those five, a few other first-degree attack vectors are discussed as follows:

- *Social Engineering:* Attackers may use social engineering techniques to deceive different stakeholders into clicking on links or downloading attachments and giving in sensitive information, such as login credentials or system details, over the phone or through other technologies. These links or attachments could contain ransomware payloads. Social engineering attack techniques may include phone conversations and chats via social media. They can even include deepfake videos and audio conversations to trick employees and customers about water management systems. Threat actors may compromise legitimate water utility websites or create counterfeit websites that appear authentic. Users visiting these websites could be exposed to ransomware through drive-by downloads or deceptive links. Threat actors may send phishing emails or messages claiming to be authenticated water utility notifications regarding billing issues, service disruptions, or payment requests. The employees might get emails regarding fake software or system upgrades. These messages trick recipients into clicking on malicious links or downloading infected attachments, leading to ransomware infections. Attackers may craft phishing campaigns related to water quality concerns, urging individuals to click on links for more information. These links could direct users to websites hosting ransomware or exploit kits. Ryuk ransomware is an example that is commonly spread via spam emails. Ryuk uses TrickBots to prepare the machines for implanting ransomware. Trickbots are the first to infect a machine and

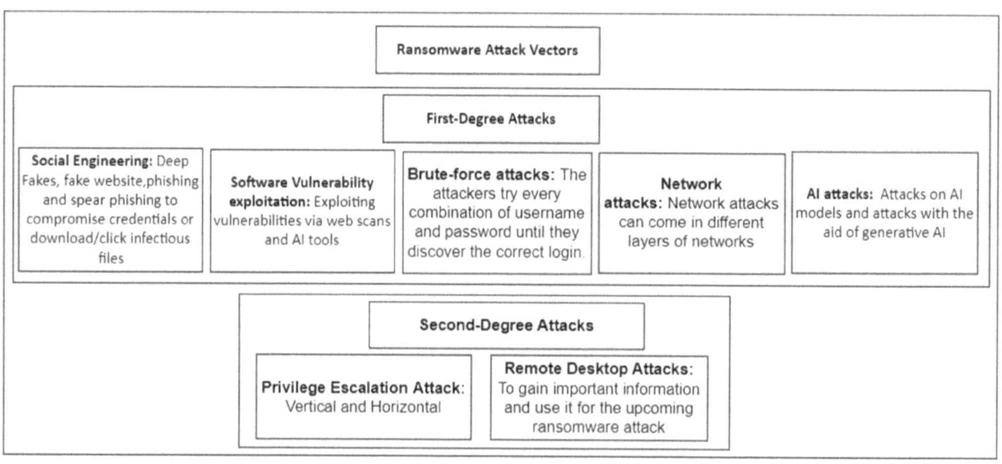

FIGURE 5.7 Ransomware Attack Vectors

steal sensitive information. Then, it starts infecting other network machines, and attackers deliver the ransomware via infected machines. Insufficient security awareness among employees and system users are two catalysts for successful social engineering attacks. For instance, without proper training, employees can easily fall for social engineering tricks such as clicking on malicious links, opening infected email attachments, or giving away information on social media.

- *Software Vulnerability Exploitation:* Exploiting vulnerabilities via web scans and AI tools is one of the most common attack vectors before launching a ransomware attack. Palo Alto's incident report notes that around 48% of ransomware attackers used vulnerable software to launch attacks. For example, attackers can use SQL injection to launch a ransomware attack. One such attack was conducted by the C10p ransomware group, which used the transfer vulnerability of MOVEit to upload a web shell to the targeted server via SQL injection (Whitefield-Madrano, 2023). Some common software vulnerabilities are outdated software and firmware, misconfigured systems, weak authentication, and unencrypted connections. IoT devices are also prone to such vulnerabilities. Sometimes, IoT devices have hard-coded passwords and can transfer data without encryption. Sometimes, the user's data is exposed via unsecured cloud storage. The attacker might find compromised credentials and use them to intrude on an IoT device, then further advance by gaining access to more devices and later infecting the whole network via ransomware. For example, a smart meter with an easily guessable or weak password can be intruded. Then, the intruder can move laterally to the whole network via this one device and infect the entire network by implanting ransomware in the SWMS. Supply chain risks arising from third-party software can also be considered vulnerabilities. For example, the attacker pretending to be a third-party vendor might launch an application for a "Water Saving Campaign" that is to check water leakage and consumption by customers and provide information on fixing and saving. The users may download this app, and the intruders may enter the device through it, steal sensitive information, and then laterally move to SWMS-related systems to infect them with ransomware.
- *Brute-Force Attacks:* The attacker launches brute-force attacks by trying every combination of username and password until they discover the correct login. The hackers might use the passwords for networked devices and systems used in the SWMS to gain unauthorized access and install ransomware.
- *Network Attacks:* Network attacks can come from different layers of networks. For example, application layer attacks may include fake applications and websites and download malicious content on the client side. Since Industry 5.0 brings the notion of personalization, the attackers may create malicious versions of water-related applications, tools, or mobile apps by giving the users an illusion of personalization. Unsuspecting users who download and install these infected applications may unknowingly introduce ransomware into the SWMS. Modern software applications are heavily dependent on APIs, which might be a target for threat vectors. The other attacks might include device spoofing and man-in-the-middle attacks, which lead to ransomware attacks. For example, the attackers might use man-in-the-middle attacks to listen to the network channel and gather sensitive information such as credentials. Then, those credentials can be used to gain unauthorized access to the systems and plant ransomware.
- *AI Attacks:* Under AI attacks, both attacks on AI models and attacks with the aid of generative AI are covered. According to Google's Cyber Security Forecast report,[7] generative AI (genAI) and large language models will aid in phishing, SMS, and other social engineering tactics to bring more flavor of legitimacy and realness. Like in the RaaS model, Gen-AI has emerged as a service. These services open the door for attackers to rent out generative AI-based potential phishing and other attack tools to launch first-degree ransomware attack vectors. AI will also make cloud attacks more sophisticated, as vulnerabilities will be easier to find using AI tools. According to the NIST report (Vassilev et al.,

2024), the most well-studied attacks in adversarial ML are evasion, poisoning, and privacy attacks, and abuse attacks are an added factor in the case of genAI. Since SWMS uses ML models to analyze data and make decisions, they are open to all these attacks. In case of a poisoning attack, the ML models in SWMS can be trained with poisonous data, which can affect the water quality.

Some of the second-degree attacks are as follows:

- *Remote Desktop Attack:* A remote desktop attack occurs when an attacker uses compromised credentials to access the system illegally. The attackers can gain access to SWMS via compromised credentials and plant ransomware.
- *Privilege Escalation Attacks:* The attackers might use the first degree of attack vectors, such as social engineering and phishing, to gain privilege escalation attacks. These attacks can take both horizontal and vertical forms. So, the attackers can use the compromised account to move into the system to install ransomware. They can also use other attack techniques, such as buffer overflow attacks, exploit vulnerabilities in the system to gain higher access, and take more information from the system before finally implanting the ransomware.

5.5.5 Identify Vulnerabilities

As discussed in the attack vector section, software vulnerabilities are one of the vital ransomware attack vectors. In the case of SWMS, vulnerabilities can arise from both software and hardware. Third-party involvement can bring new vulnerabilities. The software and hardware provided by third-party vendors can have exploitable vulnerabilities and backdoors. Misconfigurations can cause system vulnerabilities. Some such vulnerabilities are poorly configured network devices and sensors with weak and easily guessable passwords and weak access control methods. Unpatched software and remote access to update systems can also introduce vulnerabilities. If third-party vendors can access the system remotely, the attacker can use compromised accounts, or malicious third-party employees can plant ransomware in the systems. Some automated tools should be used to identify vulnerabilities in the relevant systems and software.

5.5.6 Attack Analysis

This section analyses the ransomware attack in a SWMS using four phases. Figure 5.8 shows these four phases.

The first phase is to gain initial access, which can be obtained via phishing attacks by identifying and exploiting software, networks, IoT sensors, or other devices with network access. If a phishing attack is used, a malicious payload will be sent to the victim, and social engineering tricks will be employed to make the victim execute the malicious payload. After executing the payload, the attackers will get initial access to the targeted system. Figure 5.9 shows the detailed steps of this phase.

In the second phase, the attackers install backdoors in the system and create remote access. The backdoors can be accessed both in hardware and software. In this stage, the attackers basically prepare the settings for ransomware attacks. Backdoors can be installed in SWMS applications and water monitoring devices. Then, the attackers can gain further access to the system and steal

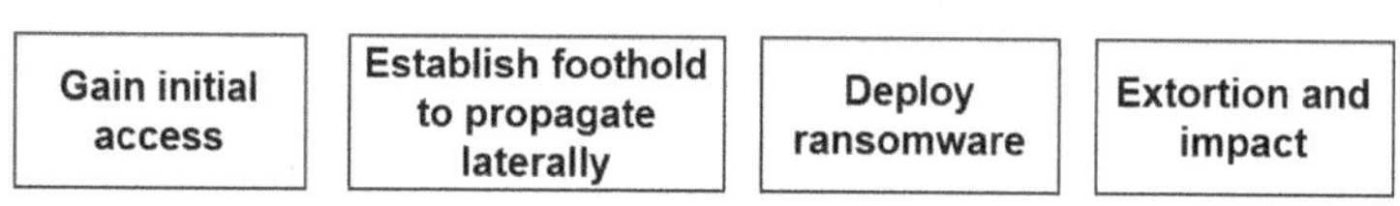

FIGURE 5.8 Four Phases of a Ransomware Attack

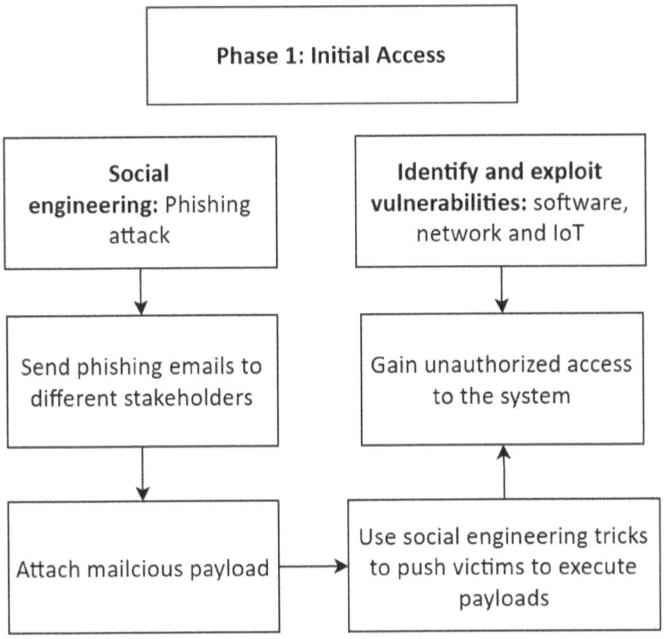

FIGURE 5.9 Initial Access

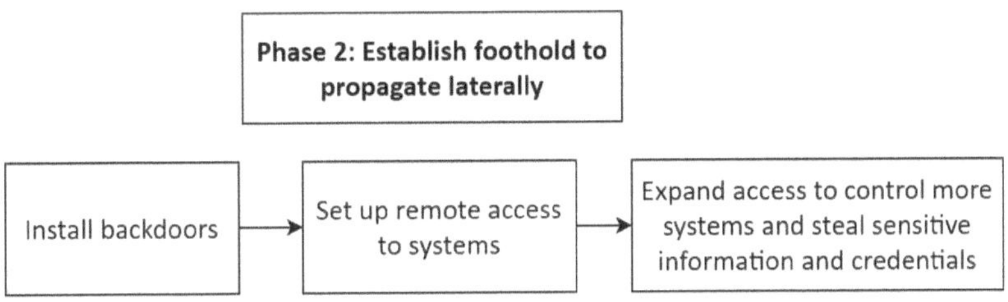

FIGURE 5.10 Establish Foothold

confidential data through different channels. They also use privilege escalation techniques by passing the standard login and getting access to more sensitive systems. Figure 5.10 shows these steps.

In phase 3, as depicted in Figure 5.11, the attackers identify critical systems with their strategies in Phase 2. They turn off backup systems to prevent easy recovery from the attack. Then, they deliver the ransomware payload, thus encrypting the systems and locking out legitimate users.

In the last phase, the attackers demand money and disrupt various operations. Here, three stages are listed before the attackers launch the final ransomware attacks. However, the attackers can avoid some steps and excel at the attack, depending on the context. For example, the attackers can even take over the system using only Phase 1 if the social engineering attacks become successful and they want to demand the ransom immediately. Figure 5.12 explains this phase.

After analyzing the four phases of ransomware attacks, the STRIDE (Jegeib, 2022) threat modeling framework is used to analyze some well-known threats that may finally end up in ransomware attacks. In STRIDE, all threats are mapped into one of the six general categories, and it is linked to the violated security attributes. The generalized categories are spoofing, tampering, repudiation, information disclosure, denial-of-service, and elevation of privilege. These security attributes are authentication, authorization, integrity, confidentiality, availability, and non-reputability. All STRIDE attacks are discussed concerning ransomware attacks on SWMS:

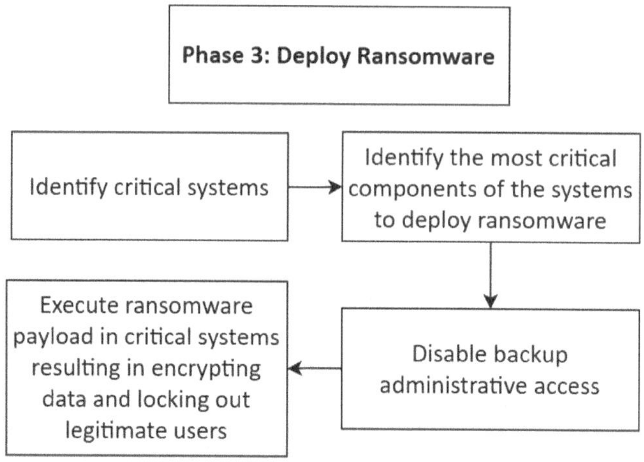

FIGURE 5.11 Deploy Ransomware

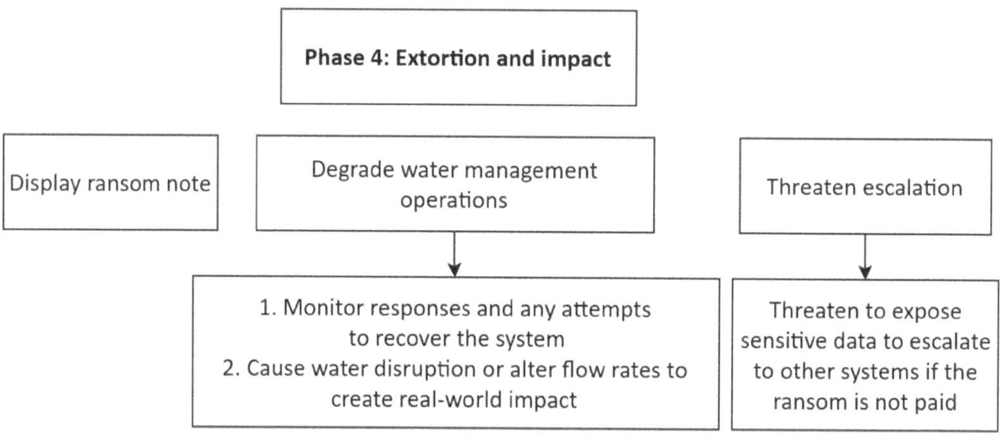

FIGURE 5.12 Extortion and Impact

- *Spoofing:* In spoofing attacks, the attacker steals the authentication token of a legitimate user and uses it to impersonate the user. The attacker uses the compromised credentials of a third-party vendor who manages the water billing system. Then, impersonating that vendor, the attacker moves laterally in the system to gather more information. The attackers steal sensitive data and implant ransomware to make the system unavailable. So, in this case, authenticity, confidentiality, and availability are violated.
- *Tampering:* Tampering attacks involve data modification by unauthorized parties. In the case of SWMS, the attackers can use phishing tactics or exploit vulnerabilities to gain unauthorized remote access to any device linked to a water treatment facility. Then, the chemical element levels of water are changed. After data poisoning, the attackers can run further havoc by implanting ransomware and demanding ransom; otherwise, the consumers will get the contaminated water. Integrity and availability are hampered in this case.
- *Repudiation:* In this case, the threat actor involved in the incident denies any role in the attack. Any malicious insider employee or third-party vendor can access the water management system and implant ransomware. Later, they can claim their account was hacked or accessed unauthorizedly by imaginative, malicious actors.

- *Information Disclosure:* Information disclosure involves publishing information to unauthorized persons. In the SWMS case, the attackers can obtain sensitive water quality information and other sensitive personal information and try to publish it if the ransomware is not paid. They can also publish partially and be threatened with publishing more if the ransomware is not paid. In this case, confidentiality and availability are affected.
- *Denial-of-Service (DoS):* This type of attack involves denying valid service to users. DoS is a network attack in which the attacker makes servers unstable by sending more traffic than they can handle. The attackers might want to run DoS attacks as a side or ransomware attacks to cause more havoc and fear among the stakeholders. In this case, availability is affected.
- *Elevation of Privilege:* An attacker uses phishing attacks or deepfakes targeting one of the water facilities to get the account details of random users. After getting information about those attacks, the attacker uses those credentials to log into systems and try to get account information for more privileged users. After getting higher-privilege account access, the attackers freely move into the system, collect data, and, in the final stage, plant ransomware.

5.5.7 Identify Impacts

Ransomware attacks in SWMS can have long-lasting impacts in different forms. Table 5.4 highlights these impacts, potential loss due to these impacts, impacted assets (if any), and impacted stakeholders.

5.5.8 Mitigation Strategies

ASD has published 16 risk mitigation strategies for smart cities in collaboration with other partner organizations.[8] These strategies focus on secure planning and design, identification and authentication management, vulnerability management, and cyber resilience. Each theme is expanded below with some examples relevant to ransomware attack analysis:

- *Secure planning and design:* These days, governments and industry alike are focusing significantly on security, privacy, and safety in the planning and designing stages of any product or service.[9] Any product and service should incorporate principles of security by design, privacy by design, and safety by design to ensure safe service delivery and protect organizations and individuals from cyberattacks. Some of the tenets of security by design principles are economy of mechanism, least privilege, separation of duties, which advocates for consensus decision-making, complete mediation, and least common mechanism. The economy of mechanism, for example, needs to be adopted to keep the design simple and complete. Mediation means that the utilization of each system/device should be checked at every entry point. These are particularly relevant to intelligent systems, which depend entirely on vulnerable connected sensors that can lead to cascade attacks. Implementing privacy by design in sensor-based systems is also crucial, as these sensors collect lots of data and are stored in the cloud. Privacy as the default is a critical principle that can be essential in mitigating the effect of ransomware attacks. One of the main tools ransomware attackers use is to threaten data leaks. If encryption is implemented correctly via privacy by default design principles, attackers would not have an easy upper hand. Maintaining supply chain privacy and security and the software supply chain are the other two critical areas for mitigating ransomware attacks and cyberattacks in general. Third parties can implement backdoors or supply compromised products if proper security and privacy are not maintained in these areas. Some mitigation strategies for supply chain attacks based on their patterns are depicted in Table 5.5.

TABLE 5.4
Impact Assessment

Areas of Impact	Potential Loss	Impacted Assets	Impacted Stakeholders
Water resource management	Water quality, water supply, service disruptions, and compromised item integrity	Physical infrastructure, applications and services, and networks	Citizens who depend on smart water systems, service providers
Environmental effect	Water quality and compromised devices in water resource management can lead to potential pollution	Water and environment	Citizens in general
Public health	Poor water quality can lead to different waterborne illnesses, burdening hospitals with more patients	Water, equipment and environment; overburdened healthcare systems	Citizens in general
Financial impact	Paying ransom, service disruptions and damage to equipment, resulting in financial loss	Physical infrastructure, applications and services, and networks	Citizens who depend on smart water systems, service providers, third-party vendors and government
Mental impact	All relevant stakeholders can go through lots of mental stress due to water supply disruption, paying ransom, and communicating with the attackers	Overburdened healthcare systems	Citizens who depend on smart water systems, service providers, third-party vendors and government
Domino effects on other systems	Other relevant systems, such as weather monitoring systems, emergency response systems, and energy management systems	System equipment for impacted systems, such as energy, weather, etc.	Service providers, governments, and citizens who are dependent on other systems, and third-party vendors
Reputation	Reputation loss due to mismanagement of smart systems, which leads to ransomware attacks and data loss.	NA	Service providers, government and third-party vendors
Data	Data loss, PII in the dark web or the hands of cybercriminals	Data flow	Service providers, governments, and citizens who are dependent on other systems, as well as third-party vendors

- *Identification and Authentication Management:* Managing identification and authentication is the key to mitigating ransomware attacks. As per the attack analysis, many attacks can happen due to weak passwords, compromised accounts, and social engineering. Enforcing multifactor authentication, zero trust architecture, and least privilege policy can significantly change cybersecurity attacks on critical infrastructure. Using biometric authentication instead of a password and continuous authentication can be a key to blocking attackers on many systems.
- *Vulnerability Management:* Vulnerable devices are another key entry point for ransomware attacks. Critical infrastructures should run vulnerability scans regularly to find vulnerable devices and apply patches where necessary.

TABLE 5.5
Supply Chain Attack Patterns in Terms of Ransomware Attacks

Pattern	Suggested Mitigation Strategy
The exploitation of third-party software: Codecov to test software vulnerability	• Secure coding by applying the security design principles
Customer company account hack: A compromised employee account at Turkish Airlines	• Use ML tools to check software vulnerabilities and any sign of intrusion continually
Zero-day vulnerability in third-party software: MOVEit Transfer, a managed file transfer (MFT)	• Use strong passwords and enforce multifactor authentication
Third-party software vulnerability: Remote code execution (RCE) vulnerability in PulseConnect Secure, a VPN software program	• Implement continuous authentication
Third-party software vulnerability: Jfrog Artifactory	

- *Cyber Resilience:* Cyber resilience is the key to mitigating cyberattacks on critical infrastructure. Systems and data should be backed up regularly, and tabletop exercises should be developed for incident response and recovery plans. Workers in critical infrastructure should have continuous cybersecurity training from top to bottom.

5.6 CONCLUSION

Smart cities are one of the growing concerns surrounding novel cyberattacks. The challenges lie in the evolving industry standards and infrastructures, which are highly technology-dependent and interconnected. The attack surface for smart cities is vast, so it is challenging to design a cybersecurity framework that integrates the needs of people, systems, and machines alike. This chapter has developed a threat modelling framework for smart city domains to deal with such challenges. Threat modeling is a vital part of the efficient design of cybersecurity frameworks. The proposed threat modeling framework starts with identifying attack entry points, individuals involved, attack vectors, threat actors, vulnerabilities, and detailed attack analysis, and concludes with some mitigation strategies. Although this chapter has used water management systems as an example, this adaptability extends beyond such systems, making it suitable for diverse systems within a smart city to study and prevent ransomware threats. Smart city stakeholders can use this framework in risk assessments and to develop long-lasting solutions.

NOTES

1 https://www.malwarebytes.com/blog/news/2021/03/150000-verkada-security-cameras-hacked-to-make-a-point.
2 https://www.cisa.gov/sites/default/files/publications/potential-threat-vectors-5G-infrastructure_508_v2_0%20%281%29.pdf.
3 https://insite.ipwea.org/intelligence-agencies-warn-on-smart-city-5g-and-iot-risks/.
4 https://trumpwhitehouse.archives.gov/articles/president-trump-unveils-americas-first-cybersecurity-strategy-15-years/.
5 https://assets.publishing.service.gov.uk/media/620131fdd3bf7f78e469ce00/national-cyber-strategy-amend.pdf.
6 https://start.paloaltonetworks.com/2022-unit42-incident-response-report.
7 https://services.google.com/fh/files/misc/google-cloud-cybersecurity-forecast-2024.pdf.
8 https://www.cyber.gov.au/resources-business-and-government/maintaining-devices-and-systems/critical-infrastructure/cybersecurity-best-practices-smart-cities?ref=search.
9 https://www.cyber.gov.au/resources-business-and-government/governance-and-user-education/secure-by-design?ref=search.

REFERENCES

Addeen, H.H., Xiao, Y., Li, J. and Guizani, M. (2021). A Survey of Cyber-Physical Attacks and Detection Methods in Smart Water Distribution Systems. *IEEE Access*, 9, pp.1–1. https://doi.org/10.1109/access.2021.3095713.

Adel, A. (2023). Unlocking the Future: Fostering Human–machine Collaboration and Driving Intelligent Automation through Industry 5.0 in Smart Cities. *Smart Cities*, 6(5), pp.2742–2782. https://doi.org/10.3390/smartcities6050124.

Albino, V., Berardi, U. and Dangelico, R.M. (2015). Smart Cities: Definitions, Dimensions, Performance, and Initiatives. *Journal of Urban Technology*, 22(1), pp.3–21. https://doi.org/10.1080/10630732.2014.942092.

Aleksandar, K. (2020). *Ransomware Statistics, Trends and Facts for 2020 and Beyond*. [online] Cloudwards. Available at: https://www.cloudwards.net/ransomware-statistics/.

Algarni, S. (2020). Cybersecurity Attacks: Analysis of 'WannaCry' Attack and Proposing Methods for Reducing or Preventing Such Attacks in Future. *Advances in Intelligent Systems and Computing*, 1270, pp.763–770. https://doi.org/10.1007/978-981-15-8289-9_73.

Bajpai, P. and Enbody, R. (2020, June). Preparing Smart Cities for Ransomware Attacks. In *2020 3rd International Conference on Data Intelligence and Security (ICDIS)* (pp.127–133). South Padre Island, TX: IEEE. https://doi.org/10.1109/ICDIS50059.2020.00023.

Beaman, C., Barkworth, A., Akande, T.D., Hakak, S. and Khan, M.K. (2021). Ransomware: Recent Advances, Analysis, Challenges and Future Research Directions. *Computers & Security*, [online] 111(1), p.102490. https://doi.org/10.1016/j.cose.2021.102490.

Bello, A., Jahan, S., Farid, F. and Ahamed, F. (2023). A Systemic Review of the Cybersecurity Challenges in Australian Water Infrastructure Management. *Water*, [online] 15(1), p.168. https://doi.org/10.3390/w15010168.

Breque, M., De Nul, L. and Petridis, A. (2021). *Industry 5.0: Towards a Sustainable, Human-centric and Resilient European Industry* (p.46). Luxembourg, LU: European Commission, Directorate-General for Research and Innovation. Available at: https://data.europa.eu/doi/10.2777/308407.

Brewer, R. (2016). Ransomware Attacks: Detection, Prevention and Cure. *Network Security*, [online] 2016(9), pp.5–9. https://doi.org/10.1016/s1353-4858(16)30086-1.

Brooks, C. (2021). The Security Challenge of Protecting Smart Cities. [online] *Forbes*. 10 Oct. Available at: https://www.forbes.com/sites/chuckbrooks/2021/10/10/the-security-challenge-of-protecting-smart-cities/ [Accessed 16 Oct. 2021].

Budde, P. (2023). *Smart Cities Vulnerable to Cyberattacks*. [online] Available at: https://independentaustralia.net/business/business-display/smart-cities-vulnerable-to-cyberattacks,17473 [Accessed 17 Apr. 2024].

Buntz, B. (2018). *Buntz*. [online] Available at: https://www.iotworldtoday.com/security/city-security-how-fort-lauderdale-deals-with-cyber-threats#close-modal [Accessed 16 Apr. 2024].

Chapple, M., Stewart, J.M., Rouse, M. and Gibson, D. (2018). Chapter 2 Personnel Security and Risk Management Concepts. In *Certified Information Systems Security Professional Official Study Guide* (8th ed., p.65). Hoboken, NJ: John Wiley & Sons.

Chen, D., Wawrzynski, P. and Lv, Z. (2021). Cyber Security in Smart Cities: A Review of Deep Learning-based Applications and Case Studies. *Sustainable Cities and Society*, 66, p.102655. https://doi.org/10.1016/j.scs.2020.102655.

Coker, J. (2022). Top 10: Deepfakes. [online] *Infosecurity Magazine*. Available at: https://www.infosecurity-magazine.com/magazine-features/top-10-deepfakes.

Czeczot, G., Rojek, I., Mikołajewski, D. and Sangho, B. (2023). AI in IIoT Management of Cybersecurity for Industry 4.0 and Industry 5.0 Purposes. *Electronics*, 12(18), p.3800. https://doi.org/10.3390/electronics12183800.

Farid, F., Bello, A., Jahan, N. and Sultana, R. (2023). Cyber Security Management in Metaverse: A Review and Analysis. In *Proceedings of the Computational Methods in Systems and Software* (pp.183–193). Cham: Springer. https://doi.org/10.1007/978-3-031-53552-9_16.

Fernando, J. (2024). *What Are Stakeholders: Definition, Types, and Examples*. [online] Investopedia. Available at: https://www.investopedia.com/terms/s/stakeholder.asp.

Gladysz, B., Tran, T.A., Romero, D., van Erp, T., Abonyi, J. and Ruppert, T. (2023). Current Development on the Operator 4.0 and Transition towards the Operator 5.0: A Systematic Literature Review in Light of Industry 5.0. *Journal of Manufacturing Systems*, 70, pp.160–185. https://doi.org/10.1016/j.jmsy.2023.07.008.

Hamid, B., Jhanjhi, N.Z., Humayun, M., Khan, A. and Alsayat, A. (2019, December). Cyber Security Issues and Challenges for Smart Cities: A Survey. In *2019 13th International Conference on Mathematics, Actuarial Science, Computer Science and Statistics (MACS)* (pp.1–7). Karachi: IEEE. https://doi.org/10.1109/MACS48846.2019.9024768.

He, Y., Aliyu, A., Evans, M. and Luo, C. (2021). Correction: Health Care Cybersecurity Challenges and Solutions Under the Climate of COVID-19: Scoping Review. *Journal of Medical Internet Research*, 23(4), p.e29877. https://doi.org/10.2196/29877.

Husar, A. (2022). *IoT Security: 5 Cyber-attacks Caused by IoT Security Vulnerabilities.* [online] www.cm-alliance.com. Available at: https://www.cm-alliance.com/cybersecurity-blog/iot-security-5-cyber-attacks-caused-by-iot-security-vulnerabilities.

Ismagilova, E., Hughes, L., Rana, N.P. and Dwivedi, Y.K. (2020). Security, Privacy and Risks Within Smart Cities: Literature Review and Development of a Smart City Interaction Framework. *Information Systems Frontiers*, [online] 24, pp.393–414. https://doi.org/10.1007/s10796-020-10044-1.

Jegeib (2022). *Threats - Microsoft Threat Modeling Tool - Azure.* [online] learn.microsoft.com. Available at: https://learn.microsoft.com/en-us/azure/security/develop/threat-modeling-tool-threats.

Kerner, S. (2022). *Colonial Pipeline Hack Explained: Everything You Need to Know.* [online] TechTarget. Available at: https://www.techtarget.com/whatis/feature/Colonial-Pipeline-hack-explained-Everything-you-need-to-know.

Kitchin, R. and Dodge, M. (2017). The (In)Security of Smart Cities: Vulnerabilities, Risks, Mitigation, and Prevention. *Journal of Urban Technology*, 26(2), pp.47–65. https://doi.org/10.1080/10630732.2017.1408002.

Kovacs, E. (2023). *Cyberattack on Irish Utility Cuts Off Water Supply for Two Days.* [online] Available at: https://www.securityweek.com/cyberattack-on-irish-utility-cuts-off-water-supply-for-two-days [Accessed 15 Apr. 2024].

Lai, B. (n.d.). *The Threat of Ransomware.* [online] www.aph.gov.au. Available at: https://www.aph.gov.au/About_Parliament/Parliamentary_Departments/Parliamentary_Library/pubs/BriefingBook47p/ThreatRansomware [Accessed 17 Apr. 2024].

Lai, T., Farid, F., Bello, A. and Sabrina, F. (2024). Ensemble Learning Based Anomaly Detection for IoT Cybersecurity via Bayesian Hyperparameters Sensitivity Analysis. *Cybersecurity*, 7(1), p.44. https://doi.org/10.1186/s42400-024-00238-4.

Lamers, C., Spoerl, E., Levey, G., Choudhury, N. and Ahmed, M. (2023). Ransomware: A Threat to Cyber Smart Cities. In Ahmed, M. and Haskell-Dowland, P. (eds.), *Cybersecurity for Smart Cities: Practices and Challenges* (pp.185–204). Cham: Springer International Publishing. https://doi.org/10.1007/978-3-031-24946-4_13.

Lyngaas, S. (2023). *Federal Investigators Confirm Multiple US Water Utilities Hit by Hackers | CNN Politics.* [online] CNN. Available at: https://edition.cnn.com/2023/12/01/politics/us-water-utilities-hack/index.html [Accessed 17 Apr. 2024].

Maigida, A.M., Abdulhamid, S.M., Olalere, M., Alhassan, J.K., Chiroma, H. and Dada, E.G. (2019). Systematic Literature Review and Metadata Analysis of Ransomware Attacks and Detection Mechanisms. *Journal of Reliable Intelligent Environments*, 5(2), pp.67–89. https://doi.org/10.1007/s40860-019-00080-3.

McDonald, J. (2022). *The Smarter the City, the Scarier the Cyber Risk.* [online] Emerging Tech Brew. Available at: https://www.emergingtechbrew.com/stories/2022/09/06/the-smarter-the-city-the-scarier-the-cyber-risk.

Mello Jr., J. (2022). *Ransomware Is Top Attack Vector on Critical Infrastructure.* [online] CSO Online. Available at: https://www.csoonline.com/article/572129/ransomware-is-top-attack-vector-on-critical-infrastructure.html [Accessed 17 Apr. 2024].

Minoli, D. and Occhiogrosso, B. (2018). Internet of Things Applications for Smart Cities. In Hassan, Q.F. (ed.), *Internet of Things A to Z: Technologies and Applications* (pp.319–358). *John* Wiley and Sons, Hoboken, New Jersey.

Mohammadpourfard, M., Khalili, A., Genc, I. and Konstantinou, C. (2021). Cyber-Resilient Smart Cities: Detection of Malicious Attacks in Smart Grids. *Sustainable Cities and Society*, 75, p.103116. https://doi.org/10.1016/j.scs.2021.103116.

Plumb, T. (2023). *Ransomware Attacks Target Critical Infrastructure – And It's Paying Off.* [online] Available at: https://www.sdxcentral.com/articles/news/ransomware-attacks-target-critical-infrastructure-and-its-paying-off/2023/03/ [Accessed 17 Apr. 2024].

Porter, S. (2020). *News.* [online] *Healthcare IT News.* Available at: https://www.healthcareitnews.com/news/europe/cyberattack-czech-hospital-forces-tech-shutdown-during-coronavirus-outbreak [Accessed 17 Apr. 2024].

Raiane, A., Gomes, R.L. and Fernandez, M.P. (2021). An Intelligent Mechanism to Detect Cyberattacks of Mirai Botnet in IoT Networks. In *2021 17th International Conference on Distributed Computing in Sensor Systems (DCOSS)* (pp.236–243). Pafos: IEEE. https://doi.org/10.1109/dcoss52077.2021.00047.

Rehman, M.U., Akbar, R., Omar, M. and Gilal, A.R. (2023, September). A Systematic Literature Review of Ransomware Detection Methods and Tools for Mitigating Potential Attacks. In *International Conference on Computing and Informatics* (pp.80–95). Singapore: Springer Nature. https://doi.org/10.1007/978-981-99-9589-9_7.

Robles, F. and Perlroth, N. (2021). 'Dangerous Stuff': Hackers Tried to Poison Water Supply of Florida Town. *The New York Times*. [online] 9 Feb. Available at: https://www.nytimes.com/2021/02/08/us/oldsmar-florida-water-supply-hack.html.

Shacklett, M.E. (2021). *What Is Attack Vector?* [online] SearchSecurity. Available at: https://www.techtarget.com/searchsecurity/definition/attack-vector.

Sidhu, K.S. (2023). *Top 5 Famous Software Supply Chain Attacks in 2023 | CloudSEK*. [online] cloudsek.com. Available at: https://www.cloudsek.com/blog/top-5-famous-software-supply-chain-cyber-attacks-in-2023.

Tatam, M., Shanmugam, B., Azam, S. and Kannoorpatti, K. (2021). A Review of Threat Modelling Approaches for APT-style Attacks. *Heliyon*, 7(1), p.e05969. https://doi.org/10.1016/j.heliyon.2021.e05969.

Teichmann, F. (2023). Ransomware Attacks in the Context of Generative Artificial Intelligence—An Experimental Study. *International Cybersecurity Law Review*, 4(4), pp.399–414. https://doi.org/10.1365/s43439-023-00094-x.

Tekinerdogan, B., Köksal, Ö. and Çelik, T. (2023). System Architecture Design of IoT-based Smart Cities. *Applied Sciences*, 13(7), p.4173. https://doi.org/10.3390/app13074173.

Ullah, Z., Al-Turjman, F., Mostarda, L. and Gagliardi, R. (2020). Applications of Artificial Intelligence and Machine Learning in Smart Cities. *Computer Communications*, 154, pp.313–323. https://doi.org/10.1016/j.comcom.2020.02.069.

Vähäkainu, P., Lehto, M. and Kariluoto, A. (2022). Cyberattacks Against Critical Infrastructure Facilities and Corresponding Countermeasures. In Lehto, M. and Neittaanmäki, P. (eds.), *Cyber Security: Critical Infrastructure Protection* (pp.255–292). Cham: Springer International Publishing. https://doi.org/10.1007/978-3-030-91293-2_11.

Vailshery, L. (2023). *IoT Connected Devices Worldwide 2019–2030*. [online] Statista. Available at: https://www.statista.com/statistics/1183457/iot-connected-devices-worldwide/.

Vassilev, A., Oprea, A., Fordyce, A. and Anderson, H. (2024). *Adversarial Machine Learning: A Taxonomy and Terminology of Attacks and Mitigations* (No. NIST Artificial Intelligence (AI) 100–2 E2023). Gaithersburg, MD: National Institute of Standards and Technology. https://doi.org/10.6028/NIST.AI.100-2e2023.

Waizel, G. (2022). Using a Modern Honeypot Model to Defend Smart Cities and Provide Early Detection to APT and Ransomware Attacks. In *Smart Cities International Conference (SCIC) Proceedings,* National University of Political Studies and Public Administration, Organized by the Smart-EDU Hub within the Faculty of Public Administration at the National University of Political Studies and Public Administration, Bucharest, Romania, (Vol. 10, pp.33–46). Available at: https://scrd.eu/index.php/scic/article/view/423.

Whitefield-Madrano, A. (2023). *What Are SQL Injections, and What Is the Risk to Businesses?* [online] Technology Solutions That Drive Business. Available at: https://biztechmagazine.com/article/2023/09/what-are-sql-injections-and-what-risk-businesses-perfcon.

Yaacoub, J.-P.A., Noura, H.N., Salman, O. and Chehab, A. (2021). Robotics Cyber Security: Vulnerabilities, Attacks, Countermeasures, and Recommendations. *International Journal of Information Security*, 21(21), pp.115–158. https://doi.org/10.1007/s10207-021-00545-8.

Part II

Novel Solutions and Applications

6 Tackling Integration Challenges of Machine Learning in Diverse Internet of Things

A Spread-Out Architectural Solution

Adarsha Bhattarai, Chathumi Samaraweera, Dongming Peng, Yutong Liu and Hamid Sharif

6.1 INTRODUCTION

The usage of Internet of Things (IoT) applications is growing every year, along with the number of devices connected to the network (Hassan, 2018). The wide and expanding use of IoT devices necessitates the holistic configuration of multiple aspects of IoT, including hardware chips, data links, edge and cloud computing, and end-user applications. This way, vulnerability points of IoT devices can be reassessed and made robust and resistant to various security threats and other design challenges (Rizvi et al., 2020; Anand et al., 2020; Raja et al., 2022; Su¨ren et al., 2023; Bhardwaj et al., 2023).

The current state-of-the-art anomaly detection algorithms proposed for Cyber-Physical Systems (CPS) must be effectively used by the various processing nodes to ensure security protection (Cook et al., 2022; Bai et al., 2022; Mansour, 2022; Zainudin et al., 2023; Jayagopal et al., 2023; Catillo et al., 2023). Particularly in a distributed network with more than one processing node, the proposed detection algorithm must match the computational complexity of the given hardware. Given multiple participants in the CPS with unique hardware specifications, the anomaly detection algorithm would be driven by artificial intelligence (AI) and appropriately distributed in the CPS. This requires the establishment of a secure communication infrastructure to support the sharing of such algorithms. Federated learning (FL) and distributed learning in IoT systems and CPS have become very popular recently (Banabilah et al., 2022; Li et al., 2022; Agrawal et al., 2022; Bousbiat et al., 2023). This is because local data is not required to be shared by nodes with a centralized server for training an AI model (Mohawesh et al., 2023; Khan et al., 2023; Pandya et al., 2023). This characteristic of FL has paved the way for safer and more innovative AI-based applications in the fields of data science, healthcare, industry, and education (Banabilah et al., 2022; Rieke et al., 2020; Lakhan et al., 2023). Furthermore, when FL leverages blockchain technology, the learning process is more secure from malicious nodes. This will prevent the manipulation of learned parameters, thereby providing a reliable learning environment for genuine nodes (Ma et al., 2022).

To enhance FL in the fields of data science, healthcare, industry, and education, it would be beneficial to explore the performance of machine learning (ML) models in depth (Li et al., 2023). This would include the study of performance metrics, such as accuracy, precision, and recall. It is observed that the unbalanced distribution of datasets or local biases causes the degradation of the ML models (Touat & Bouchenak, 2023; Wang et al., 2023). In addition, the optimization algorithms

DOI: 10.1201/9781003506638-8

in ML are of high importance when involving diverse processing nodes with unique computational complexities and communication needs (Banabilah et al., 2022; Zhou & Li, 2023; Zhu et al., 2023; Almanifi et al., 2023). The use of the same AI model hosted by the most powerful server and the same optimization algorithm might not be appropriate for heterogeneous surrounding clients. Clients with less computational power may experience crashes due to the uneven compatibility related to the hardware specifications. Hence, there is a need to continually improve the diverse/federated and distributed configuration of AI models in the edge computing networks of CPS, which consist of diverse types of sensor devices, edge devices, and servers.

In this chapter, we propose the AI spread-out model for processing platforms that have unique computational needs. In this model, the AI inference modules are partitioned and distributed among diverse processing layers of the IoT system. This, in turn, enhances the efficiency and resilience of anomaly detection. Since Artificial Intelligence of Things (AIoT) involves the use of AI models, such as deep learning models, across multiple intelligent devices (Samaraweera et al., 2025), we believe that the proposed AI spread-out would contribute to the efficient deployment of such AI models. In AI spread-out, an AI model is partitioned into multiple parts. Eventually, these parts are distributed over multiple nodes based on their computational strength. Since AI spread-out involves the distribution of AI models, a secure communication infrastructure becomes a necessity. On the other hand, FL involves multiple local nodes that remain anonymous, making it challenging to implement security measures such as traceability and accountability (Ghimire & Rawat, 2022; Chen et al., 2023; Qammar et al., 2022; Ma & Xu, 2023). Having a secure, tamper-proof record-keeping system and a malware or intrusion detection system (IDS) can collectively preserve privacy and filter out malicious nodes. Those attackers may be involved in poisoning the learning process or replaying the AI models that are yet to be delivered to a participant. In this work, we propose automatic state activation (ASA) and verification that involve multi-level processing layers to detect and verify anomalies and intrusions in CPS. This system benefits from the unique AI algorithms held by each layer.

This chapter focuses on tackling the challenges of integrating ML and AI algorithms into diverse IoT systems, particularly considering heterogeneous computing platforms and hardware resource constraints. The state-of-the-art AI techniques prominent in CPS and their security challenges will be first reviewed, and then we will present novel AI spread-out architectures and their unique features in addressing the needs for data minimization and privacy preservation. Further, details of the methodological framework used to achieve AI spread-out, including AI model partitioning, steps to form an AI spread-out chain (ASC), and methods of ASA, will be covered. We will then discuss smart power grid systems as a major application of the proposed model and how power grid systems can benefit from it. Finally, simulation results for AI spread-out, ASC, and ASA will be presented and discussed.

6.2 RELATED WORKS

Various publications have discussed methods to apply FL and distributed learning in various applications, such as CPS and other wireless sensor networks (Banabilah et al., 2022; Li et al., 2022; Agrawal et al., 2022; Bousbiat et al., 2023; Zhou & Li, 2023; Zhu et al., 2022). They have discussed the challenges of FL in terms of the computational capability of clients and optimization algorithms. The work by Zhu et al. (2022) mentioned that FL involves the selection of suitable clients by the coordinator to initiate local training based on the shared global model. Hence, it highlighted the importance of estimating the computation capability of the clients accurately. Likewise, the work of Zhou and Li (2023) presented concerns about efficient optimization algorithms in FL. It proposed both computation- and communication-efficient optimization algorithms. Further, it proposed to improve computation speed by solving sub-problems approximately and updating the client's parameters (Zhou & Li, 2023). Similarly, the work of Banabilah et al. (2022) expressed concerns about the optimization algorithms in FL for the client nodes having dynamic computational resources and communication facilities. It was noted that employing the identical optimization algorithm as in the centralized stage of the ML model might not be suitable for the client nodes in the FL setting. This approach notably decelerates the learning process and could affect the convergence speed of FL (Banabilah et al., 2022). This essential attention to the heterogeneous nature of the intelligent devices in the network is indeed a major challenge in the current AI-based edge computing network. To address this challenge, we introduce AI spread-out to various processing platforms, where the computational strength of every participating node is considered.

The compatibility of AI models with various devices and the classification accuracy of data induced by the sensors are the major challenges. In the literature, for example, several studies discussed the intersection of AI and IoT (AIoT) and aimed to ensure hardware and software components integrate well (Hu et al., 2023; Liu et al., 2024). The work published by Dong et al. (2021) introduced self-powered sensors with AI for smart homes, presented the potential of AI in home automation, and claimed to improve system performance. Within a medical domain, the approach in Pise et al. (2023) combined IoT and AI, emphasizing their proper integration to enhance the health of individuals. This study focused on patients living in rural or distant areas. However, these aforementioned studies only focused on specific cases of the intersection between AI and IoT. Hence, there has been a significant need for a systematic approach to addressing the challenges posed by the compatibility of AI models in various devices and AI model performances. In this chapter, the proposed AI spread-out approach aims to bridge this gap by allocating AI models effectively across devices and enhancing overall system efficiency through partitioning the AI inference modules themselves into multiresolution scenarios and deploying them into diverse layers of processing platforms in the IoT.

There is a serious challenge in AI-based edge computing systems other than concerns for appropriate ML models for processing platforms that have varying computational complexities. The work published by Agrawal et al. (2022) discussed FL and presented the need for a secure and efficient IDS in different environments, like CPS. With the security concern in smart grids, a combination of Transformers, FL, and Paillier cryptosystems is proposed to build the false data injection attack (FDIA) detection model (Li et al., 2022). Likewise, the FL application in smart grids is elaborated on, discussing its advantages and drawbacks (Bousbiat et al., 2023). Further, it pointed out the FDIA in FL, where a client can alternate the global model by updating malicious weights. Keeping the challenge of attacks like FDIA in mind, we propose ASA for anomaly detection and verification. This strategy involves multiple intelligent sensor devices (ISDs), intermediate platforms (IPs), and control team members that verify the state/signal by utilizing inference modules embedded in them. As the process involves multiple-tier verification, the probability of detecting anomalies increases drastically. Moreover, the work of Banabilah et al. (2022) presented concerns about raw data leakage from client nodes during the FL process. Further, the security risks of an attack on CPS devices, network attacks, firmware attacks, and cloud-based network attacks were discussed (Ghimire & Rawat, 2022). Ma and coauthors applied blockchain technology to enhance the security of a system that involved FL (Ma et al., 2022). They focused on preventing malicious nodes from poisoning the learning process (Ma et al., 2022). According to these reviews, blockchain technology can be a possible solution to provide robust security services that involve the collaboration of multiple processing layers (Zhu et al., 2022; Pulkkis et al., 2018).

The above literature review and background study present the challenges faced by heterogeneous IoT devices in AI-based edge computing environments. For example, in FL, the same optimization algorithm proposed for heterogeneous client nodes as centralized nodes led to a significant delay in the learning process and affected the convergence of AI models. In addition, the state-of-the-art research focused on specific cases like smart homes and e-health, highlighting AI and IoT integration challenges rather than the focus on developing systematic approaches to address AI and hardware compatibility issues and achieve high performance of AI across a wide range of applications, such as smart cities, smart transportation, supply chains, and agriculture. Furthermore, malicious attacks like FDIA remain a concern in CPS. Hence, a reliable anomaly detection algorithm based on the specific hardware specifications of a computing platform is necessary to detect attacks like FDIA and facilitate early intervention.

To address the aforementioned issues, we propose AI spread-out. AI spread-out involves the interaction of multiple computing platforms that share pre-trained AI algorithms through a robust security mechanism. The difference between AI spread-out and FL is that FL is a collaboration to share the parameter values after local training and updating the global model, whereas AI spread-out is a collaboration to share the AI inference algorithms that have been designed and pre-trained as per the hardware, software, and communication requirements of the node. The upcoming section will illustrate the deployment of AI model partitioning, security features, and ASA in various subsections.

6.3 SYSTEM ARCHITECTURE

The proposed AI-based architecture consists of multiple sensor nodes and edge servers in a distributed setting. This system especially fits in with those applications where signals are continuously measured and classified. For such systems, we propose the partitioning of AI algorithms, which are responsible for detecting diverse states sensed by the sensor and triggering various actions and commands based on the detected state. In these systems, sensor nodes and their corresponding edge computers and servers will be assigned the corresponding components of AI inference modules based on their computational complexity. Figure 6.1 presents the proposed system architecture in detail.

6.3.1 Intelligent Sensor Device

ISD refers to an embedded system that includes a sensing element and a processing element designed for a specific application purpose. The ISD is empowered by an embedded microprocessor-based

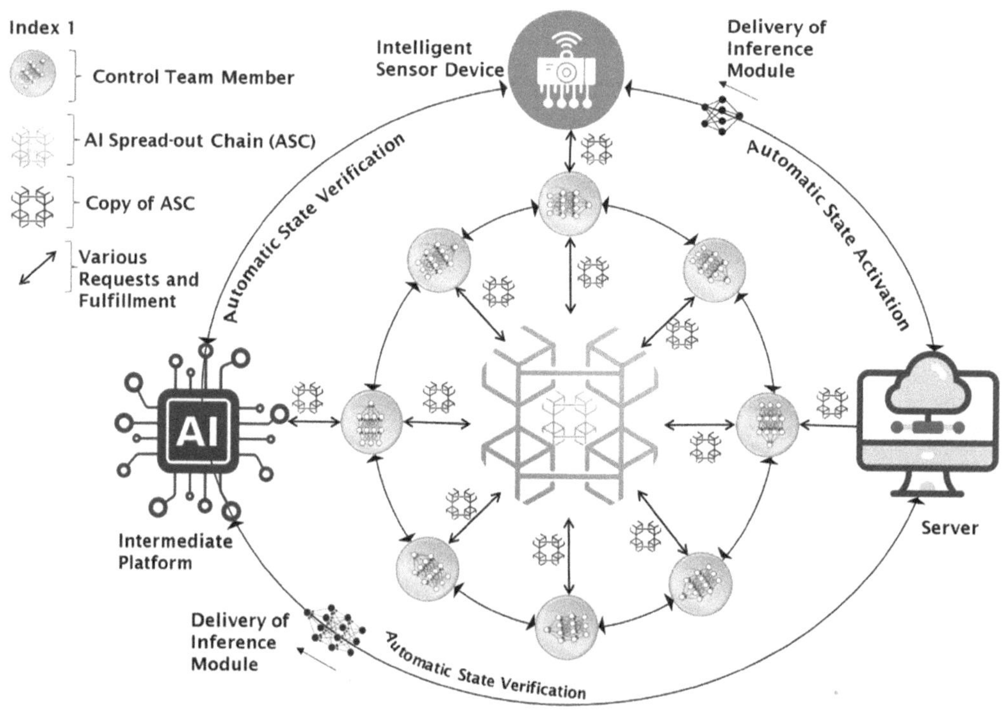

FIGURE 6.1 The Proposed System Architecture of AI Spread-Out

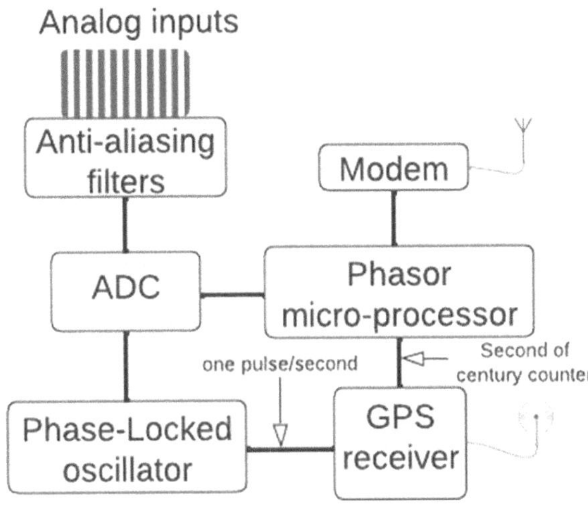

FIGURE 6.2 Configuration of a Modern Phasor Measuring Unit

computing system or high-end microcontroller-based computing system and therefore has relatively higher processing power than other regular sensor nodes. ISD could be part of many applications of CPS, such as remote anomaly detection. These ISDs might also be connected to actuators and switch breakers, depending on the type of application, to enable various physical actions. As an example, the phasor measuring unit (PMU) is a device that measures the electrical waves on a power grid system. They measure voltage and current phase angles, as well as voltage and current magnitudes. Figure 6.2 shows a typical configuration of the PMUs, which defines the internal architecture and key functional elements of the PMU. In Figure 6.2, the phasor microprocessor works with the Global Positioning System (GPS) receiver to obtain critical information, including timestamps of the second-century (SOC) counter. This counter feedbacks the count of seconds from the coordinated universal time at midnight (00:00:00) of January 1, 1970, to the current second. This device can represent an ISD that can detect anomalies in such measurements locally without dependence on the control station or server.

6.3.2 Intermediate Platforms

These platforms refer to any processing layer that resides mainly between the ISDs and servers of the system. Examples of such IPs are gateway servers or local communication network hubs. They are empowered with relatively more powerful technology compared to ISDs. The main task of the IPs is to aid the ISDs in many ways. This could be in the form of state verification, feedback, or security measures. IPs are another form of intelligent processing station dedicated to increasing the reliability of the whole system. They could do so by ensuring the accuracy of the diverse states detected by the sensors. Further, IPs become responsible for connecting to their control team members to inform them about the current scenarios or deliver requested information. For example, phasor data concentrators (PDCs) that collect phasor data from multiple PMUs of the power grid system can act as an IP. Therefore, IP in the power grid system can collect readings from multiple ISDs and verify the anomalies detected by the ISDs.

6.3.3 Control Team Members

Control team members are the human operators or the software modules interacting with the human operators. These members surround the server of the system and have a vital role in the verification of the detected state, triggering commands and actions upon emergencies, and authentication and authorization of nodes into the network. The main objective of the control team members is to form harmony with their decisions. These members could be either an automated intelligent machine connected with the system's server or a trusted authority linked to the system. The number of control team members depends upon the size of the system, i.e., the number of ISDs, IPs, and servers the system consists of.

6.3.4 The Employment of AI Spread-Out Mechanism

In FL, training is performed locally based on the initial model sent by the server to multiple local nodes. After the local training, the generated parameters are sent back to the server for aggregation. Finally, the global model is formed and distributed among the local nodes. Further, this process repeats until the optimum model is achieved. FL needs to optimize the learning delay if it is to be used for time-sensitive applications such as anomaly detection and prevention systems. Therefore, to complement FL, we present the idea of AI spread-out. AI spread-out is a major software component of the system, and the performance of the system depends upon this. In spread-out AI, a large pre-trained AI model is partitioned over multiple processing devices. That is why it is also known as AI algorithm partitioning. Here, an AI model is scaled to multiple processing platforms according to their computational strength. The novelty of this approach is that it considers the degree of strength of diverse participants or nodes present in the system's' network. This affirms that all nodes are worthy to run the algorithm and become an important part of the system in detecting various states.

6.3.5 Privacy Preservation

AI spread-out architecture necessitates a secure platform to share information and AI inference models among the processing nodes. Security services, such as authentication, authorization, and non-repudiation, are vital for AI spread-out that forms distributed nodes. In this case, AI spread-out can benefit from a secure, tamper-proof log sub-system for information and AI model sharing. In our proposed system architecture, intelligent sensors, IPs, servers, and control team members can benefit from such technology to authorize and authenticate each other. Therefore, a blockchain-based mechanism is proposed, the ASC, to ensure the security services in the proposed system.

6.4 METHODOLOGICAL FRAMEWORK

This section demonstrates the proposed methods with examples and mathematical equations. The subsections are detailed in three major areas. First, the focus will be on the schemes for achieving spread-out AI. Second, the design of the ASC, and third, the methodology for ASA.

6.4.1 Development of Inference Models in AI Spread-Out

In this step, first, an appropriate AI algorithm is considered depending on the application. Then the most powerful processing platform P_i available in the network is considered. Finally, the AI model (M_{ai}) is trained with datasets of size S_{total} and a set of training parameters T_i. S_{total} and T_i for this training are chosen in a way that matches the computational complexity of the platform P_i. The matching is done to get the optimal value of performance metrics (accuracy, precision, recall, and F1 score). We name this matching process computational complexity and parameter tuning (CCPT). When the training is complete, the inference model M_{pi} is ready to be embedded inside the processing platform P_i. After the training process, the AI model (M_{ai}) is partitioned into N shares.

The value of N is chosen as three and processing platforms P_A, P_B, and P_C are considered with their respective computational complexities, $O(P_A)$, $O(P_B)$, and $O(P_C)$, respectively. The relationship between computational complexities is given in Equation 1.

$$O(P_A) < O(P_B) < O(P_C) \tag{1}$$

We assume that the computational complexity of platform P_i is equivalent to P_C, as given in Equation 2.

$$O(P_i) \approx O(P_C) \tag{2}$$

As the computational complexity is equivalent, the same inference model M_{pi} is matched with the platform P_C, and the inference model is named M_{pC}. However, we need to train two more AI inference models dedicated to the platforms P_A and P_B. The same process used to achieve the optimal model M_{pi} mentioned in the above paragraph is repeatedly applied to obtain the models M_{pA} and M_{pB}. The iteration will run until the desired performance metrics are achieved for the inference models M_{pA} and M_{pB}. Since $O(P_A)$ and $O(P_B)$ are smaller than $O(P_C)$, CCPT should produce optimum values for the size of datasets S_{MpA} and S_{MpB} and the training parameters T_A and T_B for platforms A and B, respectively. This way, platforms P_A and P_B can run the models efficiently and effectively. The relationship between the datasets used to form the inference models M_{pA}, M_{pB}, and M_{pC} is given in Equation 3. Further Algorithm 1 talks about the procedure to partition an AI model into N shares.

$$S_{MP_A} \bigcup S_{MP_B} \bigcup S_{MP_C} = S_{MP_i} \tag{3}$$

Algorithm 1. Partitioning an AI model for a CPS application into *N* shares

Description:

- An AI model is partitioned into N shares. It is applicable for many CPS applications, like power grid systems.
- Steps include hyperparameter tuning, training, and inference, and finally, distribution of the inference models into the nodes.
- The distribution of AI models is carried out with the help of AI spread-out chain. In the process, control team members authenticate the nodes.

Data: AI algorithm M_{ai} for the CPS application, number of processing platforms N, total available dataset S_{total}, computational complexity $O(P_N)$ of all platforms.

Result: Inference AI model partitioned into N shares.

for $i = 1$ *to* N **do**

$[T_N, S_N, X_1, X_2, \ldots, X_n] \leftarrow \mathit{Hyperparameter\,Tuning}(M_{ai}, O(P_N), S_{\text{total}})$
$\mathit{inferenceModel} \leftarrow \mathit{TrainAndInfer}(M_{ai}, S_N, T_N, X_1, X_2, \ldots, X_n)$

send *inferenceModel* to nodes using AI spread-out chain

authenticate nodes

end

6.4.2 Distribution of AI Models in AI Spread-Out

In Figure 6.3, the trained AI model M_{P_i} is partitioned and spread over the ISD, IP, and server. These devices represent platforms P_A, P_B, and P_C, respectively. The server is the most powerful platform among IP and ISD. Further, as shown in Figure 6.3, the union of three inference AI models M_{PA}, M_{PB}, and M_{PC} is equivalent to the originally trained AI inference model M_{Pi}. Equation 4 illustrates the mathematical relationship among these AI models.

$$M_{P_i} = M_{P_A} \bigcup M_{P_B} \bigcup M_{P_C} \tag{4}$$

These modeled inference models are delivered to processing platforms using either offline or online methods. In the offline method, the AI model is loaded to the processing platform using an external disk without any connection to the network. This method is suitable for an application with very few nodes, i.e., small-scale applications. In contrast, the online method of embedding AI models into the processing platforms needs a connection to the network. In this process, control team members connected to the server fulfill the demand of supplying inference models to the platforms.

6.4.3 AI Spread-Out Chain

The secure sharing of inference models and other data between the main server and the processing platforms is facilitated by the ASC. The structure of the chain is built based on the application's

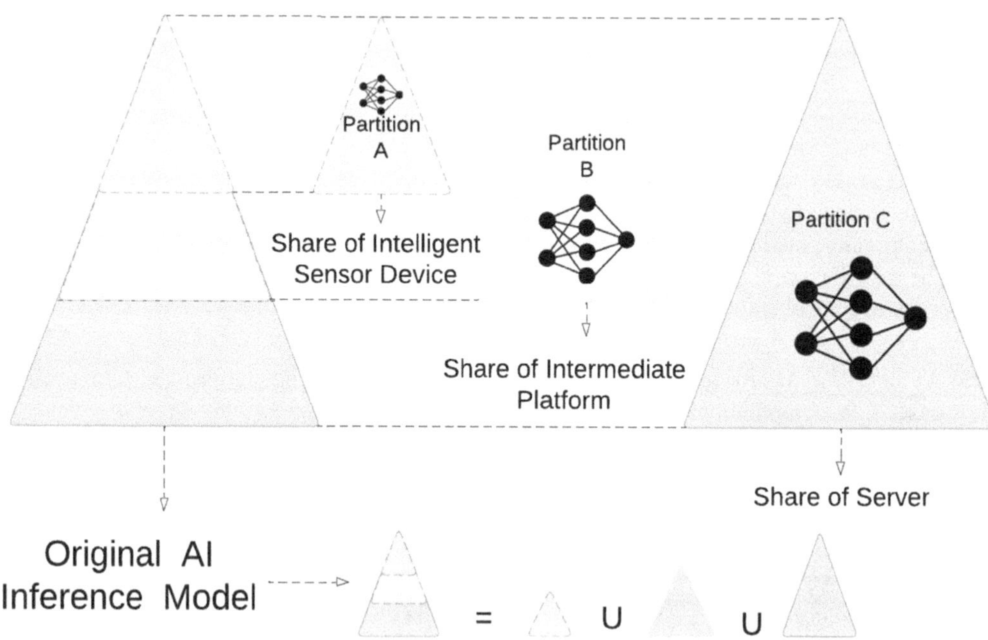

FIGURE 6.3 Partition of an AI Model into Three Parts Based on the Performance and Computational Complexity of the Processing Platforms

needs and the type of data it needs to record; the modeling of the chain acquires a chain structure through the process of linking. Considering its potential in power grid applications, a single link would record the PMU sessions and node information in encrypted form with its hash value and timestamps. For all kinds of applications, the identities of nodes in the ASC will be used to provide authentication and authorization services. This will ultimately secure the dissemination of the AI inference model from the main server to the nodes.

The scripts used in ASC are the controlling scripts written for every processing platform participating in the network. These are written by the software modules interacting with the control team members residing near the server's end and then distributed across the processing platforms. The script determines what actions to permit, such as being able to access the copy of the spread-out chain, being able to propose to add new information to the chain, and being able to update the chain with the new information. The functionality that the script provides might vary from one node to another depending upon the role and level of authorization of the node across the network. In the proposed system architecture, ISDs and IPs are authorized to obtain a copy of the ASC only if they are authenticated by the control team members. However, the control team members additionally have the authority to add new information to the ASC and update the chain with new information. The control team members mainly consist of trusted stakeholders, mostly attached to the main server of the system. The control team members play a vital role in adding new information to the AI-spread-out chain, as shown in Algorithm 2. The new information I_N, proposed to be appended to the spread-out chain, is the identity of a new node X_n in the network. And later, this new information, I_N, is used by trusted nodes to authenticate node X_n. This will result in the secure delivery of the AI inference model to X_n from some other node or server. In this process, the consensus of the control team members plays an important role. They calculate the collaborative decision (CD) factor based on their agreement and broadcast its decision to all other control team members. Finally, the CD factor is calculated once all members have given their vote. This value should be greater than the threshold value to append any identity label to the ASC. The threshold value must be greater than or equal to 0.51. This value could also be increased, making it more difficult for a group of nodes to execute a biased attack like the 51% attack, for example, by utilizing a notarization process or intelligent software agents to detect anomalies such as collusion (Aponte-Novoa et al., 2021). Furthermore, to maintain fairness and trustworthiness, control team members continuously scan the ASC for any illegal hash values. Whenever a new identity label is proposed to be appended, it is assigned a unique hash value. This hash value is also referenced to the previous identity label and will be linked to the next identity label. These hash values are generated through a highly secured cryptographic function like SHA-256 (Secure Hash Algorithm 256-bit).

Algorithm 2. Adding the identity label of a processing node to the AI spread-out chain

Data: Processing platform X_n requests control team members to update spread-out chain

Result: Information added to spread-out chain Verify X_n

// Calculate collaborative decision factor

$$CD = \frac{positive\ decisions}{total\ control\ team\ members}$$

If *CD* > *threshold* **then** Append info **end**

else Reject **end**

if *IllegalHashesFound()* **then**

exit

2. end

Update chain with I_N

6.4.4 Methodology of ASA and Verification

We propose the ASA sub-system to differentiate normal/healthy behavior from abnormal/unhealthy behaviors. The objective here is to initiate remedial action in the system with the minimum delay. This involves running partitioned AI algorithms in multiple processing layers and verifying abnormal actions or behaviors. This multi-layer state verification could give high accuracy rates with minimum false positive and false negative conditions. The ASA sub-system can bring a remarkable advancement in verifying critical cases. The ASA involves multiple processing layers running their AI inference models when an alarm is generated by the ISDs. The alarm is generated based on the state/signal sensed or measured by an ISD. In the case of normal conditions, the state is marked as safe. The safe state minimizes interaction between nodes, which leads to the conservation of power.

The ASA is powered by the partition of an AI model, M_{ai}. Here, an ISD uses an M_{pA} inference model to detect the abnormality in the system that it is connected to. When an ISD detects an abnormality, it triggers a multi-node analysis state. In this state, the ISD makes the neighboring nodes aware of the anomaly that is detected. IPs and control team members are the neighboring nodes for an ISD. So, the N_{ip} IPs and N_{pn} control team members run their respective inference models, M_{pNip} and M_{pNpn}, to analyze the abnormal signal/state initially detected by the ISD. The final decision is made by the N_{pn}, control team members by collecting all the analyses made by the control team members themselves and the IPs. Finally, control team members make a collective decision on the state by verifying the results. When a serious state of S_{Em} is detected upon verification, an emergency is called. This entire process runs automatically. When numerous nodes are verifying the abnormal state, the false positive or false negative conditions reduce significantly.

In the verification process, two different scenarios are observed: safe and abnormal states. In the safe state, the number of interactions between the nodes is reduced, which saves the transmission energy and the processing energy. On the other hand, in an abnormal state, a continual switch mode is activated, which directs the participating nodes to verify the state multiple times. Below, we present the study of ASA, including the multi-node analysis and the energy efficiency in Algorithm 3.

Algorithm 3. Automatic state verification using partitioned AI models and energy efficacy of ASA

Description: This algorithm verifies alarms sent by an ISD and activates emergencies in case of threats while conserving power otherwise.

Data: Alarm from the ISD.

Result: Verification of the alarms, abnormalities and emergency cases.

$$InitialAbnormality = ISD.DetectAbnormality\left(M_{P_A}\right)$$

foreach *IP in* N_{ip} **do**

IP.ClassifyAbnormality(M_{pNip})

3. end

foreach *CTM in* N_{pn} **do** //Control team members (CTMs)

CTM.ClassifyAbnormality(M_{pNpn})

4. end

Analyses = ISD.CollectAnalyses(N_{pn}, N_{ip})

FinalDecision = CTMs.MakeCollectiveDecision(*Analyses*)

if *FinalDecision is Threat* **then**

CallEmergency()

else

MinimizeInteractionAndConservePower()

end

6.5 APPLICATION OF THE ASC INTO POWER GRIDS

In this section, we present a case study to demonstrate the importance and relevance of our proposed novel techniques in real-world applications. These techniques focus on the development of CPS and wireless networks. Specifically, in the areas of distributed AI, cybersecurity, and ASA. We propose three important techniques elaborated below.

6.5.1 Power Grids

Power grids are managed and controlled by SCADA (Supervisory Control and Data Acquisition) systems (Keshk et al., 2021), which connect with multiple sensors, actuators, and distributed processing nodes that are part of a power grid (Sun et al., 2018). The components are responsible for transmitting the measured signal or preprocessed data for further visualization and analysis. When there are ISDs such as PMUs and PDCs distributed across a region in a power grid system, it brings challenges in the fields of cybersecurity, distributed data processing and classification, and the reliability of anomaly detection algorithms. With these challenges in mind and to improve the power grids, AI model partitioning and AI-powered ASA sub-systems are used, as demonstrated in Figure 6.4. In the AI spread-out architecture shown in Figure 6.4, there are two power generators,

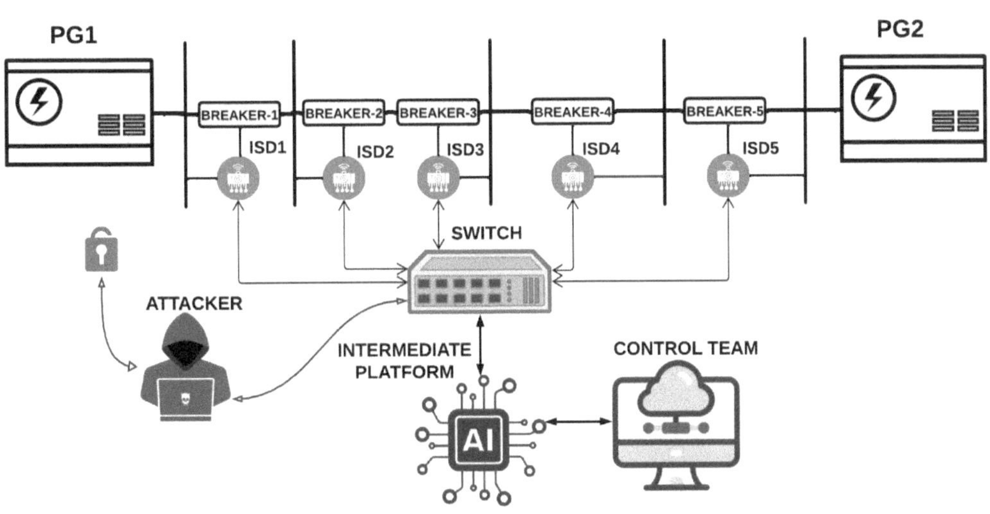

FIGURE 6.4 AI Spread-Out Architecture for Power Grid System

PG1 and PG2, five circuit breakers that can be activated in case of emergencies, five ISDs that work as smart PMUs, a switch, an IP that works as a smart PDC, and a control team. In addition, there is an attacker who can pose cyber threats, such as FDIAs.

6.5.2 AI Spread-Out in Power Grids

Dividing AI models across multiple devices can be very beneficial in smart power grid systems. For power grid systems, the computing power of the main server/SCADA system, PMUs, PDCs, switches, and other embedded systems varies a lot. Due to the varying computational power of nodes in this system, deploying an abnormality detection algorithm can be challenging. However, the nodes can benefit from an AI model that can be scaled across those processing nodes. This scaled distribution of the model will enable efficient analysis, multi-layer abnormality detection, and verification.

6.5.3 ASC in Power Grids

The large, distributed network formed by numerous processing nodes, sensors, actuators, PDCs, and ISDs in a power grid system needs to be ideally tamper-proof and hack-proof. With this objective in mind, we propose ASC for the secure authentication and authorization of any participating node in the network. ASC would work as a tamper-proof log sub-system too, where control team members of the power grid system are responsible for adding and verifying the new information. Further, ASC would allow secondary nodes such as ISDs and actuators to hold a copy of ASC so that they can identify any suspicious nodes trying to connect with them.

6.5.4 Anomaly Detection and Verification in Power Grids

We propose the ASA sub-system to differentiate normal behavior from abnormal behaviors. Then initiate the remedies in the power grid system with minimum delay. This involves running AI models in multiple nodes and verifying unusual actions or behaviors. Multi-tier state verification could yield nearly a 100% accuracy rate with zero false positive and false negative conditions. For a vulnerable system like the power grid, the ASA sub-system could bring a remarkable advancement in verifying critical cases.

6.6 SIMULATION AND DISCUSSION

In this section, Dataset 1: Power System Datasets were used from the Industrial Control System (ICS) Cyber Attack Datasets (Beaver et al., 2013; Morris & Gao, 2014). These datasets were used to perform tests to present the effectiveness of AI model partition, ASC, and ASA in a smart power grid system.

6.6.1 AI Model Partitioning

The main aim of the test here is to show that an AI model could be scaled over multiple processing nodes, sensors, actuators, PDCs, and ISDs in a power grid system. The diverse AI models were trained with varying model parameters and varying numbers of datasets. For demonstration, K-nearest neighbor (KNN) is used as an AI algorithm. Three different KNN models that vary in terms of the number of datasets and model parameters were trained. Power System Datasets consisted of 15 different .csv files, with the target column being either a natural plus normal event or a cyberattack. Each .csv file consisted of 128 feature columns. These features included 29 measurements measured by 4 different phasor measurement units (PMUs) and control panel logs, snort logs, and relay logs of the 4 PMUs. The natural plus normal event of the target class represented either fault/maintenance in the lines. This could be due to natural causes or attacks such as data injection, remote tripping command injection, and relay setting changes. Further details on datasets could be accessed from the source, ICS Cyber Attack Datasets (Beaver et al., 2013; Morris & Gao, 2014).

Three different KNN models were trained with varying sizes of datasets and unique optimal model parameters. The first KNN model used one file (data1.csv) to form KNN model 1. The dataset for KNN model 1 consisted of 128 feature columns, one target column, and 4,966 unique observations. Similarly, KNN model 2 and KNN model 3 used four files (data1.csv-data4.csv) and seven files (data1.csv-data7.csv), respectively. The dataset for KNN model 2 consisted of 128 feature columns, one target column, and 20,652 unique observations. The dataset for KNN model 3 consisted of 128 feature columns, one target column, and 36,016 unique observations. These multiple files were merged to form larger datasets compared to KNN model 1. Further details on these three models and their performances are shown in Figure 6.5 and Table 6.1.

6.6.2 Network Model Based on ASC

This simulation model was formed of 10 PMUs (PMU1–PMU10), 2 PDCs (PDC1–PDC2), and a control team with four members, as shown in Figure 6.6. Python and the Insomnia API were used to build the ASC infrastructure for the network. Members of the control team were responsible for authenticating and authorizing PMUs and PDCs into the network with their collective decision. After successful authentication, PMUs and PDCs were registered at ASC and were able to keep a

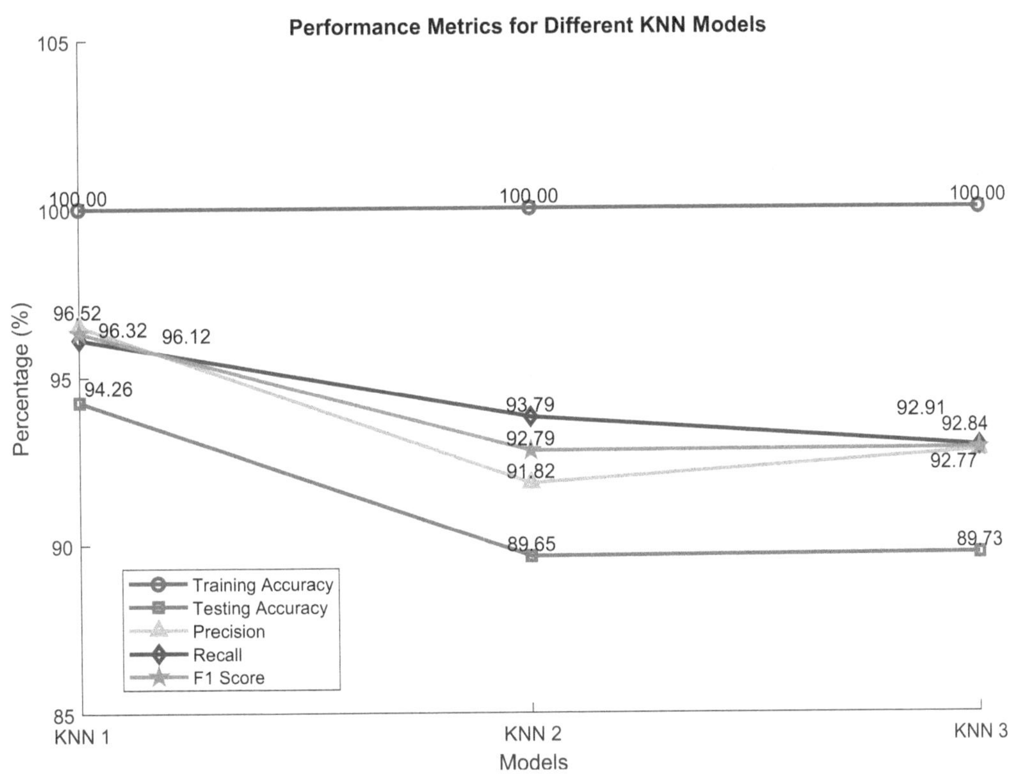

FIGURE 6.5 Performance Metrics for Different KNN Models

TABLE 6.1
Training Details for Different KNN Models

KNN Models	Feature Set	Optimal Hyperparameter	Execution Environment
KNN 1	4966 × 128	{'n neighbors':4, 'p':1, 'weights': 'distance'}	Intel Core i7-8750H, CPU @ 2.20 GHz [12 CPUs], ~2.2 GHz, and RAM 16 GB
KNN 2	20652 × 128	{'n neighbors':5, 'p':1, 'weights': 'distance'}	
KNN 3	36016 × 128	{'n neighbors':1, 'p':1, 'weights': 'uniform'}	

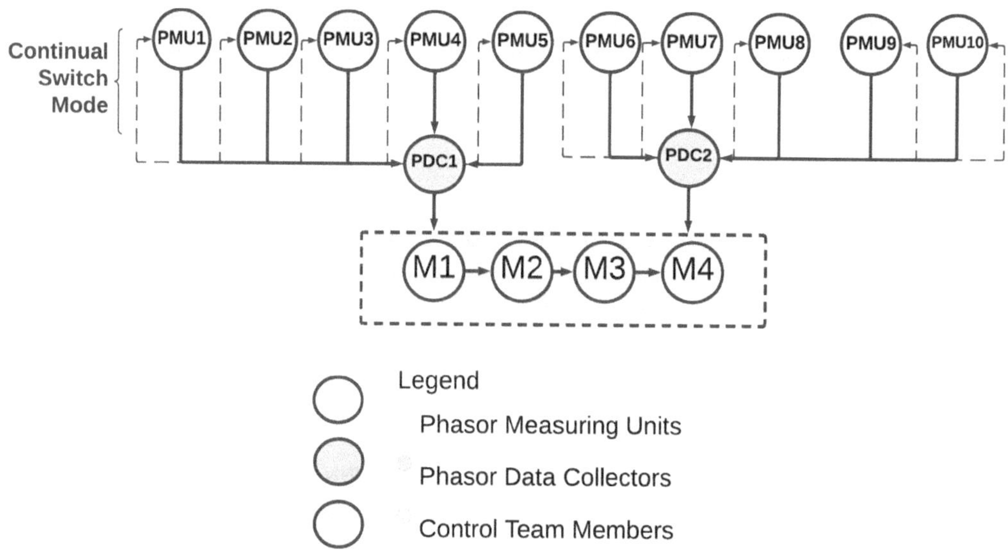

FIGURE 6.6 Simulated Network Model Based on ASC

TABLE 6.2
Observation of Connection Time in the Formation of Various Networks among PMUs, PDCs, and Control Teams in Smart Power Grid Systems

Network Formation	Time Taken (ms)
Average time for a member in the control team to connect with other member nodes and all PDCs	21.03
Average time for a PDC to connect with respective PMU and control team	4.41
Average time for a PMU to connect to respective PDC	19.68

copy of the current ASC. These copies of ASC were used by PMUs and PDCs to verify the identity of an unknown node trying to connect to the network. The time taken to form a network, request a copy of the ASC by PMUs and PDCs, verify the chain by control team members, and update the chain by control team members is shown in Tables 6.2 and 6.3.

According to the results presented in Table 6.2, it can be noted that the formation of networks between PMUs, PDCs, and control team members occurred without significant delays. The overall time taken for network formation was satisfactory. Among the different scenarios, the highest average delay of 21.03 ms was observed during the network formation involving PMU and

TABLE 6.3
Response Time (ms) for ASC Requests

Nodes	PMUs	PDCs	Control Team
Average time to obtain an ASC	22.92	13.69	20.02
Average time to add a new session to ASC	×	×	11.91
Average time to verify ASC	×	×	24.12
Average time to update ASC	×	×	44.30

its corresponding PDC. Similarly, when analyzing the average response times for various ASC requests, as depicted in Table 6.3, it can be concluded that the observed values were within acceptable ranges. The maximum average delay was found in the case of updating ASC requests initiated by the control team, with an average delay of 44.30 ms.

6.6.3 Detection of Abnormal Behavior and Verification

In this test, three KNN models (KNN 1–3) from Figure 6.5 were deployed to PMUs, PDCs, and control team members (M1–M4), respectively. These models contributed to the verification of abnormalities (i.e., attacks or natural events). State information was delivered based on the degree of threat. A higher degree of suspicion led to increased interaction between nodes and a higher chance of correctly verifying the threat state.

The initial state classification of the signals was performed by the PMU using the KNN 1 AI model. Both the signals and the classification result were transmitted to the PDC in case of attack event detection. Otherwise, only classification results were transmitted in the case of natural events. PDC also followed the same strategy as PMU, transmitting both the signals and classification results to control team members in case of an attack event. PDC used the KNN2 AI model for event classification. PDC transmitted only classification results in the case of natural event detection. The bytes of data transmitted by the PMU to the PDC and then the PDC to the respective control team members were recorded for both natural and attack events. The final state classification was performed by control team members using the KNN3 AI model to verify threat conditions such as FDIAs. The size of message delivery by the PMUs and PDCs to the control team during natural and threat scenarios is shown in Figure 6.7. A data reduction mode was initiated in cases of natural events, and the reported data sizes are demonstrated in the figure. In contrast, the original data sizes for attack events before transmission were equivalent to the reported data sizes for attack events, as denoted by a triangle and square markers. In the figure, an upward dashed slope from natural event 14 to attack event 1 denotes an emergency and continuous transmission of signals when an attack event is encountered. The slope also implied that less volume of data was transmitted in cases of natural events. However, the original sizes of the data for multiple cases of natural events ranged from 2391 to 2227 bytes, with a range of 164 bytes. It should be noted that natural events 12–14 had a relatively smaller size of the original data compared to natural events 1–11. Nonetheless, the variation in the size of the original data for natural events mattered less because of the initiation of the data reduction mode. Hence, the ASA was adaptive to the nature of events and contributed to the flexibility of having both continuous and reduced modes of data transmission.

6.7 CONCLUSION

Today's global village comprises numerous interconnected devices, from diverse IoT devices to servers with AI capabilities. The union of AI and IoT has opened the door for innovative research opportunities and progress in developing efficient IoT systems. AI has the potential to empower IoT devices to learn, reason, and make decisions based on the vast amount of data collected by the IoT sensors.

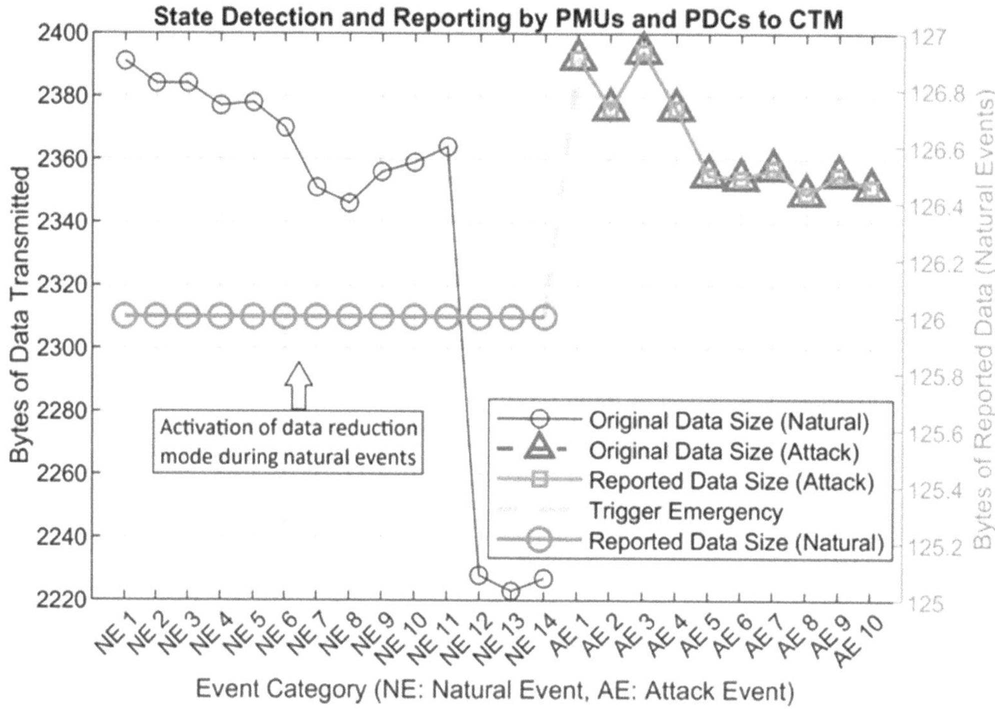

FIGURE 6.7 Automatic State Detection, Verification, and Reporting in Smart Power Grid Systems

However, the resource-constrained nature of IoT devices faces challenges in AI algorithm integration, data management, and privacy preservation. An IoT system is considered robust and resilient when its vulnerabilities and various security aspects are thoroughly studied and designed. Among the various design considerations, this chapter focused on AI and IoT integration and the corresponding challenges, particularly in CPS. Such systems are impacted by edge computing powered by innovative learning mechanisms like FL, AIoT, and distributed learning. However, the performance of AI models trained through those processes can be enhanced. In interconnected CPS, nodes with dynamic computational complexity and various communication requirements are prime concerns. Therefore, the focus on developing systematic approaches to address AI and hardware compatibility issues and achieve high performance of AI across a wide range of applications, such as smart cities, smart transportation, supply chains, and agriculture, is imperative. We proposed an AI spread-out model to deal with the diverse nature of participating nodes in the network. The proposed mechanism for AI distribution within IoT will not only facilitate learning but also enable the execution of AI processes based on that learning. Diverse computing platforms, including PMUs and PDCs of power grid systems, were studied and their corresponding AI-empowered versions, ISDs, and IPs were proposed in this study. For such applications as smart power grids, we proposed ASC to facilitate safe authentication and authorization and manage various requests. Along with the AI spread-out model and ASC, ASA was proposed to detect and verify anomalies and intrusions in the system. The simulation results of AI spread-out, ASC, and ASA prove to be very efficient and effective in terms of performance metrics of AI models, minimal delay in the ASC network, and the reduced data mode of ASA. The simulated AI spread-out model demonstrated its effectiveness through the KNN ML method. The performance metrics showed promising results for the KNN spread-out model, which achieved a maximum accuracy rate of 94.26%. This accuracy was obtained for the KNN model 1 in classifying natural events and attack events in a smart power grid system. Later, KNN models, as per hardware specification, were distributed across ISDs, IPs, and control team members that verified threats like FDIA attacks, and in safe scenarios, data reduction mode was initiated.

REFERENCES

Agrawal, S., Sarkar, S., Aouedi, O., Yenduri, G., Piamrat, K., Bhattacharya, S., Maddikunta, P. K. R. & Gadekallu, T. R., 2022. Federated Learning for Intrusion Detection System: Concepts, Challenges and Future Directions.*Computer Communications*, 195, pp. 346–361. Elsevier. Netherlands https://doi.org/10.48550/arXiv.2106.09527.

Almanifi, O. R. A., Chow, C. O., Tham, M. L., Chuah, J. H. & Kanesan, J., 2023. Communication and Computation Efficiency in Federated Learning: A Survey. *Internet of Things*, Volume 22, p. 100742, https://doi.org/10.1016/j.iot.2023.100742.

Anand, P., Singh, Y., Selwal, A., Singh, P. K., Felseghi, R. A. & Raboaca, M. S., 2020. IoVT: Internet of Vulnerable Things? Threat Architecture, Attack Surfaces and Vulnerabilities in Internet of Things and Its Applications towards Smart Grids. *Energies*, Volume 13, no. 18, p. 4813, https://doi.org/10.3390/en13184813.

Aponte-Novoa, F. A., Sandoval Orozco, A. L., Villanueva-Polanco, R. & Wightman, P., 2021. The 51% Attack on Blockchains: A Mining Behavior Study. *IEEE Access*, Volume 9, pp. 140549–140564, https://doi.org/10.1109/ACCESS.2021.3119291.

Bai, Y., Park, J., Tehranipoor, M. & Forte, D., 2022. Real-time Instruction-level Verification of Remote IoT/CPS Devices via Side Channels. *Discover Internet of Things*, Volume 2, no. 1, p. 1, https://doi.org/10.1007/s43926-022-00021-2.

Banabilah, S., Aloqaily, M., Alsayed, E., Malik, N. & Jararweh, Y., 2022. Federated Learning Review: Fundamentals, Enabling Technologies, and Future Applications. *Information Processing & Management*, Volume 59, no. 6, p. 103061, https://doi.org/10.1016/j.ipm.2022.103061.

Beaver, J. M., Borges-Hink, R. C. & Buckner, M. A., 2013. An Evaluation of Machine Learning Methods to Detect Malicious SCADA Communications. *2013 12th International Conference on Machine Learning and Applications*, Miami, FL, pp. 54–59, https://doi.org/10.1109/ICMLA.2013.105.

Bhardwaj, A., Kaushik, K., Alshehri, M., Mohamed, A. A. & Keshta, I., 2023. ISF: Security Analysis and Assessment of Smart Home IoT-Based Firmware. *ACM Transactions on Sensor Networks*, New York, NY, pp.1–19, https://doi.org/10.1145/3578363.

Bousbiat, H., Bousselidj, R., Himeur, Y., Amira, A., Bensaali, A., Fadli, F., Mansoor, W. & Elmenreich, W., 2023. Crossing Roads of Federated Learning and Smart Grids: Overview, Challenges, and Perspectives. *Machine Learning*, https://doi.org/10.48550/arxiv.2304.08602.

Catillo, M., Pecchia, A. & Villano, U., 2023. CPS-GUARD: Intrusion Detection for Cyber-physical Systems and IoT Devices Using Outlier-aware Deep Autoencoders. *Computers & Security*, Volume 129, p. 103210, https://doi.org/10.1016/j.cose.2023.103210.

Chen, J., Xue, J., Wang, Y., Huang, L., Baker, T. & Zhou, Z., 2023. Privacy-Preserving and Traceable Federated Learning for Data Sharing in Industrial IoT Applications. *Expert Systems with Applications*, Volume 213, pp. 119036–119036, https://doi.org/10.1016/j.eswa.2022.119036.

Cook, M., Paterson, C., Marnerides, A. K. & Pezaros, D., 2022. Anomaly Diagnosis in Cyber-physical Systems. *ICC 2022- IEEE International Conference on Communications, IEEE*, Seoul, Republic of Korea, pp. 5445–5450, https://doi.org/10.1109/icc45855.2022.9838968.

Dong, B., Shi, Q., Yang, Y., Wen, F., Zhang, Z. & Lee, C., 2021. Technology Evolution from Self-Powered Sensors to AIoT Enabled Smart Homes. *Nano Energy*, Volume 79, p. 105414, https://doi.org/10.1016/j.nanoen.2020.105414.

Ghimire, B. & Rawat, D. B., 2022. Recent Advances on Federated Learning for Cybersecurity and Cybersecurity for Federated Learning for Internet of Things. *IEEE Internet of Things Journal*, Volume 9, no. 11, p. 1, https://doi.org/10.1109/jiot.2022.3150363.

Hassan, Q. F. ed., 2018. *Internet of Things A to Z: Technologies and Applications*, John Wiley & Sons, Hoboken, NJ, https://doi.org/10.1002/9781119456735.

Hu, M., Cao, E., Huang, H., Zhang, M., Chen, X. & Chen, M., 2023. AIoTML: A Unified Modeling Language for AIoT-Based Cyber–Physical Systems. *IEEE Transactions on Computer-Aided Design of Integrated Circuits and Systems*, Volume 42, no. 11, pp. 3545–3558, https://doi.org/10.1109/tcad.2023.3264786.

Jayagopal, V., Elangovan, M., Singaram, S. S., Shanmugam, K. B., Subrama-Niam, B. & Bhukya, S., 2023. Intrusion Detection System in Industrial Cyber-Physical System Using Clustered Federated Learning. *SN Computer Science*, Volume 4, no. 5, https://doi.org/10.1007/s42979-023-01821-1.

Keshk, M., Sitnikova, E., Moustafa, N., Hu, J. & Khalil, I., 2021. An Integrated Framework for Privacy-Preserving Based Anomaly Detection for Cyber-Physical Systems. *IEEE Transactions on Sustainable Computing*, Volume 6, no. 1, pp. 66–79, https://doi.org/10.1109/tsusc.2019.2906657.

Khan, M., Glavin, F. G. & Nickles, M., 2023. Federated Learning as a Privacy Solution - An Overview. *Procedia Computer Science*, Volume 217, pp. 316–325, https://doi.org/10.1016/j.procs.2022.12.227.

Lakhan, A., Mohammed, M. A., Nedoma, J., Martinek, R., Tiwari, P., Vidyarthi, A., Alkhayyat, A. & Wang, W., 2023. Federated-Learning Based Privacy Preservation and Fraud-Enabled Blockchain IoMT System for Healthcare. *IEEE Journal of Biomedical and Health Informatics*, Volume 27, no. 2, p. 1, https://doi.org/10.1109/jbhi.2022.3165945.

Li, J., Mahmoodi, T. & Lam, H. K., 2023. Distributed Learning in Heterogeneous Environment: Federated Learning with Adaptive Aggregation and Computation Reduction. *ICC 2023- IEEE International Conference on Communications*, Rome, Italy, pp. 1976–1981, https://doi.org/10.1109/ICC45041.2023.10279140.

Li, Y., Wei, X., Li, Y., Dong, Z. & Shahidehpour, M., 2022. Detection of False Data Injection Attacks in Smart Grid: A Secure Federated Deep Learning Approach. *IEEE Transactions on Smart Grid*, Volume 13, no. 6, pp. 4862–4872, https://doi.org/10.1109/tsg.2022.3204796.

Liu, S., Guo, B., Fang, C., Wang, Z., Luo, S., Zhou, Z. & Yu, Z., 2024. Enabling Resource Efficient AIoT System with Cross-level Optimization: A Survey. *IEEE Communications Surveys & Tutorials*, Volume 26, no. 1, pp. 389–427, https://doi.org/10.1109/COMST.2023.3319952.

Ma, C., Li, J., Ding, M., Shi, L., Wang, T., Han, Z. & Poor, H. V., 2022. When Federated Learning Meets Blockchain: A New Distributed Learning Paradigm. *Networking and Internet Architecture*, https://doi.org/10.48550/arXiv.2009.09338.

Ma, X. & Xu, D., 2023. TORR: A Lightweight Blockchain for Decentralized Federated Learning. *IEEE Internet of Things Journal*, Volume 11, no. 1, p. 1, https://doi.org/10.1109/jiot.2023.3288078.

Mansour, R. F., 2022. Artificial Intelligence Based Optimization with Deep Learning Model for Blockchain Enabled Intrusion Detection in CPS Environment. *Scientific Reports*, Volume 12, p. 12937, https://doi.org/10.1038/s41598-022-17043-z.

Mohawesh, R., Maqsood, S., Jararweh, Y. & Salameh, H. B., 2023. Federated Learning Support for Cybersecurity: Fundamentals, Applications, and Opportunities. *2023 International Conference on Intelligent Computing Communication, Networking and Services (ICCNS)*, Valencia, Spain, pp. 50–56, https://doi.org/10.1109/iccns58795.2023.10193279.

Morris, T. & Gao, W., 2014. Industrial Control System Traffic Data Sets for Intrusion Detection Research. In J. Butts & S. Shenoi (eds.) *Critical Infrastructure Protection VIII. ICCIP 2014, IFIP Advances in Information and Communication Technology*, Springer, Berlin, Heidelberg, Volume 441, pp. 65–78, https://doi.org/10.1007/978-3-662-45355-1_5.

Pandya, S., Srivastava, G., Jhaveri, R., Babu, M. R., Bhattacharya, S., Maddikunta, P. K. R., Mastorakis, S., Piran, Md. J. & Gadekallu, T. R., 2023. Federated Learning for Smart Cities: A Comprehensive Survey. *Sustainable Energy Technologies and Assessments*, Volume 55, p. 102987, https://doi.org/10.1016/j.seta.2022.102987.

Pise, A., Yoon, B. & Singh, S., 2023. Enabling Ambient Intelligence of Things (AIoT) Healthcare System Architectures. *Computer Communications*, Volume 198, pp. 186–194, https://doi.org/10.1016/j.comcom.2022.10.029.

Pulkkis, G., Karlsson, J. & Westerlund, M., 2018. Blockchain-Based Security Solutions for IoT Systems. In Q. Hassan (ed.) *Internet of Things A to Z: Technologies and Applications*, Wiley-IEEE Press, Hoboken, NJ, pp. 255–274, https://doi.org/10.1002/9781119456735.ch9.

Qammar, A., Karim, A., Ning, H. & Ding, J., 2022. Securing Federated Learning with Blockchain: A Systematic Literature Review. *Artificial Intelligence Review*, Volume 56, pp. 3951–3985, https://doi.org/10.1007/s10462-022-10271-9.

Raja, S., Manikandasaran, S. S. & Doss, R., 2022. Threat Modeling and IoT Attack Surfaces. In S. Aurelia & S. Paiva (eds.) *Immersive Technology in Smart Cities*, Springer, Cham, Switzerland, pp. 153–165, https://doi.org/10.1007/978-3-030-66607-1_11.

Rieke, N., Hancox, J., Li, W., Milletarì, F., Roth, H. R., Albarqouni, S., Bakas, S., Galtier, M. N., Landman, B. A., Maier-Hein, K., Ourselin, S., Sheller, M., Summers, R. M., Trask, A., Xu, D., Baust, M. & Cardoso, M. J., 2020. The Future of Digital Health with Federated Learning. *npj Digital Medicine*, Volume 3, no. 119, pp. 1–7, https://doi.org/10.1038/s41746-020-00323-1.

Rizvi, S., Ahmed, S. R. & Gungor, V. C., 2020. Identifying the Attack Surface for IoT Network. *Internet of Things*, Volume 9, p. 100162, https://doi.org/10.1016/j.iot.2020.100162.

Samaraweera, C., Bhattarai, A., Peng., D., Liu, Y. & Sharif, H., 2025. Artificial Intelligence and the Internet of Things: Algorithms, Applications, and Challenges. In Q. Hassan (ed.) *Internet of Things A to Z: Technologies and Applications*, Second Edition, Wiley-IEEE Press, New Jersey, USA. In press.

Sun, C. C., Hahn, A. & Liu, C. C., 2018. Cyber Security of a Power Grid: State-of-the-Art. *International Journal of Electrical Power & Energy Systems*, Volume 99, pp. 45–56, https://doi.org/10.1016/j.ijepes.2017.12.020.

Su¨ren, E., Heiding, F., Olegård, J. & Lagerström, R., 2023. PatrIoT: Practical and Agile Threat Research for IoT. *International Journal of Information Security*, Volume 22, pp. 213–233, https://doi.org/10.1007/s10207-022–00633-3.

Touat, O. & Bouchenak, S., 2023. Towards Robust and Bias-free Federated Learning. *Proceedings of the 3rd Workshop on Machine Learning and Systems*, Rome, Italy, pp. 49–55, https://doi.org/10.1145/3578356.3592576.

Wang, G., Payani, A., Lee, M. & Kompella, R., 2023. Mitigating Group Bias in Federated Learning: Beyond Local Fairness. *Machine Learning*, https://doi.org/10.48550/arXiv.2305.09931.

Zainudin, A., Akter, R., Kim, D. S. & Lee, J. M., 2023. Federated Learning Inspired Low-Complexity Intrusion Detection and Classification Technique for SDN Based Industrial CPS. *IEEE Transactions on Network and Service Management*, Volume 20, no. 3, pp. 2442–2459, https://doi.org/10.1109/tnsm.2023.3299606.

Zhou, S. & Li, G. Y., 2023. Federated Learning Via Inexact ADMM. *IEEE Transactions on Pattern Analysis and Machine Intelligence*, Volume 45, no. 8, pp. 9699–9708, https://doi.org/10.1109/tpami.2023.3243080.

Zhu, J., Cao, J., Saxena, D., Jiang, S. & Ferradi, H., 2022. Blockchain-empowered Federated Learning: Challenges, Solutions, and Future Directions. *ACM Computing Surveys*, Volume 55, no. 11, pp. 1–31, https://doi.org/10.1145/3570953.

Zhu, J., Shi, Y., Fu, M., Zhou, Y., Wu, Y. & Fu, L., 2023. Latency Minimization for Wireless Federated Learning with Heterogeneous Local Model Updates. *IEEE Internet of Things Journal*, Volume 11, no. 1, p. 1, https://doi.org/10.1109/jiot.2023.3285937.

7 GeoSense

A Smart IoT-Driven Framework for Landslide Monitoring in the Himalayas

Praveen Kumar, Abbas Husain, Priyanka Priyanka, Akshay Kumar, Sahil Sankhyan, Kala Venkata Uday and Varun Dutt

7.1 INTRODUCTION

The ability of Internet of Things (IoT) technology to gather and share data in real time across several domains has led to its widespread adoption (Li et al., 2015; Hassan, 2018). For example, one application of IoT technology has been for assisting in monitoring and warning against disasters like landslides (Chaturvedi et al., 2018). As per Chaturvedi et al. (2018), IoT-based landslide monitoring systems (LMSs) can assist in the monitoring of landslides and in alerting people to these calamities in remote and rural locations.

Landslides frequently occur in the Himalayan region, resulting in numerous casualties and losses (Parkash, 2011). Most of these calamities are brought on by rain and happen during the monsoon season (Crosta, 1998). In the monsoon season of 2023, for instance, there was a significant flash flood and cloudburst in the Mandi district of Himachal Pradesh, India, resulting in multiple landslides and injuries.[1] In areas where landslides occur frequently, landslide detection systems (LMSs) can be installed and can assist in producing timely warnings about approaching landslides (Savvaidis, 2003; Fosalau et al., 2016; Chaturvedi et al., 2018; Van Khoa and Takayama, 2018; Pathania et al., 2020; Khan et al., 2021; Song et al., 2021). For instance, an LMS installed in a region susceptible to landslides can assist in distributing early warnings regarding slope movements through the use of pre-established thresholds based on movement values and weather conditions (Savvaidis, 2003; Fosalau et al., 2016; Chaturvedi et al., 2018; Van Khoa and Takayama, 2018; Pathania et al., 2020; Khan et al., 2021; Song et al., 2021). Yet just a handful of these LMSs can provide alerts about future slope movements in real time (Savvaidis, 2003; Fosalau et al., 2016; Chaturvedi et al., 2018; Van Khoa and Takayama, 2018; Pathania et al., 2020; Khan et al., 2021; Song et al., 2021). As a result, it is necessary to improve the current IoT-based LMSs so that they can continuously monitor landslides in order to produce forecasts about impending landslides and rainfall.

In a prior developed IoT-based LMS, Chaturvedi et al. (2018) included weather sensors to evaluate weather variables and soil sensors to measure soil variables, including slope movement (i.e., changes in soil accelerations on the hill slope). The deployed system sent weather and soil data to a cloud database. The data on the cloud was compared against pre-determined movement thresholds (determined using lab-scale setups), and early warnings were generated through SMSs when pre-defined threshold values were breached. However, this framework had certain limitations, including the absence of any local warnings in the surrounding area and the static nature of the threshold values, which did not vary based on the history of slope movements. Also, Chaturvedi et al. (2018) tested their LMS on a lab-based ramp setup, without testing it on real-world landslide

DOI: 10.1201/9781003506638-9

locations. Finally, the developed LMS was not energy efficient because the sensors were always on without exploration into sleep modes in the microcontroller. These drawbacks related to the existing IoT-based LMSs need to be overcome to provide better landslide disaster management solutions in the real world.

The main objective of this chapter is to overcome the limitations of existing LMSs. In this chapter, we introduce a "GeoSense IoT Framework," which gathers information from a number of deployed LMSs that are energy efficient. The energy efficiency is obtained by activating different sleep modes and interrupt-driven tasks. In GeoSense, a set of machine learning (ML) models are run to dynamically set thresholds for early warnings against impending landslides (Kumar et al., 2019a, 2019b; Pathania et al., 2021; Bruzón et al., 2021). For this purpose, the best-performing ML model is deployed on the cloud to predict the slope movement in real time as well as ahead of time. The framework also consists of a system to generate warnings locally using hooters and blinkers in the surrounding area using long-range radio (LoRa) technology (Zourmand et al., 2019). The GeoSense also has multiple accelerometer sensors across LMSs deployed on slopes, which help the framework sense the slope movement at multiple horizontal and sub-surface locations. Overall, the GeoSense is an advanced version of the prior IoT-based LMS (Chaturvedi et al., 2018) that provides real time monitoring of hill slopes – across energy-efficient LMSs deployed across several landslide locations, predictions of slope movements ahead of time, and local and global warning through ML models.

To explain the GeoSense IoT Framework, we first provide a literature review of the existing IoT technologies for monitoring landslides. Next, we go into the ML model development to forecast slope movements as well as the architecture of the GeoSense Framework. Additionally, we discuss the outcomes of the ML models' prediction of the slope movements within the framework. Lastly, we talk about the outcomes and practical applications of the suggested framework.

7.2 BACKGROUND

A number of technologies and equipment are described by Savvaidis (2003) as being used to monitor landslides. These include global positioning systems (GPS), photogrammetry, satellite remote sensing, and geotechnical instruments, such as tilt meters and extensometers. Landslide dynamics are better understood because of these techniques, which offer useful data on surface displacements. Their incapacity to provide real-time monitoring and prediction is one of their main drawbacks.

Fosalau et al. (2016) emphasized the critical need for functional LMSs to reduce the devastating consequences of landslides. These authors carried out their research in which they constructed wireless sensor networks (WSN) using sensitive transducers to measure slope displacement and orientation in real time. The goal was to create a system that would function as a landslide early warning system. The WSN architecture has contributed vital insights to enhance landslide mitigation measures. Nevertheless, the WSN architecture prioritizes the gathering of data in real time rather than predicting the movement of slopes based on historical data patterns. This gap may be addressed by incorporating predictive modeling techniques.

Van Khoa and Takayama (2018) created an LMS system to track landslides in Japan with the use of WSN. The Host System (HS), Cloud System (CS), and Local Sensing Node Network System (LSNNS) comprise the architecture. Nevertheless, this technology is unable to produce slope movement forecasts or real-time alerts.

In addition, an IoT infrastructure for landslide monitoring was designed by researchers (Chaturvedi et al., 2018; Pathania et al., 2020). The authors used weather monitoring sensors in these infrastructures for temperature, humidity, barometric pressure, sunlight intensity, and rainfall measurements. Furthermore, they used an accelerometer to measure the acceleration of the slope and keep track of any movement. However, the framework mentioned by Chaturvedi et al. (2018) employs only a single accelerometer sensor for surface monitoring purposes. On the other hand, the framework highlighted by Pathania et al. (2020) uses multiple accelerometer sensors to monitor subsurface movements of the slope. Furthermore, the authors built many ML models to anticipate the lateral motion of slopes based on historical data from the Gharpa landslide in Himachal Pradesh, India (Pathania et al., 2020). Nevertheless, it was difficult to confirm their predictions using historical information on the same slide events.

Khan et al. (2021) deliberated landslide monitoring via WSNs while revealing the challenge of unreliable channels of communication across tough terrains. By incorporating collaborative local data analysis, this study assessed the effect of lossy communication on the accuracy and efficiency of landslide prediction and detection models. The results emphasize the need for comprehensive strategies that combine regional and localized parameters to improve the reliability of WSN-based LMSs in practical environments. Nevertheless, the developed LMSs do not sufficiently provide continuous monitoring and lack a real-time prediction framework that is important for early recognition and timely warnings about potential landslides.

Song et al. (2021) examined the construction of the Korea Institute of Geoscience and Mineral Resources Landslide Monitoring System (KIGAM-LAMOS) for natural terrain. KIGAM-LAMOS is implemented in several national parks in Korea and combines multiple types of sensors and data acquisition equipment to monitor landslide-triggering factors, such as rainfall and soil properties, in real time. This assessment utilizes advanced cloud computing to store and manage data, increasing operational efficiency and reducing maintenance costs. This data was collected through 12 installed monitoring systems over a period of five years. Slope angles were classified using a digital elevation model. Findings indicated that there is a high probability of landslide occurrence in the study area, with 82% of the study area falling under "high" and "very high" hazard levels. Nevertheless, cloud-based ML models were not emphasized, where the system's focus was on immediate analysis based on historical information as well as current details concerning hazards.

Several researchers have developed IoT-based LMSs to monitor landslides. For example, Nguyen et al. (2024) developed a landslide monitoring technique by fusing ML with IoT systems. This approach, tested in Quang Ninh, Vietnam, combines real-time data collection with ML models to predict the likelihood of landslides. It makes use of a variety of sensors, such as soil moisture and rainfall. Although this system senses the surface conditions causing landslides, it does not provide real-time monitoring of landslides.

Sharma et al. (2019) launched cloud-based forecasting for air quality prediction using ML models. For the next 30 minutes, their method projected five meteorological parameters and seven contaminants. But this system could only predict values for one site. In order for ML models to predict values at multiple places, they need to take variability or generalization into account.

Several ML models, such as the moving average, information gain, lazy method, instance-based, autoregression, support vector machine (SVM), neural networks, and recurrent neural networks, have been developed in order to predict slope movements (Kumar et al., 2019a, 2019b, 2020, 2021a, 2021b; Pathania et al., 2022; Kumar et al., 2023). Nevertheless, the simultaneous use of cloud-based ML models for predicting slope movements at various locations concurrently may make it a time-consuming and computationally expensive task. Therefore, one requires lightweight ML models like decision trees to quickly predict multiple locations simultaneously.

Furthermore, with very few cases of movement and a high number of instances of no movement, there can be a class imbalance in the landslide dataset. Because of this mismatch, the ML model might favor more accurate predictions of no movements than one with movements. In order to solve the class imbalance problem and precisely anticipate slope movements at an Indian landslide, Kumar et al. (2019a,b) developed a Random Forest (RF) model. The RF model was compared in the study with autoregressive (AR) and SVM models. As a result, it was found that the RF model performed better than the SVM and AR models. Similar to this, Pathania et al. (2021) created a Gradient Boosting (GB) model for the Gharpa landslide in Mandi district in India. Decision Tree, RF, SVM, and Naive Bayes models were all outperformed by the GB model.

Bruzón et al. (2021) also suggested the use of the Extra Tree (ET) method for mapping the susceptibility to landslides throughout the Central and Southern Guerrero regions of Mexico. The ET model was compared to many other models. When estimating the likelihood of landslides, the ET model fared better than any other model. Building on the work of Bruzón et al. (2021), this study focuses on estimating slope changes in the GeoSense Framework utilizing ensemble models such as GB, RF, and ET.

To be effective, the deployed ML models need to generate real-time warnings. This real-time ML model deployment could be done either on the edge or on a cloud server (Hassan, 2011; Kumar et al., 2024). For example, Kumar et al. (2024) developed an LMS that made use of edge computing. Wireless edge servers were positioned close to the monitoring nodes in the LMS design to locally process real-time data from various environmental sensors and community inputs, facilitating prompt and precise decision-making. These edge servers produced adaptive thresholds for landslip notifications based on past and present data by employing ML models. However, we believe edge computing devices will consume a significant amount of energy. In edge computing, energy will be consumed by both wireless connectivity and local processing at the edge. The large energy dependence can be problematic for harsh and distant locations where regular maintenance may become difficult. Also, manual updates may be necessary for such systems, which could result in inconsistent updates and downtime. On the other hand, cloud-based predictions may likely depend upon centralized model updates, guaranteeing that all LMSs take advantage of the most recent enhancements to patterns in collected data. We believe that cloud-based LMSs are likely to be more flexible and easier to maintain due to the centralized computation in the cloud (Hassan, 2011).

While a number of LMSs have been designed to support real-time monitoring, and some of them even offer predictive analysis based on historical data, none of these LMSs have offered a comprehensive framework for producing real-time predictions. Due to this literature gap, the creation of an all-encompassing system that can continually monitor landslides, transmit data to the cloud, and produce real-time predictions concerning landslide dangers is required. The proposed paradigm is noteworthy in terms of proactive landslide control, where financial damages could be minimized and fewer lives lost.

7.3 GEOSENSE IOT FRAMEWORK

Figure 7.1 depicts an IoT framework for landslides called "GeoSense IoT Framework," which is a complete framework for real-time landslide monitoring and prediction. The data acquisition part of the framework consists of gathering weather data (rain, temperature, barometric pressure, relative humidity, and light intensity), changes in angular velocities and linear accelerations, and soil data (soil moisture and soil temperature). The GeoSense includes recording these data from all sensors, sending these data to a cloud via a 4G module, verifying whether the thresholds on slope movements are exceeded, and alerting about these slope movements. Furthermore, GeoSense utilizes LoRa (Zourmand et al., 2019) to activate the locally deployed warning poles in the valley to deliver slope movement warnings for those slope movements that exceed pre-defined thresholds. An ML model deployed in the cloud also predicts future slope movements and generates predictive warnings via SMS on mobile phones. For making the warnings understandable by disaster management officials, slope movement warnings are divided into three categories: no-risk, low-risk, and high-risk.

7.3.1 Landslide Monitoring Systems

The LMSs in the GeoSense IoT Framework utilize micro-electromechanical systems (MEMS) sensors to gather weather data (rain, temperature, barometric pressure, relative humidity, and light intensity), changes in angular velocities and linear accelerations, and soil data (soil moisture and soil temperature) at real-world landslide sites. Landslides in the Himalayan region are rainfall-induced, where increasing water in the soil could trigger landslides (Dikshit et al., 2020). Rapid changes in temperature could expand or contract the soil, which could crack the soil and weaken its strength (Yavari et al., 2016). Humidity increases the moisture content in the soil and may cause slope

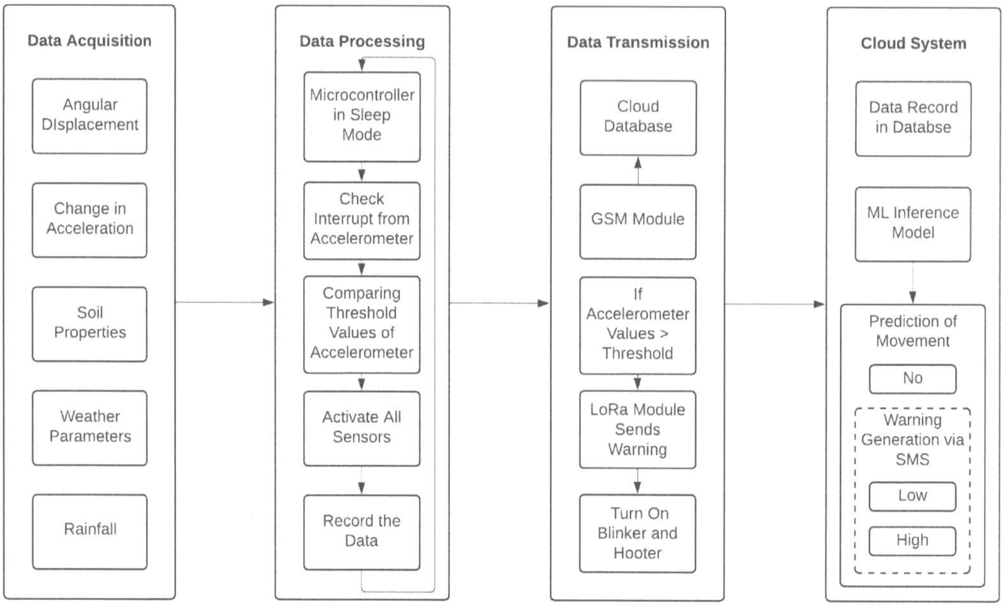

FIGURE 7.1 Different Processes in the GeoSense IoT Framework

movements (Rahardjo et al., 2001). Similarly, rainfall also increases the water content in the soil and may affect slope movements (Rahardjo et al., 2001). Lower barometric pressure conditions could cause the soil to hold water for a longer time, and this moisture could likely cause slope movements (Köhler and Schulze, 2000). Lower light intensity may likely decrease the evaporation rate of water in the soil and make the soil retain water for a longer duration, affecting landslides (Wang et al., 2017). Soil moisture measures the actual volumetric moisture content in the soil at a particular time (Wang et al., 2017). The changes in the soil linear accelerations and angular velocities allow the LMS to capture slope movements (Chaturvedi et al., 2018).

Figure 7.1 illustrates the architecture of the GeoSense IoT Framework. As mentioned above, as part of the framework, various LMS sensors evaluate slope movements, soil data, and weather data. This data is collected in LMSs from landslide locations using various sensors, such as rain gauges, accelerometers, soil moisture sensors, force resistive sensors, and weather monitoring sensors. A microcontroller with a GSM module transmits this data in real time to a cloud database every 10 minutes. The 10-minute interval is chosen based on prior work, where this interval was found to be suitable as changes in weather and soil parameters are minimal in this period (Chaturvedi et al., 2018).

The LMSs are deployed in remote areas where hardly any humans are around. If the system hangs, it is automatically restarted. To automatically restart the system if it hangs, a master-slave architecture has been developed with the help of a watchdog timer (WDT). A WDT is a hardware timer that is used to detect malfunctions and provide recovery (Chanamolu and Daim, 2020). Using the WDT, the master controller resets the non-responsive slave controller. As shown in Figure 7.2, multiple accelerometers are installed at the subsurface level or at multiple points on the outer soil surface to monitor slope movements. The master and slave run in parallel and set the WDT timer for 8 seconds. The slave initializes all the sensors and enters low-power sleep mode to conserve energy. The master also sleeps for 10 minutes and, upon waking up, sends a request to the slave for sensor data. After receiving the request, the slave wakes up, records the sensor values, and sends them to the master. The master transmits the collected data to a cloud web server using a GSM module.

Interrupt-driven events in microcontrollers are specifically built to handle interrupts only when they happen. In the absence of interrupts, the microcontroller is either inactive or performs other functions.

The processes of the master and slave systems are as follows:

- *Master:* The master system uses a LoRa module to link to a locally installed warning pole and alert valley residents of impending landslides. A 4G GSM module is also used by the master to send the recorded weather and soil data to the cloud. For battery conservation, the master system then switches on the low-power sleep mode. The WDT timer is set to run for 8 seconds, and every 8 seconds it causes an interrupt in the Interrupt Service Routine (ISR), which causes a volatile variable to increase by one. The master system awakens when the value of the volatile variable hits 75, which is equal to 10 minutes (75 × 8 seconds = 600 seconds). The GSM module is turned on by the master system when it wakes up, and it takes 30 seconds for it to connect to the Internet. To acquire data, the master interrupts the slave. The master waits for data for 8 seconds following the interrupt. It means the slave code is stuck and needs to be reset if data is not received during the waiting period. Data from the slave is transmitted to the cloud via the GSM module if it is received by the master. This pin on the accelerometer is connected to the hardware interrupt pin of the microcontroller master. The accelerometer sensors interrupt the master sleep mode instantly upon detecting movement. The master transmits a warning signal to the locally installed warning pole via the LoRa module whenever slope movement wakes it up.
- *Slave:* The WDT timer is likewise set to 8 seconds by the slave system. In ISR, the volatile variable rises by one. Up until the count of 75, the slave system waits for the master to interrupt it. The slave sends a reset signal to recover the master if this variable value hits 80, which indicates that the master's code is stuck. A multiplexer (MUX) is used to connect the slave to several accelerometers, as illustrated in Figure 7.2. The slave system is connected to rain gauges in the form of tipping buckets and weather sensors. Low-power sleep mode is the slave system's sleep mode. Both the rain interrupt from the rain sensor and the data interrupt from the master are intended to be handled by the slave system. The slave activates all of its sensors and gathers data in response to an interrupt demanding data. Serial communication is used to transmit the recorded data to the master system.

The LMS makes use of the sensors and components listed below:

- *Weather Monitoring Sensors:* To measure the temperature in degrees Celsius and the humidity in percentage, we placed a temperature sensor and a humidity sensor. We use a barometric pressure (BMP) sensor that measures the barometric pressure in millibars (mbar). To detect ambient light in lux units, we integrate a light-intensity sensor. We set up a tipping bucket rain gauge that records the amount of rain in milliliters (ml) to measure rainfall. When the bucket fills to 2.25 ml, a reed switch is activated by the tipping bucket. The reed switch interrupts the microcontroller each time the bucket tips. The number of interruptions is multiplied by 2.25 ml to obtain the total amount of rainfall. The sensor ranges are presented in Table 7.1. Figure 7.3a shows the fabricated circuit board of the LMS, and Figure 7.3b shows the weather monitoring sensors mounted on the weather nodes.
- *Force Sensitive Resistor (FSR):* We employed a variable resistor sensor as the force sensor. The sensor has a very high resistance when no force is applied to its surface. The resistance decreases with increasing force. The microcontroller reads the voltage readings through its analog pin, which is in the soil sensor port of Figure 7.3a. We use a 10K ohm resistor in a voltage divider setup and a steady voltage pushed across it to measure the force. The voltage variations are calibrated to represent the force in weight units (tons). The

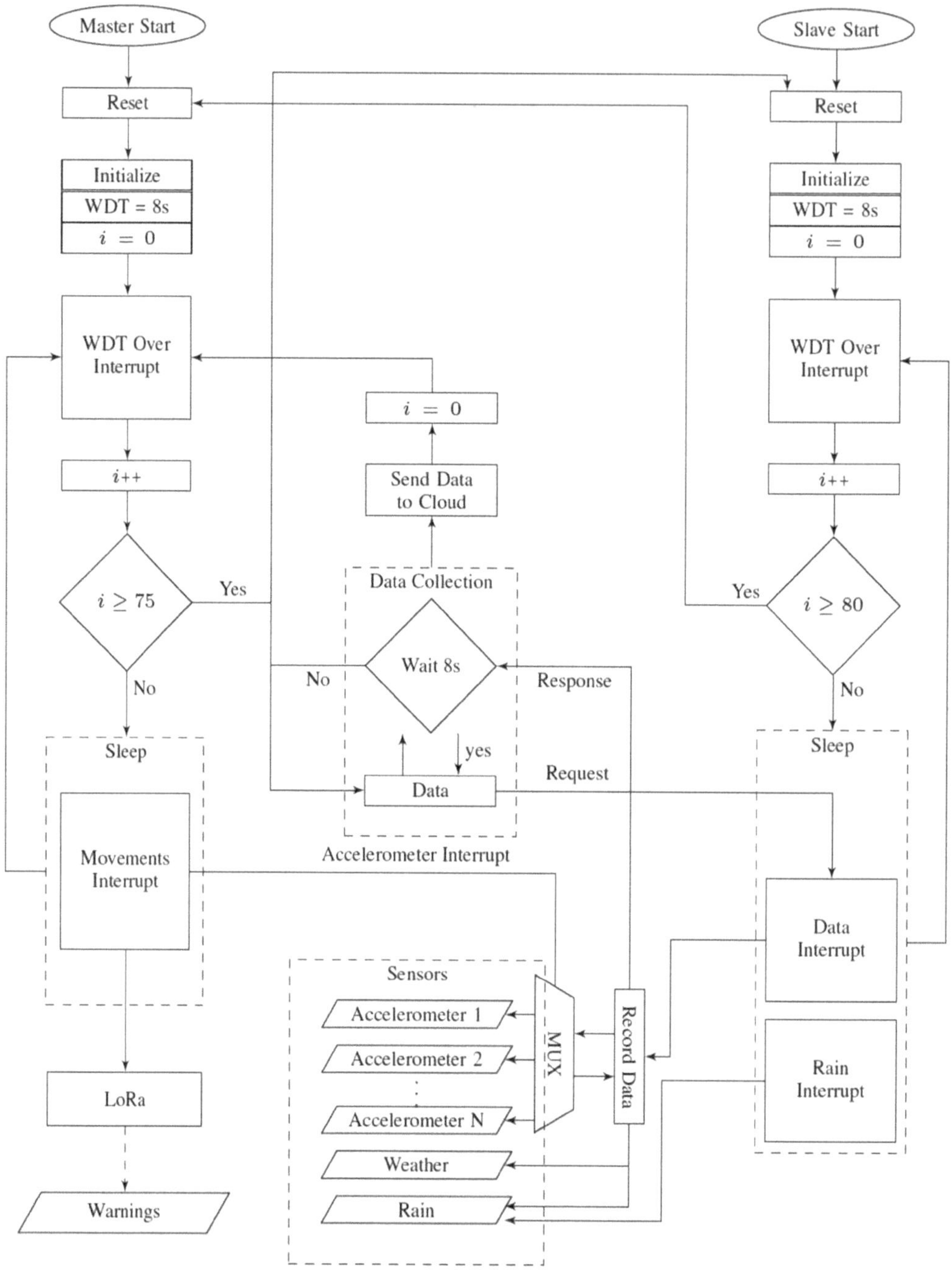

FIGURE 7.2 LMS Flowchart

microprocessor reads the force applied to the sensor by moist earth when it is buried. The FSR sensor mounted on the soil node is shown in Figure 7.3c.

- *Capacitive Soil Moisture Sensor:* The principle of capacitance underlies the operation of a capacitive soil moisture sensor (see Figure 7.3c). The sensor electronics are covered in an adhesive to ensure longevity and to make them waterproof. The dielectric characteristics of the two plates are altered when moisture is present, which increases the capacitance. This

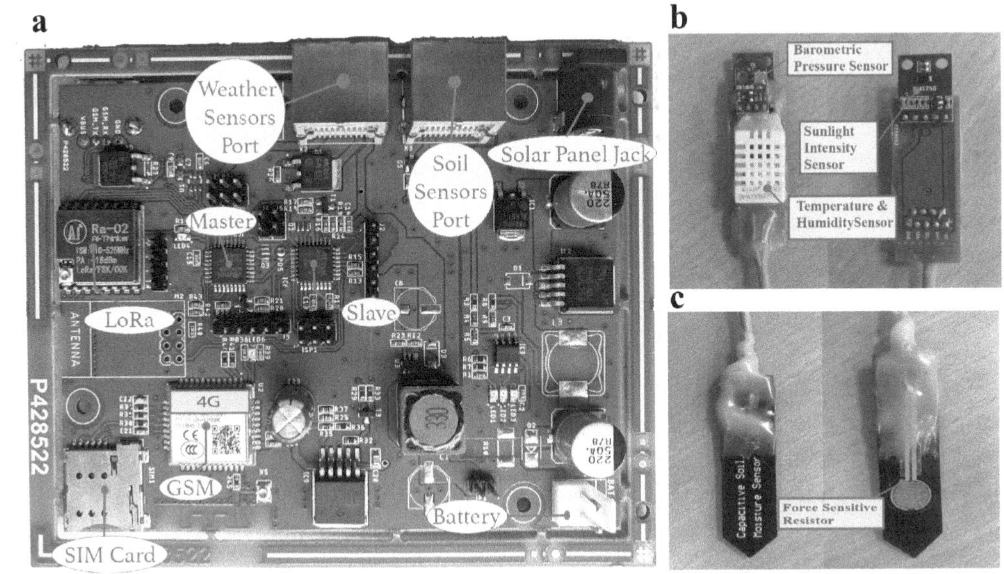

FIGURE 7.3 (a) LMS Master-slave Circuit Fabrication. (b) Weather Nodes Containing Sensors. (c) A Node with FSR and Soil Moisture Sensors

capacitance is then transformed by the sensor into a voltage value between 0 and 5 volts. Volumetric soil moisture levels ranging from 0% to 100% are detected using this sensor. This sensor is linked to the analog pin of the microcontroller, which measures the voltage.

- *Accelerometer:* Angular velocities as well as linear soil accelerations in three orthogonal directions are detected by an accelerometer during slope movements in the LMS. For slope monitoring to be effective, multiple accelerometer sensors can be located at various depths or surfaces in the LMS.
- *4G GSM Module:* This system comprises a 4G GSM module that transmits data to the cloud.
- *LoRa:* The system has a LoRa module that communicates at 433 MHz and sends alerts to a locally deployed warning pole. It has a range of 10 km from the module and requires a line of sight. When there is no mobile network coverage, the data from the LoRa module could transmit to the receiver, and then on the receiver side, a GSM could be installed to transmit the data to the Internet.
- *Warning Pole:* A warning pole (the receiving system) is made up of a steel pole containing a receiver circuit. Figure 7.4 shows the receiver circuit with its LoRa module, charging system, and hooter/blinker connections. The LoRa module on this circuit receives signals from the LMS and generates warnings via hooters and blinkers installed on the pole-top. These warning poles can be installed in the valley and villages near landslide-prone areas and beside highways, stopping traffic during slope movements.

Table 7.1 presents the different ranges and accuracy of the accelerometer, capacitive soil moisture, force-sensitive resistor, temperature, humidity, barometric pressure, and ambient light sensors.

7.3.2 Pseudocode

Algorithms 1 and 2 list the pseudocode for the master and slave microcontrollers in the LMS. Different blocks in the pseudocode include data collection, data transmission, and warning generation in the algorithms.

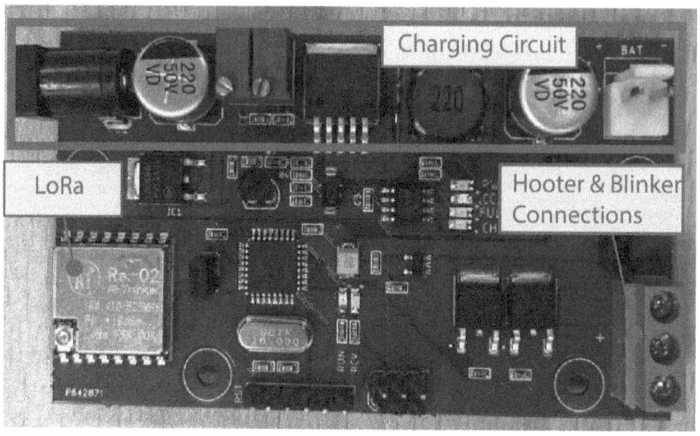

FIGURE 7.4 The Receiver's Circuit with a LoRa Module at the Warning Pole

Algorithm 1. Master operations

Description:

- The master system initializes the WDT with an 8-second timer and resets the slave system to start from a known state.
- If the master system receives a movement interrupt, immediately sends a warning to the warning pole via the LoRa module.
- The master system sends a request to get data acquired after 10 minutes and waits for 8 seconds; if the master fails to receive data in time, then it resets the slave system.
- If data is received in 8 seconds, then it transmits data to the cloud via the GSM module.

Initialization step:

- Initialize the WDT with an 8-second timer
- Reset the slave system
- $i \leftarrow 0$

While true, do

If movement interrupt is detected, **then** send a warning to the warning pole through the LoRa **end**

If the WDT timer is over, **then** $i \leftarrow i+1$

if $i \geq 75$ **then** send a request to acquire data from the slave system and wait for 8 seconds

if data is obtained from the slave system, **then** send data to the cloud via the GSM

$i \leftarrow 0$

end

else reset the slave system **end**

end

else go to sleep **end**

end

TABLE 7.1
Range, Accuracy, and Sensitivity of the Sensors

Sensors	Brand	Specifications	Values
Accelerometer	InvenSense[a]	Range	0–35°
		Accuracy	±3.0%
		Sensitivity	±0.02% full scale
Capacitive soil moisture sensor	Adafruit[b]	Range	0%–100% moisture
		Accuracy	±5.0%
		Sensitivity	±0.15%
Force sensitive resistor sensor	Adafruit[c]	Range	0.1–77.4 MPa
		Repeatability	±2.0%
		Accuracy	±10.0% full scale
		Sensitivity	0.15 MPa
		Hysteresis	+10.0%
Temperature and humidity sensors	Adafruit[d]	Humidity range	0%–100% relative humidity
		Temperature range	−40°C ~ 80°C
		Humidity accuracy	±2% Relative humidity
		Temperature accuracy	< ±0.5°C
Barometric pressure sensor	Adafruit[e]	Pressure range	300–1100 mbar
		Temperature range	−40°C ~ 85°C
		Accuracy	±12.0 millibars
		Resolution	0.0009% full scale
Ambient light sensor	Adafruit[f]	Bandwidth	400–700 nm
		Wavelength	560 nm

[a] https://shorturl.at/pOrqF
[b] https://www.adafruit.com/product/4026
[c] https://learn.adafruit.com/force-sensitive-resistor-fsr/overview
[d] https://www.adafruit.com/product/385
[e] https://www.adafruit.com/product/1603
[f] https://learn.adafruit.com/adafruit-bh1750-ambient-light-sensor/overview

Algorithm 2. Slave operations

Description:

- The slave system initializes the WDT with an 8-second timer and initializes variables $rain \leftarrow 0$.
- If the slave system receives a data request, it immediately records the data from sensors and transmits it to the master system.
- The slave system receives interrupts from the rain sensor, and then it stores the rain values.
- If the variable i value reaches 80, then reset the master.

Initialization step:

- Initialize the WDT with an 8-second timer
- $i \leftarrow 0$
- $rain \leftarrow 0$

While true, do

If interruption from the rain sensor, **then** $rain \leftarrow rain + 1$ **end**

If a request for data is received from the master system, **then**

1. Turn on all the weather and soil moisture sensors
2. Record the weather parameters and soil moisture
3. Record the acceleration and angular velocity values from several accelerometer sensors using a multiplexer

$$[Ax,\ Ay,\ Az,\ Wx,\ Wy,\ Wz,\ soil\ \ temperature] \leftarrow Accelerometer$$

4. Fetch stored rain value from *rain* variable
5. Transmit data to the master system
6. Turn off all the weather and soil moisture sensors
7. *rain*
8. $i \leftarrow 0$

end

If the WDT timer is over, **then** $i \leftarrow i+1$

if $i \geq 80$ **then** reset the master system and $i \leftarrow 0$ **end**

else go to sleep **end**

end

7.3.3 Cloud System

A cloud-based system was created, in which backend processing and data management were managed by a PHP server. A relational database was created to store processed data coming from different LMSs situated at various places. Each of these LMSs sending data possessed a unique identification number that was stored in a table in the SQL database. An ML model deployed on the CS used this new data and predicted the slope movement as 0 (no movement), 1 (low movement), or 2 (high movement).

Figure 7.5a shows the real-time data received from a selected location, where the user can see the weather data, soil data, and real-time predictions by the ML models. Figure 7.5b shows all LMSs installed as part of the GeoSense Framework in the Kinnaur district. A set of dots represents every location in Figure 7.5b. In Figure 7.5b, the ML model turns the points from blue to orange or red if an area is prone to landslides. A color scheme from yellow to deep red signifies a range between low and high probabilities of landslide hazards. Moreover, the registration menu has a link that directs users to subscribe for SMS alerts.

7.3.4 Study Area and Data Description

Figure 7.6a and b shows the study area and the locations of the 15 LMSs in Kamand Valley, Himachal Pradesh, India, 16 km from Mandi, India. Figure 7.6c shows an actual LMS installed at a landslide site with a circuit board, weather sensors, soil node, rain gauge, and a solar panel to power the system. Figure 7.6d shows the warning pole equipped with a hooter, blinker, solar panel, and a warning board to alert nearby people to landslide hazards. The warning pole was installed near the highway to warn vehicular traffic about landslide threats. Communication between the LMS and the warning pole is achieved through a LoRa module.

The dataset contained features such as temperature (°C), humidity (%), light intensity (lux), barometric pressure (in Pascal), rainfall (in mm), acceleration (Ax, Ay, and Az) in three directions

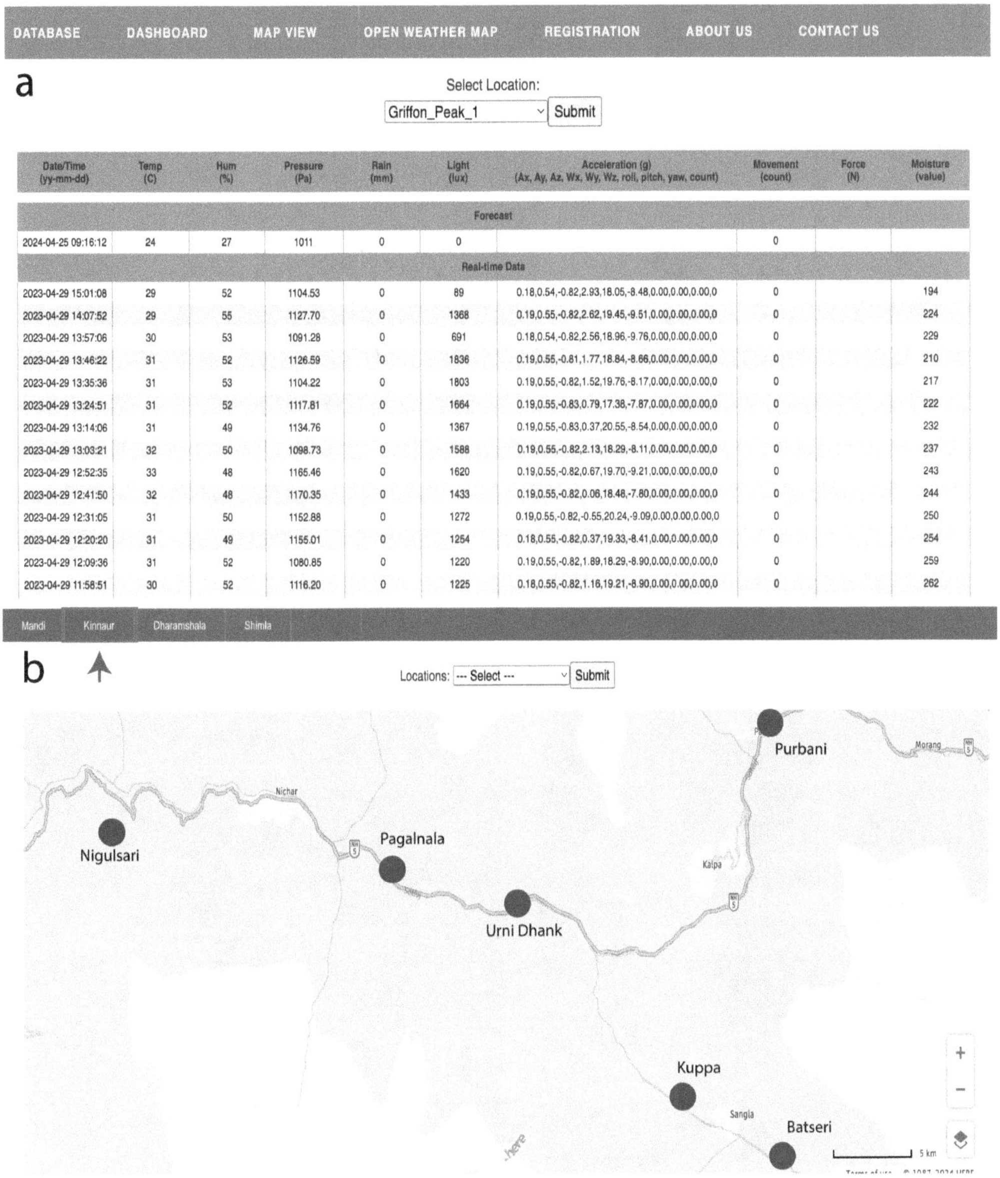

Date/Time (yy-mm-dd)	Temp (C)	Hum (%)	Pressure (Pa)	Rain (mm)	Light (lux)	Acceleration (g) (Ax, Ay, Az, Wx, Wy, Wz, roll, pitch, yaw, count)	Movement (count)	Force (N)	Moisture (value)
Forecast									
2024-04-25 09:16:12	24	27	1011	0	0		0		
Real-time Data									
2023-04-29 15:01:08	29	52	1104.53	0	89	0.18,0.54,-0.82,2.93,18.05,-8.48,0.00,0.00,0.00,0	0		194
2023-04-29 14:07:52	29	55	1127.70	0	1368	0.19,0.55,-0.82,2.62,19.45,-9.51,0.00,0.00,0.00,0	0		224
2023-04-29 13:57:06	30	53	1091.28	0	691	0.18,0.54,-0.82,2.56,18.96,-9.76,0.00,0.00,0.00,0	0		229
2023-04-29 13:46:22	31	52	1126.59	0	1836	0.19,0.55,-0.81,1.77,18.84,-8.66,0.00,0.00,0.00,0	0		210
2023-04-29 13:35:36	31	53	1104.22	0	1803	0.19,0.55,-0.82,1.52,19.76,-8.17,0.00,0.00,0.00,0	0		217
2023-04-29 13:24:51	31	52	1117.81	0	1664	0.19,0.55,-0.83,0.79,17.38,-7.87,0.00,0.00,0.00,0	0		222
2023-04-29 13:14:06	31	49	1134.76	0	1367	0.19,0.55,-0.83,0.37,20.55,-8.54,0.00,0.00,0.00,0	0		232
2023-04-29 13:03:21	32	50	1098.73	0	1588	0.19,0.55,-0.82,2.13,18.29,-8.11,0.00,0.00,0.00,0	0		237
2023-04-29 12:52:35	33	48	1165.46	0	1620	0.19,0.55,-0.82,0.67,19.70,-9.21,0.00,0.00,0.00,0	0		243
2023-04-29 12:41:50	32	48	1170.35	0	1433	0.19,0.55,-0.82,0.06,18.48,-7.80,0.00,0.00,0.00,0	0		244
2023-04-29 12:31:05	31	50	1152.88	0	1272	0.19,0.55,-0.82,-0.55,20.24,-9.09,0.00,0.00,0.00,0	0		250
2023-04-29 12:20:20	31	49	1155.01	0	1254	0.18,0.55,-0.82,0.37,19.33,-8.41,0.00,0.00,0.00,0	0		254
2023-04-29 12:09:36	31	52	1080.85	0	1220	0.19,0.55,-0.82,1.89,18.29,-8.90,0.00,0.00,0.00,0	0		259
2023-04-29 11:58:51	30	52	1116.20	0	1225	0.18,0.55,-0.82,1.16,19.21,-8.90,0.00,0.00,0.00,0	0		262

FIGURE 7.5 Graphical User Interface of the LMS's Cloud Platform. (a) A Chosen Location's Data Is Shown in Real Time. (b) A Summary of Every LMS in a Chosen District, with Various Dot Colors Denoting the Likelihood of a Landslide Hazard

(in m/s²), angular velocity (Wx, Wy, and Wz), soil moisture content (in %), and movement classification (0 for no movement, 1 for small movement, and 2 for high movement). Movement labels were calculated by setting a threshold for the change in acceleration and angular velocity values, such as within ±2 SD labeled 1 and beyond ±2 SD labeled 2.

The dataset comprised 218,882 samples and was split in an 80:20 ratio. The first 80% was utilized for training and the remaining 20% for testing. The dataset possessed a class imbalance as there were very few slope movement points and a large number of no movement points. The accuracy measure is not a good measure in the class imbalance dataset, as the ML model may show high accuracy due to the prediction of no movement class (Folleco et al., 2008). Precision in landslide prediction refers to the accuracy of the model in predicting slope movement events when it predicts

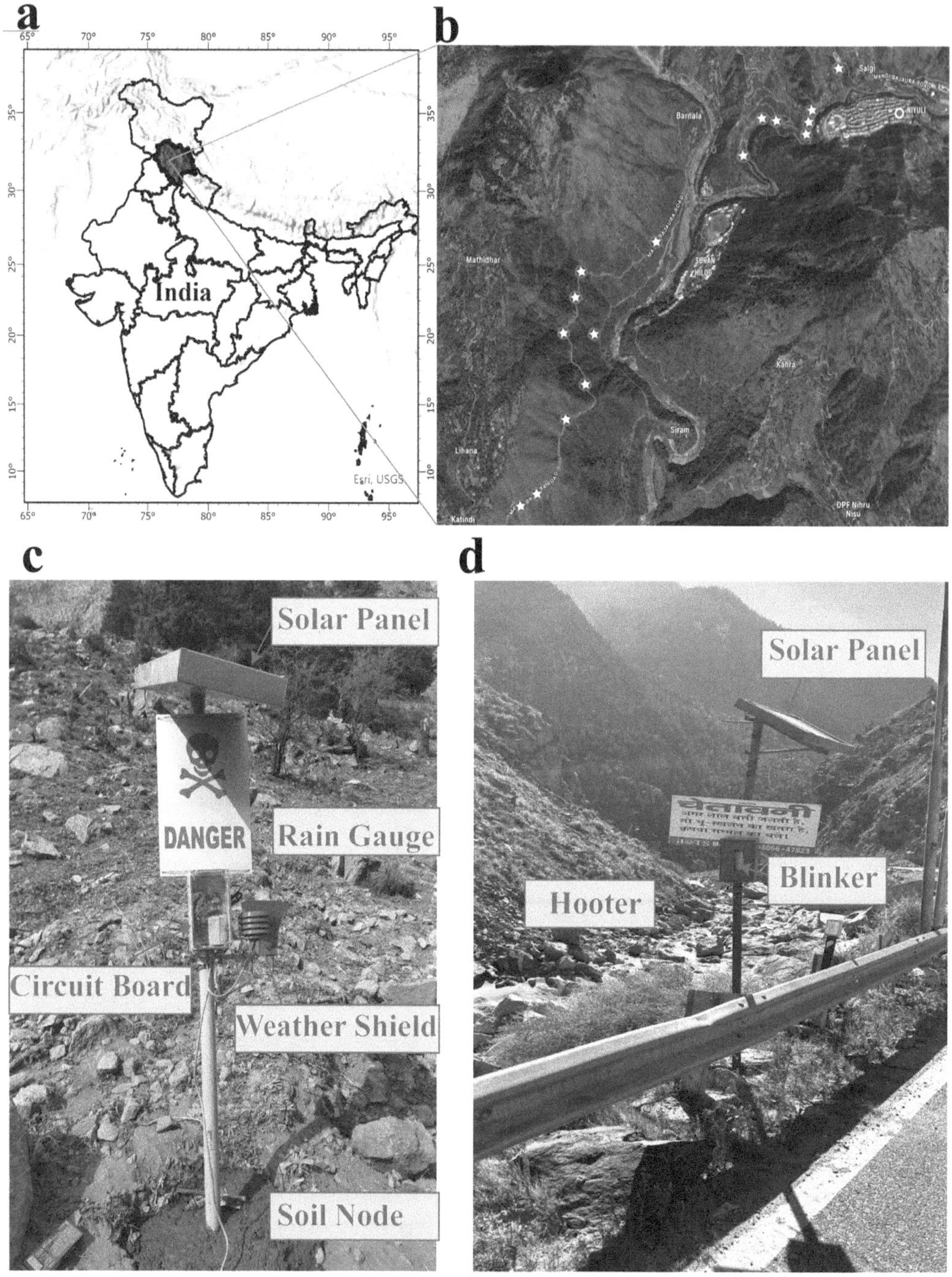

FIGURE 7.6 LMSs Installed in the Himalayan Mountains. (a) The Study Area in the Himalayan Region. (b) 15 Landslide Locations Indicated by star symbols on Google Maps within Kamand Valley. (c) An LMS Installed on a Real-World Landslide. (d) A Warning Pole Installed Near the Highway Running below the Mountain Where the LMS Was Installed

a slope movement. Recall in slope movement prediction measures the model's ability to detect actual slope movement events. The F1 score, which is a harmonic mean of precision and recall, is a good measure for a class imbalance dataset.

7.3.5 ML Models

From the literature, we found that tree-based ML models are the best choice for deployment on the CS. Tree-based ML models are lightweight and quickly predict multiple landslide location values. In this study, we selected RF, ET, and GB ML models to develop and predict slope movement. We developed these models using the scikit-learn 1.5.0[2] library in Python 3.0.[3] First, we trained these ML models individually using training data and later tested them using unseen test data. The input features for the ML models were temperature, humidity, light intensity, barometric pressure, rainfall, acceleration, angular velocity, and soil moisture content. The output of the ML models was the next movement classification, either 0, 1, or 2. The best model was selected based on the testing results and deployed on the CS for real-time prediction. The details of the ML models and their hyperparameters that we varied in this study are the following:

- *RF:* It is a collection of different decision trees that produces the average results for all trees (Kumar et al., 2019b). Because of averaging, it overcomes the overfitting problem. The overfitting problem arrives when a single decision tree memorizes the training data and predicts the training results very well but fails on the test data. Table 7.2 discusses the hyperparameters of RF, such as the number of estimators or decision trees used for averaging. The maximum depth determines the depth or height, of each decision tree. Next, max features hyperparameters of the RF model determine how many input features are considered for node splits of the decision tree. Furthermore, the max samples hyperparameters determine the number of samples used to train each decision tree. Next, RF trains each decision tree with random sampling and the bootstrap hyperparameter value (true or false) determines whether sampling is with or without replacement.
- *ET:* Like RF, it averages many decision trees (Bruzón et al., 2021). Instead of RF's greedy strategy, ET randomly selects features at tree split points. Overfitting is avoided because each tree is random. Table 7.2 covers ET hyperparameters, such as the number of

TABLE 7.2
The Model Hyperparameters and their Ranges

Modal	Hyperparameters	Range
RF	Number of estimators	[5, 10, 50, 100, 500]
	Max depth	[1, 2, 4, 6, 8, 10, 12, 14, 16, 18, 20]
	Max features	[1 to 15] with step size 2
	Max samples	[0.1 to 1] with step size 0.2
	Bootstrap	[False, True]
ET	Number of estimators	[5, 10, 50, 100, 500]
	Max features	[1 to 15] with step size 2
	Min samples split	[2 to 15] with step size 2
GB	Number of estimators	[5, 10, 50, 100, 500]
	Max depth	[2 to 10] with step size 2
	Max samples	[0.1 to 1] with step size 0.2
	Learning rate	[0.1 to 1] with step size 0.2
	Max features	[1 to 15] with step size 2

estimators, max features, and min samples split (the minimum amount of samples needed to split a node).

- *GB:* The GB model is composed of different weak learners (Pathania et al., 2021). The weak learners are the simple decision trees with only one split. The data passed through weak learners, where each weak learner gave more weight to incorrectly classified data than to correctly classified data. The following weak learner receives this weighted output using the same methodology. The data could be correctly classified after passing through several weak learners. The GB hyperparameters are listed in Table 7.2. For example, the number of estimators, maximum depth, maximum samples, learning rate, and maximum features varied in this model.

The ranges for several hyperparameters in this study were determined based on the ranges used by researchers who have developed equivalent models in the literature. This means that when we enlarged the range, we ensured some hyperparameters had optimized values that were not beyond the limits of the boundaries. We made use of grid search to overcome this problem of multiple hyperparameter combinations. Using it, all possible combinations of setting the hyperparameters are fitted to our ML model and their resulting accuracy metrics are recorded. The corresponding maximum F1 score measure was used as a criterion for determining optimum hyperparameters.

7.4 RESULTS

In this study, we used grid search strategies to pass various combinations of hyperparameters to the RF, ET, and GB models. Tables 7.3 and 7.4, respectively, illustrate the models' performance during training and testing. With a testing F1 score of 0.895, the ET model performed better than the other models. This could be the case since ET splits nodes arbitrarily, in contrast to other models that do so using a greedy strategy. It appeared that the ET was a smart choice for slope movement prediction. We deduced that all three models demonstrated a training F1 score better than 0.9 and a test F1 score greater than 0.87 based on the training and test results of RF, ET, and GB. This suggests that the models are general in nature. We chose the ET model and installed it on the CS to forecast slope movements in real time based on the model's highest F1 score. The optimum values for each ML model's hyperparameters are displayed in Table 7.5. For instance, 50 estimators were taken into account for prediction in all three ML models. Subsequently, the RF and GB decision trees have

TABLE 7.3
Models' Performance in the Training Dataset

Model	F1 Score	Precision	Recall	Accuracy
ET	0.921	0.922	0.922	0.922
GB	0.917	0.919	0.917	0.918
RF	0.906	0.908	0.907	0.907

TABLE 7.4
The Performance Measures of the Models in Testing

Model	F1 Score	Precision	Recall	Accuracy
ET	0.895	0.974	0.846	0.846
GB	0.891	0.974	0.839	0.840
RF	0.871	0.973	0.809	0.809

TABLE 7.5
Best Value of the Hyperparameters in ML Models

Model	Hyperparameters	Value
ET	Number of estimators	50
	Max features	7
	Min samples split	3
GB	Number of estimators	50
	Max depth	4
	Max samples	0.6
	Learning rate	0.3
	Max features	7
RF	Number of estimators	50
	Max depth	8
	Max features	5
	Max samples	0.2
	Bootstrap	True

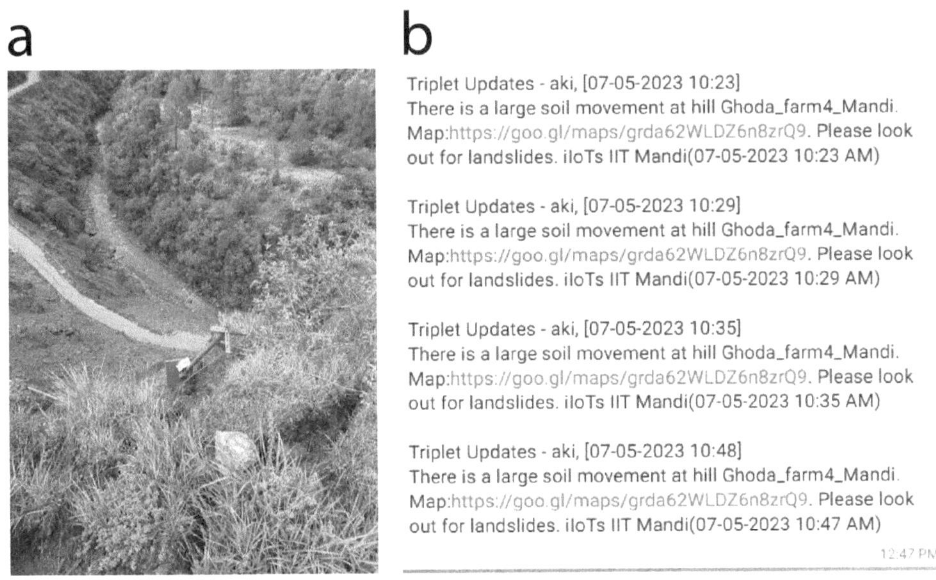

FIGURE 7.7 A Damaged LMS at a Landslide Site and Its Alerts. (a) Images of the LMS after a Landslide Event; (b) Messages Triggered by GeoSense through SMSs

maximum depths of 8 and 4, respectively. Additionally, the models' maximum feature counts were five for RF, seven for ET, and seven for GB.

The real-time data section in Figure 7.5a shows live updates of the weather and soil data every 10 minutes. Whenever our CS obtains data from an LMS, it starts with a Python script. This script, along with the ET model, retrieves the current and previous data from the database to predict future values. These predicted values are shown in the forecast section of Figure 7.5a, which helps us to see what we might expect in the future.

Several LMSs are deployed in the Kamand valley, including one at the Ghora Farm location in the valley. As shown in Figure 7.7a, a large slope movement occurred at Ghora Farm (Kamand Valley) on 7th May 2023, where the deployed LMS was damaged. A number of movement messages were triggered from the GeoSense Framework during the large slope movement event via the LMS before it was damaged (see Figure 7.7b). The framework, upon receiving movement data from the LMS, was able to trigger messages on registered mobile numbers. Overall, this slope event showed how well the GeoSense Framework worked, where it predicted a landslide 10 minutes before it occurred.

7.5 CONCLUSION

A major problem faced by the Himalayan region of India is slope failures and landslides owing to unpredictable weather changes (Parkash, 2011; Crosta, 1998). To alert people about impending landslides, we developed a GeoSense IoT Framework using low-cost LMSs together with advanced ML models. The framework helps provide real-time predictions and alerts regarding slope movements to disaster management authorities and to people living in landslide-prone areas.

The GeoSense IoT Framework has made a significant leap forward in landslide detection and monitoring. This is through the use of an early warning system that is fitted with strategically placed sensors in areas prone to landslides. It measures weather and soil parameters. All these elements are analyzed by GeoSense to know how they contribute to the movements of slopes, hence making it easy to predict what will happen next.

GeoSense Framework's real-time data analysis is one of its main features. The framework assessed RF, GB, and ET classifiers as ML models for predicting slope movements. While there was promise for all three models, the ET classifier outperformed the other two in terms of its ability to anticipate slope movements. In the training dataset, the ET classifier's F1 score was 0.921, while in the test dataset, it was 0.895. It might have done so by using a random split method as opposed to a greedy strategy.

On May 7, 2023, the accuracy of the GeoSense Framework's prediction of a landslide in Ghora Farm, Kamand Valley, proved its worth. The framework automatically delivered warnings to the appropriate stakeholders. These incidents demonstrate why GeoSense is a useful foundation for community safety assurance and risk reduction in landslide-prone areas.

The low-priced sensing technology of the GeoSense Framework and real-time forecasting of slope movements have shown that it is useful for disaster management. The application has been validated by its ability to predict slope movement at Ghora Farm in real time and prior to its actual occurrence.

We intend to keep enhancing the GeoSense IoT Framework so that it is easier to use and can be installed in as many locations as feasible where landslides are likely to occur. But there are still a lot of obstacles to overcome, like keeping Internet connections up and deploying LMSs in really remote locations. We intend to overcome the existing obstacles by putting in place an edge-based monitoring system that enables local data processing and forecasting. Furthermore, future improvements to the suggested ML models are anticipated. To increase the accuracy and dependability of landslide predictions, we also intend to focus on improving sensor technologies. Because of the trade-off between covered area and energy efficiency, one might consider developing low-power networks that can handle more significant areas. In conclusion, putting this approach into practice with the aid of IoT has the potential to significantly alter people's resistance to the hazards of landslides in their communities. The difficulty here is to further develop this framework so that it can function effectively across wide variations in technology in various contexts and become extremely sensitive to changes related to monitoring and predicting landslides.

ACKNOWLEDGMENTS

The authors would like to thank the district disaster management authority of Mandi, Kangra, and Kinnaur of Himachal Pradesh for providing the funds for this research. They also thank IIT Mandi for providing the space and computing facilities for this research work.

NOTES

1 https://en.wikipedia.org/wiki/2023_North_India_floods.
2 https://scikit-learn.org/stable/.
3 https://www.python.org/download/releases/3.0/.

REFERENCES

Bruzón, A. G., Arrogante-Funes, P., Arrogante-Funes, F., Martín-González, F., Novillo, C. J., Fernández, R. R., Vázquez-Jiménez, R., Alarcón-Paredes, A., Alonso-Silverio, G. A., Cantu-Ramirez, C. A., and Ramos-Bernal, R. N., 2021. Landslide Susceptibility Assessment Using an AutoML Framework. *International Journal of Environmental Research and Public Health*, 18(20), 10971. https://doi.org/10.3390/ijerph182010971

Chanamolu, S. R., and Daim, T., 2020. Technical Transformation: Internet of Things. In S. P. Vadera (Ed.), *Digital Transformation: Evaluating Emerging Technologies* (pp. 83–139). World Scientific Publishing, Singapore. https://doi.org/10.1142/9789811214639_0004

Chaturvedi, P., Thakur, K. K., Mali, N., Kala, V. U., Kumar, S., Yadav, S., and Dutt, V., 2018. A Low-cost IoT Framework for Landslide Prediction and Risk Communication. In Hassan, Q. F. (ed.), *Internet of Things A to Z: Technologies and Applications* (pp. 593–610). John Wiley & Sons, New Jersey, https://doi.org/10.1002/9781119456735.ch21

Crosta, G., 1998. Regionalization of Rainfall Thresholds: An Aid to Landslide Hazard Evaluation. *Environmental Geology*, 35(2), 131–145. https://doi.org/10.1007/s002540050300

Dikshit, A., Sarkar, R., Pradhan, B., Segoni, S., and Alamri, A. M., 2020. Rainfall Induced Landslide Studies in Indian Himalayan Region: A Critical Review. *Applied Sciences*, 10(7), 2466. https://doi.org/10.3390/app10072466

Folleco, A., Khoshgoftaar, T. M., and Napolitano, A., 2008, December. Comparison of Four Performance Metrics for Evaluating Sampling Techniques for Low Quality Class-imbalanced Data. In *2008 Seventh International Conference on Machine Learning and Applications* (pp. 153–158). San Diego, CA: IEEE. https://doi.org/10.1109/icmla.2008.11

Fosalau, C., Zet, C., and Petrisor, D., 2016. Implementation of a Landslide Monitoring System as a wireless sensor network. In *2016 IEEE 7th Annual Ubiquitous Computing, Electronics & Mobile Communication Conference (UEMCON)* (pp. 1–6). New York, NY: IEEE. https://doi.org/10.1109/uemcon.2016.7777813

Hassan, Q. F., 2011. Demystifying Cloud Computing. *Journal of Defense Software Engineering*, 24, 16–21. Available at https://www.researchgate.net/publication/215795101_Demystifying_Cloud_Computing

Hassan, Q. F., 2018. *Internet of Things A to Z: Technologies and Applications*. John Wiley & Sons, New Jersey, https://doi.org/10.1002/9781119456735

Khan, R., Yousaf, S., Haseeb, A., and Uddin, M. I., 2021. Exploring a Design of Landslide Monitoring System. *Complexity*, 2021, 1–13. https://doi.org/10.1155/2021/5552417

Köhler, H. J., and Schulze, R., 2000, November. Landslide Triggering Induced by Barometric Pressure Changes. In *ISRM International Symposium. ISRM-IS-2000–465*. Available at https://onepetro.org/ISRMIS/proceedings-pdf/IS00/All-IS00/ISRM-IS-2000-465/1908920/isrm-is-2000-465.pdf

Kumar, P., Priyanka, P., Dhanya, J., Uday, K. V., and Dutt, V., 2023. *Analyzing the Performance of Univariate and Multivariate Machine Learning Models in Soil Movement Prediction: A Comparative Study. IEEE Access*, 11, 2368–62381. https://doi.org/10.1109/access.2023.3287851

Kumar, P., Priyanka, Pathania, A., Agarwal, S., Mali, N., Singh, R., Chaturvedi, P., Uday, K. V., and Dutt, V., 2020. Predictions of Weekly Slope Movements Using Moving-average and Neural Network Methods: A Case Study in Chamoli, India. In *Soft Computing for Problem Solving 2019: Proceedings of SocProS 2019* (Volume 2, pp. 67–81). Singapore: Springer. https://doi.org/10.1007/978-981-15-3287-0_6

Kumar, P., Sihag, P., Pathania, A., Agarwal, S., Mali, N., Singh, R., Chaturvedi, P., Uday, K. V., and Dutt, V., 2019a. Predictions of Weekly Soil Movements Using Moving-average and Support-vector Methods: A Case-study in Chamoli, India. In *International Conference on Information technology in Geo-Engineering* (pp. 393–405). Cham: Springer International Publishing. https://doi.org/10.1007/978-3-030-32029-4_34

Kumar, P., Sihag, P., Pathania, A., Agarwal, S., Chaturvedi, P., Mali, N., Singh, R., Uday, K. V., and Dutt, V., 2019b. Landslide Debris-flow Prediction Using Ensemble and Non-ensemble Machine-learning Methods: A Case-study in Chamoli, India. In *Contributions to Statistics: Proceedings of the 6th International Conference on Time Series and Forecasting (ITISE)* (pp. 614–625). Granda, Spain: Springer. Available at https://itise.ugr.es/ITISE2019_Vol1.pdf

Kumar, P., Sihag, P., Pathania, A., Chaturvedi, P., Uday, K. V., and Dutt, V., 2021a. Comparison of Moving-average, Lazy, and Information Gain Methods for Predicting Weekly Slope-Movements: A Case-study in Chamoli, India. In *Understanding and Reducing Landslide Disaster Risk*, Volume 3 Monitoring and Early Warning 5th (pp. 321–330). Cham: Springer. https://doi.org/10.1007/978-3-030-60311-3_38

Kumar, P., Sihag, P., Sharma, A., Pathania, A., Singh, R., Chaturvedi, P., Mali, N., Uday, K. V., and Dutt, V., 2021b. Prediction of Real-world Slope Movements via Recurrent and Non-recurrent Neural Network Algorithms: A Case Study of the Tangni Landslide. *Indian Geotechnical Journal*, 51(4), 788–810. https://doi.org/10.1007/s40098-021-00529-4

Kumar, S., Panigrahi, R. K., and Kanungo, D. P., 2024, January. Development of IoE-Edge-AI Architecture For Landslide Early Warning System. *In 2024 IEEE Applied Sensing Conference (APSCON)* (pp. 1–4). Goa, India: IEEE. https://doi.org/10.1109/apscon60364.2024.10465809

Li, S., Xu, L. D., and Zhao, S., 2015. The Internet of Things: A Survey. *Information Systems Frontiers*, 17, 243–259. https://doi.org/10.1007/s10796-014-9492-7

Nguyen, V. T., Nguyen, Q. A., and Nguyen, N. K. (2024). Method for Monitoring and Forecasting Landslide Phenomenon Based on Machine Learning. *MethodsX*, 12, 102797. https://doi.org/10.1016/j.mex.2024.102797

Parkash, S., 2011. Historical Records of Socio-economically Significant Landslides in India. *Journal of South Asia Disaster Studies*, 4(2), 177–204. Available at: https://www.researchgate.net/publication/272905239_Historical_Records_of_Socio-economically_Significant_Landslides_in_India

Pathania, A., Kumar, P., Priyanka, Maurya, A., Kumar, M., Chaturvedi, P., Singh, R., Uday, K. V., and Dutt, V., 2022. Predictions of Soil Movements Using Persistence, Auto-regression, and Neural Network Models: A Case-study in Mandi, India. *International Journal of Swarm Intelligence*, 7(1), 94–109. https://doi.org/10.1504/ijsi.2022.121100

Pathania, A., Kumar, P., Priyanka, Maurya, A., Uday, K. V., and Dutt, V., 2021. Development of an Ensemble Gradient Boosting Algorithm for Generating Alerts about Impending Soil Movements. In *Machine Learning, Deep Learning and Computational Intelligence for Wireless Communication: Proceedings of MDCWC 2020* (pp. 365–379). Singapore: Springer. https://doi.org/10.1007/978-981-16-0289-4_28

Pathania, A., Kumar, P., Sihag, P., Chaturvedi, P., Singh, R., Uday, K. V., and Dutt, V., 2020. A Low-cost, Sub-surface IoT Framework for Landslide Monitoring, Warning, and Prediction. In *Proceedings of 2020 International Conference on Advances in Computing, Communication, Embedded and Secure Systems*. Available at https://www.researchgate.net/publication/339850328_A_Low_Cost_Sub-Surface_IoT_Framework_for_Landslide_Monitoring_Warning_and_Prediction

Rahardjo, H., Li, X. W., Toll, D. G., and Leong, E. C., 2001. The Effect of Antecedent Rainfall on Slope Stability. In Toll, D. G. (ed.), *Unsaturated Soil Concepts and their Application in Geotechnical Practice* (pp. 371–399). Dordrecht: Springer. https://doi.org/10.1007/978-94-015-9775-3_8

Savvaidis, P. D., 2003. Existing Landslide Monitoring Systems and Techniques. *From Stars to Earth and Culture*, pp. 242–258. Available at https://www.academia.edu/10602541/Existing_Landslide_Monitoring_Systems_and_Techniques

Sharma, R., Saini, T., Kumar, P., Pathania, A., Chitineni, K., Chaturvedi, P., and Dutt, V., 2019. An Online Low-cost System for Air Quality Monitoring, Prediction, and Warning. In *International Conference on Distributed Computing and Internet Technology* (pp. 311–324). Cham: Springer International Publishing. https://doi.org/10.1007/978-3-030-36987-3_20

Song, Y. S., Chae, B. G., Kim, K. S., Park, J. Y., Oh, H. J., and Jeong, S. W., 2021. A Landslide Monitoring System for Natural Terrain in Korea: Development and Application in Hazard Evaluations. *Sensors*, 21(9), 3040. https://doi.org/10.3390/s21093040

Van Khoa, V., and Takayama, S., 2018. Wireless Sensor Network in Landslide Monitoring System with Remote Data Management. *Measurement*, 118, 214–229. https://doi.org/10.1016/j.measurement.2018.01.002

Wang, J. D., Gu, T. F., and Xu, Y. J., 2017. Field Tests of Expansive Soil Embankment Slope Deformation under the Effect of the Rainfall Evaporation Cycle. *Applied Ecology & Environmental Research*, 15(3). https://doi.org/10.15666/aeer/1503_343357

Yavari, N., Tang, A. M., Pereira, J. M., and Hassen, G., 2016. Effect of Temperature on the Shear Strength of Soils and the Soil–structure Interface. *Canadian Geotechnical Journal*, 53(7), 1186–1194. https://doi.org/10.1139/cgj-2015-0355

Zourmand, A., Hing, A. L. K., Hung, C. W., and AbdulRehman, M., 2019, June. Internet of Things (IoT) using LoRa Technology. In *2019 IEEE International Conference on Automatic Control and Intelligent Systems (I2CACIS)* (pp. 324–330). Selangor, Malaysia: IEEE. https://doi.org/10.1109/I2CACIS.2019.8825008

8 IoT-MediGuard
An Advanced Monitoring and Predictive Analytics System for Transportation of Medical Supplies

Surbhi Ranga, Tushar Vasudev, Praveen Kumar, Sahil Sankhyan, Yashwant Kumar, Kala Venkata Uday and Varun Dutt

8.1 INTRODUCTION

Due to a lack of appropriate medical supplies, including blood, vaccinations, and other medications, India has been dealing with a number of issues.[1] For instance, the nation can only donate 11 million of the 15 million units of blood that are needed each year. Regretfully, studies suggest that 12,000 fatalities per day are attributed to a shortage of high-quality blood. A further issue is that every year, over 0.65 million blood units become useless as a result of unmonitored temperatures during transportation.[2] Furthermore, in underdeveloped nations, vaccinations need to be transported at low temperatures (Kartoglu and Ames, 2022). All things considered, it is crucial to keep a watch on the temperatures at which pharmaceuticals, blood, and vaccines are carried in order to prevent waste and guarantee that patients can get them. In distant, mountainous regions of India, where traffic and weather patterns are unpredictable and medical supplies may be transported over extended distances, this kind of monitoring is extremely important (Aggarwal et al., 2023). For example, natural events such as landslides may cause road blockages and traffic disruptions for long periods, leading to delays in the delivery of medical supplies.

Sensors, microcontrollers, and cloud-based modules are objects that may connect with one another because of the Internet of Things (IoT) technology (Schoder, 2018). IoT technology has been applied in many fields, including healthcare. IoT technologies, for instance, are thought to be able to accomplish real-time health management, enhance diagnostic accuracy, and monitor patients remotely (Sagahyroon et al., 2017; Kapur et al., 2023). One possible use case for the IoT is the intelligent conveyance of medical supplies. For instance, a Global System for Mobile Communications (GSM) module, a microcontroller, and IoT sensors may be installed in a medical supply box. When moving a box containing medical supplies from one area to another, one may use these components to keep track of the box's position and the ambient temperatures both inside and outside the container. To the best of the authors' knowledge, however, there is currently no use case for a smart box for the delivery of medical supplies, particularly in developing nations like India, where the demand for medical goods frequently outpaces the supply.

In the past years, several machine learning (ML) algorithms have been developed to indicate how patterns might be learned from historical data to predict future outcomes. For example, some researchers have made advanced predictions of landslides and soil movements using environmental and weather data (Kumar et al., 2019a). Some other ML algorithms for the prediction of future data

DOI: 10.1201/9781003506638-10

based on historical data that have been applied include making predictions about blood sugar levels; it could predict hypoglycemia by analyzing breath-sensed volatile organic compounds using time series up to 20 minutes before onset (Kapur et al., 2024). However, to the best of the authors' knowledge, the literature lacks information on implementing ML models to predict future temperature values for a confined space, such as a smart box. Thus, there is still a void in the literature about the application of IoT for predictive analytics in the transportation of medical supplies, even though earlier research has examined several facets of IoT applications in healthcare.

In this chapter, we address the gaps in the literature highlighted above, and we propose a new IoT smart box technology named IoT-MediGuard that offers a complete solution to the critical problem of unmonitored temperatures of medical supplies during transportation. The device uses continuous internal temperature and humidity monitoring and is intended to safely transport medical supplies, including blood, vaccines, and other medicines, which require low temperatures. The IoT-MediGuard system fitted in an icebox uses Global Positioning System (GPS) to track the exact location of the box in real time when it is transported. In addition, the GSM module in the smart box offers cellular networks that the box optimally uses to transmit data to the cloud, allowing immediate access to crucial temperature and humidity information. When temperature and humidity levels exceed the permissible ranges, alerts are sent to selected phones via SMS. Also, the local team carrying the box is alerted via a buzzer placed on the box. The IoT-MediGuard system enhances transportation of medical supplies by monitoring the temperature, humidity, and live location of the icebox and alerting the stakeholders 30 minutes ahead of time if internal temperatures are about to exceed the permissible range. This optimization is achieved through circuit design and sleep modes, which make the system sleep for 30 minutes and only wake up for 2–3 minutes to measure temperature and transmit these data to the cloud. In the cloud, a pre-trained autoregressive integrated moving average (ARIMA) model predicts future temperatures inside the box. Due to these features, the IoT-MediGuard system not only increases access to healthcare but also plays a critical role in saving lives in neglected areas.

We first review the literature and discuss how the IoT-MediGuard offers new capabilities for monitoring and forecasting internal box temperatures in medical iceboxes. Next, we go into great depth about the IoT-MediGuard system's architecture and development. We also describe the outcomes of utilizing ML approaches to forecast future temperatures and data collection via IoT-MediGuard. We conclude this chapter with a discussion of our findings and the practical ramifications of using the IoT-MediGuard system to carry medical supplies in isolated, mountainous regions of India.

8.2 LITERATURE REVIEW

Studies on real-time monitoring and storage systems have demonstrated the need for maintaining optimal conditions to ensure the safety and efficacy of medical supplies (Afreen and Bajwa, 2021; Aswath et al., 2021; Salah et al., 2020). For example, Afreen & Bajwa (2021) devise a real-time IoT cold storage system designed specifically for monitoring ambient conditions in cold storage environments. Storage item safety is guaranteed by the system's real-time temperature deviation alert mechanism. The method primarily addresses the storage of medical supplies rather than their transportation. By including medical supply quality management, Aswath et al. (2021) broadened the use of real-time monitoring. Power and connectivity issues during transit were solved by the suggested solution. The system enables continuous environmental monitoring as it ensures the effectiveness and safety of medical products throughout transit.

Additionally, studies have demonstrated how crucial it is to monitor medical supplies throughout transportation (Cil et al., 2022; Bapatla et al., 2022; Kiruthika et al., 2021). To facilitate comprehensive monitoring of blood banks in India, Kiruthika et al. (2021) developed an random forest (RF) tag system that incorporates infrared temperature sensors. The technology provides automated updates and ensures a safe blood supply with continuous tracking and cloud-based time monitoring. The method is designed to identify blood types within hospital blood banks rather than to concentrate on blood during transportation. For container ports especially, Cil et al. (2022) suggested a real-time cold chain monitoring system with IoT capabilities. This technology continually monitors the temperature and alerts users to ensure that perishable goods—including medications—remain within the proper temperature range whether in storage or transit. In order to safely monitor medical supply conditions throughout transportation, Bapatla et al. (2022) describe PharmaChain 2.0, a system that leverages blockchain, cloud computing, and IoT. This methodology enhances the medical supply chain's dependability by guaranteeing high-quality supplies to pharmaceutical customers.

Reliability of predictive analytics to anticipate environmental elements that might affect medical supply quality has attracted attention in the field of medical supply transportation (Ahmadimanesh et al., 2023; Sandaruwan et al., 2020; Tao et al., 2019). One such system that Tao et al. (2019) present is a cold chain monitoring system for pharmaceuticals based on LoRa wireless technology. This system continuously collects and transmits temperature data and emphasizes real-time monitoring of drug efficacy throughout storage and transit by utilizing LoRa's extended communication range and low power draw. Sandaruwan et al. (2020) suggested using blockchain technology in combination with k-means clustering, long short-term memory (LSTM), and geographic information systems to enhance blood bank management. However, blood transportation issues are not adequately addressed by their system, which forecasts blood demand and recommends blood donation sites. Ahmadimanesh et al. (2023) developed an optimal model for blood logistics management in Iran's hierarchical blood transfusion network at the local, regional, and national levels. By enhancing blood distribution and preservation via the use of neural networks and simulation techniques, this method improves the dependability and efficiency of blood supply management. Nonetheless, blood monitoring during transit is not addressed in Ahmadimanesh et al.'s (2023) work.

Through the use of sophisticated ML techniques like LSTM, random forest (RF), support vector machine, and multilayer perceptron (MLP), as well as statistical models like ARIMA, environmental parameters like temperature and humidity have been predicted in the literature (Ghide et al., 2022; Kumar et al., 2019a, 2019b, 2021, 2023; Li et al., 2019; Pathania et al., 2022; Rodrigues et al., 2023). Li et al. (2019), for instance, highlighted the effectiveness of statistical models like ARIMA in predicting future values from historical data in their comparison of several environmental prediction models. With smaller datasets, ARIMA models have been shown to perform better than traditional techniques. Pathania et al. (2022) concentrated on one area, where they use ARIMA models to predict soil changes. This approach bears some resemblance to the prediction of temperature in enclosed spaces, such as delivery boxes containing medical supplies. In other fields, such as landslide and meteorological forecasting, Ghide et al. (2022), Rodrigues et al. (2023), and Kumar et al. (2019a, 2019b, 2021, 2023) have also utilized comparable models. Research such as the aforementioned studies continuously highlights the effectiveness of ARIMA models in predicting environmental factors, underscoring its significance for the delivery of medical supplies.

Building on these previous research works, this study presents the IoT-MediGuard system, an intelligent and effective medical supply transit solution that combines real-time monitoring and predictive analytics, building on earlier research works as previously mentioned. Since hospital-to-hospital transfers are given priority in the architecture of IoT-MediGuard, it is possible to continuously monitor the temperature and relative humidity both inside and outside the transport box. The IoT-MediGuard system uses a local SMS alert system in conjunction with location tracking to ensure timely notifications in the event that the specified conditions are not met. The IoT-MediGuard system also uses the cloud-based ARIMA model to anticipate interior temperatures in transportation. When combined with the real-time monitoring aspects of the system, this predictive capability offers a strong means of guaranteeing the effectiveness and security of medical supplies, especially in areas where transportation is difficult.

8.3 IOT-MEDIGUARD MONITORING SYSTEM

In this section, we will discuss the IoT-MediGuard system's design, working, and installation.

8.3.1 System Design

The monitoring of medical supplies during transportation is the design objective of the IoT-MediGuard system. For this purpose, continuous temperature and humidity monitoring of boxes carrying medical supplies in the system is done. Figure 8.1 shows the IoT-MediGuard system's design and explains the system's operational methodology. The first unit (sensor unit) comprises various sensors that gather data. After the data is extracted from these sensors, the GSM module facilitates its transmission to the cloud via an HTTP GET Request made to an API. The captured data is shown on the website for tracking and monitoring (the data logging and prediction unit). The ML model can use this collected data to forecast future interior temperatures and produce alerts if needed.

Furthermore, the collected readings are compared to the predefined internal box temperature and humidity permissible ranges. The alert system is activated when the temperature or humidity values exceed the thresholds. This turns the SMS notification module on for those registered to receive such alerts via SMS. A built-in alarm system also alerts the person carrying the box anytime the temperature or humidity exceeds the defined permissible ranges (the alert system).

A web interface has been designed to make things easier for users to understand. This interface shows the box's internal and external temperatures. Additionally, the web interface integrates the ARIMA model to predict the future values of internal temperatures and display them on the webpage.

The safety of blood and vaccinations during transportation is the design objective of this system. The sensors chosen for this system closely follow the prescribed guidelines. As per the recommended guidelines,[3] the temperature inside a blood center refrigerator should be maintained at 4°C ± 2°C. Continuous temperature monitoring of boxes carrying medical supplies should be done every 4 hours. As per the guidelines, the medical supplies box is recommended to have an alarm system and a power supply. The different sensors and their ranges used in the IoT-MediGuard Box are displayed in Table 8.1. For instance, the two temperature sensors that measure the box's interior and exterior temperatures range from −40°C to 80°C.

Similarly, humidity sensors record interior and outdoor relative humidity data, ranging from 0% to 100%. In addition, a warning system with an 85 dB sound capability is employed. A GPS module is embedded to track the box's location with ±2.5 meters accuracy. These sensor ranges meet the National Blood Transportation Corporation's (NBTC) standards, where NBTC is the apex body setting standards for blood handling in India.[4]

Every sensor in this system communicates with the microcontroller, which retrieves data and displays it on the LCD panel. SPI, I2C, and UART are some of the widely used protocols in communication (Subero et al., 2021). These communication protocols help the sensors and microcontrollers in the box to communicate. The GSM module transmits data using an HTTP GET request method.

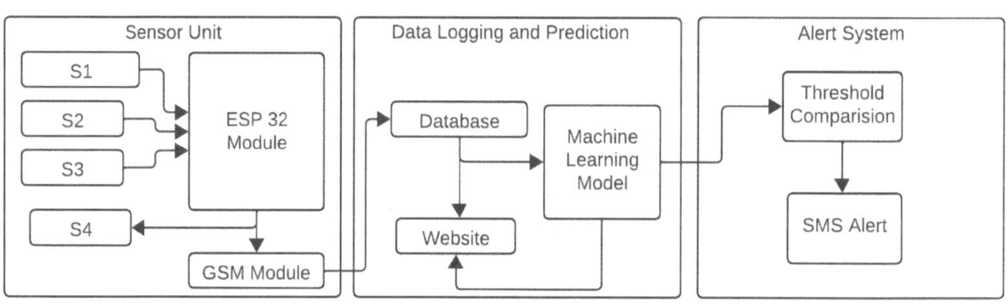

FIGURE 8.1 IoT-MediGuard's System Design

TABLE 8.1
Different Sensors, Manufacturers, Cost, and Range of Operations

Sensors	Manufacturer	Cost (US$)	Range
Temperature sensor	Adafruit[a]	3.95	−40°C to 80°C
Humidity sensor	Adafruit[b]	9.95	0%–100%
Alarm system	Adafruit[c]	1.50	85 dB
GPS sensor	Adafruit[d]	29.95	±2.5 m accuracy

[a] https://www.adafruit.com/product/3721;
[b] https://www.adafruit.com/product/385;
[c] https://www.adafruit.com/product/160;
[d] https://www.adafruit.com/product/746

We have created a cloud system via Firebase,[5] a NoSQL, cloud-hosted database. The data collected from multiple iceboxes used for transporting medical supplies is stored in the Firebase database. When the GPS module is powered on, the microcontroller records the data the GPS module obtains from the satellites, including latitude, longitude, date, time, and speed. The collected data is transmitted to the Firebase cloud via the GSM module. The Firebase cloud stores the received data in its NoSQL database.

The GPS sensor does not require Internet connectivity to gather data from the satellite. However, Internet connectivity is needed when the system sends data to the database (via SIM card). If there is no Internet connectivity, the GPS sensor will store the values, and once the network is available, it will send them to the database. In the absence of the network, a local alarm system has been added to provide alerts in such situations. The permissible range for the local alarm system is set to be between 2°C and 5°C. If the temperature is outside this range, then the alert system activates appropriate alert messages via SMS as well as a sound alarm via a buzzer on the box.

The sensors are connected to an ESP32 microcontroller, which is powered by a rechargeable 10 Ah battery using a boost converter, raising the voltage from 3.3 V to 5 V. ESP32 requires a continuous 5 V power supply for uninterrupted operation. The system contains an integrated battery charger to ensure that batteries are regularly recharged. The temperature and humidity within the IoT-MediGuard are displayed in real-time on the LCD. When the temperature within the box rises, an alert system notifies the user.

8.3.2 The Smart Box in the IoT-MediGuard System

Figure 8.2 shows the smart box in the IoT-MediGuard system. The system requires this custom-made container to house and transport medical supplies. In addition, temperature and humidity sensors in the box monitor its internal environment. Another set of sensors is responsible for the external environmental parameters of this box. Hence, the rest of the circuitry is found outside, sealed off by a smaller cover from dust, rain, or mist.

The exterior casing of the electronic circuit has been carefully designed to have important features, such as a large battery holder for the system's power source. Moreover, it also has an LCD to show the temperature, humidity, and warnings. After testing several prototypes, we developed one final design that best suited the application. The specific design was chosen because it was the most suitable for meeting safety standards and ruggedness requirements in remote hilly areas.

The IoT-MediGuard system's prototype circuit board is depicted in Figure 8.3. The different components, like the microcontroller, GPS sensor, GSM module, LCD screen, buzzer, and temperature and humidity sensors, that were described above are shown on the circuit board. A booster circuit is also added to the board to boost the voltage from 3.3 V to 5 V. The prototype circuit board was created during the early phases of the IoT-MediGuard system's design. The validation of the system's fundamental features, including GPS tracking, GSM connectivity, and temperature and

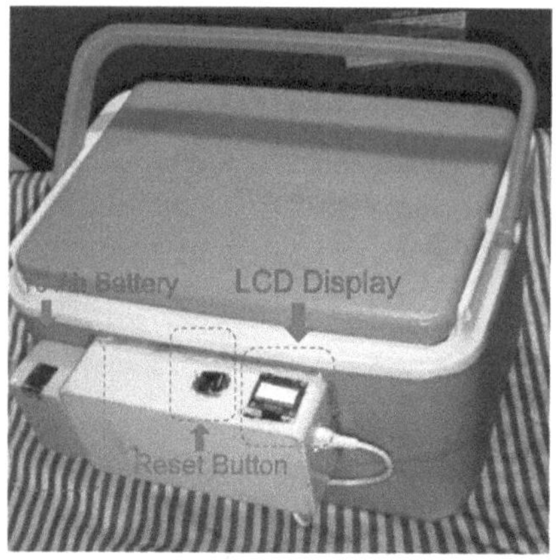

FIGURE 8.2 Robust IoT-MediGuard System

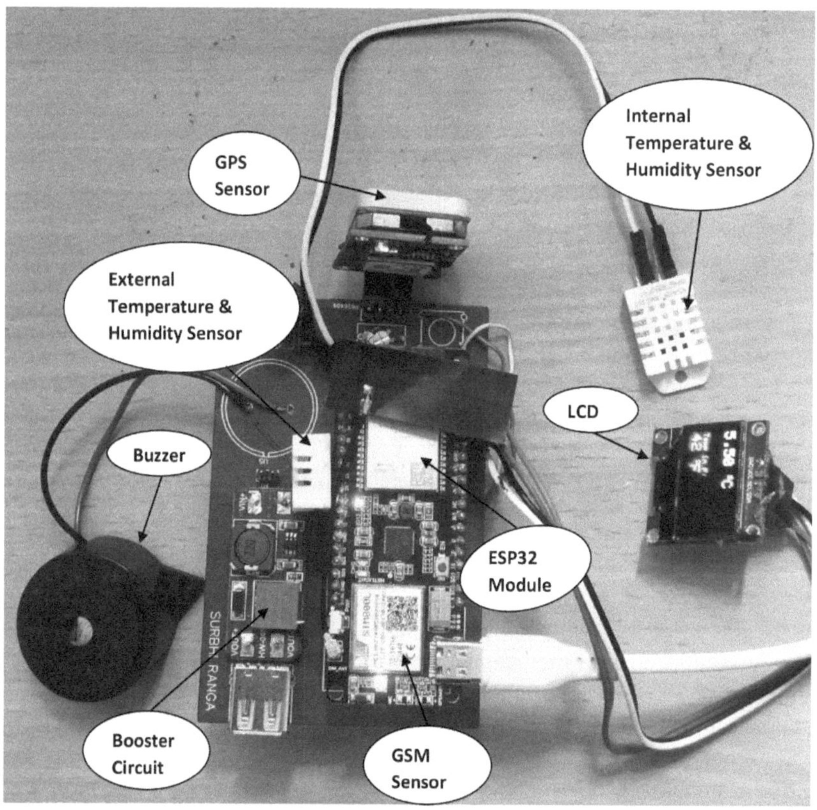

FIGURE 8.3 A Prototype Circuit Board of IoT-MediGuard

humidity sensors, was greatly assisted by this prototype. We were able to improve the way these parts interacted and ensure the system could consistently monitor and send data in a variety of scenarios by testing them in the prototype.

It is crucial to note that this prototype is only a first step, and it may not accurately depict the IoT-MediGuard system as it is in its finished form. The prototype, in particular, concentrates on the electronic parts and their interplay. It does not incorporate the final housing, shielding, or power management solutions that are essential for practical implementation. These electronics are protected from physical damage, moisture, and dust by being contained in a weather-resistant, tough enclosure that houses the whole system. The final design of the IoT-MediGuard also includes a separate power supply, which includes circuitry for power management and a rechargeable battery to guarantee that the system functions even in distant areas without access to external power sources. The final IoT-MediGuard system has these extra characteristics to suit the rigorous needs of carrying medical supplies, especially in difficult and isolated areas. The lessons acquired from this prototype were crucial in the development of this system.

8.3.3 Microcontroller Programming

Figure 8.4 shows a flowchart of the system's working. On system boot-up, libraries are initialized, pins are set, and modules are configured in the initialization phase. After this step, the software runs in a loop that performs other activities repeatedly.

The flow chart is as follows:

- *Initialization Phase:* When the system is powered on, it initializes and configures its peripheral hardware, such as sensors and actuators, a GSM module, a GPS module, and an LCD.
- *Sensor Data Acquisition:* These sensors' temperature and humidity readings start after setting up the initialization phase. Then, GPS data collection starts at the same time as the above process. Even if it may take some time to connect to satellites, the GPS sensor waits for data to be collected to proceed with more accurate position fixing. It will wait until the exact GPS data is acquired again.
- *Display and Alarm System:* After collecting GPS and sensor data, the system displays the temperature value on the LCD sourced from the temperature sensor. Additionally, the alarm is triggered if the temperature or relative humidity breaches predefined permissible ranges.

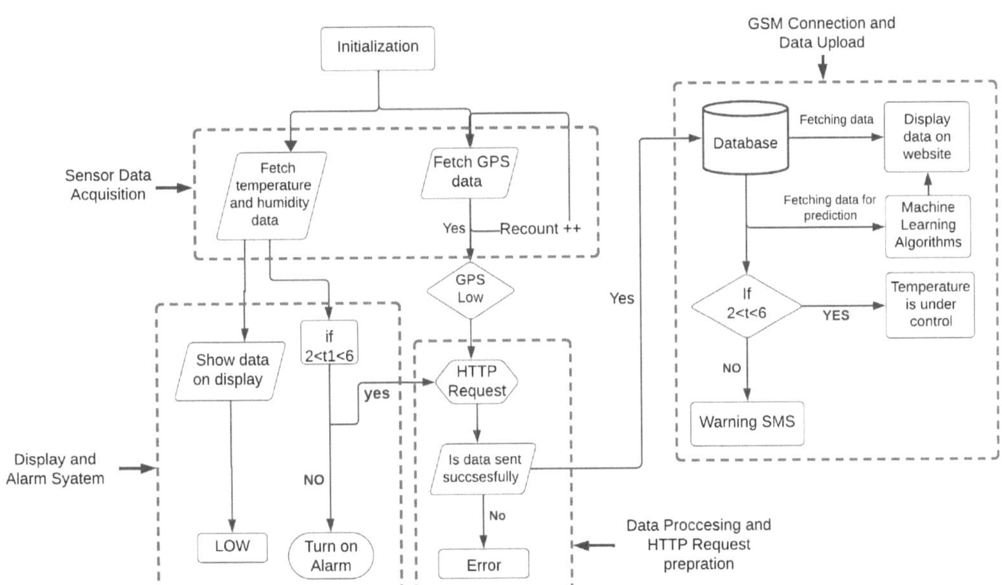

FIGURE 8.4 Flow Chart of the System's Working

- *Data Processing and HTTP Request Preparation:* After handling the alarm, the software processes the collected data and shows all sensor data. Then, it prepares an HTTP request to send the data collected from the different sensors and GPS modules to the database.
- *GSM Connection and Data Upload:* The GSM module uses an HTTP GET request to send the collected data to the server once a successful connection is established. In case of failure to establish the connection, the application will log the error and attempt to reconnect.
- *Deep Sleep Mode and Timer Wake-Up:* The system remains in deep sleep to save power and only wakes up to sense and upload the data to the cloud system. Before going into sleep mode, the system sets a built-in timer (Timer Wake-Up), and after this timer expires, the system wakes up.

8.3.4 PSEUDOCODE

The code listing below shows the pseudocode for the microcontroller in the IoT-MediGuard system. This pseudocode explains the microcontroller's working.

Algorithm 1

Description:

1. Initialize the system with sensors and reset the modem.
2. Fetch the data from temperature, humidity, and location sensors.
3. Every time data is fetched, the threshold is checked.
4. Prepare the HTTP GET request for sending data to the cloud.
5. Make a connection to the server if the connection fails and retry three times.
6. After sending data to the cloud, go to sleep for 10 minutes.

While true do:

- While the latitude and longitude are zero and the retry count1 < 6:
  If retry count1 >= 0
  retry count++
  Try to fetch GPS data
  If valid data is obtained:
  Collect this data and send it to the server
  Retry count1 = 0
  Exit loop
- Initialize t1 = 0°C
  If temperature t1 > 2°C && t1 < 5°C:
  Temperature t1 is in the normal range, buzzer remains deactivated and no SMS notification
  Else If temperature t1 > -5°C && t1 < = 2°C:
  Activate buzzer for sending alerts for 10 seconds as well as send low-temperature alert via SMS notification
  Else If temperature t1 < = -5°C:
  Activate buzzer for sending alerts for 10 seconds as well as send very low-temperature alerts via SMS notification
  Else If temperature t1 > = 5°C:

Activate buzzer for sending alerts for 10 seconds as well as send high-temperature alert via SMS notification

- While not connected to the server and the retry count2 < 3:
  If retry count2 >= 0
  Retry count++
  Try to connect with the server
  If connection with the server successful:
  Send the data to the server
  Retry count2 = 0
  Exit loop
- Goto sleep for 10 minutes

8.3.5 Cloud System

Each medical supply box is assigned a unique identification number so that distinct data may be stored in the Firebase cloud in a data structure known as a "collection," which is essentially a table. Using the current internal temperature of the box, the Python ML model predicts the future temperature 30 minutes ahead of time. We have also developed a homepage for our system that shows the temperature values predicted by the ML model together with other prevailing parameter values (e.g., humidity, latitude, and longitude) for the convenience of the user.

Figure 8.5 displays the real-time data of a specific icebox in tabular form. The user may view the many metrics coming from the box in the database, including its latitude and longitude, temperature, and humidity levels. The user can also view the data for a certain system ID by selecting it. The system also has a dashboard, which is shown in Figure 8.6. The dashboard provides statistics for a given period as well as a 30-minute forecast for the temperature[6].

8.3.6 Power Consumption

The use of the IoT-MediGuard system is intended for remote or resource-constrained locations where dependable power sources might not be available. We implemented a number of power-saving techniques to guarantee the constant monitoring of medical supplies while they are being transported. The ESP32 microcontroller of the system runs in two main modes: deep sleep mode, which drastically lowers power consumption while the system is idle, and active mode, which

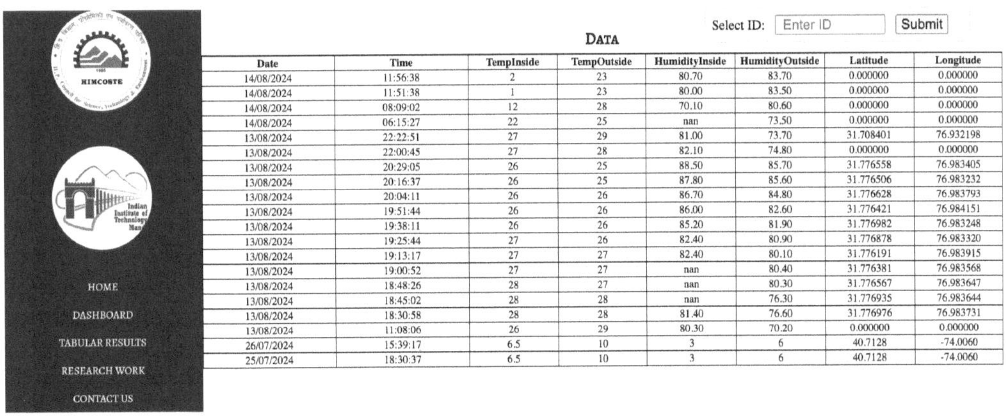

Date	Time	TempInside	TempOutside	HumidityInside	HumidityOutside	Latitude	Longitude
14/08/2024	11:56:38	2	23	80.70	83.70	0.000000	0.000000
14/08/2024	11:51:38	1	23	80.00	83.50	0.000000	0.000000
14/08/2024	08:09:02	12	28	70.10	80.60	0.000000	0.000000
14/08/2024	06:15:27	22	25	nan	73.50	0.000000	0.000000
13/08/2024	22:22:51	27	29	81.00	73.70	31.708401	76.932198
13/08/2024	22:00:45	27	28	82.10	74.80	0.000000	0.000000
13/08/2024	20:29:05	26	25	88.50	85.70	31.776558	76.983405
13/08/2024	20:16:37	26	25	87.80	85.60	31.776506	76.983232
13/08/2024	20:04:11	26	26	86.70	84.80	31.776628	76.983793
13/08/2024	19:51:44	26	26	86.00	82.60	31.776421	76.984151
13/08/2024	19:38:11	26	26	85.20	81.90	31.776982	76.983248
13/08/2024	19:25:44	27	26	82.40	80.90	31.776878	76.983320
13/08/2024	19:13:17	27	27	82.40	80.10	31.776191	76.983915
13/08/2024	19:00:52	27	27	nan	80.40	31.776381	76.983568
13/08/2024	18:48:26	28	27	nan	80.30	31.776567	76.983647
13/08/2024	18:45:02	28	28	nan	76.30	31.776935	76.983644
13/08/2024	18:30:58	28	28	81.40	76.60	31.776976	76.983731
13/08/2024	11:08:06	26	29	80.30	70.20	0.000000	0.000000
26/07/2024	15:39:17	6.5	10	3	6	40.7128	-74.0060
25/07/2024	18:30:37	6.5	10	3	6	40.7128	-74.0060

FIGURE 8.5 Data of the IoT-MediGuard System in Tabular Form

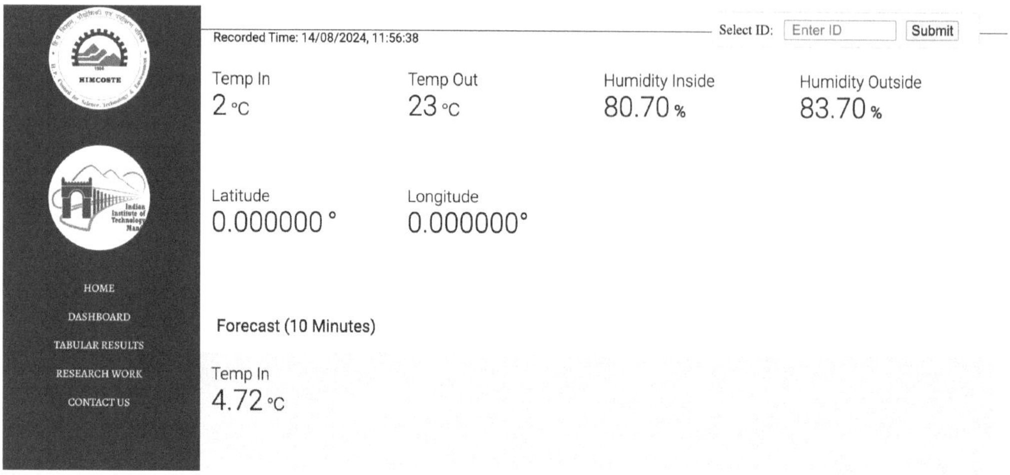

FIGURE 8.6 Dashboard of the Website

gathers sensor data and sends it to the cloud. To be more precise, the ESP32 draws about 280 mA for its 3-minute active intervals and just 5.2 mA during deep sleep, which is controlled by a timer that periodically awakens the device to carry out monitoring duties.

We also use optimal power management in the programming of the system. To save energy, this power management involves turning off certain non-essential parts, such as the LCD display and some sensors, only when they are not in use (Hakim et al., 2022; Pathania et al., 2019). The IoT-MediGuard system runs continuously for up to 72 hours on a single charge thanks to a 10 Ah lithium-ion battery. This longer battery life is essential to guarantee that the system can continue to monitor for lengthy periods of time during transportation. Overall, the IoT-MediGuard system provides dependable monitoring during the transit phase while maintaining energy efficiency by utilizing deep sleep modes and innovative power management.

8.4 ALERTS VIA ML MODEL

We integrated the ARIMA model into the IoT-MediGuard system to establish a reliable temperature forecast system for medical supply preservation during transit. Three popular time series forecasting methods (autoregression, differencing, and moving average) are utilized to predict future data points based on historical values (Kumar et al. 2020). Along with a few essential modules, Python 3.0 is utilized for both data preparation and model training. The Python libraries used are numpy[7] for numerical calculations, Pandas[8] for data analysis and manipulation, and matplotlib[9] for visualizing data. The ARIMA model is built using the statsmodel[10] package, and pmdarima[11] is used to assist in choosing the best model parameters for ARIMA. Sklearn[12] is used to evaluate the model performance using measures like mean squared error. Scipy[13] is also used for optimization.

Following the procedures of Kumar et al. (2023) and utilizing a comparable statistical ML model for soil movement prediction, we adjusted this model to meet our needs for temperature forecasting. We can forecast inside box temperatures with good accuracy by optimizing the parameters of ARIMA. Three parameters define the ARIMA model: moving average order (q) to determine the relationship between an observation and a residual error; differencing order (d) to eliminate seasonality and trends from the data; and autoregressive order (p) to determine the relationship between the number of past observations and the current observation. We determine the ideal parameters, which may vary based on trend, seasonality, and residual error, in order to recalibrate the model. For instance, based on the input characteristics, the autoregressive order in ARIMA may change depending on the number of prior values that correlate with the present value. This adjustment

TABLE 8.2
Range of Different Parameters of the ARIMA Model

Parameters	Range
p	0–5
d	0–5
q	0–5

ensures that the model remains flexible and responsive to changes in how past temperatures influence the present. By dynamically recalibrating the autoregressive order, the ARIMA model can better capture underlying patterns and improve forecasting accuracy. To meet the particular forecasting requirements of our research, a grid search was performed on the training set of data to determine the values of p, d, and q. In order to thoroughly investigate and enhance the performance of the model, we set the range for these parameters to be between 0 and 5. In order to reduce the root mean squared error between the predicted and actual values, a grid search process was used to determine the range of the parameters (p, d, and q) within the range of variation. The accuracy and reliability of time series forecasts were increased when the optimal ARIMA model parameters were found using this grid search technique (Table 8.2).

Carrying an icebox on a bus allowed for the collection of data, which was then preprocessed to train the ML model. We collected statistics from 17 distinct travel scenarios to investigate the icebox's interior and exterior environments. The icebox was transported between two places on the Mandi campus of the Indian Institute of Technology (IIT) in Himachal Pradesh, India, as well as between Mandi town and the IIT Mandi campus. Every 30 minutes, the humidity and temperature were measured and recorded. 80% of the gathered datasets were designated as test datasets, and the remaining 20% as training datasets. By computing the root mean square error (RMSE) using internal temperature data, we were able to assess the model's performance.

8.5 RESULTS

Figure 8.7a shows the IoT-MediGuard's implementation within the experimental setup, and Figure 8.7a provides additional information on a randomly selected dataset used to train the ML model. We designed a small external box to hold the battery and all integrated electronics (ICs). So live tracking, continuous monitoring, and system functionality were ensured through extensive testing. Figure 8.7b shows that when the system temperature exceeded the set predefined threshold, an SMS alert was sent to a designated number as soon as possible. Also, a local alarm connected to the box gave active warnings via a buzzer and the LCD also displayed temperature-breaching warnings when the predefined temperature threshold was breached.

Figure 8.8 illustrates the temperature change over time both within and outside the box in a randomly selected scenario. The temperature outside is indicated by a dotted line, and the temperature inside is by a solid line. NBTC guidelines suggest that if the permissible threshold temperature of 6°C is breached, it can spoil the medical supplies (e.g., in the case of blood and vaccines). As shown in Figure 8.8, the outside temperature (dotted line) was relatively constant, whereas the inside temperature (solid line) increased over time. The scenario started at 5 pm on April 27th, 2023, and the internal temperature breached the permissible threshold at 12 pm on April 28th, 2023, about 19 hours after the start of the scenario. Furthermore, it took four more days for the internal icebox temperature to come close to the ambient temperature outside the box.

For the purpose of predicting interior temperature, the ARIMA model was created and calibrated using five randomly chosen transport scenarios. These 5 transport scenarios are chosen randomly out of the 17 collected transport scenarios, where each scenario involved the transport

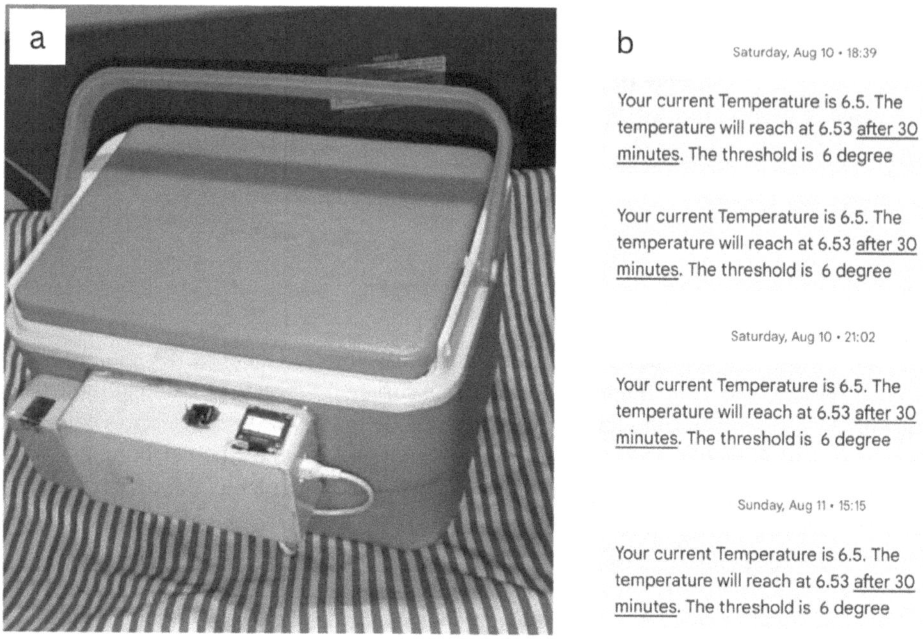

FIGURE 8.7 (a) Real-time System Monitoring Displaying Live Temperature Data. (b) Temperature Warnings Are Provided by the System to Alert Users of Critical Temperature Fluctuations

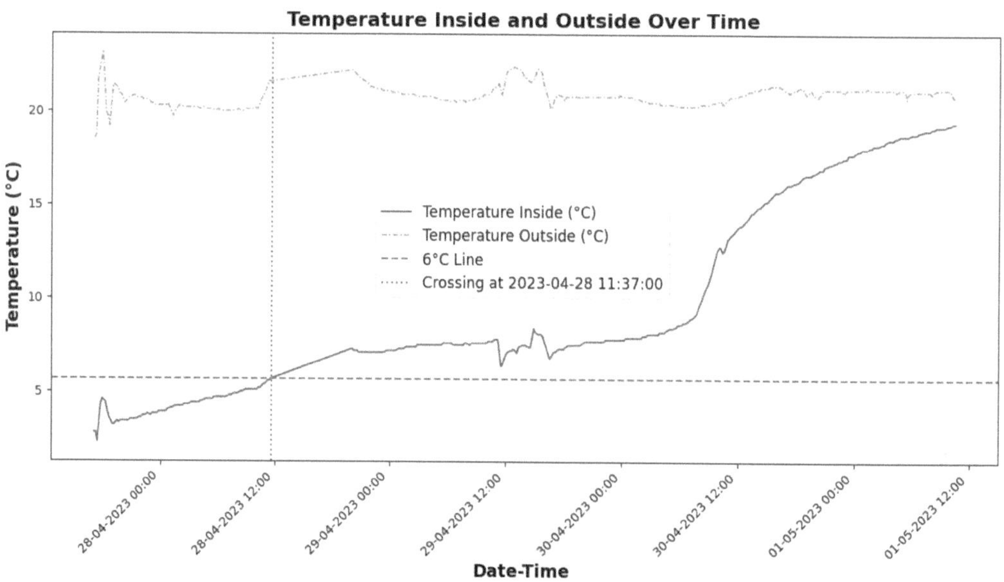

FIGURE 8.8 Variation in Temperature over Time

of medical supplies from a start location to an end location in a vehicle. We chose a 30-minute look-ahead projection because it gave us enough time to take the appropriate precautions in the event that the temperatures were expected to rise over the predetermined allowable limit. To find the best ARIMA configuration with the lowest RMSE calculated on internal temperatures, a grid search was employed. $p = 2$, $d = 2$, and $q = 4$ were the best ARIMA models found using the 80% training data from all five situations. With these settings, the training average RMSE was 0.12°C, while the test average RMSE was 0.24°C.

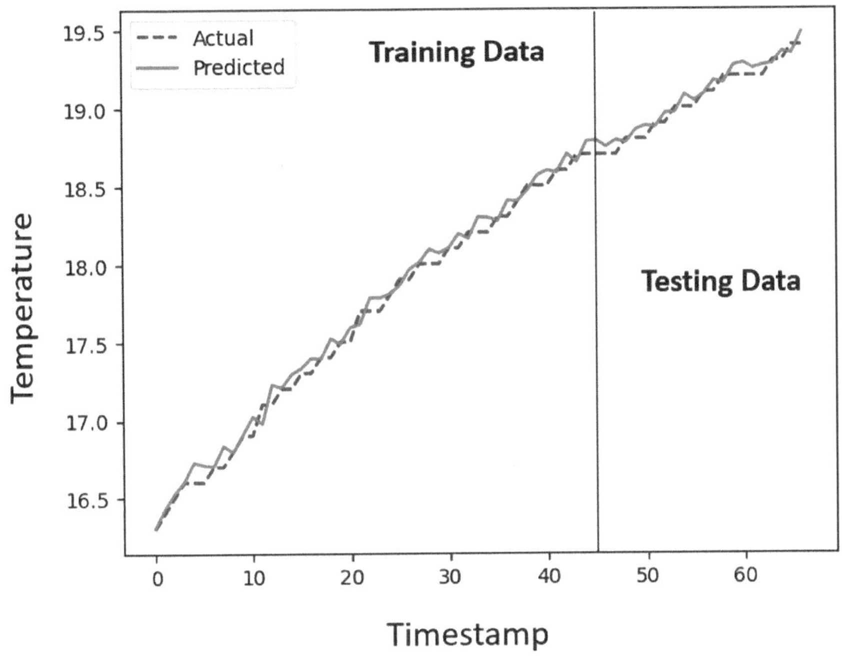

FIGURE 8.9 Comparison of ARIMA Model Predictions with Actual Temperature Values over Time

Two low-power design elements of the IoT-MediGuard system enable it to perform well with little energy consumption: deep sleep modes and selective activation of components, such as the LCD (Pathania et al., 2019; Sharma et al. 2019). With the ESP32 microcontroller, the system may operate for up to 72 hours on a 10 Ah lithium-ion battery. It consumes just 5.2 mA when in deep sleep and 280 mA during brief active periods. The energy-efficient architecture of the system ensures that medical supplies can be continually monitored for extended periods of time, even in circumstances when power is scarce (Hakim et al., 2022).

Figure 8.9 illustrates how well the ARIMA model predicts interior temperatures. This offers vital early warnings so that appropriate action may be taken to preserve the integrity of the medical supplies. For the safe delivery of necessary medical supplies, IoT-MediGuard is a dependable choice because of its energy-efficient architecture and accurate ARIMA model predictions.

8.6 CONCLUSION

If medical supplies are exposed to temperatures above what is recommended, they may deteriorate and have dangerous, even lethal, effects. Dependable systems that can keep an eye on and preserve the integrity of medical supplies throughout transit are desperately needed in India, especially in the country's remote and mountainous areas. In response to this demand, the IoT-MediGuard system was created to offer real-time temperature and humidity monitoring both inside and outside of iceboxes that are used to carry medical goods. Because of the system's power-efficient architecture, which incorporates sensors for temperature, humidity, GPS, and GSM, it is possible to monitor continuously for up to 72 hours, even under difficult circumstances.

Additionally, the ARIMA model in the IoT-MediGuard system achieves an amazing RMSE of 0.07°C when predicting interior temperatures 30 minutes in advance. This predictive capacity is essential for facilitating prompt responses and reducing the possibility that medical supplies would deteriorate while en route. The device underwent extensive testing on the IIT Mandi campus, effectively adjusting to external environmental parameters, including temperature variations. These experiments confirmed the system's practical effectiveness and demonstrated its ability to maintain medical supply integrity, particularly in difficult-to-reach places.

Anticipating subsequent deployments in pharmacies, blood banks, and hospitals, further data will be gathered to confirm the system's functionality. The system will be improved by using sophisticated ML models, such as Transformer, LSTM, or MLP, to raise the efficiency and accuracy of predictions (Moise et al., 2023). Furthermore, edge computing will improve real-time decision-making, decrease latency, and process data closer to the source, all of which will increase the system's responsiveness (Kumar et al., 2024). These developments will further guarantee the timely delivery of essential medical supplies, improving the efficiency and sustainability of healthcare logistics.

To sum up, the IoT-MediGuard system provides a reliable and effective way to track and forecast interior icebox temperatures in real time, greatly enhancing healthcare logistics' dependability in maintaining the effectiveness and security of medicinal supplies. This device is a significant development that will help to guarantee that vital medical supplies reach their destinations in optimal condition, especially in areas where transportation is difficult.

ACKNOWLEDGMENT

The Himachal Pradesh Council for Science, Technology, and Environment (HIMCOSTE) funded this study for Dr. Varun Dutt under project number IITM/HIMCOSTE/VD/416. We thank HIMCOSTE and IIT Mandi for their financial and computation support.

NOTES

1 https://health.economictimes.indiatimes.com/news/industry/india-needs-better-network-to-avoid-blood-wastage/92200637.
2 https://www.businessinsider.in/science/health/news/nearly-12000-individuals-die-every-day-due-to-lack-of-quality-blood/articleshow/92196089.cms
3 https://naco.gov.in/sites/default/files/Standards%20for%20Blood%20Banks%20and%20Blood%20Transfusion%20Services.pdf
4 https://www.adafruit.com/product/385.
5 https://firebase.google.com/.
6 Among internal temperature and humidity, the system predicts future internal temperature and not future internal humidity. That is because the packaging of the medical supplies generally protects them from varying humidity conditions.
7 numpy: https://github.com/numpy/numpy.
8 numpy: https://github.com/numpy/numpy.
9 matplotlib: https://github.com/matplotlib/matplotlib.
10 statsmodels: https://github.com/statsmodels/statsmodels.
11 pmdarima: https://github.com/alkaline-ml/pmdarima.
12 scikit-learn: https://github.com/scikit-learn/scikit-learn.
13 scipy: https://github.com/scipy/scipy.

REFERENCES

Afreen, H. and Bajwa, I.S., 2021. An IoT-based real-time intelligent monitoring and notification system of cold storage. *IEEE Access*, 9, pp.38236–38253. IEEE. USA. https://doi.org/10.1109/ACCESS.2021.3056672

Aggarwal, S., Gupta, P., Mahajan, N., Balaji, S., Singh, K., Bhargava, B. and Panda, S., 2023. Implementation of drone-based delivery of medical supplies in North-East India: Experiences, challenges, and adopted strategies. *Frontiers in Public Health*, 11, p. 1128886. https://doi.org/10.3389/fpubh.2023.1128886

Ahmadimanesh, M., Safabakhsh, H.R. and Sadeghi, S., 2023. Designing an optimal model of blood logistics management with the possibility of return in the three-level blood transfusion network. *BMC Health Services Research*, 23(1), p.1304. https://doi.org/10.1186/s12913-023-10240-0

Aswath, M.B., Kamalesh, V., Koushik, M. and Prema, R., 2021. February. Medicine parameters monitoring using IoT. In *2021 Third International Conference on Intelligent Communication Technologies and Virtual Mobile Networks (ICICV)* (pp.349–353). Tirunelveli, India: IEEE. https://doi.org/10.1109/ICICV50876.2021.9388623

Bapatla, A.K., Mohanty, S.P., Kougianos, E. and Puthal, D., 2022, December. PharmaChain 2.0: A blockchain framework for secure remote monitoring of drug environmental parameters in pharmaceutical cold supply chain. In *2022 IEEE International Symposium on Smart Electronic Systems (iSES)* (pp.185–190). Warangal, India: IEEE. https://doi.org/10.1109/iSES54909.2022.00046

Cil, A.Y., Abdurahman, D. and Cil, I., 2022. Internet of Things enabled real-time cold chain monitoring in a container port. *Journal of Shipping and Trade*, 7(1), p.9. https://doi.org/10.1186/s41072-022-00110-z

Ghide, L., Wei, S. and Ding, Y., 2022. Comparative study of Wavelet-SARIMA and EMD-SARIMA for forecasting daily temperature series. *International Journal of Analysis and Applications*, 20, pp.17–17. https://doi.org/10.28924/2291-8639-20-2022-17

Hakim, G.P.N., Hajar, M.H.I., Firdausi, A. and Ramadhan, E., 2022. Benchmarking in microcontroller development board power consumption for low power IoT WSN application. *JurnalTeknologiElektro*, 13(1), pp.25–28. https://doi.org/10.22441/jte.2022.v13i1.005

Kiruthika, S., Sakthi, P., Kaviya, M. and Vishnupriya, S., 2021. Blood bank monitoring and blood identification system using IoT device. Annals of the Romanian Society for Cell Biology, 25(6), pp. 182–192. Association of Cell Biology Romania. Romania. https://www.researchgate.net/publication/360385949_Blood_Bank_Monitoring_and_Blood_Identification_System_Using_Iot_Device

Kapur, R., Kumar, Y., Sharma, R., Sharma, S., Singh, E., Rohilla, D., Kanwar, V., Kumar, B., Bhavsar, A. and Dutt, V., 2024. GlucoBreath: An IoT, ML, and breath-based non-invasive glucose meter. *IEEE Access*, 12, pp.59346–59360. https://doi.org/10.1109/ACCESS.2024.3392015

Kapur, R., Kumar, Y., Sharma, S., Rastogi, V., Sharma, S., Kanwar, V., Sharma, T., Bhavsar, A. and Dutt, V., 2023. DiabeticSense: A non-invasive, multi-sensor, IoT-based pre-diagnostic system for diabetes detection using breath. *Journal of Clinical Medicine*, 12(20), p.6439. https://doi.org/10.3390/jcm12206439

Kartoglu, U. and Ames, H., 2022. Ensuring quality and integrity of vaccines throughout the cold chain: the role of temperature monitoring. Expert Review of Vaccines, 21(6), pp. 799–810. Taylor and Francis. USA. DOI: https://doi.org/10.1080/14760584.2022.2061462

Kumar, P., Priyanka, P., Dhanya, J., Uday, K.V. and Dutt, V., 2023. Analyzing the performance of univariate and multivariate machine learning models in soil movement prediction: A comparative study. *IEEE Access*, 11, pp. 62368–62381. IEEE. USA. https://doi.org/10.1109/ACCESS.2023.3287851

Kumar, P., Priyanka, Pathania, A., Agarwal, S., Mali, N., Singh, R., Chaturvedi, P., Uday, K.V. and Dutt, V., 2019a, September. Predictions of weekly soil movements using moving-average and support-vector methods: A case-study in Chamoli, India. In *International Conference on Information Technology in Geo-Engineering* (pp.393–405). Cham: Springer International Publishing. https://doi.org/10.1007/978-3-030-32029-4_34

Kumar, P., Priyanka, Pathania, A., Agarwal, S., Mali, N., Singh, R., Chaturvedi, P., Uday, K.V. and Dutt, V., 2020. Predictions of weekly slope movements using moving-average and neural network methods: A case study in Chamoli, India. In *Soft Computing for Problem Solving 2019: Proceedings of SocProS 2019* (Volume 2, pp.67–81). Singapore: Springer. https://doi.org/10.1007/978-3-030-32029-4_34

Kumar, P., Sihag, P., Pathania, A., Agarwal, S., Mali, N., Chaturvedi, P., Singh, R., Uday, K.V. and Dutt, V., 2019b, September. Landslide debris-flow prediction using ensemble and non-ensemble machine-learning methods. In *International Conference on Time Series and Forecasting* (pp.25–27). Granada, Spain: IAEA. Springer. Switzerland. https://www.researchgate.net/publication/336241696_Landslide_debris-flow_prediction_using_ensemble_and_non-ensemble_machine-learning_methods_a_case-study_in_Chamoli_India

Kumar, P., Sihag, P., Pathania, A., Chaturvedi, P., Uday, K.V. and Dutt, V., 2021. Comparison of moving-average, lazy, and information gain methods for predicting weekly slope-movements: A case-study in Chamoli, India. In Casagli, N., Tofani, V., Sassa, K., Bobrowsky, P. T. and Takara, K. (eds.) *Understanding and Reducing Landslide Disaster Risk: Volume* 3 *Monitoring and Early Warning* (pp.321–330). Cham: Springer. https://doi.org/10.1007/978-3-030-60311-3_38

Kumar, S., Panigrahi, R.K. and Kanungo, D.P., 2024, January. Development of IoE-Edge-AI architecture for landslide early warning system. In *2024 IEEE Applied Sensing Conference (APSCON)* (pp.1–4). IEEE. https://dx. doi.org/10.1109/APSCON60364.2024.10465809 Li, Z., Zou, H. and Qi, B., 2019, October. Application of ARIMA and LSTM in relative humidity prediction. In *2019 IEEE 19th International Conference on Communication Technology (ICCT)* (pp.1544–1549). Xi'an, China: IEEE. https://doi.org/10.1109/ICCT46805.2019.8947142

Moise, M.V., Pantazica, M. and Otoiu, M., 2023, May. IoT portable patient health monitoring system using MEMS sensors. In *2023 46th International Spring Seminar on Electronics Technology (ISSE)* (pp.1–4). Timisoara, Romania: IEEE. https://doi.org/10.1109/ISSE57496.2023.10168525

Pathania, A., Kumar, P., Kesri, J., Priyanka, Agarwal, S., Mali, N., Singh, R., Chaturvedi, P., Uday, K.V. and Dutt, V., 2019, September. Reducing power consumption of weather stations for landslide monitoring. In *International Conference on Information Technology in Geo-Engineering* (pp.144–158). Cham: Springer International Publishing. https://doi.org/10.1007/978-3-030-32029-4_13

Pathania, A., Kumar, P., Priyanka, Maurya, A., Kumar, M., Chaturvedi, P. and Dutt, V., 2022. Predictions of soil movements using persistence, auto-regression, and neural network models: A case-study in Mandi, India. *International Journal of Swarm Intelligence*, 7(1), pp.94–109. Inderscience Publishers. Switzerland. https://doi.org/10.1504/IJSI.2022.121100

Rodrigues, R.K., Torné, I.G., de Sousa Cardoso, F., Sobrinho, A.M.F., Printes, A.L. and Cisneros, E.A.G., 2023. Real-time monitoring and alert system in pharmaceutical products transportation using IoT: Sistema de monitoramento e alertaem tempo real no transporte de produtosfarmacêuticosutilizando IoT. *Concilium*, 23(6), pp.212–231. https://doi.org/10.53660/CLM-1122-23D24C

Sagahyroon, A., Aburukba, R. and Aloul, F., 2017. The Internet of Things and e-Health: Remote patients monitoring. In Hassan, Q. F., Khan, A. R. and Madani, S. A. (eds.) *Internet of Things* (pp.303–319). Chapman and Hall/CRC. USA.

Salah, K., Alfalasi, A., Alfalasi, M., Alharmoudi, M., Alzaabi, M., Alzyeodi, A. and Ahmad, R.W., 2020, January. IoT-enabled shipping container with environmental monitoring and location tracking. In *2020 IEEE 17th Annual Consumer Communications & Networking Conference (CCNC)* (pp.1–6). Las Vegas, NV: IEEE. https://doi.org/10.1109/CCNC46108.2020.9045495

Sandaruwan, P.A.J., Dolapihilla, U.D.L., Karunathilaka, D.W.N.R., Wijayaweera, W.A.D.T.L., Rankothge, W.H. and Gamage, N.D.U., 2020, December. Towards an efficient and secure blood bank management system. In *2020 IEEE 8th R10 Humanitarian Technology Conference (R10-HTC)* (pp.1–6). Kuching, Malaysia: IEEE. https://doi.org/10.1109/R10-HTC49770.2020.9356980

Schoder, D., 2018. Introduction to the Internet of Things. In Hassan, Q. F. (ed.) *Internet of Things A to Z: Technologies and Applications* (pp.1–50). Hoboken, NJ: Wiley. https://doi.org/10.1002/9781119456735.ch1

Sharma, R., Saini, T., Kumar, P., Pathania, A., Chitineni, K., Chaturvedi, P. and Dutt, V., 2019, December. An online low-cost system for air quality monitoring, prediction, and warning. In *International Conference on Distributed Computing and Internet Technology* (pp.311–324). Cham: Springer International Publishing. https://doi.org/10.1007/978-3-030-36987-3_20

Subero, A., 2021. *Programming Microcontrollers with Python: Experience the Power of Embedded Python*. Berkeley, CA: Apress. https://doi.org/10.1007/978-1-4842-7058-5_4

Tao, T., Bin, C. and Yunxiang, L., 2019, November. A medicine cold chain monitor system based on LoRa wireless technology. In *2019 International Conference on Intelligent Informatics and Biomedical Sciences (ICIIBMS)* (pp.93–98). Shanghai, China: IEEE. https://doi.org/10.1109/ICIIBMS46890.2019.8991458

9 IoTMaxHRV
A Low-Cost IoT Device for Heart Rate Variability Assessment

Yashwant Kumar, Ritu Kapur, Sahil Sankhyan, Premchand, Shivani Sharma, Arnav Bhavsar and Varun Dutt

9.1 INTRODUCTION

The Internet of Things (IoT) is transforming the healthcare business by enabling the creation of smart devices that can collect, evaluate, and share health data in real time (Hassan, 2018). These technologies are especially essential for bettering patient care because of remote patient monitoring (RPM) (Russo and Eriksson, 2018). Since RPM allows medical practitioners to closely monitor their patients' health outside of official clinical settings, it provides various advantages for managing chronic diseases, recovering from surgery, and caring for the elderly. Wearable sensors and smart implants are examples of IoT-enabled devices that healthcare professionals can use to monitor vital indications, including blood pressure, glucose levels, and heart rate. Here, by lowering the need for frequent hospital visits and aiding in the early diagnosis of any health issues, continuous data streams may likely help improve patient outcomes and save healthcare costs (Russo and Eriksson, 2018; Sagahyroon et al., 2017).

Cardiovascular diseases account for approximately 17.9 million annual deaths that are expected to increase further in both developing and developed countries, as forecasted by the World Health Organization[1] (WHO) and the American Heart Association (AMA) (Benjamin et al., 2017; Virani et al., 2021). The major factors leading to cardiovascular diseases include population growth, unhealthy lifestyle choices, obesity, diabetes, and hypertension (Benjamin et al., 2017; Virani et al., 2021). Heart rate variability (HRV) is closely related to cardiovascular activity and the autonomic nervous system (Valenza et al., 2020). It is often the most relevant biomarker for assessing cardiovascular health status and autonomic nervous system function (Xhyheri et al., 2012). HRV could be used to check the possibility of people getting a cardiovascular disease by learning from their cardiovascular health history (Abdelnabi, 2019; Thayer et al., 2010).

The human body contains numerous individual cells and extracellular fluids containing substances, such as water, proteins, and electrolytes (Sampson and McGrath, 2015). The cells exhibit permeability to electrolytes to pass through the membrane in and out. The electrolyte concentration inside and outside the cell and flow through the membrane are responsible for its electrical activity. The electrolytes sodium and potassium are responsible for the heart's electrical activity measured using the electrocardiography (ECG) method with the help of electrochemical electrodes. On the other hand, Allen (2007) states that photoplethysmography (PPG) utilizes optoelectronic hardware with a light source to detect the blood volume changes in tissues under the skin. The optoelectronic detector outputs a sinus-pulsating waveform considered to be synchronized with the heart-blood pumping cycle. Nowadays, various wearable devices like smartwatches and oximeters utilize the PPG-based method for various heart-related measurements, including HRV.

The HRV analysis is a non-invasive technique to assess the cardiac and autonomic nervous system function status related to human heart health (Shaffer and Ginsberg, 2017). HRV measurements can be used to predict adverse cardiac events. The ECG signal-measuring method, introduced in 1995,

DOI: 10.1201/9781003506638-11

has been a notable advancement in signal processing (Shaffer et al., 2014). HRV is semi-periodic; successive heartbeats lead to an ECG signature. ECG signals possess sharp peaks, known as R peaks in ECG signals, due to depolarization of the heart ventricles. An effective measure of autonomic function and total heart rate is the mean of R-R intervals, MeanRR, or the average interval between consecutive R peaks in the ECG signal. Decreasing values of the MeanRR measure might indicate autonomic dysregulation, heart rate elevation, or cardiac strain. Poorer autonomic control and a higher risk of cardiac events are linked to lower levels of MeanRR. Comparably, the variance in R-R intervals is measured by the standard deviation of NN intervals (SDNN; Shaffer and Ginsberg, 2017). Although HRV is reflected by the standard deviation of heart rate (SDHR), elevated mean heart rate (MeanHR) over time can indicate significant stress on the heart, raising the likelihood of adverse events (Rajendra Acharya et al., 2006). Lower values of the SDHR suggest reduced vagal activity and increased cardiac risk. Additionally, MinHR and MaxHR may be used to infer unfavorable cardiovascular events. According to Munoz et al. (2015), excessive MaxHRs can suggest tachycardia or arrhythmias, whereas too low MinHRs could suggest bradycardia. The difference between consecutive R peaks is squared, and the resulting root mean square of successive differences (RMSSD) is determined by averaging the differences. The RMSSD is the square root of the mean. RMSSD may allow researchers to assess parasympathetic activity and short-term HRV. As per Thayer et al. (2010), there is a likely correlation between a lower RMSSD value and an increased risk of consequences, including sudden cardiac death. In order to improve cardiac health conditions in the general population, monitoring patients' HRV may be an effective approach for early identification of heart diseases.

Also, these measures may help in real-time treatment plan adjustments based on cardiac reconditioning program results, which are intended to improve client health management. In HRV detection techniques, the most widely used electrical and photoelectric methods are ECG and PPG. While PPG measures changes in blood volume inside the microvascular bed of tissues, ECG uses electrochemical electrodes to directly measure the heart's electrical activity.

Max30003[2] and AD8232[3] are some commonly used ECG detection devices, and emWave Pro[4] is a commercially available PPG-based device commonly used to drive HRV parameters. Max30003 has been reported for long-term ECG monitoring (Cinotti et al., 2024), while the AD8232 has been proposed for its real-time monitoring capabilities, often coupled with data transmission functionalities (Güvenç, 2020). However, accurate PPG devices from established manufacturers are expensive, which makes them out of reach of many, especially for the economically weaker sections in developing countries like India. Thus, there is an urgent need for the development of portable, low-cost HRV measurement devices. Precise HRV evaluation devices can help detect sudden heart activity variations, indicating the risk of potential heart conditions. Such low-cost devices can be set up at various public healthcare centers and help with accurate HRV measurements, especially in remote areas in the Himalayas with limited medical facilities.

Furthermore, as per WHO[5] and AMA, transportation is a critical barrier to HRV assessment in the elderly (Kotval et al., 2020). The inability of the elderly to travel to hospitals or medical centers for timely HRV assessment may result in many undetected heart diseases, which could lead to sudden heart attacks (Hancock et al., 2013). Older individuals with disabilities travel less, and the effect of disability on travel increases more in patients aged 65 or older (Mackett, 2017). Thus, we propose a novel HRV assessment device, IoTMaxHRV, which can help quickly detect sudden cardiac problems. The device could also be used in remote areas for HRV assessment. The IoTmaxHRV device provides an easy-to-use solution for cardiac health management by facilitating the remote interface and ECG signal processing algorithm for HRV parameter estimation. The IoTMaxHRV is compared for its performance with available HRV assessment solutions: AD8232 (based on an ECG sensor) and a commercial device emWave Pro (based on a PPG sensor). To perform this comparison, the following HRV metrics are assessed from all devices: MeanRR, SDNN MeanHR, SDHR, MinHR, MaxHR, and RMSSD.

To explain the HRV assessment with the IoTMaxHRV device, we first provide the background on existing hardware solutions for HRV monitoring systems. Next, we provide the literature review of ECG-and-PPG-based heart health management systems. After that, the IoTMaxHRV device formation and remote interface development are explained in detail. Lastly, the experiment outcomes, IoTMaxHRV's utilization in HRV assessment, and future work are discussed.

9.2 BACKGROUND

HRV has been identified as a critical biomarker for improving the health condition of the autonomic and cardiovascular systems. Accurate HRV measurement is necessary for improved HRV estimation or incorporation of adequate cardiac health evaluation into medical devices such as PPG- and ECG-based ones. This review of the literature examines studies on portable ECG and PPG sensors, identifies knowledge gaps, and highlights advancements with the suggested IoTMaxHRV device for low-cost, real-time HRV measurements.

9.2.1 Portable PPG and ECG Equipment

Currently, various ECG- and PPG-based portable hardware solutions are available for HRV-related measurements. This study used Max30003 and AD8232 sensors for ECG detection. Max30003 provided a digital output of the measured data, and the AD8232 ECG sensor provided an analog output signal. An emWave Pro commercial PPG-based device was used to detect HRV in this study using the PPG technique. The sensors used for HRV parameter assessment in this study are explained below:

- A medical-grade ECG-detecting chip, the Max30003 sensor has an integrated R-R peak detector, a serial peripheral interface (SPI), and an analog-to-digital converter (ADC). Wearable measurement applications for ECG signal detection make use of the Max30003 sensor's low power consumption and broad detection range.
- Using an electrochemical gel electrode, the AD8232 sensor is a semiconductor chip for ECG detection that generates an analog ECG signal. Additionally, it features a high-pass filter to eliminate low-frequency noise and an integrated instrument amplifier to augment weak ECG readings. As per Kumar et al. (2020), the gadget has the ability to monitor cardiac abnormalities in real time and can be easily integrated with data transport capabilities using microcontroller systems.
- HRV characteristics can also be estimated by the emWave Pro PPG device, which is sold commercially. This PPG sensor measures HRV parameters using a photodetector and an infrared light source. But more research needs to be done to determine how useful it is in comparison to ECG-based sensors (Singstad et al., 2021).

Research comparing PPG and ECG sensors has produced different findings. Bhowmick et al. (2020) suggested low-cost equipment for remote monitoring and emphasized the dependability of HRV parameters from combined ECG and PPG data. Plews et al. (2017) established that smartphone PPG technology and ECG, especially for athletes, agree reasonably when assessing HRV.

To show that time-frequency power spectral density analysis for HRV is feasible, Chou et al. (2011) created a portable ECG processor and successfully executed real-time HRV analysis using a Lomb-Scargle periodogram. Similarly, Esgalhado et al. (2022) discovered that, under some circumstances, PPG might function as a suitable substitute for ECG in HRV extraction.

Even with these developments, real-time remote evaluation and reliable HRV measurement in motion situations continue to be difficult tasks. To improve measurement reliability, algorithms to remove muscle and motion abnormalities in ECG and PPG data are still required (Pinheiro et al., 2016).

9.3 LITERATURE GAPS AND THE PROPOSED IOTMAXHRV DEVICE

For the purpose of detecting PPG and ECG signals, modern devices such as the Max30003, AD8232, and emWave Pro work well. However, in order to log data in real time to a cloud server, these devices need a remote interface. Nevertheless, these devices frequently lack the integration required for affordable, user-friendly real-time data storage and continuous HRV monitoring. This need is filled by the IoTMaxHRV gadget, which offers a novel solution that blends continuous HRV monitoring with a remote connection. Because it reduces the need for frequent trips to medical institutions, this function is especially helpful for seniors and people with mobility challenges. It also lessens the financial and physical stress on patients.

The IoTMaxHRV device's usage of the clinical-grade ECG sensor Max30003, which guarantees excellent accuracy in HRV readings, is one of its primary characteristics. The IoTMaxHRV's ECG-based architecture makes it resilient against noise and motion aberrations, unlike many portable PPG-based devices that might have trouble with them, especially while engaging in physical activity. This makes it a dependable alternative for both clinical and personal use. Healthcare professionals may continually watch their patients' heart health with the capacity to remotely monitor and record HRV data in real time. This enables them to make prompt interventions and improve the management of cardiovascular disorders. For patients with chronic cardiac disorders, this is essential since it allows for the early identification of any problems, thereby avoiding major implications.

Furthermore, the IoTMaxHRV device is intended to be affordable, which means that people from a wider range of socioeconomic backgrounds may use it. Because of the device's mobility and simplicity of use, it may be used in a variety of contexts, from private homes to remote healthcare facilities in rural locations. This allows people who might not otherwise have access to such technology to monitor their heart rates. By enabling proactive treatment of heart health, this can enhance health outcomes, especially in underprivileged populations.

ECG and PPG methods have been proposed to assess the HRV. For example, various PPG-based wrist-worn devices are utilized for HRV analysis due to ease of use. However, the ECG measurement is still considered the gold standard for HRV analysis (Jeyhani et al., 2015). Bhowmick et al. (2020) found a good correlation between ECG and PPG HRV parameters in spite of that study indicating that the PPG estimated HRV parameters should be considered with caution for measurement during sports activity, and ECG is a must for reliable R-R peak detection. Tobón et al. (2017) proposed a spectro-temporal analysis method for robust noise removal for extremely noisy ECG signals. Esgalhado et al. (2022) proposed a PPG signal processing algorithm using the Hilbert double envelope method and suggested that the PPG can be used as an alternative to ECG with good measurement accuracy.

All things considered, previous studies show that although PPG and ECG techniques are both appropriate for HRV study, portable PPG measuring instruments have distinct drawbacks in specific situations. In order to analyze HRV for both clinical and personal usage, it is necessary to investigate a portable and reliable remote HRV monitoring system, such as the IoTMaxHRV. The IoTMaxHRV is a useful tool for enhancing heart health management in a variety of populations since it offers an inexpensive, dependable device with a remote interface for HRV monitoring, therefore solving the issues of motion artifacts and the requirement for real-time remote examination.

9.4 MATERIALS AND METHODS

In this work, we created the IoTMaxHRV device for HRV measurement and evaluated its efficacy against two other HRV measuring devices, namely, emWave Pro (PPG-based) and AD8232 (ECG-based). The IoTMaxHRV device consists of an ECG sensor interfaced with a Wi-Fi-enabled microcontroller for ECG detection and remote data logging. The embedded C code was utilized to link the ECG sensor with the microcontroller, and the ESP 32 microcontroller's built-in Wi-Fi performed data logging to a distant server. An algorithm was utilized to extract the R-R interval sequence from the unprocessed ECG data to obtain the HRV features. HRV parameters were also obtained for every ECG sample using a PPG device for comparison. Several HRV parameters were collected for the study from three different HRV measuring devices (Max30003, AD8232, and emWave Pro), and a comparative analysis was carried out. R-R intervals were extracted from the raw ECG data to compute HRV parameters. The ECG monitoring devices' operating and construction details are given below.

9.4.1 IoTMaxHRV ECG Device

IoTMaxHRV includes hardware and software components. The device utilized an ECG sensor and Wi-Fi-enabled microcontroller for data transmission and ECG recording of the study participants. The ESP 32 microcontroller accomplishes the task of data transmission to the remote server using the built-in Wi-Fi feature. In Table 9.1, we list the components necessary to build the IoTMaxHRV device, including their manufacturer's details and prices.

The Max30003 ECG sensor is a single-channel integrated biopotential heart rate monitor for 3/5 electrode configuration. The other components were selected for their ease of integration and the lowest acquisition cost possible to support the development of an efficient, reliable HRV monitoring device with a compact design.

Figure 9.1a shows the overall workflow of the IoTMaxHRV device. The IoTMaxHRV ECG sensor-based remote HRV monitoring system comprises a 3-lead single-channel ECG lead, a Max30003 sensor (ECG), a Wi-Fi-based microcontroller, a cloud server, and a database for data storage (Figure 9.1a).

The Max30003 sensor is our choice for the development of IoTMaxHRV because of its features like low power consumption and Bluetooth wireless serial communication interface. The IoTMaxHRV uses a single-channel, three-electrode ECG lead to detect ECG data from the human body with the help of ECG gel electrodes attached to the body (Figure 9.1b). The ESP 32 microcontroller was integrated with the ECG sensor module connected to the human body for ECG detection. Two electrodes corresponding to the left-hand and right-hand, and a third right-foot electrode served as the reference electrode for ECG measurements (Figure 9.1b). The Max30003 ECG sensor reads the ECG signal and transmits its digital representation to the ESP32 microcontroller unit (MCU) over the SPI interface bus. The ESP32 MCU serves as the SPI master. The Max30003 operates as an SPI slave device. Thus, the ESP32 oversees the clock and chip select (C/S) signals to select the slave device and communicate with it. The ECG system architecture utilized an ESP 32 as a remote client to transmit data to a distant cloud server. The InfluxDB,[6] a time series database, stores the ECG raw data streamed over the IoT interface (Figure 9.1a). An open-source Python-based R peak detection algorithm[7] is used to find R peaks in raw ECG data, where these R peaks can then be utilized to compute HRV parameters like R-R intervals (Figure 9.1a).

Figure 9.1b shows that the ESP 32 microcontroller and the Max30003 sensor require seven wiring connections to power and communicate with each other. The Max30003 sensor power supply and ground pins were connected to the 3.3V and ground pins of the ESP32 microcontroller board. The Max30003 ECG sensor has a wide input range of sigma-to-delta type 18-bit ADC to read analog ECG signals. Therefore, the Max30003 ECG sensor digitally represents the measured ECG raw data over the SPI interface, as shown in Figure 9.1a.

TABLE 9.1
The Bill of Materials Used to Construct the IoTMaxHRV Device

Component Name	Manufacturers	Price (USD)
Max30003 (ECG sensor)[a]	Analog Electronics	30.95
Breadboard[b]	ROBU India	0.6
Male to male jumper wires[c]	ROBU India	0.96
ESP 32-S3[d]	Espressif	15.12
Micro USB Type B cable[e]	pTron	1.2

[a] https://shorturl.at/Xzhvu
[b] https://shorturl.at/llrqH
[c] https://shorturl.at/lrhI8
[d] https://shorturl.at/xriZG
[e] https://shorturl.at/RACH6

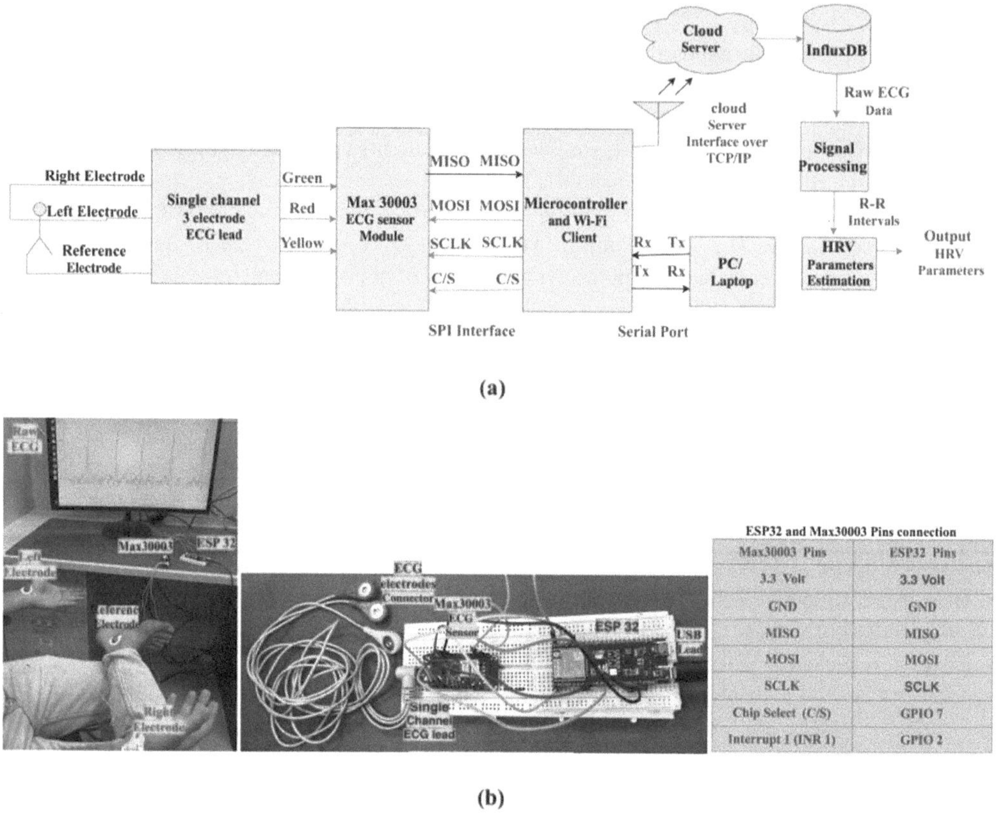

FIGURE 9.1 (a) The IoTMaxHRV ECG Monitoring Device's Block Diagram. (b) The IoTMaxHRV ECG Monitoring Device's Hardware Implementation

9.4.1.1 Master-Slave Working of the SPI Interface

The SPI interface uses four pins to communicate between the SPI master (ESP 32) and the SPI slave (Max30003) devices. The four pins of the SPI communication protocol were defined as master in slave out (MISO),[8] master out slave in (MOSI),[9] system clock (SCLK),[10] and C/S.[11] The MISO pin receives serial data from the slave device, and each bit is synchronized with the SCLK clock signal. Similarly, the MOSI pin sends serial data from the slave to the master device. Both the MISO and MOSI can run simultaneously in synchronization with SCLK. The master device generates the SCLK clock signal and controls the data transmission speed accordingly. The C/S pin enabled the slave device to communicate by activating the C/S pin from the master device. SPI enables bidirectional communication between the master and slave devices; we only utilized it for data reception from the Max30003 slave device.

The interrupt pin of the Max30003 is connected to GPIO 2 of the ESP32. A USB lead connected to the PC/laptop was utilized for code flashing to the microcontroller, and it worked as a debug probe during system development to check the device's response, as shown in Figure 9.1b.

9.4.2 AD8232 ECG Device

In Table 9.2, we detail a bill of materials for the various components employed in the AD8232 device development buffer table, along with the manufacturer and pricing details. The major parts are the AD8232 ECG sensor from Analog Electronics, the breadboard and male-to-male jumper wires from ROBU India, the ESP 32-S3 microcontroller board by Espressif, and a Micro USB Type B cable from pTron. We selected these components to be compatible and efficient in building an ECG monitoring system.

TABLE 9.2
The Bill of Materials Used to Construct the AD8232 Device

Component Name	Manufacturers	Price (USD)
AD8232 (ECG sensor)[a]	Analog Electronics	24
Breadboard	ROBU India	0.6
Male to male jumper wires	ROBU India	0.96
ESP 32-S3	Espressif	15.12
Micro USB Type B cable	pTron	1.2

[a] https://rb.gy/jcg04w

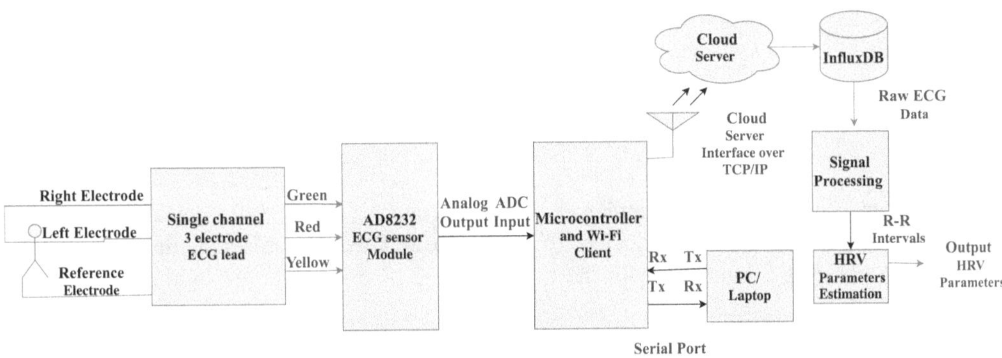

(a)

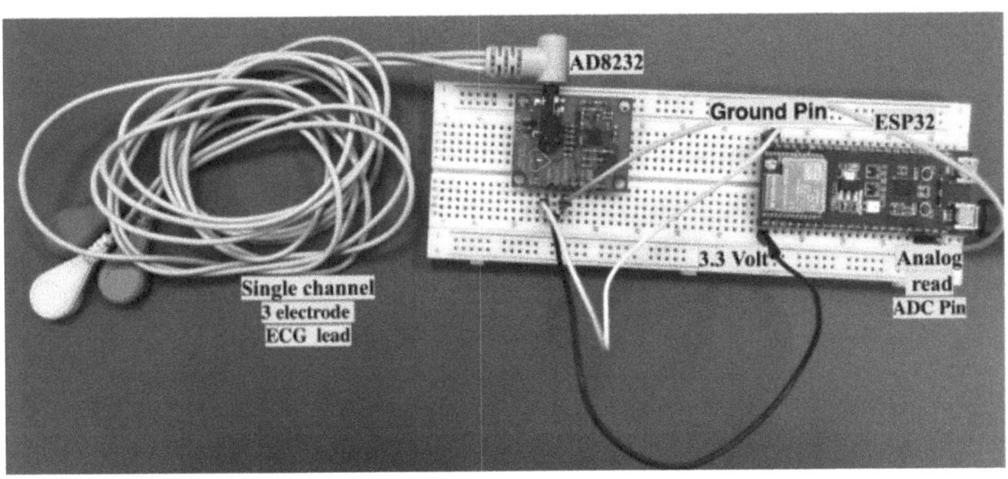

(b)

FIGURE 9.2 (a) The AD8232 ECG Monitoring Device's Block Diagram. (b) The AD8232 ECG Monitoring Device's Hardware Implementation

Figure 9.2a and b shows the AD8232 device's block diagram and hardware integration to implement the ECG monitoring system. Figure 9.2a shows that the AD8232 device connected with ECG lead and ECG gel electrodes was integrated with a microcontroller to read the analog ECG signal from participants. The ESP 32 utilizes an inbuilt 12-bit successive approximation type ADC converter to represent the analog ECG signal digitally. The AD8232 ECG sensor power supply and ground pins were connected to the 3.3V and ground pins of the ESP 32 microcontroller board. Figure 9.2a and b shows the pin mappings for the wired interface between the AD8232 and ESP32 microcontroller to communicate with the sensor.

9.4.3 emWave Pro PPG Device

The emWave Pro by HeartMath uses optical sensors to detect blood volume changes in the microvascular bed of tissue in a way other than the ECG-based devices we discussed earlier that can measure HRV. Figure 9.3 shows emWave Pro's elegant design and user-friendly interface, demonstrating ease in measuring HRV. The system is lightweight and captures PPG signals. It is costly (it costs USD 315 at the HeartMath store).

The optical sensor in emWave Pro detects the pulse wave that occurs from blood flow in skin tissues. Proprietary algorithms process the captured PPG data and estimate HRV parameters, yielding insights into a user's autonomic nervous system function. This device allows us to measure HRV via PPG in a noninvasive way and can be used for consecutive, practical, and comfortable monitoring.

9.4.4 Software Implementation

Table 9.3 displays the code that is implemented to integrate the ESP32 ECG detecting device with the Max30003 and AD8232 ECG devices (the code for emWave Pro is not available because it is private). The libraries compatible with the target ESP 32 board are chosen by the "WiFiMulti.h"

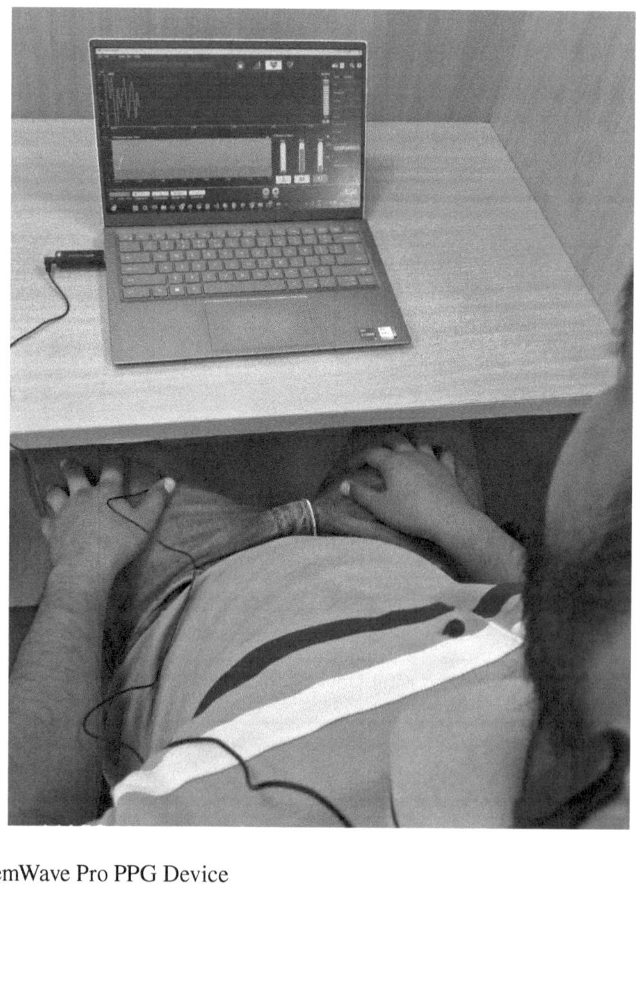

FIGURE 9.3 The emWave Pro PPG Device

TABLE 9.3
ESP 32 Code Blocks for Max30003 and AD8232 Sensors

SN	Max30003	AD8232
1	#if defined(ESP32) #include <WiFiMulti.h> WiFiMulti wifiMulti; #define DEVICE “ESP32” #elif defined(ESP8266) #include <ESP8266WiFiMulti.h> ESP8266WiFiMulti wifiMulti; #define DEVICE “ESP8266” #endif #include <InfluxDbClient.h> #include <InfluxDbCloud.h> #include<SPI.h> #include “protocentral_Max30003.h”	#if defined(ESP32) #include <WiFiMulti.h> WiFiMulti wifiMulti; #define DEVICE “ESP32” #elif defined(ESP8266) #include <ESP8266WiFiMulti.h> ESP8266WiFiMulti wifiMulti; #define DEVICE “ESP8266” #endif #include <InfluxDbClient.h> #include <InfluxDbCloud.h>
2	MAX30003 max30003; float ECG_s = 0;	const int analogInPin = A0; int sensorValue = 0;
3	#define WIFI_SSID “SSID” #define WIFI_PASSWORD “PASSWORD” #define INFLUXDB_URL “XXXXX” #define INFLUXDB_TOKEN “XXXXX” #define INFLUXDB_ORG “XXXXXX” #define INFLUXDB_BUCKET “XXX” #define TZ_INFO “XXXX” InfluxDBClient client(INFLUXDB_URL, INFLUXDB_ORG, INFLUXDB_BUCKET, IN- FLUXDB_TOKEN, InfluxDbCloud2CACert); Point sensor(“wifi_status”);	#define WIFI_SSID “SSID” #define WIFI_PASSWORD “PASSWORD” #define INFLUXDB_URL “XXXXXX” #define INFLUXDB_TOKEN “XXXXX” #define INFLUXDB_ORG “XXXX” #define INFLUXDB_BUCKET “XXX” #define TZ_INFO “XXXX” InfluxDBClient client(INFLUXDB_URL, INFLUXDB_ORG, INFLUXDB_BUCKET, INFLUX- DB_TOKEN, InfluxDbCloud2CACert); Point sensor(“wifi_status”);

4	void setup() {
	Serial.begin(57600);
	WiFi.mode(WIFI_STA);
	wifiMulti.addAP(WIFI_SSID, WIFI_PASS-
	WORD);
	Serial.print(“Connecting to wifi”);
	while (wifiMulti.run() != WL_CON-
	NECTED) {
	Serial.print(“.”);
	delay(100);}
		void setup() {
		Serial.begin(115200);
		WiFi.mode(WIFI_STA);
		wifiMulti.addAP(WIFI_SSID, WIFI_PASS-
		WORD);
		Serial.print(“Connecting to wifi”);
		while (wifiMulti.run() != WL_CONNECT-
		ED) {
		Serial.print(“.”);
		delay(100);}

5	pinMode(MAX30003_CS_PIN,OUTPUT);
	digitalWrite(MAX30003_CS_PIN,HIGH);
	SPI.begin();
	SPI.setBitOrder(MSBFIRST);
	SPI.setDataMode(SPI_MODE0);
	bool ret = max30003.max30003Read-
	Info();
	if(ret){
	Serial.println(“Max30003 read ID
	Success”);
	}else{
	while(!ret){
	ret = max30003.max30003ReadInfo();
	Serial.println(“Failed to read ID,
	please make sure all the pins are
	connected”);
	delay(10000); }}
	Serial.println(“Initialising the
	chip ...”);
	max30003.max30003Begin(); }	

continued

TABLE 9.3
continued

SN	Max30003	AD8232

6	`timeSync(TZ_INFO, "pool.ntp.org",`
	`"time.nis.gov");`
	`if (client.validateConnection()) {`
	`Serial.print("Connected to InfluxDB:`
	`");`
	`Serial.println(client.getServer-`
	`Url());}`
	`else {`
	`Serial.print("InfluxDB connection`
	`failed:");   Serial.println(client.`
	`getLastErrorMessage());}`
	`sensor.addTag("device", DEVICE);`
	`sensor.addTag("SSID", WiFi.SSID());}`
		`timeSync(TZ_INFO, "pool.ntp.org",`
		`"time.nis.gov");`
		`if (client.validateConnection()) {`
		`Serial.print("Connected to InfluxDB:`
		`");`
		`Serial.println(client.getServer-`
		`Url());}`
		`else {`
		`Serial.print("InfluxDB connection`
		`failed: ");`
		`Serial.println(client.getLastEr-`
		`rorMessage());}`
		`sensor.addTag("device", DEVICE);`
		`sensor.addTag("SSID", WiFi.SSID());}`

7	`void loop() {  max30003.getEcgSam-`
	`ples();   Serial.println(max30003.`
	`ecgdata);`
	`ECG_s =max30003.ecgdata;`
	`sensor.clearFields();`
	`sensor.addField("rssi", ECG_s );`
	`Serial.print("Writing: ");`
	`Serial.println(sensor.toLineProto-`
	`col());`
	`if (wifiMulti.run() != WL_CONNECTED)`
	`{`
	`Serial.println("Wifi connection`
	`lost");}`
	`if (!client.writePoint(sensor)) {`
	`Serial.print("InfluxDB write failed:`
	`");`
	`Serial.println(client.getLastEr-`
	`rorMessage());}`
	`delay(8);}`
		`void loop() {`
		`sensorValue = analogRead(analogInPin);`
		`sensor.clearFields();`
		`sensor.addField("rssi", sensorValue);`
		`Serial.print("Writing: ");`
		`Serial.println(sensor.toLineProto-`
		`col());`
		`if (wifiMulti.run() != WL_CONNECTED) {`
		`Serial.println("Wifi connection`
		`lost");}`
		`if (!client.writePoint(sensor)) {`
		`Serial.print("InfluxDB write failed:`
		`");`
		`Serial.println(client.getLastEr-`
		`rorMessage());}`
		`delay(8);}`

library. Communication between ESP 32 and the InfluxDB server is established using the "InfluxDbClient.h" and "InfluxDbCloud.h" libraries. SPI connectivity between the ESP 32 and Max30003 is established via the SPI library. Communication between the ESP32 and the Max30003 sensor is made possible via the "protocentral Max30003.H" library. First, the Max30003 device initialization and SPI interface setup between ESP32 and Max30003 are accomplished by means of the library. The sensor's data is then read using the library via the SPI interface. In Table 9.3, row 1 displays the libraries AD8232 and Max30003 for ESP 32 communication. The code for initializing the Max30003 sensor SPI interface using ESP32 is displayed in row 2 of the table. The analog pin could read the ECG signal's AD8232 output. The unified resource locator (URL), security token, organization, and bucket name are the codes that were entered to connect to the InfluxDB server over the Internet. To connect to the ESP 32 Wi-Fi, credentials such as the service set identification and password are needed. The code for Max30003 is displayed in row 5, and it initializes the SPI data bus prior to the sensor's continuous data reading. The code to get various time zone statuses for connecting to the InfluxDB server is displayed in row 6. The code to read and store sensor data via the SPI bus is displayed in row 7. Following that, the measured data is entered into the InfluxDB database. After that, the ESP32 is successfully used to establish continuous ECG detection and data logging from the ECG sensor.

Table 9.4 shows how R-R intervals are extracted from raw data stored in the database and then used to estimate HRV parameters. As a result, the proposed system controls the process of identifying

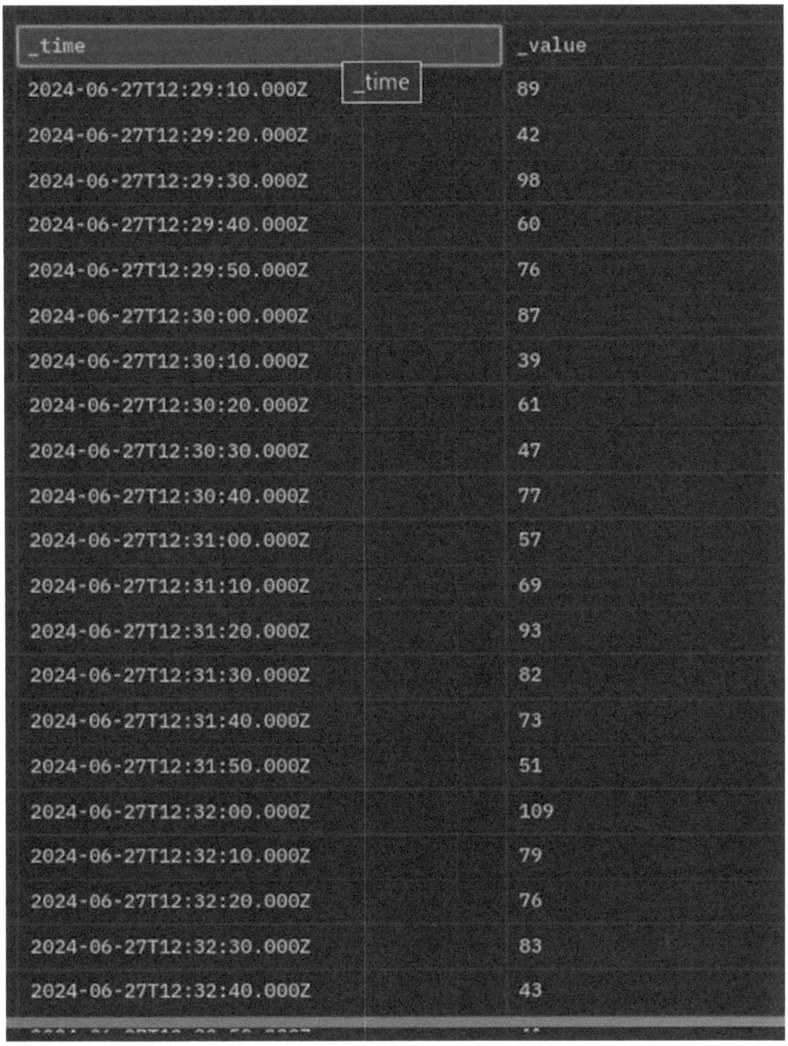

_time	_value
2024-06-27T12:29:10.000Z	89
2024-06-27T12:29:20.000Z	42
2024-06-27T12:29:30.000Z	98
2024-06-27T12:29:40.000Z	60
2024-06-27T12:29:50.000Z	76
2024-06-27T12:30:00.000Z	87
2024-06-27T12:30:10.000Z	39
2024-06-27T12:30:20.000Z	61
2024-06-27T12:30:30.000Z	47
2024-06-27T12:30:40.000Z	77
2024-06-27T12:31:00.000Z	57
2024-06-27T12:31:10.000Z	69
2024-06-27T12:31:20.000Z	93
2024-06-27T12:31:30.000Z	82
2024-06-27T12:31:40.000Z	73
2024-06-27T12:31:50.000Z	51
2024-06-27T12:32:00.000Z	109
2024-06-27T12:32:10.000Z	79
2024-06-27T12:32:20.000Z	76
2024-06-27T12:32:30.000Z	83
2024-06-27T12:32:40.000Z	43

FIGURE 9.4 InfluxDB Time Series Data Log Table

R-R intervals in unprocessed ECG data and provides HRV values as the output. Figure 9.4 displays raw ECG data that has been logged into an InfluxDB database table along with the time stamp.

9.4.5 Experimental Setup

The study's main purpose was to test the efficacy of the IoTMaxHRV device. Our study comprised 44 subjects randomly split into two groups based on the devices used for HRV measurements: (a) the IoTMaxHRV group comprising 22 participants for whom the HRV measurements (Table 9.4) were performed using IoTMaxHRV and emWave Pro and (b) the AD8232 group comprising the remaining 22 participants for whom the HRV measurements were performed using AD8232 and emWave Pro.

9.4.5.1 HRV Parameters Used in Data Collection

Table 9.4 describes the HRV parameters evaluated using ECG and PPG sensors in this study. The efficacy of IoTMaxHRV and AD8232 devices against the commercially available emWave Pro was evaluated by comparing their respective estimates of the HRV parameters.

TABLE 9.4
HRV Parameters Considered in Our Study

HRV Parameter	Description
MeanRR	The mean value of the RR intervals between successive R peaks.
SDNN	The standard deviation of NN intervals discussed by Shaffer et al. (2014) in successive heartbeats over the observation time.
MeanHR	The mean of HR readings.
SDHR	The standard deviation of the HR readings.
MinHR	The minimum HR reading from the readings collected.
MaxHR	The maximum HR reading from the readings collected.
RMSSD	The root mean square of successive differences between normal heartbeats.

As shown in Table 9.4, the MeanRR defines the average of R-R intervals over the recording period. An abnormally elevated MeanRR may likely indicate tachycardia (i.e., an increase in heart rate) and an abnormally low MeanRR indicates bradycardia (i.e., a decrease in heart rate). SDNN represents the overall HRV over the recording period. A lower value of SDNN may likely indicate less autonomic flexibility prone to cardiovascular disease and sudden cardiac death. RMSSD represents the mean of successive squared differences between NN intervals sensitive to high-frequency variations in HR. A lower value of RMSSD may likely indicate stress and autonomic dysfunction, and elevated RMSSD indicates good autonomic balance. MeanHR defines the average heart rate over the measurement period. An abnormally high MeanHR may likely indicate stress, anxiety, hyperthyroidism, or cardiovascular diseases. SDHR defines the standard deviation over the observation period. Elevated SDHR may likely indicate good autonomic nervous system and cardiac health. A lower SDHR may likely indicate autonomic dysfunction and chronic stress. The least value of HR during the observation period is represented as MinHR. The lower MinHR may likely indicate bradycardia with symptoms like dizziness and fatigue. Highly fit people often have lower MinHR. MaxHR represents the highest HR over the observation period. Elevated MaxHR under rest conditions may likely be an indicator of cardiovascular problems like tachycardia, acute stress, and anxiety.

9.4.6 Procedure for HRV Analysis

Participants were comfortably seated during data collection, with sensors positioned on the bottom part of the ear for PPG measurement and on the wrists and legs for ECG recording. An ECG lead with connectors in the colors red, yellow, and green equipped the electrochemical gel electrodes. We attached the red electrode to the left hand's wrist, the yellow electrode to the right hand's wrist, and the green electrode to the right foot's ankle as a reference electrode. We positioned the PPG sensor at the inferior section of the ear lobe to acquire PPG data. Data collection for each participant lasted 5 minutes. For IoTMaxHRV and AD8232 ECG devices and emWave Pro PPG measurements, we calculated the mean, minimum, maximum, and 95% confidence intervals, as well as lower and upper bounds for the captured biosignals. Due to the intricate nature of assessing HRV parameters using PPG and ECG sensors, the present study employed human data collection under rest conditions.

9.4.6.1 Detecting R-R Peaks Using the Peak Detection Algorithm

Python-based open-source preprocessing and a peak detection algorithm were employed to remove distortion and detect R-R peaks from the recorded biosignals. The detected R-R peaks were then used to derive R-R intervals between the detected peaks. The R-R intervals were used to calculate various HRV parameters.

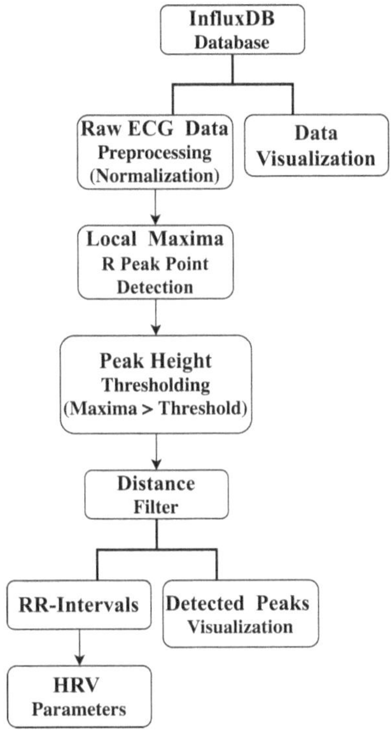

FIGURE 9.5 Flowchart of the Algorithm to Derive R-R Intervals from Raw ECG Data

As shown in Figure 9.5, raw ECG biosignals collected for 5 minutes were provided as input for the peak detection algorithm. The raw ECG signal sometimes contains an error introduced during signal acquisition, and this error was removed before further processing. After that, the preprocessed signal was analyzed to find local maxima, i.e., the R peak points. A height-thresholding hyperparameter was adjusted until the optimal threshold for peak detection was found. Only the peaks whose amplitude was higher than the threshold were considered valid. Only the peaks whose distances from the previous R peak were larger than the minimum distance threshold of successive peaks were considered valid R peaks. This systematic approach made peak detection more reliable and accurate in ECG signal processing. Due to its proprietary nature, the PPG signals recorded by the emWave Pro were processed using the Kubios software. Kubios software is an HRV analysis tool that estimates HRV parameters independent of measuring methods, including raw signal preprocessing. This software is recommended to process emWave Pro data.

9.4.6.2 Determining the R-R Interval

Figure 9.6 shows the R-R intervals between successive peaks, where the peaks were identified using a thresholding hyperparameter tuned to obtain optimal peak detection from the raw ECG data.

9.4.6.3 Outlier Correction

An outlier correction was applied to refine the R-R peaks. This correction used an R-R interval distribution with two standard deviation boundaries. The points beyond two standard deviations of the R-R interval mean were considered outliers. R-R intervals after outlier correction were used to estimate HRV parameters.

9.5 EXPERIMENTAL RESULTS

Figure 9.7 shows the measurements of HRV parameters (listed in Table 9.1) for the IoTMaxHRV and the AD8232 group. Both groups contain 22 participants each. The measured HRV parameters

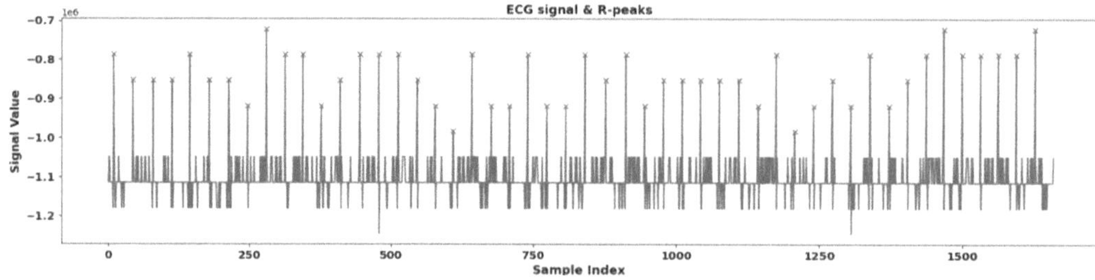

FIGURE 9.6 ECG Signal with R-R Detected Peak Pointers

are as follows: MeanRR, SDNN, MeanHR, SDHR, MinHR, MaxHR, and RMSSD. The error bars display a 95% confidence interval around the point estimate. Figure 9.7a represents the MeanRR readings (measured in milliseconds, ms) for both groups considered in our experiments. The ECG sensors performed very closely with the commercial device emWave Pro for the measurement of MeanRR parameter estimation. However, the IoTMaxHRV device observations were much closer to the emWave Pro than the AD8232 group. Figure 9.7b represents the SDNN parameter (ms) obtained from the two groups of subjects. Figure 9.7b shows that the emWave Pro sensor has a slightly larger SDDN than both ECG sensors. The Max30003 sensor performed more closely with the emWave Pro than the AD8232 sensor for the following parameters: MeanRR, MeanHR, and MinHR. Figure 9.7c represents the MeanHR (measured in beats per minute, beats/min) obtained from the two groups.

Figure 9.7d represents the SDHR (beats/min) obtained from two groups of subjects. The emWave Pro sensor observed a larger value than ECG sensors in both groups (Figure 9.7d). Figure 9.7f shows the MinHR obtained from two groups of subjects. ECG sensors observed a slightly higher value of MinHR than emWave Pro. However, the Max30003 sensor performed more closely with emWave Pro than the AD8232, as shown in Figure 9.7f. Figure 9.7e shows the MaxHR (beats/min) obtained from the IoTMaxHRV and AD8232 groups. The emWave Pro shows a slightly larger value than both groups' ECG sensors. Figure 9.7g represents the RMSSD (ms) obtained from two groups of subjects. The emWave Pro observed a higher value of RMSSD than the Max30003 and AD8232 sensors, as shown in Figure 9.7g. Conversely, the Max30003 sensor was much closer to the emWave Pro than the AD8232, as shown in Figure 9.7g.

Figure 9.7 shows that the Max30003 performed closely to the commercial device emWave Pro. The results show good alignment between the emWave Pro device and the IoTMaxHRV device for HRV parameter estimation. Parameters, such as MeanRR, MeanHR, MinHR, and MaxHR, were well aligned between IoTMaxHRV and emWave Pro.

Figure 9.8 depicts the root mean square error (RMSE) value for the different data measurements collected from the IoTMaxHRV and AD8232 groups. The RMSE in HRV parameter measurement for both ECG sensor groups was quite balanced for most HRV parameters, except for HRV parameter RMSSD, in which almost 62.13 milliseconds were the largest difference in RMSE for both groups. In contrast, MeanHR demonstrates the minimum difference of 0.3 beats/min in RMSE between the groups. The RMSE for HRV parameters, such as MeanHR, SDNN, and MinHR, was lower for IoTMaxHRV than for the AD Group. The RMSE for HRV parameters was lower for the AD Group than for the IoTMaxHRV Group for SDHR, MaxHR, and RMSSD. The HR-driven HRV parameters tend to have a lower RMSE value than the RR-driven ones for both groups, such as SDNN and RMSSD. Considering everything, both ECG sensors give relatively similar results to the measurement of the commercial PPG device.

In this study, 5 minutes of ECG data was collected in lab settings under rest conditions. The participants were requested to sit for 15 minutes on a comfortable chair. The proposed IoTMaxHRV device was not tested for ECG recording during activities like sports, exercise, and other activities

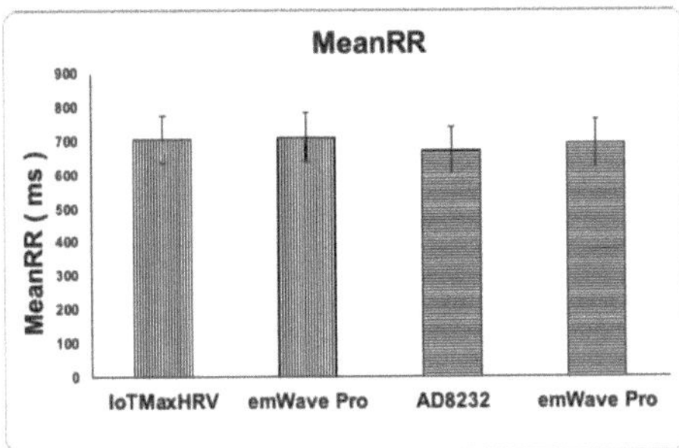

(a) MeanRR

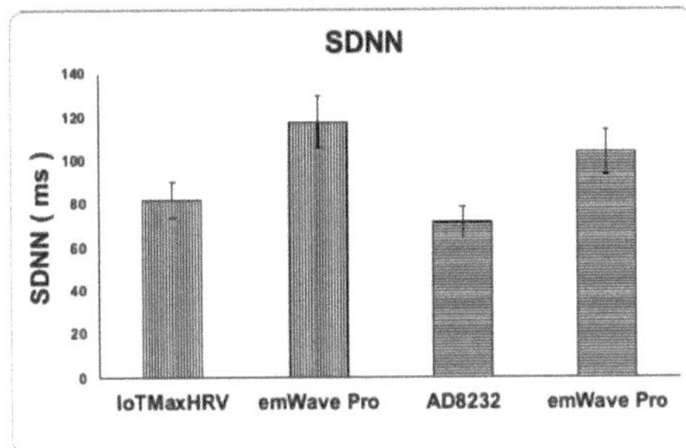

(b) SDNN

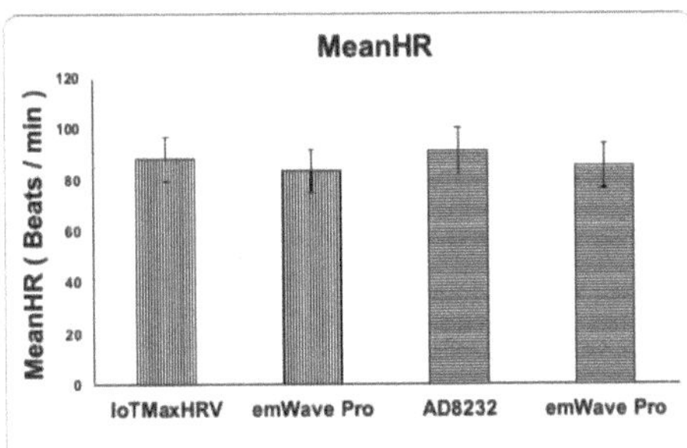

(c) MeanHR

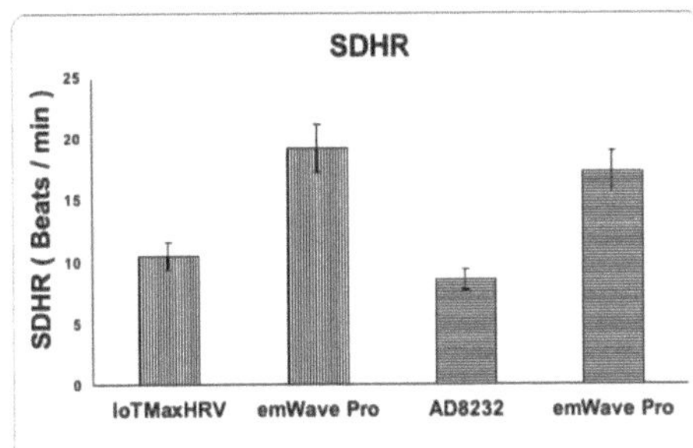

(d) SDHR

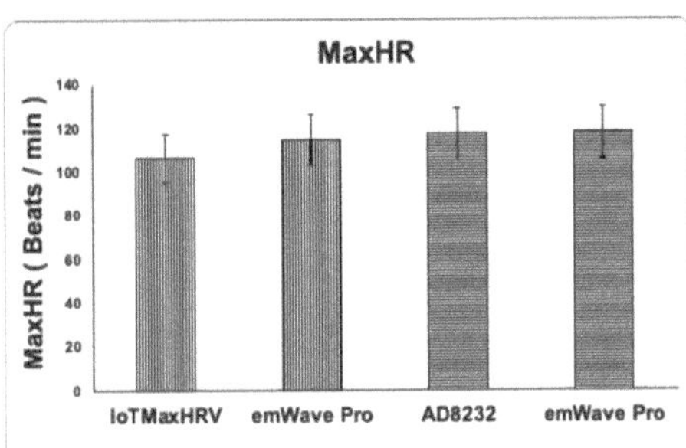

(e) MaxHR

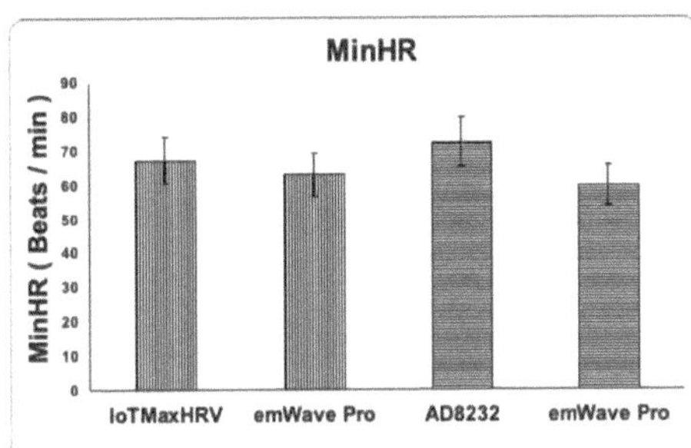

(f) MinHR

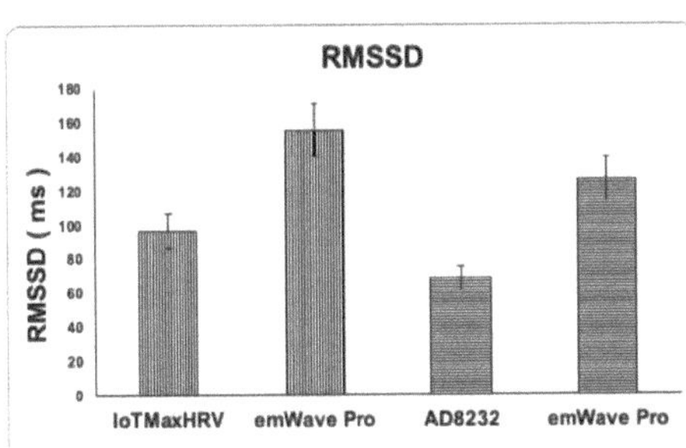

(g) RMSSD

FIGURE 9.7 HRV Parameter for Two Sensor Groups: (a) MeanRR, (b) SDNN, (c) MeanHR, (d) SDHR, (e) MaxHR, (f) MinHR, and (g) RMSSD

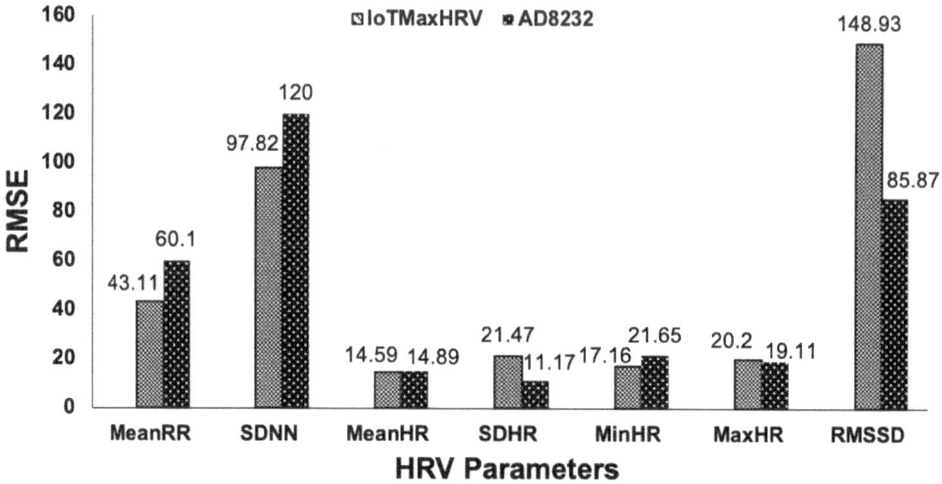

FIGURE 9.8 RMSE between ECG and PPG HRV Parameter Measurements for Two ECG Sensor Groups

and was only tested when the participants were in a rest condition. The data collected in rest conditions in this study was less prone to muscle and motion artifacts. However, sports and other fitness activities in real-world applications may incur larger muscle artifacts than rest conditions. Nevertheless, we believe IoTMaxHRV to perform accurately when tested on participants performing real-world activities and plan to validate the same in the future. The data collected by the proposed device can be utilized in the development of artifacts (that emerge during physical activity) rejection ML approaches for actual world fitness activities like sports and exercise. Additionally, the ECG is a well-established approach to assessing cardiac health parameters as compared to other measurement approaches. Therefore, the device can be further improved and tested with specific fitness activities in the future. The device-collected ECG data can be utilized for further signal processing and analysis approaches development to improve the overall measurement accuracy in HRV estimation during sports and exercise.

9.6 CONCLUSION

The HRV is an important biomarker related to cardiac activity utilized to evaluate patient's cardiovascular health and the working of the autonomic nervous system. While clinical health management is the primary use of HRV, it also plays a crucial role in monitoring both clinical and personal health. This chapter investigates the performance of PPG and ECG devices in HRV assessments. In conclusion, this chapter introduces an affordable IoTMaxHRV device developed using low-cost ECG sensors. The proposed device utilized a low-cost microcontroller with Wi-Fi functionality for the remote interface of the offline ECG monitoring sensor for healthcare applications. Then, the device logged data stored on the cloud used for HRV parameter estimation and further use in artificial intelligence (AI)-based healthcare applications and other clinician utilization. This study concludes that HRV sensor validation is essential when considering the enormous clinical and personal health monitoring applications. The work employed the two critical low-cost HRV sensors, Max30003 and AD8232, from Analog Devices. Results indicated consistent HRV parameter estimation between the commercial and costly emWave Pro PPG-based and low-cost ECG-based devices. The IoTMaxHRV proved to be a reasonable device for HRV measurements. IoTMaxHRV's use is likely significant for physically challenged people, elderly patients, and patients in medical facilities devoid of remote hilly regions, as it is a low-cost solution to assess the HRV parameters to assess cardiovascular health remotely. In addition, the IoTMaxHRV utilizes the Max30003 sensor for its power efficiency compared to other devices critical for portable healthcare applications. Besides, the IoTMaxHRV device can assess people with heart conditions, such as arrhythmia, previous heart attack, and chest pain, caused by blocked or narrowed heart arteries. Overall, IoTMaxHRV is an effective HRV measurement device (also a remote HRV assessment alternative) that can help detect cardiac issues due to sudden heart rate changes, especially in remote areas without proper medical facilities and for the elderly who cannot visit medical centers regularly. Our results can also help people choose the most accurate HRV measurement devices. The ECG sensors (IoTMaxHRV, AD8232) and commercial PPG device emWave were very close in HRV parameter estimation for most of the parameters. However, the RMSSD showed the largest RMSE between the ECG (IoTMaxHRV, AD8232) and PPG sensors. A maximum RMSE difference of 62.13 milliseconds between both groups was observed for RMSSD parameters. The MeanHR, MinHR, MaxHR, and SDHR parameter estimates showed a lower RMSE difference for both the ECG and PPG methods for both groups, with a minimum RMSE difference of 0.3 beats/min for the SDHR parameter. The IoTMaxHRV device, with a small additional cost, performed better than the AD8232 device on four out of the seven HRV measures (MeanRR, SDNN, MeanHR, and MinHR) and similar to the AD8232 device on one HRV measure (MaxHR). Thus, the IoTMaxHRV device provides measurement reliability across five out of seven HRV measures. For the RMSSD and SDHR parameters, the IoTMaxHRV device showed a higher deviation than the AD8232. The high RMSE for RMSSD and SDHR with the IoTMaxHRV device implies that this device may be less reliable for these two measures than the AD8232 device. Our findings provide a detailed performance comparison of commonly available HRV assessment alternatives for selecting suitable sensors based on measurement requirements. The sample size used in this study is small; therefore, the device may need to be tested for different age groups and large populations. However, it might be sufficient to detect large effect sizes.

The findings of this study can be confirmed with the findings of future studies. Also, the duration of HRV measurements (5 minutes) may fail to capture long-term variability trends and require investigation on extended monitoring periods. Moreover, using the emWave Pro sensor required minimal head movement and participant interaction during the measurement process to avoid artifacts and interferences. Repetitive recordings may be needed to achieve quality measures with less distortion if movements cause malfunctioning of PPG-based sensors. Future research should include these advances to understand more about what these kinds of affordable monitoring systems can do and where they have their limitations.

Various PPG-based low-cost wearable devices are used to estimate HRV parameters. However, the PPG sensor output signal needed to be further processed to predict the location of the R peak in successive heartbeats. ECG is still considered for precise R peak detection due to the direct measure of heart electric activity and easily differentiable R peaks in the ECG data. Additionally, the PPG signals are considered more susceptible to artifacts and noise than ECG signals; therefore, repetitive measures for reliable results may be required. Robust signal processing algorithms are proposed to deal with highly noisy ECG data and can be utilized to remove the noise and artifacts effectively. The proposed IoTMaxHRV is a low-cost and optimal power consumption device for portable HRV assessment applications. Future research may investigate new methods of motion artifact rejection, complexity, user-friendliness, and compatibility of IoTMaxHRV with existing and new AI and IoT-based healthcare and telemedicine applications.

ACKNOWLEDGMENTS

This work was supported by a grant from the IIT Mandi iHub and HCI Foundation to Prof. Varun Dutt (IITM/iHub and HCIF-IIIT Mandi/VD/412). The authors are also grateful to the Indian Institute of Technology Mandi for providing lab support for the project.

NOTES

1 https://www.who.int/news-room/fact-sheets/detail/cardiovascular-diseases-(cvds).
2 https://www.analog.com/en/products/max30003.html.
3 https://www.analog.com/en/products/ad8232.html.
4 https://www.heartmath.co.uk/product/emwavepro/.
5 https://www.who.int/publications/i/item/9789241565042.
6 https://www.influxdata.com.
7 https://peakutils.readthedocs.io/en/latest/.
8 MISO: Master in slave out pin is used to transmit data from slave to master device.
9 MOSI: Master out slave in pin is used to transmit data from master to slave device.
10 SCLK: data communication between the master and slave is synchronized with the clock signal.
11 C/S – Chip select signal is used to select the slave device for communication.

REFERENCES

Abdelnabi, M.H., 2019. Cardiovascular clinical implications of heart rate variability. *International Journal of Cardiovascular Academy*, 5(2), p.37. https://doi.org/10.4103/IJCA.IJCA_36_18

Allen, J., 2007. Photoplethysmography and its application in clinical physiological measurement. *Physiological Measurement*, *28*(3), p.R1. https://doi.org/10.1088/0967-3334/28/3/R01

Benjamin, E.J., Blaha, M.J., Chiuve, S.E., Cushman, M., Das, S.R., Deo, R., De Ferranti, S.D., Floyd, J., Fornage, M., Gillespie, C. and Isasi, C.R., 2017. Heart disease and stroke statistics—2017 update: A report from the American Heart Association. *Circulation*, *135*(10), pp.e146–e603. https://doi.org/10.1161/CIR.0000000000000485

Bhowmick, S., Kundu, P.K. and Mandal, D.D., 2020, February. HRV performance analysis in photoplethysmography and electrocardiography. In *2020 IEEE Calcutta Conference (CALCON)* (pp.342–345). IEEE. https://doi.org/10.1109/CALCON49167.2020.9106418

Cinotti, E., Centracchio, J., Parlato, S., Andreozzi, E., Esposito, D., Muto, V., Bifulco, P. and Riccio, M., 2024. A Narrowband IoT Personal Sensor for Long-Term Heart Rate Monitoring and Atrial Fibrillation Detection. *Sensors*, *24*(14), p.4432. https://doi.org/10.3390/s24144432

Chou, C.C., Tseng, S.Y., Chua, E., Lee, Y.C., Fang, W.C. and Huang, H.C., 2011, September. Advanced ECG processor with HRV analysis for real-time portable health monitoring. In *2011 IEEE International Conference on Consumer Electronics-Berlin (ICCE-Berlin)* (pp.172–175). IEEE. https://doi.org/10.1109/ICCE-Berlin.2011.6031850

Esgalhado, F., Batista, A., Vassilenko, V., Russo, S. and Ortigueira, M., 2022. Peak detection and HRV feature evaluation on ECG and PPG signals. *Symmetry*, 14(6), p.1139. https://doi.org/10.3390/sym14061139

Güvenç, H., 2020, November. Wireless ECG device with Arduino. In *2020 Medicalechnologies Congress (TIPTEKNO), Turkey,* (pp.1–4). IEEE. https://doi.org/10.1109/TIPTEKNO50054.2020.9299248

Hancock, H.C., Close, H., Mason, J.M., Murphy, J.J., Fuat, A., Singh, R., Wood, E., de Belder, M., Brennan, G., Hussain, N. and Kumar, N., 2013. High prevalence of undetected heart failure in long-term care residents: Findings from the Heart Failure in Care Homes (HFinCH) study. *European Journal of Heart Failure*, 15(2), pp.158–165. https://doi.org/10.1093/eurjhf/hfs165

Hassan, Q.F. ed., 2018. *Internet of Things A to Z: Technologies and Applications*. John Wiley & Sons.

Jeyhani, V., Mahdiani, S., Peltokangas, M. and Vehkaoja, A., 2015, August. Comparison of HRV parameters derived from photoplethysmography and electrocardiography signals. In *2015 37th Annual International Conference of the IEEE Engineering in Medicine and Biology Society (EMBC)* (pp.5952–5955). IEEE. https://doi.org/10.1109/EMBC.2015.7319747

Kotval, K.Z., Keilman, L. and Wang, W., 2020. Transportation services for older adults and preventive health-care attainment. *Urban Science*, 4(3), p.38. https://doi.org/10.3390/urbansci4030038

Kumar, S., Srivastava, R., Pathak, S. and Kumar, B., 2020. Cloud-based computer-assisted diagnostic solutions for e-health. In A.K. Singh and M. Elhoseny (Eds.), *Intelligent Data Security Solutions for e-Health Applications* (pp.219–235). Academic Press. https://doi.org/10.1016/B978-0-12-819511-6.00012-1

AmitKumar Singh and Mohamed Elhoseny

Mackett, R., 2017. Older people's travel and its relationship to their health and wellbeing. In C. Musselwhite (Ed.), *Transport, Travel and Later Life* (Vol. 10, pp.15–36). Emerald Publishing Limited. https://www.emerald.com/insight/content/doi/10.1108/s2044-994120170000010001/full/html

Munoz, M.L., Van Roon, A., Riese, H., Thio, C., Oostenbroek, E., Westrik, I., de Geus, E.J., Gansevoort, R., Lefrandt, J., Nolte, I.M. and Snieder, H., 2015. Validity of (ultra-) short recordings for heart rate variability measurements. *PLoS One*, *10*(9), p.e0138921. https://doi.org/10.1371/journal.pone.0138921

Pinheiro, N., Couceiro, R., Henriques, J., Muehlsteff, J., Quintal, I., Goncalves, L. and Carvalho, P., 2016, August. Can PPG be used for HRV analysis?. In *2016 38th Annual International Conference of the IEEE Engineering in Medicine and Biology Society (EMBC)* (pp.2945–2949). IEEE. https://doi.org/10.1109/EMBC.2016.7591347

Plews, D.J., Scott, B., Altini, M., Wood, M., Kilding, A.E. and Laursen, P.B., 2017. Comparison of heart-rate-variability recording with smartphone photoplethysmography, polar H7 chest strap, and electrocardiography. *International journal of sports physiology and performance*, *12*(10), pp. 1324–1328. https://doi.org/10.1123/ijspp.2016-0668

RajendraAcharya, U., Paul Joseph, K., Kannathal, N., Lim, C.M. and Suri, J.S., 2006. Heart rate variability: A review. *Medical and Biological Engineering and Computing*, 44, pp.1031–1051. https://doi.org/10.1007/s11517-006-0119-0

Russo, N.L. and Eriksson, J., 2018. The Internet of Things and people in health care. In Hassan, Q.F. (ed.), *Internet of Things A to Z: Technologies and Applications* (pp.447–474). Wiley. https://doi.org/10.1002/9781119456735.ch16

Sagahyroon, A., Aburukba, R. and Aloul, F., 2017. The Internet of Things and e-Health: Remote patients monitoring. In Hassan, Q.F., Khan, A. and Madani, S. (eds.), *Internet of Things* (pp.303–319). Chapman and Hall/CRC.

Sampson, M. and McGrath, A., 2015. Understanding the ECG. Part 1: Anatomy and physiology. *British Journal of Cardiac Nursing*, *10*(11), pp.548–554. https://doi.org/10.12968/bjca.2015.10.11.548

Shaffer, F. and Ginsberg, J.P., 2017. An overview of heart rate variability metrics and norms. *Frontiers in Public Health*, *5*, p.290215. https://doi.org/10.3389/fpubh.2017.00258

Shaffer, F., McCraty, R. and Zerr, C.L., 2014. A healthy heart is not a metronome: An integrative review of the heart's anatomy and heart rate variability. *Frontiers in Psychology*, *5*, p.108292. https://doi.org/10.3389/fpsyg.2014.01040

Singstad, B.J., Azulay, N., Bjurstedt, A., Bjørndal, S.S., Drageseth, M.F., Engeset, P., Eriksen, K., Gidey, M.Y., Granum, E.O., Greaker, M.G. and Grorud, A., 2021. Estimation of heart rate variability from finger photoplethysmography during rest, mild exercise and mild mental stress. *Journal of Electrical Bioimpedance*, 12(1), pp.89–102. https://doi.org/10.2478/joeb-2021-0012

Thayer, J.F., Yamamoto, S.S. and Brosschot, J.F., 2010. The relationship of autonomic imbalance, heart rate variability and cardiovascular disease risk factors. *International Journal of Cardiology*, 141(2), pp.122–131. https://doi.org/10.1016/j.ijcard.2009.09.543

Tobón, D.P., Jayaraman, S. and Falk, T.H., 2017. Spectro-temporal electrocardiogram analysis for noise-robust heart rate and heart rate variability measurement. *IEEE Journal of Translational Engineering in Health and Medicine*, 5, pp.1–11. https://doi.org/10.1109/jtehm.2017.2767603

Valenza, G., Faita, F., Citi, L., Saul, J.P., Bruno, R.M. and Barbieri, R., 2020, September. Validation of sympathetic activity index from heart rate variability series: A preliminary muscle sympathetic nerve activity study. In *2020 Computing in Cardiology* (pp.1–4). IEEE. https://doi.org/10.22489/CinC.2020.365

Virani, S.S., Alonso, A., Aparicio, H.J., Benjamin, E.J., Bittencourt, M.S., Callaway, C.W., Carson, A.P., Chamberlain, A.M., Cheng, S., Delling, F.N. and Elkind, M.S., 2020. Heart disease and stroke statistics-2021 update: A report from the American Heart Association. *Circulation*, *141*(9). p. e254. https://doi.org/10.1161/CIR.0000000000000757

Xhyheri, B., Manfrini, O., Mazzolini, M., Pizzi, C. and Bugiardini, R., 2012. Heart rate variability today. *Progress in Cardiovascular Diseases*, 55(3), pp.321–331. https://doi.org/10.1016/j.pcad.2012.09.001

Part III

Emerging Technologies

10 Split Learning and Synergetic Inference
When IoT Collaborates with the Cloud-Edge Continuum

Joana Tirana and Dimitris Chatzopoulos

10.1 INTRODUCTION

Modern mobile and Internet of Things (IoT) devices feature different sensors that produce voluminous and multivariate data, which can be used to train powerful neural network (NN) models. A typical way to train NN models or process the produced data is by gathering the data on one machine (server) and starting to train the model there. As shown in Figure 10.1, the model is trained in *epochs*, which is the period when the model is updated using all the data in the dataset. Specifically, the model is updated in batches (i.e., the dataset is split into multiple batches of data). First, a batch of data is inputted into the model and is forward-propagated through the whole model. During forward propagation, each layer produces a vector of activations that is passed to the subsequent layer. Then, the last layer will compute the predicted output. This output is used to compute the loss (i.e., quantifying how far the predicted output is from the real target), which is also back-propagated through the model to update the model parameters accordingly. During back-propagation, each layer produces a vector of gradients, which is used for the weight update operation.

However, as the demands for processing more data and training larger models are growing, we moved to distributed training techniques, in which multiple processing units collaborate to accelerate the training process via techniques that allow parallel processing. But still, the orchestration and administration of the training process are server-based (i.e., all data is gathered from the IoT devices at a centralized machine and processed there). On the other hand, the technological development that has been seen in IoT devices in the past years[1] (equipped with processing units, not just microcontrollers) allows parts of the processing and storage burdens to move to these devices using *client-based distributed training techniques*. This enables data processing and orchestration to take place at the very edge of the system. Such approaches, apart from their performance benefits, can also protect clients' data, since personal devices do not share user data with other entities.

The first proposed client-based distributed training technique is federated learning (FL) from McMahan et al. (2017), which allows a machine learning (ML) model (usually a NN) to be trained in a distributed manner while keeping the data decentralized. In other words, the raw data does not leave the device. In FL, training is implemented in global training rounds, also known as *communication rounds*. The steps of this process are shown in Figure 10.2. In each communication round, a central entity, known as the *aggregator*, selects a subset of the clients. The size of the subset may vary from a small fraction to the whole set of available clients. Next (step F1), the clients perform on-device training (i.e., training the model locally using their data) for a predetermined number of local epochs. One local epoch is completed when the device has trained the local model using all data samples from its dataset. Then (step F2), the clients send their updated version of the model to the aggregator. The aggregator uses an aggregation algorithm like FedAvg (McMahan et al., 2017)

DOI: 10.1201/9781003506638-13

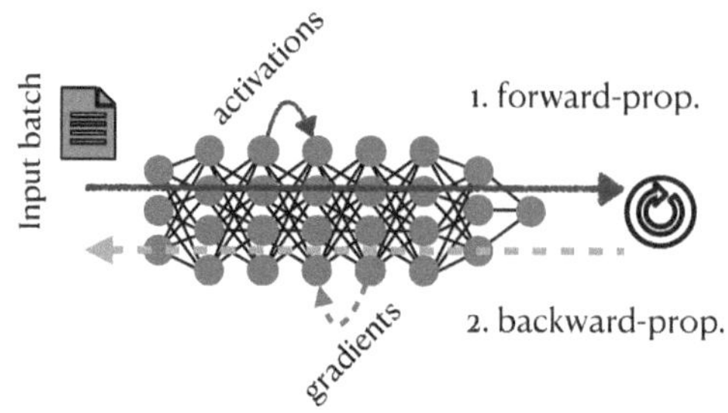

FIGURE 10.1 Modern Mobile and Internet of Things (IoT) Device Trained in Epochs

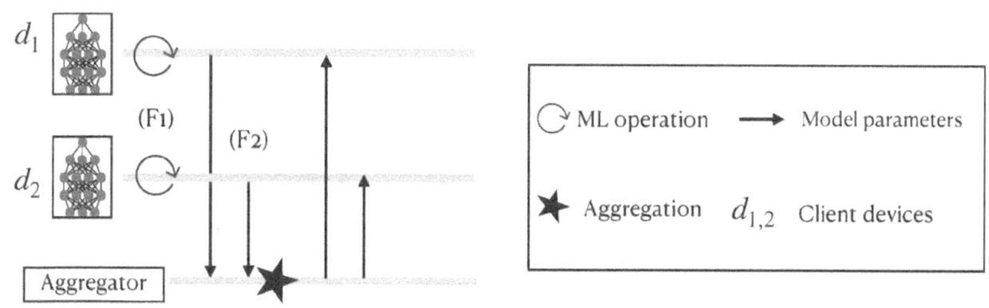

FIGURE 10.2 Steps of Training with FL When There Are Two Clients and One Aggregator

to generate a new global model. Finally (step F3), the aggregator dispatches the new global model to all clients. These steps will be repeated until the model has converged or the maximum number of training rounds has been reached.

Moreover, there is an alternative configuration of FL, named *decentralized FL,* which allows the implementation of collaborative learning without the need for another third party. Specifically, the clients complete the aggregation phase using peer-to-peer algorithms or utilizing other decentralized tools, such as blockchain (Ramanan, Nakayama and Sharma, 2020) or interplanetary file system networks (Pappas et al., 2022). The benefits behind this approach are: (i) clients become fully independent; (ii) it becomes an inevitable approach in cases where the aggregator cannot be trusted. However, usually, decentralized syncing protocols require additional communication overhead in order to converge into one solution. As a result, the aggregation phase requires more steps compared to the centralized FL.

Notably, both centralized and decentralized FL can be quite demanding in terms of computing, memory resources, and communication load. To begin with, one must consider the memory wall given that deep NNs require a large batch size (i.e., samples of the clients' data per batch update) to ensure good accuracy and convergence speed. Thus, the model will generate large vectors containing the produced activations during forward propagation (see Figure 10.1). Depending on the size of the model, this may require the device to allocate a significant amount of memory to be able to train the model. For instance, training a ResNet-101 (He et al., 2016) model with an input size of 6 MB requires, at any time during the training process, at least 872 MB of the main memory. However, we cannot assume that all resource-constrained devices are capable of storing such amounts of data. Furthermore, training can be quite computationally intensive. Although recently released mobile GPUs can support the training of deep NNs, they still perform poorly. Therefore, even if a device can conduct on-device training, its constrained computing resources will result in the *stragglers effect*, wherein a solitary slow device can significantly prolong training delays. A way to tackle this is by using techniques that can improve the performance of the device by converting the NN model into a more lightweight model. Such approaches are pruning (Liu et al., 2020), quantization (Zhou et al., 2018), and knowledge distillation (Bucilŭa, Caruana and Niculescu-Mizil, 2006). However,

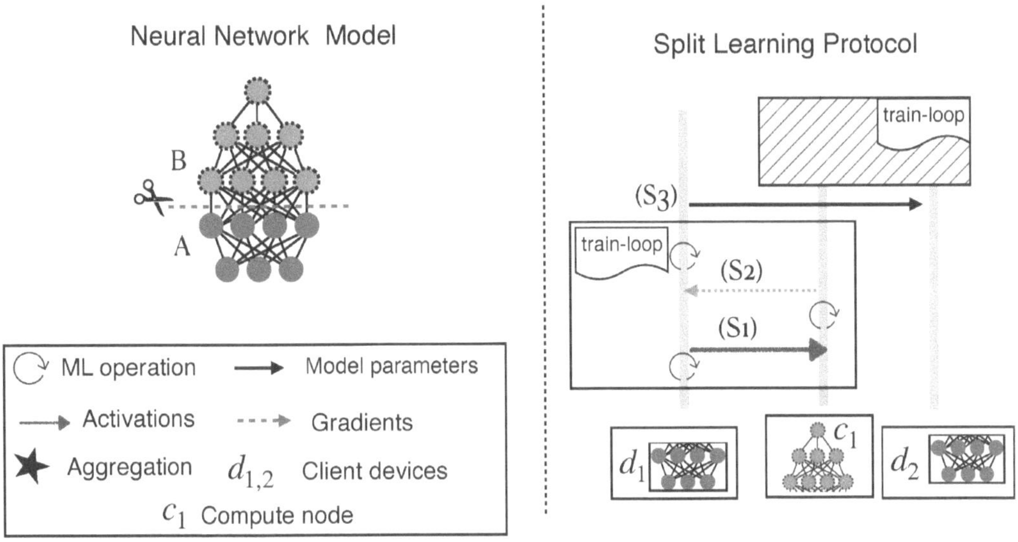

FIGURE 10.3 Steps of Training One Model with One Split (i.e., One Helper) and Two Clients with SL

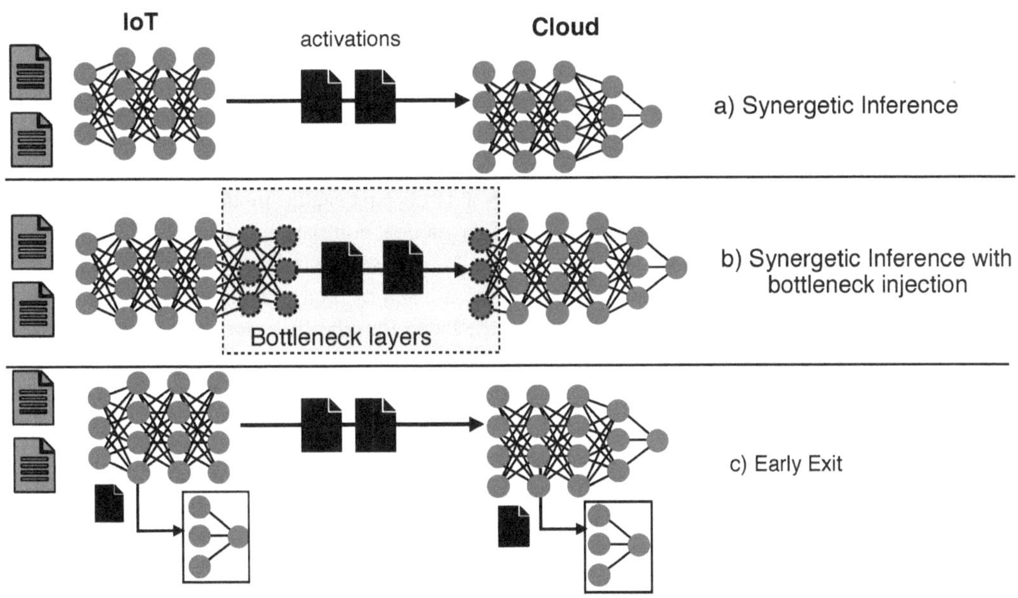

FIGURE 10.4 Indicative Cases of SI: (a) SI without Changing the Model; (b) SI with Injecting a Bottleneck Model Part; and (c) Early Exit. Note that the Dotted Circles in (b) Represent the Inserted Bottleneck Layers. In All Cases, the Grey File-Shaped Icons Represent the Query Data That Is Used as the Input Data Samples, while the Dark File-Shaped Icons Are the Intermediate Activations Shared with the Cloud

these approaches may reduce the precision of training operations, and hence, there is a tradeoff between reducing training overhead and model accuracy, which makes these approaches practically challenging when aiming for high model performance. Therefore, exclusively on-device training is not always feasible and cannot be supported by all IoT systems. Also, for the reasons that were mentioned above, a cloud-only approach is not acceptable.

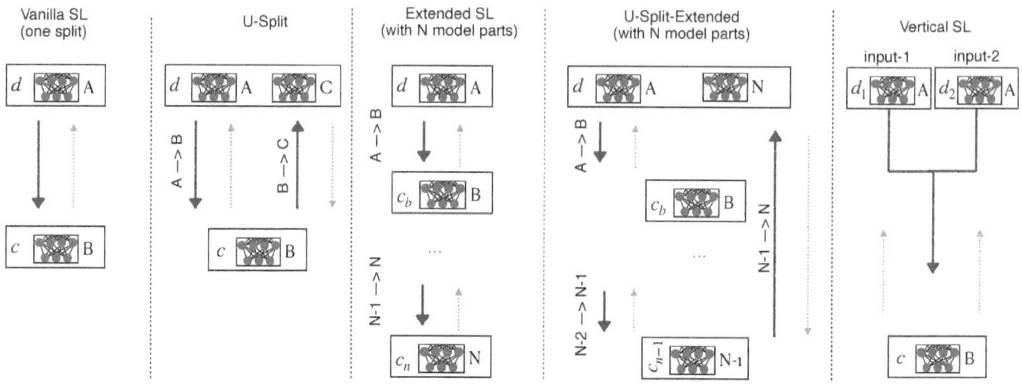

Figure 10.5 **Types of SL Split Policies**

As a result, an intermediate solution is *split computing* (SC), in which the client devices collaborate with devices located at the cloud-edge continuum to get support in processing their ML tasks. SC started with the emergence of *synergetic inference* (SI), where the model is split into a head and a tail using a *cut layer,* as shown in the left plot of Figure 10.3. The head contains the first layers of the model, while the tail contains the rest. To reduce energy consumption on client devices, the tail is offloaded to an edge server. When an inference task starts, the client forwards the input samples into the head model and sends the intermediate activations to the edge server to compute the output using the tail model.

Going a step further, split learning (SL) (Vepakomma et al., 2018; Palanisamy et al., 2021) enables resource-constrained devices to collaboratively train state-of-the-art models in an acceptable time while the data remains decentralized. Like in SI, the model is split, and the most computationally demanding model parts are assigned to a powerful *compute node* (see Figure 10.3) that performs the respective training process. While clients only keep a small part of the model. This makes it possible for resource-constrained clients to train far bigger models compared to FL. The rightmost plot in Figure 10.3 shows the steps of SL, where the compute node sequentially communicates with each client, and they collaboratively update the model using a batch of data each time. First, the client d_1 starts the forward propagation phase by forwarding the batch into the first model part (i.e., head) and sending the intermediate activations to the compute node (step S1). Then, the compute node starts the back-propagation phase and updates the second model part (i.e., the tail). Once the compute node has computed the gradients of the cut layer, it sends them to the client (step S2). Finally, the client, upon receiving the gradients, updates the head model. Steps S1–S2 are repeated for all batches at the device. Then, in step S3, the client d_1 sends the updated model part to the next client d_2 to proceed with training. The compute node will serve the clients in a round-robin order. Notably, SC is the intermediate solution, which allows IoT client devices to train or use an ML model in a decentralized manner while enabling them to support models with millions of parameters as large parts of the model are offloaded onto the cloud. Also, the burden of computation is proportionally distributed to the edge and cloud.

In this chapter, we will discuss the synergy between IoT and the cloud-edge continuum for supporting ML applications and operations, such as inference with SI and training with SL. We mainly focus on SL by presenting a generic and abstract mathematical model of the systems that allows the optimization and tuning of various system factors. Also, it examines the main challenges of SL (i.e., data heterogeneity and privacy concerns). Finally, emerging trends and future research ideas and directions are discussed.

10.2 CHALLENGES OF CROSS-DEVICE DECENTRALIZED AI

There are mainly two settings of client-based distributed training: (i) cross-silo and (ii) cross-device. The cross-silo setting involves training ML models between large institutions, like hospitals, where data is combined to enhance model accuracy. When referring to cross-silo systems, we can assume that the involved devices are resourceful, without any computation or memory constraints. On the other hand, the cross-device setting focuses on collaboration among individual small devices (e.g., IoT and mobile devices) to train ML models. Notably, when working with cross-device settings, one has to come across the challenges of such environments, some of which are listed below:

- *Constrained Computational Resources:* These devices are usually resource-constrained, preventing them from completing tasks with heavy computing demands. Note that modern ML tasks involve training large NN models with hundreds of parameters, which both have computational and memory demands. But if we consider the specification of common IoT devices, like the Raspberry Pi B+,[2] it is not feasible for such devices to support these types of operations.
- *System Heterogeneity:* Decentralized AI requires the collaboration and synchronization of multiple nodes. However, it is expected that the participating devices will have resource discrepancies, with some of them being more powerful than others. Additionally, computing or network fluctuations can be noticed for the same device throughout an ML task.
- *Communication Overheads:* As the training procedure becomes distributed, the entities will communicate with each other more frequently. Moreover, in IoT systems, the connection between the nodes is expected to have slow network speed (e.g., 3G, LTE, etc.), or the network interface and inbound/outbound capabilities of the devices are not powerful.
- *Dynamic Availability:* The availability of personal devices depends on the user's behavior, their battery autonomy (in the case of IoT devices), and their connectivity to the Internet. Also, IoT devices tend to experience CPU thermal throttling, or they are using unreliable network connections, in addition to the dynamic nature of the network. Hence, the system should be able to easily adapt to such fluctuations without causing significant overhead or negatively affecting the accuracy of the model.
- *Statistical Heterogeneity:* Data is not evenly distributed among the clients. Some of the clients have larger datasets than others, causing data imbalance, or the type of data each client has access to or gathers may differ essentially from client to client. As a result, when considering collaborative training operations, we need to create training algorithms that will be robust to non-independent and identically distributed (non-IID) datasets.
- *Privacy and Security:* ML models trained in such systems use sensitive personal data. Even though the raw format of the data will never leave the devices, there are other forms of data shared with third-party entities (e.g., model updates, ML configurations, intermediate results, etc.) that can expose the user.

Notably, cross-device configurations are more challenging than cross-silo. In cross-device configurations, there is no limit to the number of participating clients (e.g., in applications for agriculture, there are large fields surveilled by IoT devices or drones that constantly collect data from their environment), and it is not always feasible to establish centralized authority to orchestrate these entities (e.g., applications for smart cities, which have a more arbitrary structure and consist of mobile devices). Considering that, we believe that the primary focus of this chapter should be on cross-device distributed learning, as more issues and approaches provide research interest to discuss. Note that throughout the chapter, when we refer to collaborative learning or decentralized AI, we imply cross-device settings, unless specified otherwise. Considering cross-device SL and SI, which mainly employ IoT devices with the characteristics described above, we will explore how SC has been developed to overcome their constraints.

10.3 SYNERGETIC INFERENCE

The clients can input into the model their data and receive a prediction of the corresponding task. Depending on the design of the application, the model can be deployed in the cloud, or the clients can locally download a copy of the model (if their resources can support such action). In the latter case, the client needs to transfer their data to the cloud and receive (download) the results. Whereas, when the model is already installed on the device, the delay of a local inference operation depends on the device's resources and temporal load. In fact, a study of inference's latency cost showed that IoT devices with sufficient computing capabilities (e.g., GPU-enabled) can implement inference tasks faster and with smaller energy consumption than completely offloading tasks onto the cloud (Kang et al., 2017). This fact is valid when the network speed between the cloud and IoT is slow (i.e., 3G, LTE, etc.) or when the cloud has already received a large number of requests. Nevertheless, these assumptions cannot always be valid (as discussed in Section 10.2). An intermediate solution is SI, which, as shown at the top of Figure 10.4, mainly splits the model at a layer that will reduce the amount of data transferred to the cloud (i.e., bottleneck layer), and at the same time, it evenly distributes the computing overhead among IoT and the cloud. Moreover, taking advantage of the edge (i.e., nodes that are located between IoT clients and the cloud), one can split the model using one more cut layer (i.e., three model parts) and offload the intermediate part onto the edge node. As a result, the clients will send their activations to the edge, which is usually closer to them, and then the edge, after forward-propagating through some layers, sends the corresponding activation to the cloud (Liu et al., 2023). SI has been studied in numerous research works; hence, in this section, we study more techniques that support SI. Finally, we discuss the differences in implementing and designing a system that will support the operation of an inference task vs. a learning task (i.e., SL).

10.3.1 Synergetic Inference Approaches

To evenly distribute the computing cost between IoT and the cloud while keeping the communication cost low, one must carefully determine the cut layer, as it directly affects these two metrics. As mentioned above, Neurosurgeon (Kang et al., 2017) searches for the bottleneck layer (i.e., cuts the model once). Another approach is to decide for each layer independently whether it should be offloaded or not (Eshratifar, Abrishami and Pedram, 2019). To do so, Eshratifar, Abrishami, and Pedram (2019) formulate an integer linear programming (ILP) problem with objectives set to minimize latency or reduce the client's energy consumption. Yet, searching for the bottleneck layer(s) in a model implies that this is an inhering property in NN models. However, this is not always the case because there are models without natural bottlenecks (e.g., residual models, RNNs, etc.). Therefore, more sophisticated approaches should be considered. Some examples are discussed below:

- *Embedding Bottleneck Layers:* Instead of searching for the optimal bottleneck layer, one can inject one such layer. As shown in the middle of Figure 10.4, additional layers are added to the model, which will reduce the dimension of the activations transferred over the network. Specifically, the embedded layers act as encoders and decoders. The encoder part is added after the last layer on the client device, while the decoder sits before the first layer on the cloud. BottleNet, as introduced by Eshratifar, Esmaili, and Pedram (2019), is one of the first of such approaches, in which a learnable bottleneck unit is added. The client device utilizes both spatial and channel-wise reduction units, followed by a compressor unit, to generate a condensed representation of the feature tensor. This compressed tensor is then transmitted to the cloud. However, this comes at a cost. First, the addition of the bottleneck layers alters the intermediate activation, which will result in a noisy output with the risk of degrading accuracy. Also, before deployment, the injected layers need to be trained as well. Notably, the training approach will determine the level of accuracy degradation.
- *Early Exit (EE) Design:* Considering the heterogeneous nature of cross-device systems, EE (Laskaridis, Kouris and Lane, 2021) is a more adaptive approach in which the inference process is adjusted according to the input, device, and network characteristics. Concretely, as is shown at the bottom of Figure 10.4, along the depth of the model, earlier exit gateways are added, which can compute the output in an earlier stage of the model. The intuition behind this is that not all inputs need the same processing; there are easier inputs that can achieve good accuracy using fewer layers. As a result, the output can be computed locally

without passing through the whole model. In such implementation, some of the challenges one needs to consider are (i) the design of the EE layers, (ii) the training of the model, and (iii) the confidence function that will evaluate whether an EE's output is satisfactory or not.

10.3.2 Synergetic Inference vs. Split Learning

SL could be considered an extension of SI, which supports training tasks that are implemented with the synergy of IoT and the cloud-edge continuum. However, SL is naturally a more complex process because of the following characteristics:

- *Larger Computing Demands:* A training operation is more challenging as it consists of multiple repetitions of forward- and backward-propagation operations. One batch update (i.e., one forward propagation followed by one backward propagation operation) is repeated for each batch of data and multiple epochs. Moreover, the forward propagation must be followed by the corresponding back propagation (persevering the order); hence, they are co-dependent operations. Also, the batch updates themselves are independent of each other; no concurrent batch updates are allowed. Hence, there is no freedom in decoupling these operations and implementing them in parallel.
- *Larger Memory Demands:* During forward propagation, the model generates activations crucial for computing gradients during back-propagation. These activations must be stored until the back-propagation stage is completed and the corresponding gradients are generated. Afterwards, these gradients are utilized to update the model's weights. Depending on the complexity of the model's architecture, the generated activations may require significant memory storage, often amounting to hundreds of megabytes.
- *Time-Consuming Operation:* It may take hundreds of training rounds to complete the training of the models, which may take hours or days to complete.
- *Synchronization Complexity:* In the case of multiple clients (i.e., collaborative learning), we *should* consider the complexity and overhead of client synchronization. FL has solved this using aggregation functions, which adds a large communication cost.

Hence, even though there has been a great breakthrough in improving SI, these existing approaches cannot be applied directly to SL because of the reasons mentioned above. As will be presented in the following sections, many aspects of SL need to be approached that essentially lead to new research directions.

10.4 EVOLUTION OF SPLIT LEARNING

SL was first introduced in Vepakomma et al. (2018), the protocol of which was described in Section 10.1. This initial work inspired several research studies to expand this first approach and hence, this section will present an overview of SL applications, case studies, as well as works that enhance the SL protocol by considering the challenges of cross-device systems (see Section 10.2).

10.4.1 Types of Splits

Figure 10.5 shows some of the main designs for SL regarding the split policy. First, it shows the vanilla SL, where there is one split and the interaction between the compute node and client is the one that was discussed in Section 10.1. However, in the one-split configuration, the compute node is the one that will generate the predicted output and will receive the values of the real labels in order to compute the loss and initiate the back-propagation. This can be avoided with the U-Split or no-label-sharing configuration. As is shown by the second plot in Figure 10.5, in U-Split, the model is split into three parts, and the client keeps locally the first and the last part of the model.

This will guarantee that the client will have full control over the raw format of the data samples. In that case, there is an extra communication step, where the compute node sends to the client the activation and receives the gradients of the layer that connects the second and last model parts (i.e., second cut layer).

One step further, there is SL with multiple model parts (i.e., more than two for vanilla SL and more than three for U-Split), which are offloaded to different compute nodes while the first and/or last model parts stay in the client (third and fourth plots, respectively). To achieve that, the client and the compute nodes, which have successively modeled parts, communicate and transmit the activation/gradients of the corresponding cut layers. This configuration is useful for systems where the compute nodes are cheap or have limited memory capacity and cannot store the whole intermediate model part. Moreover, as we will discuss further, this multi-hop configuration allows pipeline parallelism, which provides acceleration and scalability. Finally, the last plot shows an indicative example of how SL can support training for clients with vertically split datasets, which is one of the common applications that FL supports, and we refer to this as *vertical FL* because each of the clients acquires only a subset of the features for each data sample. As a result, these clients must combine their data in order to make a single prediction. This can be achieved by concatenating their activations, as shown in Figure 10.5, and other techniques, as we will discuss in the following section.

10.4.2 Split Learning Cases and Applications

SL was initially introduced as a training tool that can be used for medicine and healthcare applications. Many works extended this direction, like in Ayad, Frei and Schmeink (2022), where they proposed an alternative SL for ECG classification. Specifically, their approach targets accelerating the training delay and minimizing the communication overhead by adding autoencoder (AE) layers that will compress the load of activations/gradients between the cut layers. In the following part of this section, we will examine additional techniques that target reducing the communication cost in SL. Moreover, in Ayad, Frei and Schmeink (2022), a threshold policy is applied to decide when a batch input should be used for updating the model or not in order to avoid unnecessary updates that add additional communication and computing costs to the system.

Furthermore, the SL architecture can naturally support a vertically split dataset, which is a very common application in healthcare systems. For instance, assume that one hospital holds X-rays for a patient while the patient's blood test results are stored in another hospital. In that case, the two hospitals cannot share their knowledge due to privacy reasons; however, combining this information will boost the predictive and modeling tools of the hospitals. In Ceballos et al. (2020), the authors study the performance of vertical SL for five different pooling mechanisms to aggregate the partial output from all clients. As we discussed, the most straightforward approach is concatenation, but it has one limitation. Namely, it requires all clients to send their intermediate activations before proceeding into the second model part (combine and then predict). This is sensitive to the straggler effect, as slower clients will slow down the whole process. Other approaches studied in Ceballos et al. (2020) are elementwise averaging, addition, summary, etc. All methods have similar performance; hence, the selection depends on the system's preference. For instance, elementwise can be preferred for more privacy-preserving systems because it can support the technique of secure aggregation. Moreover, the works in Allaart et al. (2022) and Ads, Alfares and Salem (2021) apply vertical SL for several medical and healthcare datasets and their results verify the performance of this approach.

Directly related to this, the research work in Ni et al. (2024) presents an application of Internet of Medical Things (IoMT), in which wearable devices can train deep NNs by applying a SL protocol. Also, apart from the healthcare system, SL has been supporting other IoT applications. One such example is the work of Wang et al. (2024), which builds an intelligent smart meter system with the use of NN models. These models are trained by edge devices using SL. Additionally, SL has been implemented for intelligent transportation systems (Otoum et al., 2022) and mmWave power prediction (Koda et al., 2019). Moreover, SL, apart from convolutional neural networks (CNN)-based models that are usually used, can be used for recurrent neural networks (RRNs) (Zhou et al., 2022) by utilizing the split policy of multi-hop (as shown in Figure 10.5).

10.4.3 Split Learning Protocols and Approaches

As described in Section 10.1, the SL protocol sequentially iterates through the available clients. This subsequential attribute of SL can result in a prohibited training delay when the system scales as more clients are added to it. Moreover, in FL, a single communication action (i.e., a single message) is significantly more costly than in SL. In SL, an individual message only contains the vector of activations/gradients of a single layer, whereas in FL, a single message contains the whole NN model that will be sent to the aggregator. However, in SL, the clients communicate with the compute nodes more frequently—in every batch update. As a result, the total communication overhead in SL may exceed that in FL.

Therefore, a great deal of research has been targeted to tackle the issues of the conventional SL protocol. This section presents such works and taxonomizes them according to their approach. First, we start with works that focus on the training protocol and target to accelerate the training time, and then we proceed with works that endeavor to minimize the communication overhead.

10.4.3.1 Enhancing Split Learning

One approach that enhances the performance of SL is the combination of SL with FL, also referred to as *parallel SL*. On the one hand, there is FL, which allows the clients to train the model in parallel, but it requires them to have the appropriate resources. And on the other hand, SL drops this requirement since clients will handle a small portion of the computing load, but the training will run consecutively. Hence, the motivation for combining these two protocols is to take advantage of their benefits and minimize their weaknesses. This approach was first introduced in the SplitFed framework by Thapa et al. (2022), providing two algorithms named *splitfedv1* and *splitfedv2*. In the former, for every client, the compute node manages a different version of the model part that oversees. In the latter, all clients share the same version. In parallel SL, the clients can concurrently send the activations to the compute node and be served in parallel. During the aggregation phase, the compute node is in charge of aggregating the model parts that are handled locally, while the client's model part will be sent to the aggregator. In Gao et al. (2021), the authors propose a general split federated learning (SFL) protocol named SFLG. They try to combine splitfedv1 and splitfedv2 by grouping the clients and allowing the compute node to only maintain a model part for each group.

Moreover, apart from the system performance, the combination of SL and FL can boost the personalization capabilities of the model while remaining generalized. We have seen personalized FL (Fallah, Mokhtari and Ozdaglar, 2020), but they perform poorly on non-IID data and because of the FL's nature, such approaches cannot be supported by all devices (i.e., resource constraints). But SplitGP (Han et al., 2023), which also combines SL with FL, provides a training and inference method for a generalized and personalized model. Additionally, the combination of SL and FL is used as a method to support transfer learning and pre-train the models using large-scale training data. The FedGKT (He et al., 2020) is a framework that combines FedAvg (McMahan et al., 2017) with SL to train large CNN models even with non-IID data distribution. Furthermore, Gao et al. (2020) employ a technique named *ensemble learning* to accelerate the training. That is, the nodes train other versions of the model in parallel in order to utilize idle clients. At the same time, the compute node interacts with the clients in a round-robin order for each ensemble model (i.e., version of the model). Note that, although each version of the model is trained sequentially, the models are updated in parallel by a different client each time. This is different from SFL, as all versions of the model belong to all clients, and the clients sequentially train each version of the model. Whereas in SFL, each client has its own instance of the model (i.e., they do not share versions) and only trains the corresponding instance. Alternatively, SL and FL can be combined to create a hybrid approach (Liu, Deng and Mahmoodi, 2022a,2022b). The devices may select between the FL or the SL protocol; this decision is made during the client selection phase. To optimize this decision, they formulate a multiple-choice knapsack problem (MCKP) and a multi-arm bandit (MAB) selection scheme that minimizes total energy consumption.

10.4.3.2 Reducing Communication Overhead

There are research studies in the literature that propose approaches to improve the performance of SL by reducing communication overhead. We categorize them into two main approaches, as discussed in the following:

- *Local Updates:* mainly use techniques from local-loss-based training (Nøklan and Eidnes, 2019). Specifically, Han et al. (2021) extend the SplitFed protocol (Thapa et al., 2022) by introducing two different local loss functions, one on the client side and one on the compute-node side. The clients send to the compute node the intermediate activations, but the two model parts (i.e., the one on the clients and the one on the compute nodes) are updated in parallel using the respective local-loss functions. This way, the nodes do not send the intermediate gradients to each other for back-propagation. Additionally, Mu and Shen (2023) integrate the local loss function into the parallel SL protocol, which not only reduces the communication overhead but also diminishes the storage burden at the compute node. Note that in parallel SL, the compute node stores an instance of the model part for every client it serves. This restricts the system from scaling; hence, Mu and Shen (2023) use a single model training strategy for the compute nodes, while the clients send activations to the compute node less frequently because they use a local auxiliary model to update their model part.
- *Data Compression:* reduces the communication delay by scaling down the data load transferred between the clients and the compute nodes (i.e., activations/gradients). However, note that this comes with a risk of accuracy loss. A first such approach is introduced by Chen, Li and Chakrabarti (2021), which reduces the load of the transmitted messages by changing the variable precision from a 32-bit floating point to an 8-bit floating point. This is achieved by using a hybrid 8-bit floating point representation in conjunction with a search-based quantization method that fixes the number of exponent bits and exponent bias across epochs. Alternatively, in Ayad, Renner and Schmein (2021), the authors add an AE model after the first cut layer (like techniques that are applied in SI). The AE is an unsupervised NN that is used to reduce the data load sent from the clients to the compute node. The client stores the encoder part, while the compute node has the decoder part of the AE.

10.4.4 Comparison of SL and FL

After having discussed various SL approaches and considered their challenges, we can have an overview of the two most known client-based distributed ML protocols, SL and FL. First, there is a significant difference training-wise between FL and conventional SL. First, in FL, the clients work on their own version of the model during training and synchronize their results. While in SL, each client applies its update directly to the model, similarly, to training the model in a centralized manner, and there is no need for aggregation. This similarity promises that the performance of the model will be close to that of centralized training. In fact, empirical evaluations in Gao et al. (2020) and Gao et al. (2021) show that the SL is more robust than FL for imbalanced clients (i.e., the number of data samples in each client differs significantly) but FL performs better than SL when the data distribution is extremely non-IID. However, the performance of SL is sensitive to the order in which the compute node will iterate the clients (see discussion of heterogeneous data in Section 10.6).

Resource-wise, in SL, clients consume less energy for computing than in FL. As a result, a higher peak of power and temperature and much more memory are used in FL than in SL (Gao et al.,

2020, 2021). However, in SL, the communication operations are more frequent than in FL (i.e., in SL, clients communicate with compute nodes in every batch update). Yet, the per-communication data load is less than in FL, where clients share the whole model with the aggregator. Moreover, this communication overhead in SL can be controlled with split selection or by applying any of the techniques discussed in Section 4.3.2. On the other hand, there is parallel SL, whose training protocol is the same as in FL (training phase followed by an aggregation phase), while preserving the energy consumption of the clients at the same level as in SL. However, in parallel SL, the computing demands for the compute nodes are larger and require more memory because the compute node stores a different instance of the intermediate model part for each client it is in charge of.

A new upcoming approach that seems to be a middle solution between parallel SL and FL is Hybrid FL-SL (Liu, Deng and Mahmoodi, 2022a, 2022b). The motivation is that not all clients in the system need to offload their model parts to a compute node. There are client devices capable enough to implement on-device training without delaying the whole training process. Hence, by keeping these clients in a FL configuration, we can reduce the load on the compute nodes, while clients with constrained resources can use a SL configuration. This approach essentially will lead to a faster training process, as the queue on the compute nodes will be smaller while the on-device delay is overlapped by the SL operations.

10.5 MODELING AND OPTIMIZING COLLABORATIVE LEARNING SYSTEMS

One main research focus in SL is on mathematically modeling the system. This formalization allows the construction of optimization problems, which will define key factors across various aspects of the system (e.g., splitting policy, task scheduling, etc.). First, in this section, we will provide a general structure of a model for a system that supports parallel SL. Afterwards, we will discuss how several research studies model the cost of training as a parameter of latency or energy consumption and how they formulate the entire optimization problem.

10.5.1 Modeling Split Learning Systems

We will formulate a baseline and abstract model of a system that uses parallel SL. First, this section presents the main components of the system and then formally describes the workflow of the training protocol that uses parallel SL. Note that the description of the model will be as generic as possible, in order to describe as many setups as possible.

10.5.1.1 Entities

A system for parallel SL is comprised of the following entities:

- A set of clients $c_k \in C$ where $|C| = C$. The clients have access to a local dataset D_c, which contains a set of $|D_c|$ training samples, $\{x_i, y_i\}_{i=1}^{|D_c|}$ (i.e., input data and target class). The client c_k can be a small device (e.g., IoT or mobile device) with computing capacity, denoted as $cap(c_k)$ floating point operations per second (flops) and memory capacity as $mem(c_k)$ GB. Note that in SL it is expected that the $(cap(c_k), mem(c_k))$capabilities not to be sufficient for training the whole model of the training task.
- A set of compute nodes $n_k \in N$, where $|N| = N$, which assist the clients throughout the training process. The capabilities of the compute nodes $(cap(n_k), mem(n_k))$ are more powerful than those of the client machines and can support the whole training of the model.

- The aggregator node n_{aggr}, which is in charge of aggregating the clients' models at the end of each training round. This can be a separate node (i.e., $n_{aggr} \notin N$), or one of the compute nodes.

We consider that each client c_k can communicate with all compute nodes (and the aggregator node) via a network link net_{c_i} $\forall$ N.

10.5.1.2 Training Workflow of Parallel Split Learning

The clients and the compute nodes collaborate to train one NN model, annotated as M, which consists of L layers; specifically, the model contains L vectors with trainable weights, i.e., $M = (w_1; w_2; \ldots; w_L)$. The model is split into S model parts, where each model part contains a subset of sequential layers, i.e., $M_s = (w_i; w_{i+1}; \ldots; w_j)$. Therefore, following the *splitting policy*, the model can be represented by $M = \bigcup_{s=1}^{S} M_s$. Additionally, recall that in parallel SL, each client c_i trains a different version of the main model; hence, $M^i = \bigcup_{s=1}^{S} M_s^i$, $\forall i \in C$. Moreover, following the *allocation policy*, let $\pi(M_s^i)$ be the node (client c_i or one of the compute nodes) that will be in charge of the model part M_s^i.

As discussed in Section 10.4, in parallel SL, each client trains its model part independently of the other clients. The clients train the model for $r \in R$ local epochs, before sending its model updates to the aggregator to complete one communication round. Let $e \in E$ be the number of communication rounds. The clients cluster the data of their local dataset, D_c, into $b_k \in B_c$ batches of data, which will be used to update the model. The batch update starts by forwarding the input batch into the model parts (forward-propagation) as,

$$fwd\left(b_k^{re}\right) = M_S^{cre}\left(M_{S-1}^{cre} \ldots \left(M_1^{cre}\left(b_k^{cre}\right) \ldots\right)\right) = \hat{Y}_k^{re} \tag{1}$$

Each model part generates a vector of the intermediate activations that are forwarded to the subsequent mode part. Let $a_{i \to j}$ be the activations that connect the layers i and j. Of course, $j = i + 1$, as the forward-propagation starts from the first layers and completes at the final one. Furthermore, if the model part $w_i^c \in M_s^c$, and $w_j^c \in M_{s'}^c$ belong to different nodes, then the node $\pi(M_s^c)$ will transfer the activation to node $\pi(M_{s'}^c)$. Next, using $\hat{Y}_k^{re}$ and the real targets, Y_k^{re} from the local dataset, the client can compute the loss function.

Afterwards, when back-propagating the loss throughout the model, the gradients of each layer are computed, and the model weights are updated. We denote this operation as

$$bck\left(b_k^{re}\right) = M\grave{}_1^{cre}\left(M\grave{}_2^{cre} \ldots \left(M\grave{}_S^{cre}\left(loss_k^{cre}\right) \ldots\right)\right) = \hat{Y}_k^{re} \tag{2}$$

Similarly, to forward-propagation, in back propagation, each model part generates a vector of the intermediate gradients along subsequent model part. Let $g_{i \to j}$ be the gradients that connect the layers i and j, of course, $j = i - 1$ as back-propagation follows the inverse path to forward-propagation. Correspondingly, if model parts $\pi(M_s^c) \neq \pi(M_{s'}^c)$, then $\pi(M_s^c)$ will transfer the gradients to $\pi(M_{s'}^c)$. Finally, the computing round is completed when the aggregator generates the global model of the next training round as

$$M^{e+1} = \frac{1}{C}\sum_{c=1}^{C} \mathrm{M}^{ce} \tag{3}$$

10.5.1.3 Training Cost Analysis

Having provided a general description of the system model and the training steps, we will discuss how practically the latency and energy cost are modeled. Starting with ARES (Samikwa, Maio and Braun, 2022), which models a system with one compute node (N = 1) and multiple clients working in parallel. There is a single split, and hence there are two model parts, i.e., $M = (M_1, M_2)$, where $\pi(M_1^i)$ is c_i (i.e., the client itself) and $\pi(M_2^i)$ is the compute node n_1. Note that each client has a different splitting policy, which is also updated (if needed) in every communication round. In ARES, the communication delay is computed using the formula $\frac{|\text{data}|}{\text{throughput}}$, where the throughput is defined by the network line between the client and compute node. Additionally, the computing cost for one batch update is equal to $\sum_{i=1}^{2} 2\text{cap}\left(\pi\left(M_i^{ce}\right)\right)$, where the cost of the forward- and backward-propagation is determined by the machine's capacity, which is responsible for the corresponding model part. In ARES, the objective function is to find the optimal split that will jointly minimize the cost of the client with the maximum training delay and energy consumption inside the communication round. In other words, decreasing the training duration of the slowest device (the straggler) results in a reduction of the overall training time for the training round.

Considering larger-scale scenarios, CoopFL (Wang et al., 2023) models a system with multiple compute nodes and uses a similar latency estimation to ARES. However, the aforementioned models do not take into account the fact that multiple clients compete for the same resources on the edge device (compute node). They assume that when a client sends a task to the compute node, it will be served without any delay, whereas Tirana et al. (2024) consider the queuing delay on the compute nodes, which is a key overhead in parallel SL. In detail, Tirana et al. (2024) model the system as a parallel machine scheduling problem and construct an ILP optimization problem for minimizing the latency of a client with the largest batch delay. Finally, an alternative system is multi-hop Parallel Split Learning (MP-SL) (Tirana, Lalis and Chatzopoulos, 2024). In this technique, the compute nodes handle different model parts, supporting pipelined parallelism. The system model of MP-SL focuses on the pipeline's delay, which is defined by the compute node that takes the longest time to complete its task.

10.5.2 Optimizing System Aspects

Using model construction, latency estimation, and the optimization problem definition, the next discussion is about the aspects of the system that can be optimized.

10.5.2.1 Splitting Policy

As discussed in Section 10.3, deciding the splitting points in SC is not trivial, especially when considering systems of collaborating training. A typical approach is to define the splitting points during the offline period (i.e., before training starts) (Tirana et al., 2024). Alternatively, ARES (Samikwa, Maio and Braun, 2022) is an adaptive protocol for parallel SL in which each client minimizes its own training time and energy consumption by optimizing its splitting policy. However, these two objectives are contradictory, so it is up to the user to decide the balance between acceleration and energy. Similarly, in Kim et al. (2023), the researchers focus on the trade-off between energy consumption and selecting more privacy-preserving splits. The deeper the cut, the less information becomes available to the compute node. However, the client will get a heavier computing load, which leads to more energy consumption. They model the system as a multiplayer bargaining optimization problem and solve it with the Kalai-Smorodinsky bargaining solution. In a different set-up, MP-SL (Tirana, Lalis and Chatzopoulos, 2024) implements a pipelined parallel SL protocol in which the ML model is split multiple times and uses a different compute node for each model part.

In MP-SL, the model parts are defined by solving an optimization problem that minimizes the delay of the pipeline of compute nodes.

On the other hand, some studies do not use conventional mathematical tools to optimize the splitting decision and employ stochastic methods. Like FedAdapt (Wu et al., 2022), in which the splitting point changes to adapt to the changes in the network bandwidth between the clients and the compute node. FedAdapt decides the offloading point (i.e., cut layer) with the help of a trained reinforcement learning (RL) agent. However, a significant limitation of this work is that a single RL agent cannot be used for a large number of clients, and one should also consider the overhead of training the RL agent.

10.5.2.2 Client Allocation and Task Scheduling

The memory and computing demands for the compute nodes are enormous and are increasing as more clients participate in the training procedure. A natural solution to this is to scale the resources horizontally. Even though the significance of system models involving multiple compute nodes is acknowledged, such systems have not been broadly explored. In this case, apart from the splitting policy, another key factor is allocating the clients to the available compute nodes. However, considering the special features of cross-device systems, the decision of allocation should be decisive and carefully designed. For instance, consider a system where there are two compute nodes n_a, n_b and a 100 clients. Let us assume that 70% of the clients have a faster link to n_a than n_b. Additionally, n_b is computationally more powerful than n_a. A question that needs to be answered is: how should the clients be allocated to the compute nodes. If we decide to map them according to their network links, then the node n_a will be overutilized, and clients will end up waiting for other clients to complete. Neither is fair; allocating clients equally. Note that n_b is more powerful and can serve more clients in parallel, but at the same time, network limitations should also be considered. Allocation and scheduling of clients are multiparameter problems.

CoopFL (Wang et al., 2023) is an example of research work that jointly makes the decision of allocation and split policy by minimizing the summary of all clients' training delays. The formulation leads to an NP-hard optimization problem, which is solved by using a heuristic algorithm based on exhaustive searching. However, the model of CoopFL does not capture the queueing delay at the compute nodes. On the contrary, Tirana et al. (2024) is the only work that models the whole workflow of a horizontally scaled parallel SL system with a U-Split SL (i.e., there are three model parts, ensuring that the first and last layer will be processed only by the client) and jointly optimizes the client allocation and task scheduling. In Tirana et al. (2024), the splitting policy is user-defined, but they support clients having different splitting definitions tuned to the individual device characteristics.

10.6 DATA HETEROGENEITY

Cross-device collaborative learning protocols provide a promising approach to training ML models on a large scale (i.e., many client devices can participate) while being a privacy-preserving solution since the data is decentralized. However, it is common for such systems to suffer from the negative effects of non-IID (i.e., training data that is not evenly distributed across devices), which can significantly deteriorate the performance.

10.6.1 Non-IID Types

There are many categories of non-IID. Some representative examples are:

- *Label Distribution Skew:* The label distribution among the clients is different. An instance of this category could be when the clients have access to data samples that correspond to a subset of the set of possible labels. Another example is when the clients have access to data samples that correspond to the full set of possible labels but the portion of data samples for each label follows a non-IID distribution.

- *Feature Distribution Skew:* The feature distribution among the clients differs. In other words, the clients have a different view of the same thing. For instance, assume there are two clients. Both of them have a dataset that contains figures that correspond to a "dog," but they have figures from different breeds.
- *Quantity Skew:* The number of data samples for each client is different. Some clients may have more samples than other clients (i.e., not all clients contribute the same during the training process).

10.6.2 Tackling Non-IID in Client-Based Distributed Training

As SL is an extension of FL, we will start our discussion with mechanisms and algorithms that tackle non-IID in FL. This section taxonomizes these approaches into categories based on their approach. Afterwards, we will present related works that use SL in highly heterogeneous systems. Not surprisingly, the SL approaches seem to perfectly fit into the FL categories.

10.6.2.1 Non-IID in Federated Learning

Starting with FL protocols, there are a plethora of approaches that deal with data heterogeneity. These approaches fall into the following categories:

- *Data-based:* This group of approaches consists of techniques that aim to "fix" the non-IID distribution of the clients' local datasets. Such techniques are data sharing (Zhao et al., 2018) and data augmentation (Cheng, Zhang and Li, 2023). In the former, before the actual training starts, the aggregator trains the model using an IID set of samples. This IID set can be found in a globally available repository, or clients can share their data (Yoshida et al., 2020). Even though this is a straightforward and promising approach, it violates one of the fundamental properties of FL since clients share their data. On the other hand, data augmentation can repair the clients' data distribution in a more privacy-preserving manner. Data augmentation can be improved using generative adversarial networks (GANs)[3] (Cheng, Zhang and Li, 2023). Namely, clients collaboratively train a GAN, and then the GAN's generator is used to produce synthetic data that will be used by the clients to align their data distribution.
- *Algorithm-based:* More popular approaches focus on the FL protocol itself. For instance, FedProx (Li et al., 2019) focuses on the local-update level and introduces an additional regulation term in the clients' local objective function to limit the distance between the local and global models. Another example is FedNova (Wang et al., 2020), which adds a regulation term to the aggregation function that is applied in the aggregator.
- *Cluster-based:* This group of approaches mainly focuses on the idea of personalized FL, in which the system does not force incompatible clients to train one single model. As cross-device systems commonly exhibit data heterogeneity, assuming all users are IID samples may hinder the feasibility of personalized predictions (Ghosh et al., 2022). Hence, instead, we can group clients that have similar data distributions and train personalized models for each such group of clients.

10.6.2.2 Non-IID in Split Learning

Similarly, the performance of SL protocols degrades when the distribution of clients' data is non-IID (Gao et al., 2020, 2021). In the same vein as in FL, some approaches propose strategies to mitigate this problem. However, the research area toward non-IID and SL is not as broadly explored as in FL. Starting with the data-driven approaches, there is IoTSL (Feng et al., 2023), an approach that uses data augmentation. It combines GANs and a generator-sharing strategy. IoTSL uses the technique of transfer learning, in which each client pre-trains the model locally using GANs. Then, clients' generators are combined to generate data from the missing classes. As a result, this will construct an IID data distribution for every client. An additional data-based approach is LocFedMix-SL (Oh et al., 2022), in which the compute node, receives a mix of activations from the client as a data augmentation technique; thus, they use the local loss to maximize the information between the input data and the smashed data (i.e., activations that will be mixed with other client's activations). Also, the technique of mixed activations is used in MergeSFL (Liao et al., 2023), in which clients share with the compute node activations from a micro-batch (i.e., a smaller set than the batch of data). All the micro-batches are combined to construct one full batch of samples. This approach can encounter the forgetting effect of SL (Madaan, Kulkarni and Pant, 2021), which can significantly harm convergence as the updates of one client are overwritten by other clients' updates. Another example of an algorithm-based approach is Cai and Wei (2022), which is based on an incentive mechanism and inspired by the Stackelberg Game. Specifically, in this approach, clients are motivated to select data samples from their local dataset that will satisfy the compute node's utility function. This utility function forces clients to use data samples that have an even distribution. However, note that clustering-based algorithms that focus on personalized training have not yet been explored in an SL context.

10.7 PRIVACY AND SECURITY

One of the primary motivations for using client-based distributed ML is the privacy-preserving attribute. The clients involved in the training process desire to remain the owners of their data and do not wish to disclose it. However, for the training to be feasible, some knowledge needs to be shared. This may allow some intruders to threaten the client's privacy. Depending on what and how much knowledge is revealed to third parties, this categorizes the attack conducted by an intruder into one of the following categories: *black-box*, *white-box*, or *partial white-box* (Soykan et al., 2022). Specifically, for client-based distributed ML, a black-box attack is when the attacker uses techniques that are model- and ML-task-agnostic, whereas in a white-box attack, the intruder knows the model architecture, parameters, hyper-parameters, and the ML-task and settings. Notably, the partial white-box attacker has more knowledge than a black-box attacker but less than a white-box attacker. Moreover, typically, the attackers fall into two categories of adversarial behavior, namely, the *honest-but-curious* attackers and the *malicious* ones. The honest-but-curious entities do not harm the procedure of training; they follow the protocol properly. They do, however, have access to sensitive data and records and use them for their own interests, whereas the malicious attackers interfere with the process and sabotage the training. In a client-based distributed ML, the adversary can be the aggregator, the compute nodes (in the case of SL), or even the clients.

Consider that in SL, the insight and knowledge that can be revealed are the following:

- Model architecture (similar to FL)
- Training hyper-parameters (similar to FL)
- Parameters of a model part M_s^c and their intermediate updates. This information is disclosed to the node in charge of the corresponding model part, that is, $\pi(M_s^c)$.
- The intermediate activations and gradients. Recall that if $\pi(M_s^c) \neq \pi(M_{s'}^c)$ so, these two nodes need to exchange this information among themselves.
- The clients' model is updated at the end of a communication round (similar to FL). This information is disclosed to the aggregator during the aggregation phase.

Notably, in SL, there is additional information disclosed to third parties compared to FL. The focus of this section will be on the vulnerabilities and threat models for SL; hence, the discussion will be about the attacks that are based on the information being disclosed due to the SL protocol and not the union of them. Therefore, there will not be any discussion about attacks conducted by adversary aggregators or attacks happening during the inference phase (e.g., membership inference, property inference). Such attacks have been well covered in many surveys (Soykan et al., 2022).

10.7.1 Vulnerabilities of Split Learning

Table 10.1 shows the SL works that attempt to threaten and attack SL clients. The attacks are categorized according to their behavior and the methodology they use. Starting with one of the most popular SL attacks, which uses a technique called the Feature-Spaces Hijacking attack (FSHA). This attack was first seen in Pasquini et al. (2020) and can be characterized as a Model Inversion Attack (Soykan et al., 2022), as it uses deductive knowledge from the training phase and builds a reverse model that can be used in the inference/testing phase to reconstruct the input data samples. Specifically, in Pasquini, Ateniese and Bernaschi (2020), the attacker is the compute node that has access to one of the intermediate model parts (i.e., $s > 1$ and $s \geq S$).[4] The FSHA requires access to a set of public input samples X_{pub} that captures the same domain of the clients' training sets. But other than that, the attacker is model- and ML-task-agnostic and takes place during the training phase. Specifically, the adversary compute node, while training the model part M_s^c, also trains two adversary model parts $M_s^c, M_s^{c^{-1}}$ using the intermediate activations form M_{s-1}^c and the X_{pub}. In detail, $M_s^{c^{-1}}$ is the inverse model that will be used in the testing phase for reconstructing the input data, while M_s^c is used auxiliary during the training phase to alter the M_s^c to produce activations toward a target feature space, which is defined by the X_{pub}. Essentially, Pasquini, Ateniese and Bernaschi (2020) train $M_s^{c^{-1}}$ using an adversarial loss function like in GANs.[33] This attack is categorized as malicious as it implicitly poisons the M_s^c; without this being the main purpose of the attack. The performance of the attack does not depend on the number of clients or splits.

However, in Gao and Zhang. (2023), the authors noticed that the success of FSHA is owed to the change of the loss function when applying the adversarial objective function. But, when using U-Split SL (i.e., ensuring that the first and last model parts are exclusively handled by the client), the loss calculation process is on the client side, which will not allow FSHA to train the inverse model. Additionally, as will be shown in the following section, malicious attacks are detectable and can be interrupted once the client notices them. Therefore, Gao and Zhang (2023) propose a new honest-but-curious FSHA attack named PCAT, which does not depend on the activation but on the model part that the compute node is in charge of. It is a partial white-box attack, as it requires minimal knowledge. PCAT, like in FSHA, requires access to a small public dataset to train a client-copy model, which will then be used to train a reverse model. Similarly, Unsplit (Erdoğan, Küpçü and Çiçek, 2022) is an honest-but-curious attack that reconstructs the client's model part and the input data. But Unsplit is based on the received activations and requires much more knowledge than PCAT.

The work in Liu et al. (2023) and Li et al. (2021) considers a different configuration in which the definition of client and compute node alters. Specifically, there is a *non-label* party and a *label* party, which essentially means that the no-label party has access only to X_{priv} data and the model part that is in charge contains the first layer, i.e., M_1. Whereas the label party has access to Y_{priv} (label values) and is in charge of the last model part, i.e., M_2, which contains the output layer. In Liu et al. (2023) and Li et al. (2021), the attacker is the no-label party whose goal is to inspect the label of the input data. Liu and Lyu (2023) propose a distance-based online label inference attack in which the no-label party analyzes the intermediate activations and uses the Euclidean distance metric to find the closest class to which the input data may belong. This attack can be applied to binary or multi-class tasks. Whereas the attack in Li et al. (2021) is designed for binary classification problems. However, note that these attacks are designed for a restrictive SL setup, only for one client, which also does not have access to the real label values.

10.7.2 Privacy-Preserving Split Learning

Attempts have been made to design SL systems that consider the privacy vulnerabilities of the protocol. Therefore, some works propose defensive mechanisms for enhancing the privacy of SL systems. However, when selecting a privacy enhancement mechanism, we should first study the characteristics of the system and the client-based distributed protocol and then decide which defense mechanisms will best fit. There is not a panacea mechanism that can fully safeguard the system for every possible threat model. In fact, the research work in Soykan et al. (2022) provides a guideline for choosing one of the privacy-preserving technologies. The most common techniques found in distributed ML are: *differential privacy (DP)*, *homomorphic encryption (HE)* (Melchor, Castagnos and Gaborit, 2008), *secure multi-party computation (MPC)* (Goldreich, Micali and Wigderson, 2019), *functional encryption (FE)*, and *confidential computing techniques (CCT)* like *trusted execution environments (TEE)*. In this section, we will present works that use such approaches to protect the privacy of SL clients.

First, we consider an attacker with malicious behavior (i.e., the attacker will harm the training procedure and poison the model). In this case, Erdogan et al. (2023) have developed a protocol that can detect an FSHA attack during training. To do so, the clients periodically input batches with randomized labels and create fake batches. Then, they analyze the gradients that they receive from the compute node. Theoretically, if the server follows the proper protocol, then the following facts should be valid:

- The angle distance between fake and correct gradients will be larger than the distance between two random subsets of regular gradients.
- The fake gradients will have a higher magnitude than regular gradients.

If these two facts are not valid, then the client can safely conclude that the compute node is not properly following the training protocol, which may be injecting an FSHA attack.

Moreover, when considering honest-but-curious attacks, one of the proposed defence mechanisms is SplitHe (Pereteanu, Alansary and Passerat-Palmbach, 2022), which allows the implementation of HE in SL. Specifically, with HE, one can apply computation operations to ciphertexts (i.e., encrypted data). Hence, the client can send encrypted activations/gradients to the compute node and the compute node can process them without learning their real values. However, supporting HE can be extremely computationally heavy for a system. Another approach that has been broadly used in FL is applying DP to the model parameters before sending them to the aggregator. DP is a lightweight process and can be easily implemented by client devices. In fact, the DP scheme has also been applied in the SL context; such examples are noPeek (Vepakomma et al., 2020; Wu et al., 2023), in which the clients add noise into their activations/gradients before sharing them with the compute node. DP can potentially protect clients from model and input inversion attacks. However, conventional DP is not suitable for label inference attacks because both the no-label and label parties are aware of the identity of which example has participated in a given gradient update (Li et al., 2021). Therefore, they developed a defense mechanism named Marvell, which searches for the optimal structure of the noise for the gradients to minimize the amount of label leakage.

10.8 FUTURE TRENDS AND CHALLENGES

SL is an important technique as it enables resource-constrained devices to participate in a collaborative learning process. We can benefit from clients' data without violating their privacy. However, it is still a recent approach and there are still many open research questions to be answered. We encourage more researchers to work on this topic, as it can lead to new research areas. Here, we provide a high-level discussion of the future directions and the open questions for SL.

10.8.1 Scalability

Considering the parallel SL protocol, it effectively combines the advantages of SL and FL. This protocol provides scalability as clients can perform the training phase in parallel, similar to FL. However, for that to be true, we assume that the machine(s) that host the compute nodes to assist the clients are very powerful machines with multiple processing units to work in parallel with each client. Moreover, recall that the compute node needs to store the training state for every client. The training state contains the parameters of the model part that the compute node is in charge of and the produced activations (i.e., activations are stored during the forward propagation because they will be used at the backward propagation phase to compute the gradients). Hence, in summary, we assume that these machines have high computing and memory capacities. Furthermore, recall that the duration of a training process may take hours or days to complete. As a result, if we utilize the pay-as-you-go billing system and rent virtual machines (VMs) from public cloud providers, which is a common practice in such systems, this will significantly increase the cost of the training, as the cost is proportional to the resources of the VMs (i.e., memory and CPU cores).

We discussed earlier that a natural direction is the horizontal scaling of the entities of the compute nodes. This will ensure that the system can support even more clients if needed. But still, even in this case, we should consider that the model part that the compute nodes are in charge of is usually comprised of hundreds of parameters. Thus, to support a sufficiently large number of clients, the memory and computation demands remain large, even for the horizontally scaled compute nodes. On the other hand, we have seen in server-based distributed learning the technique of pipelined parallelism, which can relax the resource demands from the compute node as the model is split into even more parts, and for each of them, a separate compute node is used. Consequently, each compute node will process a part of the original task. MP-SL has adopted this concept into the SL scheme and has shown that it can significantly drop the cost with a small increase in the training delay. Yet, the combination of horizontal and pipelined scaling has not been explored. Multiple cheap compute nodes will work in parallel for the same sub-part of the model but assist different sets of clients. However, such a system requires careful orchestration that can dynamically be adjusted to the system's demands without overloading or underutilizing the resources of the compute nodes.

10.8.2 Privacy

As discussed in Section 10.7, the SL protocol, compared to FL, exposes more information about the clients. As a result, there are several attacks that can take place during the training process, which could expose the participating clients to privacy threats. On the other hand, as we also showed in Section 10.7, several defense mechanisms have been developed, mainly based on HE or altering the source data by adding noise (e.g., DP). However, following the paradigm of FL, there have not yet been any examples that leverage other privacy enhancement techniques, such as MPC, FE, or CCT. For instance, we could consider the CCT, which provides secure settings for users to run applications in remote environments by exploiting hardware security tools. Specifically, in modern computing, there have been platforms that offer advanced isolation and privacy support through so-called TEEs, like Intel Software Guard Extension (SGX) (Costan and Devadas, 2016). So, a potential approach could be to use such platforms for the compute nodes and safeguard the system from malicious or honest-but-curious acts. However, the main issue with TEEs is that they have a limited memory size and thus cannot be used to train large models. But, by splitting the model into smaller and more lightweight parts (like in MP-SL), it becomes possible to fit each part on a separate TEE (node), thereby achieving fully privacy-preserving training even for very large models.

Moreover, based on Agora (Koutsos et al., 2021), a blockchain-based data marketplace, we see that FE schemes could perfectly support privacy-preserving SC applications. In detail, Agora integrates a zero-knowledge proof (ZKP) protocol that is compatible with their proposed FE scheme to ensure output verifiability. However, to make feasible the combination of these two protocols, Agora

TABLE 10.1
SL Attacks Are Characterized by: the Adversary's Behavior

Attack	Behavior	Methodology	#Clients	Adversary Knowledge		
				Model	Data Values	Other Properties
Unleashing the tiger (Pasquini, Ateniese and Bernaschi, 2020)	Malicious	FSHA	Multiple*	✗	✗	X_{pub}
PCAT (Gao and Zhang, 2023)	Honest	Model inversion, input reconstruction	Multiple*	✗	✗	X_{pub}, Y_{pub}
UnSplit (Erdoğan, Küpçü and Çiçek 2022)	Honest	Model inversion, input reconstruction	Multiple*	✔	✔	Target values
Two-party label leakage (Li et al., 2021)	Honest	Label inference	One	✔	✔	Input values
Distance-based online label inference (Liu and Lyu, 2023)	Honest	Label inference	One	✗	✔	Input values

Note: (Malicious, honest-but-curious), the methodology used, how many clients can support it, and the knowledge the attacker has. Note that for short, instead of honest-but-curious, we write "honest." Additionally, regarding the number of clients, we indicate what is written in the corresponding paper (description of attack), but we use the symbol "*" to mark the works that have not tested the proposed attack in larger scaled systems.

builds upon the multi-client functional encryption (MCFE) scheme of Chotard et al. (2018). In this paper, we see a working example of SI for a trainable NN model that contains only fully connected layers (i.e., neurons implement a weighted average operation). Of course, the support of further types of operations that are usually found in NN (e.g., convolutional layers, back-propagation operations, etc.) will be challenging as FE operations are costly and complex. However, such techniques can ensure the clients' protection under a threat model that only needs to trust the authority that generates the cryptographic keys.

10.8.3 Personalized Training

As we discussed, inherently, the data that the clients gather cannot be guaranteed to be IID. But even in this case, the importance of collaborative learning is essential, as the number of training samples per user is not sufficient to train individual ML models in isolation. Therefore, a new direction in FL is personalized FL, which elaborates on a different concept from conventional FL. Specifically, in personalized FL, instead of training one global model, which should be general enough to capture all clients' behavior, it makes more sense to train multiple models that represent different groups of clients (Ghosh et al., 2022). As an alternative to conventional FL, instead of having a whole different model for each group of clients, one can train personalized layers. The model is split into a base and a personalized part. The personalized part is different for each client and is trained exclusively, whereas the base part is a generalized part that is trained by multiple clients. This approach is very close to SL, and hence a future research direction would be to explore how SL could contribute to more personalized-driven training.

10.9 CONCLUSION

The importance of decentralized AI lies in the need to allow clients to be the owners of their data and guarantee that their data will remain private. On the other hand, developing ML applications that are based on such decentralized systems can be very challenging, as the devices participating in the system are computationally constrained, with dynamic availability, and geographically distributed. In this chapter, we highlighted the challenges IoT devices face when utilizing ML models and the motivation behind SI and SL with the assistance of resources placed at the cloud-edge continuum. Also, we discussed the issues IoT devices face when training ML models, particularly due to their limited computational capabilities, that motivate the need for distributed ML model training protocols that are executed between IoT devices and powerful servers on the cloud-edge continuum. Finally, we have dived more into the challenges of cross-device distributed ML and discussed data heterogeneity and privacy concerns. To sum up, SC for ML applications can be very beneficial and can be applied to many IoT-based applications, but it is still an emerging research area that needs to be further explored.

NOTES

1 https://www.nvidia.com/en-us/autonomous-machines/embedded-systems/jetson-xavier-series/.
2 https://www.raspberrypi.com/products/raspberry-pi-3-model-b-plus/.
3 GANs models are trained using unsupervised learning and are mainly used to generate synthetic data. A GAN model consists of two NNs: a generator and a discriminator. These two models are trained concurrently using adversarial training. In detail, the generator receives random noise as input and its task is to learn to generate data instances that approximate the distribution of the original (true) dataset by trying to minimize the number of times the discriminator identifies fake data instances. Whereas, the discriminator is a binary classifier trained on a mixture of real and fake data, trying to maximize accuracy in recognizing whether the input comes from the true dataset or the synthetic one.
4 In the paper, the demonstrated attack is conducted by the compute node, which is in charge of the last model part, i.e., $s = S$.

REFERENCES

Ads, O. S., Alfares, M. M. and Salem, A. M. (2021), Multi-limb split learning for tumor classification on vertically distributed data, in *2021 Tenth International Conference on Intelligent Computing and Information Systems (ICICIS)*, IEEE, Cairo, Egypt, pp. 88–92. https://doi.org/10.1109/ICICIS52592.2021.9694163.

Allaart, C. G., Keyser, B. Bal, H. and van Halteren, A. (2022), Vertical Split Learning-an exploration of predictive performance in medical and other use cases, in *2022 International Joint Conference on Neural Networks (IJCNN)*, IEEE, Padua, Italy, pp. 1–8. https://doi.org/10.1109/IJCNN55064.2022.9891964.

Ayad, A., Frei, M. and Schmeink, A. (2022), Efficient and private ecg classification on the edge using a modified split learning mechanism, in *2022 IEEE 10th International Conference on Healthcare Informatics (ICHI)*, IEEE, Rochester, MN, pp. 1–6. https://doi.org/10.1109/ICHI54592.2022.00048.

Ayad, A., Renner, M. and Schmein, A. (2021), Improving the communication and computation efficiency of split learning for IoT applications, in *2021 IEEE Global Communications Conference (GLOBECOM)*, IEEE, Madrid, Spain, pp. 1–6. https://doi.org/10.1109/GLOBECOM46510.2021.9685493.

Buciluˇa, C., Caruana, R. and Niculescu-Mizil, A. (2006), Model compression, in *Proceedings of the 12th ACM SIGKDD International Conference on Knowledge Discovery and Data Mining*, Association for Computing Machinery (ACM), Philadelphia, PA, pp. 535–541. https://doi.org/10.1145/1150402.1150464.

Cai, Y. and Wei, T. (2022), Efficient split learning with non-iid data, in *The 23rd IEEE International Conference on Mobile Data Management (MDM)*, IEEE, Paphos, Cyprus, pp. 128–136. https://doi.org/10.1109/MDM55031.2022.00039.

Ceballos, I., Sharma, V., Mugica, E., Singh, A., Roman, A., Vepakomma, P. and Raskar, R. (2020), Splitnn-driven vertical partitioning, in arXiv preprint arXiv:2008.04137. https://doi.org/10.48550/arXiv.2008.04137.

Chen, X., Li, J. and Chakrabarti, C. (2021), Communication and computation reduction for split learning using asynchronous training, in *2021 IEEE Workshop on Signal Processing Systems (SiPS)*, IEEE, Coimbra, Portugal, pp. 76–81. https://doi.org/10.1109/SiPS52927.2021.00022.

Cheng, Y., Zhang, L. and Li, A. (2023), GFL: Federated learning on non-IID data via privacy-preserving synthetic data, in *2023 IEEE International Conference on Pervasive Computing and Communications (PerCom)*, IEEE, Atlanta, GA, pp. 61–70. https://doi.org/10.1109/PERCOM56429.2023.10099110.

Chotard, J., Dufour Sans, E., Gay, R., Phan, D. H. and Pointcheval, D. (2018), Decentralized multi-client functional encryption for inner product, in *Advances in Cryptology–ASIACRYPT 2018: 24th International Conference on the Theory and Application of Cryptology and Information Security*, Springer, Brisbane, QLD, Australia, pp. 703–732. https://doi.org/10.1007/978-3-030-03329-3_24.

Costan, V. and Devadas, S. (2016), Intel SGX explained, in Cryptology ePrint Archive. https://eprint.iacr.org/2016/086.

Erdoğan, E., Küpçü, A. and Çiçek, A. E. (2022), Unsplit: Data-oblivious model inversion, model stealing, and label inference attacks against split learning, in *Proceedings of the 21st Workshop on Privacy in the Electronic Society*, Association for Computing Machinery (ACM), Los Angeles, CA, pp. 115–124. https://doi.org/10.1145/3559613.3563201.

Erdogan, E., Teksen, U., Celiktenyildiz, M. S., Kupcu, A. and Cicek, A. E. (2023), SplitOut: Out-of-the-box training-hijacking detection in split learning via outlier detection, in arXiv preprint arXiv:2302.08618. https://doi.org/10.48550/arXiv.2302.08618.

Eshratifar, A. E., Abrishami, M. S. and Pedram, M. (2019), JointDNN: An efficient training and inference engine for intelligent mobile cloud computing services, in *IEEE Transactions on Mobile Computing* 20(2), pp. 565–576. https://doi.org/10.1109/TMC.2019.2947893.

Eshratifar, A. E., Esmaili, A. and Pedram, M. (2019), Bottlenet: A deep learning architecture for intelligent mobile cloud computing services, in *2019 IEEE/ACM International Symposium on Low Power Electronics and Design (ISLPED)*, IEEE, Lausanne, Switzerland, pp. 1–6. https://doi.org/10.1109/ISLPED.2019.8824955.

Fallah, A., Mokhtari, A. and Ozdaglar, A. (2020), Personalized federated learning with theoretical guarantees: A model-agnostic meta-learning approach, in *Advances in Neural Information Processing Systems* 33, pp. 3557–3568. https://proceedings.neurips.cc/paper_files/paper/2020/file/24389bfe4fe2eba8bf9aa9203a44cdad-Paper.pdf.

Feng, X., Luo, C., Chen, J., Huang, Y., Zhang, J., Xu, W., Li, J. and Leung, V. C. (2023), IoTSL: Towards efficient distributed learning for resource-constrained Internet of Things, in *IEEE Internet of Things Journal* 10(11), pp. 9892–9905. https://doi.org/10.1109/JIOT.2023.3235765.

Gao, X. and Zhang, L. (2023), {PCAT}: Functionality and data stealing from split learning by {Pseudo-Client} attack, in *32nd USENIX Security Symposium (USENIX Security 23)*, USENIX Association, USA, pp. 5271–5288. https://www.usenix.org/conference/usenixsecurity23/presentation/gao.

Gao, Y., Kim, M., Abuadbba, S., Kim, Y., Thapa, C., Kim, K., Camtepe, S. A., Kim, H. and Nepal, S. (2020), End-to-end evaluation of federated learning and split learning for internet of things, in arXiv preprint arXiv:2003.13376. https://doi.org/10.48550/arXiv.2003.13376.

Gao, Y., Kim, M., Thapa, C., Abuadbba, A., Zhang, Z., Camtepe, S., Kim, H. and Nepal, S. (2021), Evaluation and optimization of distributed machine learning techniques for internet of things, in *IEEE Transactions on Computers* 71(10), pp. 2538–2552. https://doi.org/10.1109/TC.2021.3135752.

Ghosh, A., Chung, J., Yin, D. and Ramchandran, K. (2022), An efficient framework for clustered federated learning, in *IEEE Transactions on Information Theory* 68(12), pp. 8076–8091. https://doi.org/10.1109/TIT.2022.3192506.

Goldreich, O., Micali, S. and Wigderson, A. (2019), How to play any mental game, or a completeness theorem for protocols with honest majority, in *Providing Sound Foundations for Cryptography: On the Work of Shafi*, Association for Computing Machinery, New York, NY, pp. 307–328. https://doi.org/10.1145/3335741.3335755.

Han, D. J., Bhatti, H. I., Lee, J. and Moon, J. (2021), Accelerating federated learning with split learning on locally generated losses, in *ICML 2021 Workshop on Federated Learning for User Privacy and Data Confidentiality*, ICML Board. https://fl-icml.github.io/2021/papers/FL-ICML21_paper_6.pdf.

Han, D. J., Kim, D. Y., Choi, M., Brinton, C. G. and Moon, J. (2023), Splitgp: Achieving both generalization and personalization in federated learning, in *IEEE INFOCOM 2023-IEEE Conference on Computer Communications*, IEEE, New York City, NY, pp. 1–10. https://doi.org/10.1109/INFOCOM53939.2023.10229027.

He, C., Annavaram, M. and Avestimehr, S. (2020), Group knowledge transfer: Federated learning of large CNNs at the edge, in *Advances in Neural Information Processing Systems* 33, pp. 14068–14080.

He, K., Zhang, X., Ren, S. and Sun, J. (2016), Deep residual learning for image recognition, in *Proceedings of the IEEE Conference on Computer Vision and Pattern Recognition*, pp. 770–778. IEEE. USA. https://doi.org/10.1109/CVPR.2016.90

Kang, Y., Hauswald, J., Gao, C., Rovinski, A., Mudgeason, T., Mars, J. and Tang, L. (2017), Neurosurgeon: Collaborative intelligence between the cloud and mobile edge, in *ACM SIGARCH Computer Architecture News* 45(1), pp. 615–629. https://doi.org/10.1145/3093337.3037698.

Kim, M., DeRieux, A. and Saad, W. (2023), A bargaining game for personalized, energy efficient split learning over wireless networks, in *2023 IEEE Wireless Communications and Networking Conference (WCNC)*, IEEE, Glasgow, pp. 1–6. https://doi.org/10.1109/WCNC55385.2023.10118601.

Koda, Y., Park, J., Bennis, M., Yamamoto, K., Nishio, T. and Morikura, M. (2019), One pixel image and RF signal based split learning for mmWave received power prediction, in *Proceedings of the 15th International Conference on Emerging Networking EXperiments and Technologies*, Association for Computing Machinery, Orlando, FL, pp. 54–56. https://doi.org/10.1145/3360468.3368176.

Koutsos, V., Papadopoulos, D., Chatzopoulos, D., Tarkoma, S. and Hui, P. (2021), Agora: A privacy-aware data marketplace, in *IEEE Transactions on Dependable and Secure Computing* 19(6), pp. 3728–3740. https://doi.org/10.1109/TDSC.2021.3105099.

Laskaridis, S., Kouris, A. and Lane, N. D. (2021), Adaptive inference through early-exit networks: Design, challenges and directions, in *Proceedings of the 5th International Workshop on Embedded and Mobile Deep Learning*, Association for Computing Machinery, Virtual, WI, pp. 1–6. https://doi.org/10.1145/3469116.3470012.

Li, O., Sun, J., Yang, X., Gao, W., Zhang, H., Xie, J., Smith, V. and Wang, C. (2021), Label leakage and protection in two-party split learning, in arXiv preprint arXiv:2102.08504. https://doi.org/10.48550/arXiv.2102.08504.

Li, X., Huang, K., Yang, W., Wang, S. and Zhang, Z. (2019), On the convergence of fedavg on non-iid data, in arXiv preprint arXiv:1907.02189. https://doi.org/10.48550/arXiv.1907.02189.

Liao, Y., Xu, Y., Xu, H., Wang, L., Yao, Z. and Qiao, C. (2023), MergeSFL: Split federated learning with feature merging and batch size regulation, in arXiv preprint arXiv:2311.13348. https://doi.org/10.48550/arXiv.2311.13348.

Liu, G., Dai, F., Xu, F., Fu, X., Dou, W., Kumar, N. and Bilal, M. (2023), An adaptive DNN inference acceleration framework with end–edge–cloud collaborative computing, in *Future Generation Computer Systems* 140, pp. 422–435. https://doi.org/10.1016/j.future.2022.10.033.

Liu, J. and Lyu, X. (2023), Distance-based online label inference attacks against split learning, in *ICASSP* 2023-2023 *IEEE International Conference on Acoustics, Speech and Signal Processing (ICASSP)*, IEEE, Rhodes Island, Greece, pp. 1–5. https://doi.org/10.1109/ICASSP49357.2023.10096955.

Liu, N., Ma, X., Xu, Z., Wang, Y., Tang, J. and Ye, J. (2020), Autocompress: An automatic DNN structured pruning framework for ultra-high compression rates, in *Proceedings of the AAAI Conference on Artificial Intelligence* 34(4), pp. 4876–4883. https://api.semanticscholar.org/CorpusID:202558809.

Liu, X., Deng, Y. and Mahmoodi, T. (2022a), A novel hybrid split and federated learning architecture in wireless UAV networks, in *ICC 2022-IEEE International Conference on Communications*, IEEE, Seoul, Republic of Korea, pp. 1–6. https://doi.org/10.1109/ICC45855.2022.9838867.

Liu, X., Deng, Y. and Mahmoodi, T. (2022b), Energy efficient user scheduling for hybrid split and federated learning in wireless uav networks, in *ICC 2022-IEEE International Conference on Communications*, IEEE, Seoul, Republic of Korea, pp. 1–6. https://doi.org/10.1109/ICC45855.2022.9882277.

Madaan, H., Kulkarni, M. G. V. and Pant, A. (2022), Vulnerability due to training order in split learning, in *ICT Systems and Sustainability: Proceedings of ICT4SD 2021*, Volume 1, Springer, Singapore, pp. 103–112. https://doi.org/10.1007/978-981-16-5987-4_11.

McMahan, B., Moore, E., Ramage, D., Hampson, S. and Arcas, B. A. (2017), Communication-efficient learning of deep networks from decentralized data, in *Artificial intelligence and statistics*, PMLR, pp. 1273–1282. https://proceedings.mlr.press/v54/mcmahan17a?ref=https://githubhelp.com.

Melchor, C. A., Castagnos, G. and Gaborit, P. (2008), Lattice-based homomorphic encryption of vector spaces, in *2008 IEEE International Symposium on Information Theory*, IEEE, Toronto, ON, Canada, pp. 1858–1862. https://doi.org/10.1109/ISIT.2008.4595310.

Mu, Y. and Shen C. (2023), Communication and storage efficient federated split learning, in *ICC 2023-IEEE International Conference on Communications*, IEEE, Rome, Italy, pp. 2976–2981. https://doi.org/10.1109/ICC45041.2023.10278891.

Ni, W., Ao, H., Tian, H., Eldar, Y. C. and Niyato, D. (2024), FedSL: Federated split learning for collaborative healthcare analytics on resource-constrained wearable IoMT devices, in *IEEE Internet of Things Journal* 11(10), pp. 18934–18935. https://doi.org/10.1109/JIOT.2024.3370985.

Nøkland, A. and Eidnes, L. H. (2019), Training neural networks with local error signals, in *International Conference on Machine Learning*, PMLR, pp. 4839–4850.

Oh, S., Park, J., Vepakomma, P., Baek, S., Raskar, R., Bennis, M. and Kim, S. L. (2022), Locfedmix-sl: Localize, federate, and mix for improved scalability, convergence, and latency in split learning, in *Proceedings of the ACM Web Conference 2022*, Association for Computing Machinery, Virtual Event, Lyon, pp. 3347–3357. https://doi.org/10.1145/3485447.3512153.

Otoum, S., Guizani, N. and Mouftah, H. (2022), On the feasibility of split learning, transfer learning and federated learning for preserving security in ITS systems, in *IEEE Transactions on Intelligent Transportation Systems* 24(7), pp. 7462–7470 https://doi.org/10.1109/TITS.2022.3159092.

Palanisamy, K., Khimani, V., Moti, M. H. and Chatzopoulos, D. (2021), Spliteasy: A practical approach for training ml models on mobile devices, in *Proceedings of the 22nd International Workshop on Mobile Computing Systems and Applications*, Association for Computing Machinery, Virtual, pp. 37–43. https://doi.org/10.1145/3446382.3448362.

Pappas, C., Papadopoulos, D., Chatzopoulos, D., Panagou, E., Lalis, S. and Vavalis, M. (2022), Towards efficient decentralized federated learning, in *IEEE 42nd International Conference on Distributed Computing Systems Workshops (ICDCSW)*, IEEE, Bologna, Italy, pp. 79–85, https://doi.org/10.1109/ICDCSW56584.2022.00023.

Pasquini, D., Ateniese, G. and Bernaschi, M. 2020. Unleashing the tiger: Inference attacks on split learning, in *Proceedings of the 2021 ACM SIGSAC Conference on Computer and Communications Security*, Association for Computing Machinery, Virtual Event, Republic of Korea, pp. 2113–2129. https://doi.org/10.1145/3460120.3485259.

Pereteanu, G. L., Alansary, A. and Passerat-Palmbach, J. (2022), Split HE: Fast secure inference combining split learning and homomorphic encryption, in arXiv preprint arXiv:2202.13351. https://doi.org/10.48550/arXiv.2202.13351.

Ramanan, P., Nakayama, K. and Sharma, R. (2020), BAFFLE: Blockchain based aggregator free federated learning, in *IEEE International Conference on Blockchain (Blockchain)*, IEEE, Rhodes, Greece, pp. 72–81, https://doi.org/10.1109/Blockchain50366.2020.00017.

Samikwa, E., Maio, A. D. and Braun, T. (2022), Ares: Adaptive resource-aware split learning for internet of things, in *Computer Networks* 218, p. 109380. https://doi.org/10.1016/j.comnet.2022.109380.

Soykan, E. U., Karacay, L., Karakoc, F. and Tomur, E. (2022), A survey and guideline on privacy enhancing technologies for collaborative machine learning, in *IEEE Access* 10, pp. 97495–97519. https://doi.org/10.1109/ACCESS.2022.3204037.

Thapa, C., Mahawaga Arachchige, P. C., Camtepe, S. and Sun, L. (2022), Splitfed: When federated learning meets split learning, in *Proceedings of the AAAI Conference on Artificial Intelligence* 36(8), pp. 8485–8493. https://doi.org/10.1609/aaai.v36i8.20825.

Tirana, J., Lalis, S. and Chatzopoulos, D. (2024), MP-SL: Mul-tihop parallel split learning, in arXiv preprint arXiv:2402.00208.

Tirana, J., Tsigkari, D., Iosifidis, G. and Chatzopoulos, D. (2024), Workflow optimization for parallel split learning, in *IEEE INFOCOM 2024-IEEE Conference on Computer Communications,* Vancouver, BC, pp. 1331–1340. IEEE. doi: 10.1109/INFOCOM52122.2024.10621348.

Vepakomma, P., Gupta, O., Swedish, T. and Raskar, R. (2018), Split learning for health: Distributed deep learning without sharing raw patient data, in arXiv preprint arXiv:1812.00564. https://doi.org/10.48550/arXiv.1812.00564.

Vepakomma, P., Singh, A., Gupta, O. and Raskar, R. (2020), NoPeek: Information leakage reduction to share activations in distributed deep learning, in *2020 International Conference on Data Mining Workshops (ICDMW)*, IEEE, Sorrento, Italy, pp. 933–942. https://doi.org/10.1109/ICDMW51313.2020.00134.

Wang, J., Liu, Q., Liang, H., Joshi, G. and Poor, H. V. (2020), Tackling the objective inconsistency problem in heterogeneous federated optimization, in *Advances in Neural Information Processing Systems* 33, pp. 7611–7623.

Wang, Y., Li, Y., Qin, D. and Poor, H. V. (2024), Introducing edge intelligence to smart meters via Federated Split Learning. https://doi.org/10.1038/s41467-024-53352-9.

Wang, Z., Xu, H. Xu, Y., Jiang, Z. and Liu, J. (2023), Coopfl: Accelerating federated learning with dnn partitioning and offloading in heterogeneous edge computing, in *Computer Networks* 220, p. 109490. https://doi.org/10.1016/j.comnet.2022.109490.

Wu, D., Ullah, R. Harvey, P., Kilpatrick, P., Spence, I. and Varghese, B. (2022), Fedadapt: Adaptive offloading for IoT devices in federated learning, in *IEEE Internet of Things Journal* 9(21), pp. 20889–20901. https://doi.org/10.1109/JIOT.2022.3176469.

Wu, M., Cheng, G., Li, P., Yu, R., Wu, Y., Pan, M. and Lu, R. (2023), Split learning with differential privacy for integrated terrestrial and non-terrestrial networks, in *IEEE Wireless Communications* 31(3), pp. 177–184. https://doi.org/10.1109/MWC.015.2200462.

Yoshida, N., Nishio, T., Morikura, M., Yamamoto, K. and Yonetani, R. (2020), Hybrid-FL for wireless networks: Cooperative learning mechanism using non-IID data, in *ICC 2020-2020 IEEE International Conference on Communications (ICC)*, IEEE, Dublin, Ireland, pp. 1–7. https://doi.org/10.1109/ICC40277.2020.9149323.

Zhao, Y., Li, M., Lai, L., Suda, N., Civin, D. and Chandra, V. (2018), Federated learning with non-iid data, in arXiv preprint arXiv:1806.00582. https://doi.org/10.48550/arXiv.1806.00582.

Zhou, W., Qu, Z., Zhao, Y., Tang, B. and Ye, B. (2022), An efficient split learning framework for recurrent neural network in mobile edge environment, in *Proceedings of the Conference on Research in Adaptive and Convergent Systems*, Association for Computing Machinery, Virtual Event, pp. 131–138. https://doi.org/10.1145/3538641.3561500.

Zhou, Y., Moosavi-Dezfooli, S.-M., Cheung, N.-M. and Frossard, P. (2018), Adaptive quantization for deep neural network, in *Proceedings of the AAAI Conference on Artificial Intelligence* 32(1), 4596–4604. https://api.semanticscholar.org/CorpusID:19227870.

11 Smart Radio Environments
An Emerging Paradigm for the Internet of Things

Akaa Agbaeze Eteng

11.1 INTRODUCTION

The Internet of Things (IoT) has emerged as a critical ecosystem enabling the interconnection of objects and devices for the collection and exchange of data and the automation of interventions in the physical world. Arguably, all aspects of modern society have been impacted by the emergence of IoT technology, ranging from business to industry, healthcare, education, and governance (Hassan, 2018). Estimates put the global number of connected IoT devices in the range of tens of billions, all connected through various network access technologies.

The various IoT application domains can be roughly categorized into four broad segments, namely massive, critical, broadband, and industrial IoT (IIoT). While massive IoT refers to those applications characterized by a massive deployment of low-power network nodes (e.g., environment-monitoring sensors), critical IoT applications are characterized by fewer nodes (e.g., autonomous vehicles) and real-time transmission of much higher data volumes. Broadband IoT is realized by the integration of broadband Internet connectivity in IoT devices (e.g., advanced wearables), while IIoT combines aspects of massive, broadband, and critical IoT with AI-driven automation in industrial settings.

Wireless connectivity is integral to facilitating IoT networks in the various service domains, and in many cases, it is the only way to establish connectivity between devices. The quality of the radio environment is therefore a critical consideration for establishing and maintaining connectivity. Wireless access technologies have traditionally been designed from the perspective of an adversarial and uncontrollable radio environment. Conventionally, signaling and information encoding schemes, as well as receiver and transmitter configurations, are optimized to overcome constraints presented by the wireless environment.

An alternative approach, however, has recently begun to gain significant traction in the wireless community. The idea is to deploy programmable artifacts that can be used to alter the electromagnetic (EM) state of the environment in the operational area. Thus, the radio environment is converted into a smart space that can be manipulated on demand to support wireless connectivity. This new direction is a consequence of the convergence of research in multiple disciplines, providing novel hardware and software enablers for realizing smart radio environments.

The rest of this chapter provides an overview of the technology for realizing smart radio environments. It also describes some application scenarios that are relevant to next-generation IoT. Finally, emerging trends and directions in the realization of smart radio environments are examined.

11.2 TECHNOLOGY OVERVIEW

The vision for implementing smart radio environments is based on strategically installed smart surfaces, which interact with ambient radio waves in such a manner as to positively alter the wireless

 DOI: 10.1201/9781003506638-14

environment. The responses of these smart surfaces can be dynamically programmed so as to optimally condition the propagation of transmitted radio waves to meet high-level wireless network objectives.

Various nomenclatures have been used in literature to describe these smart surfaces. Usually, these surfaces are referred to as large intelligent surfaces (LIS) when they are implemented as large planar antenna arrays, where each array element is connected to a dedicated radio frequency (RF) amplification circuitry. If the smart surface is fully passive and re-engineers the radio environment by reflecting impinging radio waves, the surface is often called an intelligent reflecting surface (IRS). In contrast, a reconfigurable intelligent surface (RIS) usually refers to a smart surface that is able to manipulate the radio environment in more ways than just reflection and whose functionality can be reconfigured by higher-level network intelligence during its operation. The RIS designation encompasses a wider range of functionality and will be used in the rest of the chapter as the general designation for smart surfaces used in manipulating EM waves.

11.2.1 System Architecture

Smart radio environments are physically actuated by hardware in direct contact with the radio environment, as illustrated in Figure 11.1. The facility for intelligent interaction with incident EM waves is provided by software-driven coordination of this hardware to enable the customization of wireless propagation to serve operational objectives. Hardware and software components that enable smart radio environment functionalities are discussed in the subsequent subsections.

11.2.2 Hardware Components

The fundamental hardware enabler of smart radio environments is the RIS. Typically, an RIS is composed of an EM surface and various components that enable it to function as a reconfigurable and intelligent structure for modifying EM propagation in a wireless network environment (Di Renzo et al., 2020b). RIS implementations are generally designed to be nearly passive, with minimal energy requirements sustained by non-grid sources. The hardware components associated with an RIS are shown in Figure 11.2 and discussed in the following subsections.

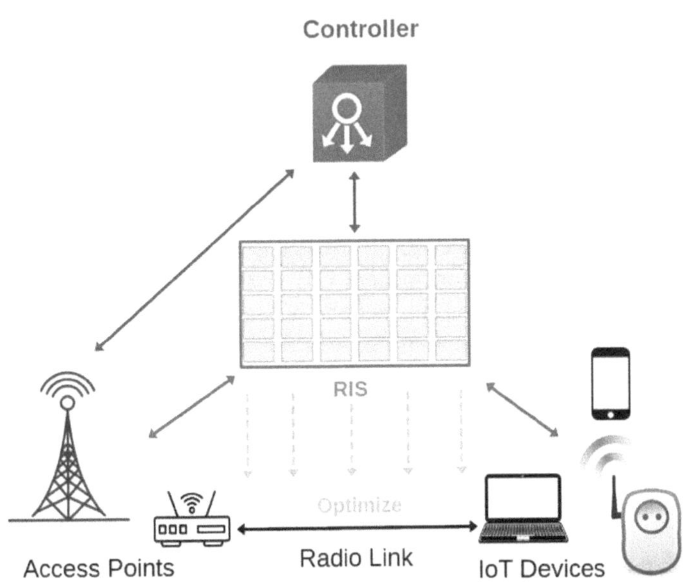

FIGURE 11.1 Smart Radio Environments

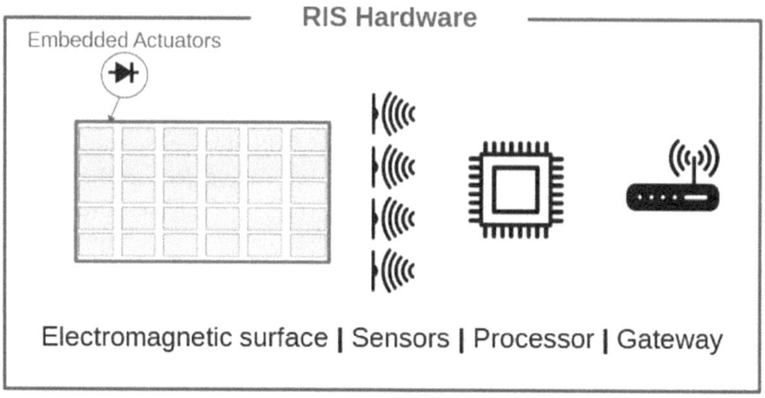

FIGURE 11.2 RIS Hardware Components

11.2.2.1 The EM Surface

The core component of an RIS is an EM surface, whose thickness is considerably smaller than its transverse dimensions. Usually, the physical dimensions of this surface are characterized in terms of the fundamental wavelength of the interacting radio waves, such that while the surface may be many wavelengths in transverse size, its thickness is a fraction of a wavelength. Consequently, it is often convenient to electromagnetically characterize the EM surface as a zero-thickness surface.

The EM surface is a composite pattern of conductive or dielectric patches printed on a thin substrate, usually backed by another conducting sheet, which constitutes a ground plane. Depending on their sizes relative to the wavelength of the impinging wave, the patterned patches can either be viewed as a two-dimensional array of planar antennas or as a planar aggregate of meta-atoms, collectively forming an EM metasurface. Antenna array implementations of RIS, also known as reflectarrays, are typified by half-wavelength array element sizes and spacing. On the other hand, EM metasurface characterization is used when the sizes and spacing between the meta-atoms are less than one-tenth of the wavelength of the interacting radio wave (Glybovski et al., 2016).

Reflectarrays are the structurally simpler but functionally less flexible of both alternatives for implementing the RIS EM sheet. The adopted half-wavelength spacing between the array elements ensures negligible mutual coupling between them. This simplifies the design of the overall EM response of the total surface to an incident radio wave. The individual array elements act as reflectors, which alter the phase of the impinging wave. By carefully coordinating the resulting phase shifts from individual elements, the reflectarray can shape an incident radio wave. This can be done by superposing these phase shifts to realize an overall wavefront response.

On the other hand, EM metasurfaces are two-dimensional equivalents of EM metamaterials, which are structures with artificially engineered EM properties. The overall response of an EM metasurface depends both on localized interactions between individual meta-atoms and the incident EM wave and on the mutual coupling between the closely spaced meta-atoms. Other kinds of metasurfaces exist, such as mechanical metasurfaces and acoustic metasurfaces. However, EM metasurfaces function within the domain of radio waves and will subsequently be referred to simply as metasurfaces without any loss of meaning.

Various conductive patterns have been developed as implementations of meta-atoms. This imbues the metasurface with non-naturally occurring EM properties, such as simultaneously negative values of permittivity and permeability. Some of the more widely used meta-atom designs include the split-ring resonator, the complementary split-ring resonator, and the electric-field coupled resonator, in addition to several other non-canonical geometries engineered to provide specific responses.

Figure 11.3 illustrates some of the EM wave manipulations that can be provided by metasurfaces within specified frequency ranges (Di Renzo et al., 2020b):

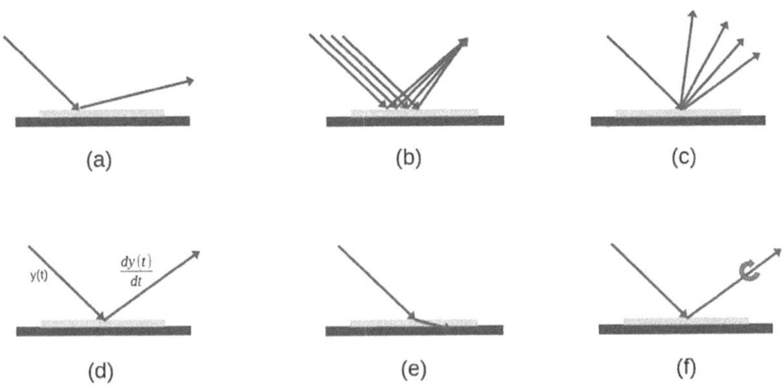

FIGURE 11.3 Examples of EM Wave Manipulations: (a) Anomalous Reflection; (b) Focusing; (c) Collimation; (d) Signal Processing; (e) Perfect Absorption; (f) Polarization Conversion

- *Anomalous Reflection and Refraction*: where an impinging radio wave is reflected at a different angle from its angle of incidence or is transmitted by the metasurface at an angle of opposite sign.
- *Focusing and Collimation:* where an EM wave is reflected either with its energy directed toward a specific location in space or dispersed over a region of space.
- *Signal Processing:* in which the metasurface acts as a spatial analog processor such that the reflected or refracted wave is a mathematically transformed version of the incident wave.
- *Perfect Absorption:* realized by the surface suppressing both the reflection and refraction of an incident radio wave, which also provides a basis for harvesting energy from the radio wave.
- *Polarization Conversion:* where the spatial orientations of the field components of the reflected or refracted EM wave are altered, often also leading to anomalous reflective or refractive responses.

11.2.2.2 Sensing and Actuation

Early implementations of metasurfaces were designed to provide static EM responses to incident waves with specific characteristics. However, practical control of EM waves requires that the maintenance be reconfigurable, either so that its dynamic response can enable maintaining a specific functionality irrespective of variations in the EM environment or so that different functionalities can be provided as requested for specific services. It is therefore important that the EM surface be equipped with a means to sense the EM environment. This is achieved by equipping the EM surface with low-power sensors to receive pilot signals from network access points or base stations, from which the channel state information (CSI) can be extracted. To maintain the nearly passive nature of the RIS, energy harvesting sensors, which are powered by harvested ambient EM energy, are typically proposed for sensing implementations (Di Renzo et al., 2020b).

Global actuation, in which all meta-atoms are controlled in concert, was the first proposed approach toward reconfiguring metasurface platforms (Tsilipakos et al., 2020). This is facilitated through inclusions into the metasurface structure, whose EM characteristics are altered by applying appropriate stimuli, leading to a modification of the overall response to an impinging wave. Various forms of stimuli have been suggested for global actuation of metasurfaces, depending on the frequencies of interest in the EM environment, such as thermal, electrical, magnetic, and optical stimuli.

Electrical actuation is the more widely used tuning mechanism within microwave frequencies. It requires the use of electronic components whose input impedances can be altered using DC bias voltages. The components are embedded in each meta-atom so that when they are collectively stimulated by an appropriate bias voltage, the surface impedance of the metasurface is altered, thereby tuning it to any of a variety of pre-set operational modes. Typical examples of such electronic tuning components include variable resistors, varactor diodes, and PIN diodes (Rana et al., 2023).

In contrast with global actuation, being able to tune individual meta-atoms provides an added dimension of flexibility, as it enables switching among multiple functionalities when a metasurface is in operation. In addition, a local actuation approach enables the implementation of additional functionalities, especially those that rely on dynamic control of the spatial distribution of the surface impedance over a metasurface. This enables greater levels of wavefront control that can be employed for controlled anomalous wave reflection and diffraction, which can be used for practical beam steering (Vasic et al., 2020). Local actuation at the meta-atom level can be achieved using the same tuning mechanisms used to realize global tuning of the whole structure. However, tuning implementations need to adopt specific measures to achieve a desired EM effect on a target group of meta-atoms without interfering with the operation of others. For this reason, local actuation of meta-atoms within microwave frequencies is often realized using lumped electronic components.

The nature of control exerted over a metasurface, whether digital or analog, depends on the nature of the tuning components and materials incorporated into the structure. Components that offer a continuous variation of response to the external tuning stimulus enable analog reconfiguration of the metasurface. On the other hand, digital tuning requires the use of tuning components with discrete states. For example, electronic tuning using PIN diodes (R. Chen et al., 2022) provides two discrete states, ON and OFF, which can be mapped to bits. Metasurfaces with digital actuation mechanisms are also known as coding metasurfaces and are the basis for software-driven implementations.

Digital control can also be used to approximate the continuous control offered by analog schemes. Such quasi-analog control can be realized by grouping combinations of discrete responses from digital tuning elements in a manner that creates more intermediate states between those offered by single elements. For instance, by grouping N binary-state meta-atoms into a sub-wavelength super-cell, the metasurface would now comprise super-cell building blocks, each capable of offering 2^N different responses.

11.2.2.3 Local Processing Hardware

Coordinated control of the actuation and sensing elements on an RIS platform is achieved through local processing hardware. The most basic processing involves mapping EM states, such as impedances and phases, to corresponding actuator states. By integrating processors into metasurfaces, programmed logic can be used for the dynamic reconfiguration of EM functionalities while the surface is in operation. Programmable control of the metasurface is often realized using a single embedded chip, with implementations ranging from field-programmable gate arrays (Zhang et al., 2020) and microcontrollers (Chen et al., 2018) to application-specific integrated circuits (Petrou et al., 2022). The programmable controller, together with the network of connected actuators, can also be viewed as forming the control network of the RIS.

11.2.2.4 Communications Hardware

Additional capabilities required for real-time reconfiguration of the wireless radio environment include the ability to communicate with other devices in the wireless network. It is therefore practical that an RIS be equipped with communications hardware that enables it to collaborate with other network nodes in order to satisfy high-level operational objectives. This also introduces opportunities for the remote coordination of installed RIS, either through centralized or distributed network coordination methodologies. It also implies software-defined RIS functionalities, enabling the provision of various RIS-driven wireless network functionalities as a wireless service (Strinati et al., 2021). Such software-defined metasurfaces (Tsilipakos et al., 2020) are a logical extension of programmable metasurfaces and constitute the basis for the practical realization of a smart and programmable radio environment.

Basically, networking capabilities require that an RIS be equipped with a gateway, which enables communication between it and the external wireless network. Network protocols can then be used to address and configure the RIS as a node within the wireless network while abstracting the EM details of the wireless channel. The gateway also enables the formation of inter-RIS networks, which leverage machine-type communications to facilitate the cooperative use of multiple RIS structures within a wireless network scenario.

11.2.3 Software Enablers

Although there has been broad interest in realizing hardware platforms for smart radio environments, there has been comparatively less activity on the software enablers for integrating and orchestrating these devices within a wireless network. A notable exception is the European-funded VisorSurf project,[1] which has developed a full suite of software components to accompany its hardware implementation of smart and programmable radio environments. The documented vision of a programmable radio environment envisions two categories of software enablers, namely the middleware and the application programming interface (API) (Liaskos et al., 2018, 2019, 2022; Tsioliaridou et al., 2020). This framework ensures that the underlying EM interactions at the metasurfaces are transparent to the software for orchestrating a set of networked RISs. Central to this arrangement is the alignment of the end-to-end system architecture with the software-defined network (SDN) paradigm, which, among other benefits, enables compatibility with existing wireless network infrastructure and IoT integration (Liaskos et al., 2022). Figure 11.4 illustrates the system model.

The roles of these software enablers, as discussed in Tsioliaridou et al. (2020), are summarized here. An SDN controller contains the server to centrally monitor and configure a group of networked RISs. The middleware is linked to a unified southbound API for all IoT devices, which is also responsible for the orchestration of deployed RIS platforms, abstracted as parameterized routing nodes among other network devices. The southbound interface is the bridge through which API callbacks are translated into hardware directives for the actuation of deployed RISs. The northbound API is responsible for the provision of high-level operational objectives, such as energy efficiency, coverage extension, security, or localization, as a service. An API callback is a functionality request either initiated by an external device (e.g., PC, smartphone, etc.) through a graphic user interface or internally by the central network controller in an attempt to maintain operational objectives in reaction to unexpected changes in the radio environment.

The middleware has two components, the first of which is offline and encapsulates the calibration and evaluation of the accompanying RIS hardware. It provides a database containing all configurations that can be actuated on the RIS, as specified by API callbacks. The second part, the online component, handles the real-time actuation of RIS hardware by employing the configuration database as a codebook. For each specific API callback, the online component of the middleware retrieves the corresponding RIS configurations from the codebook. Complex configurations without direct codebook entries are actualized using an interleaving algorithm to combine settings in order to provide the required hardware actuation.

11.2.4 Network Deployment Modes

IoT networks can broadly be categorized into short-range, wide-area, cellular, and non-terrestrial networks (Vaezi et al., 2021). Short-range implementations are established using such wireless technologies as Wi-Fi, RFID, Bluetooth, and Zigbee over relatively short distances. Wide-area IoT networks, on the other hand, employ such technologies as LoRa and Sigfox to facilitate IoT connections between devices that are more physically dispersed. Cellular IoT leverages the wide coverage provided by mobile networks, mainly fourth-generation (4G) and fifth-generation (5G) networks, and is based on cellular network technologies, such as narrowband IoT (NB-IoT) and 4G Long-Term Evolution (LTE). Non-terrestrial networks are a relatively new class of IoT networks

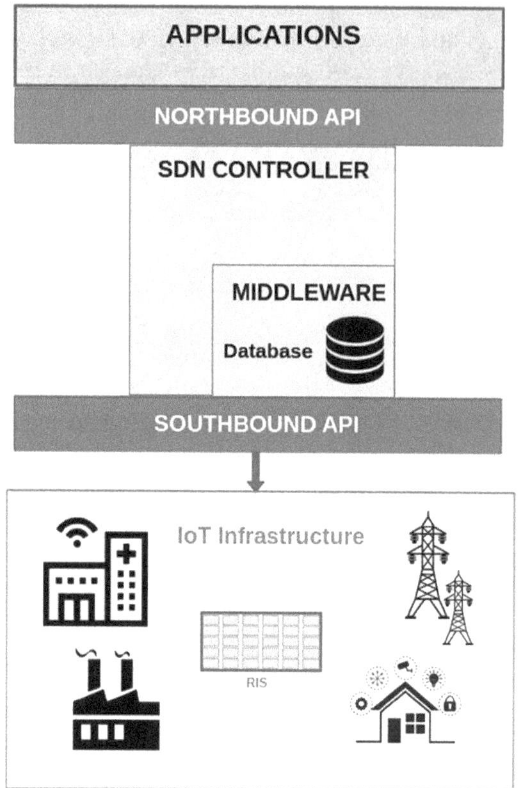

FIGURE 11.4 SDN-Based Smart Radio Environment Model with Software Enablers

based on unmanned aerial vehicles (UAVs) and satellite networks, which provide an extended spatial dimension to network coverage.

Three broad categories of RIS deployments in the highlighted network classes can be identified (Di Renzo et al., 2020b). The first, a centralized implementation, requires a central controller to orchestrate a network of deployed RISs within the operational area. The earlier discussed software framework for smart radio environments supports a centralized network architecture. In this architecture, RISs are abstracted as EM relays and centrally coordinated through the southbound interface of an SDN controller. In contrast, standalone implementations grant autonomy to deployed RISs. In this case, RISs are fully powered platforms that are able to sense channel conditions and apply the optimal functionalities required by service objectives. It may also be possible to have a single autonomous RIS shared by multiple networks (Alexandropoulos et al., 2023). Although standalone deployments can be employed in unlicensed spectra to provide localized coverage, optimal scaling to accommodate multi-user access by heterogeneous devices still remains an open issue. A third deployment option is a hybrid strategy that deploys both autonomous and centrally controlled RIS platforms within an operational environment.

Given the growing importance of cellular IoT deployments, the practical integration of RIS platforms into already existing cellular networks is receiving special consideration. Like other radio infrastructure, the coverage area of an RIS platform is a crucial consideration when planning its deployment over an area serviced by licensed spectrum. However, native support for the IoT, which is provided in 5G networks and is expected to deepen in 6G, could be leveraged to provide a framework for deploying RISs. For instance, in a cellular network scenario with RIS platforms deployed to influence the quality of connectivity between a gNodeB (the next-generation equivalent of a base station) and user equipment, novel radio interfaces and radio channels will be required to provide an avenue for orchestrating the smart radio environment (Alexandropoulos et al., 2023).

The cellular network integration scenario most often envisaged in available literature is one where RISs are physically deployed close to distributed users. While this strategy is effective in providing alternative line-of-sight (LOS) links to improve radio coverage, a central coordination of such deployments is expected to result in a comparatively large signal overhead. Infrastructure-sided deployment, for instance, close to a base station, would expand the influence of the RIS to cover more network nodes, but at the cost of less effective radio coverage within critical zones. In an attempt to trade off these two extremes, recent studies have begun to examine hybrid strategies, seeking to leverage the gains provided by both user-side and infrastructure-side RIS deployments (You et al., 2022).

11.3 APPLICATION SCENARIOS IN NEXT-GENERATION IOT

Although short-range implementations presently account for a sizeable portion of IoT connections, cellular IoT connections are increasing at a significant rate.[2] This trend is expected to continue, given that research into crystallizing a vision for 6G has commenced in earnest. 6G is expected to be AI-driven, complemented by non-terrestrial networks, and IoT-dominant in the sense of device-to-device connections being the most prevalent type (Vaezi et al., 2021). Current trends in IoT development suggest a movement toward a more integrated and seamless global ecosystem, with a significantly higher density and variety of interlinked devices. The following sub-sections highlight some ways in which smart radio environments can enable the realization of this vision.

11.3.1 Coverage, Connectivity, and Reliability Enhancement

One of the key network enhancements that can be provided in a smart radio environment is the improvement of network coverage in areas where LOS wireless links between user devices and access points are not possible. This is especially useful in indoor scenarios where obstructions in the indoor space create dead zones without direct wireless links between network nodes. Moving toward millimeter-wave (mm-wave) frequencies in 5G and next-generation networks makes the condition even worse. These frequencies, notwithstanding the higher capacities they provide, are severely impacted in the absence of a direct signal path. Sporadic outages can also occur due to fading phenomena arising from multiple interfering signal paths reflected from obstacles in the environment. The reliability of such wireless links can be improved by a carefully planned deployment of RISs, which can be orchestrated to dynamically program alternative non-LOS signal paths (Khawaja et al., 2020). Depending on the operational environment, any of the previously mentioned deployment strategies, such as a single RIS deployment, multiple autonomous RISs, or multiple networked RISs, can be employed to this end.

A related application is one where RISs are used to improve connectivity in wireless environments that are highly susceptible to polarization mismatches between connected devices (Chen et al., 2021). While higher-complexity user devices, such as mobile phones, have the luxury of sophisticated antenna systems, low-cost IoT nodes often employ simple single-polarization antennas. A change in the physical orientation of these devices may misalign their antenna polarizations compared to access point antennas, leading to a significant degradation in link quality. For instance, this challenge occurs when low-cost nodes are attached to moving machinery. A strategically installed RIS platform can be programmed to re-orient the polarization of signals that reflect off its surface to match the device antennas within a serviced coverage area.

Within the ambit of enabling smart cities, there are a range of scenarios in which RIS platforms can prove beneficial for extending coverage and maintaining reliable connectivity. Indoor RIS deployments can aid in the facilitation of smart homes, factories, and hospitals. RIS platforms could also be deployed on the façade of large buildings to enable interfacing between indoor and outdoor network infrastructure. Large billboards also present an opportunity for outdoor RIS installations to improve connectivity for vehicle-to-vehicle (V2V) IoT interactions. An interesting scenario is one where RIS platforms are attached to UAVs as part of UAV-IoT deployments to increase their area of influence over the radio environment (Kisseleff et al., 2020).

11.3.2 Sensing, Localization, and Positioning

Large-scale deployments of RISs can also turn the wireless environment into a distributed sensing platform, where sensors embedded within the EM surfaces can be used to characterize and map the environment. Accurate contextual and location awareness is a key requirement for many IoT-enabled services that depend on the positioning and tracking of target nodes or devices within an operational area. The environmental mapping provided through such services is critical for the navigation of robots and other autonomous vehicles. It also finds use in services that employ remote actuation, as may be used in smart industry, healthcare, logistics management, and transportation.

Conventional techniques for localization often employ range-based measurements of some parameters of propagated wireless signals. Designated nodes within the network, such as base stations, serve as anchor points that provide a frame of reference for these measurements, which are then fed to localization and mapping algorithms. The success of a localization scheme depends on the accuracy and resolution of the obtained measurements. Although accuracy implies a minimal deviation between the determined signal parameter of interest and its true value, resolution refers to being able to discriminate between multiple signals within the same operational environment.

Generally, smart radio environments can enable localization where it would have otherwise been a daunting task without deliberate control of the wireless environment. For instance, where there are no base station anchor points in the vicinity to aid in the process. An RIS platform whose location and orientation are known a priori can serve as an anchor point and enable improved accuracy and extended coverage. If the target object is within the near-field of the RIS, localization can also be achieved using the physics of waveform curvature that occurs within this region (Alexandropoulos et al., 2022). Also, where there is no direct LOS path to the target device being tracked, RIS platforms can be programmed to provide alternative signal paths. Similarly, in IIoT scenarios typified by high densities of mobile and stationary objects, RIS platforms can be used to re-engineer these usually harsh propagation environments to improve the resolution and accuracy of localization signal measurements. Physically, the flat-surface topology of an RIS implies the provision of a maximum of 180° coverage by each platform. Consequently, providing localization coverage over an indoor or outdoor operational environment requires a carefully planned installation of these surfaces.

11.3.3 Security

Smart radio environments can introduce new physical layer security capabilities by manipulating the physical characteristics of the wireless channel to prevent information leakage to unintended users. The radio environment can be configured to minimize or eliminate radiation in unintended directions that could be exploited by malicious users in the IoT network. Deployed RIS platforms can be programmed to focus the propagated signal on the intended receiving devices or, conversely, to steer the signal away from located eavesdroppers. Another proposed mechanism for ensuring secure communication randomizes the weights of reflection coefficients of RIS elements such that only an intended user can decrypt the RIS-reflected signal (Luo et al., 2021).

Using signal-theoretic methodologies, various studies have demonstrated physical layer security gains that can be achieved by deploying RIS platforms in various wireless network scenarios (Chu et al., 2020, 2021; Feng et al., 2020). Often, these performance gains are characterized in terms of a secrecy rate, usually formulated as capacity differences between legitimate and eavesdropper channels.

However, the novel physical layer security capabilities introduced by smart radio environments will be compromised without a framework for securing the enabling infrastructure. Different attack vectors aimed at smart radio infrastructure and some proposed defense strategies have been explored in the literature (Liaskos et al., 2022). At the physical hardware level, RIS platforms are susceptible to physical tampering. To counter this, a likely solution may be found in the integration of physically unclonable functions in RIS hardware implementations. Also, networked deployments of RIS platforms are susceptible to man-in-the-middle attacks, allowing intruders to tamper with communications between the network controller and RIS platforms. In this case, the security of controller-RIS links may be enhanced by the use of cryptographic and blockchain solutions. The centralized controller is also susceptible to denial-of-service attacks, leading to outages. Mitigating this security threat would require the deployment of redundant controllers with appropriate load-balancing measures.

11.3.4 Wireless Power Transfer and Ambient Backscatter

The beam-steering capabilities of RIS platforms can be used to assist in the provision of over-the-air energy for recharging the batteries of low-power IoT devices. A typical wireless power transfer system has an independently powered transmitter, which generates an EM field toward the receiver. Electrical power is extracted from this field and supplied to an appropriate load connected to the receiver. The application of this concept within IoT scenarios has attracted significant interest due to the limited network lifetime occasioned by the use of batteries in IoT devices. Unfortunately, many such devices have to be deployed in environments that preclude subsequent access during their operational lifetime for the replacement of batteries. It has therefore become imperative to investigate wireless approaches to recharging IoT device batteries, including potential applications of RIS platforms in these circumstances (Tran et al., 2022).

The impact of RIS-assisted wireless power delivery has been studied under two broad network categories. The first, wireless powered communication networks, assumes the deployment of dedicated power beacons within the network operational area, whose major utility is to transmit EM energy to power appropriate network devices. The other scenario, simultaneous wireless information and power transfer (SWIPT), describes wireless networks where devices extract electrical power from the same propagated EM wave as was used for information transfer. These scenarios are often modeled as optimization problems, and with appropriate algorithms, the resulting analyses have demonstrated performance gains from the deployment of RIS platforms (Tran et al., 2022; Zheng et al., 2020).

An alternative approach to circumventing the limited network lifetime of sensor deployments in IoT scenarios is the idea of recycling propagated EM waves within the network operational area. Information transmission, in this case, is realized through passive modulation and reflection of ambient radio waves among network devices without generating new transmissions. For instance, a heterogeneous assortment of sensors can be attached to an RIS, which can be used to monitor various aspects of the surrounding environment without needing an independent power source to transmit sensed data. Rather, sensed data can be encoded onto reflections of incident waves impinging on the RIS, thereby effectively reusing ambient EM energy (Di Renzo et al., 2020b).

11.3.5 Multi-access Edge Computing

Multi-access edge computing (MEC) is an architecture that aims to move cloud computational and storage resources to the edge of the wireless network, close to the user devices (Sabella et al., 2016). This allows analytics to be performed close to the data source, which reduces the communication and computational overhead associated with accessing centralized clouds. Such network architectures allow connected devices to offload computational tasks to edge servers placed close to access points, which perform the required computations on behalf of the devices. In IoT contexts, MEC enables the provision of reliable, real-time data analytics even where terminals have severe hardware constraints, such as limited computational abilities and low-power requirements.

Offloading computational and storage tasks to edge servers is, however, subject to the vagaries of the radio propagation environment, especially at high data-rate mm-wave frequencies.

RIS platforms deployed in the wireless environment can be used to offset the unpredictability of such connections and secure the benefits of MEC systems (Alexandropoulos et al., 2022). Through smart beamforming to negate multi-path and non-line-of-sight propagation conditions, data-offloading to edge servers can be realized at high data rates.

11.4 OPEN CHALLENGES, EMERGING TRENDS, AND OUTLOOK

Although RISs have widely been touted as the most promising technology for future wireless networks, these expectations are somewhat dampened by some fundamental open challenges that need to be resolved prior to their mainstream practical adoption. These problems and the new research directions they have engendered are briefly highlighted in this section.

11.4.1 Fundamental Performance Limits

The introduction of RIS structures to aid coverage extension in harsh wireless environments bears some similarities with an already established method for extending coverage in harsh radio environments, which is the use of relays. In cases where time-averaged channel conditions are fairly stable, RIS functionality can be viewed as roughly equivalent to amplify-and-forward relaying, albeit without overt power amplification (Di Renzo et al., 2020a). The question, therefore, arises as to what threshold conditions warrant the use of RIS-assisted systems rather than relays. A related problem is the issue of a clear characterization of the fundamental performance limits achievable by using RISs to enable a smart radio environment. Knowledge of these performance bounds not only provides clarity with respect to whether the gains from deploying the technology are significant enough to warrant the necessary investments, but it also provides insight on how to appropriately scale deployments of RIS structures over a coverage area.

These questions have garnered significant research interest, and numerous efforts are currently underway to provide elaborate characterizations of performance bounds of RIS-supported radio environments in various network scenarios. Progress in this regard is tied to the development and experimental validation of accurate and EM-consistent channel models.

11.4.2 RIS Hardware Development

Some of the concerns about smart radio environments border on hardware implementation issues. The practical realization of smart radio environments in any of the potential use cases is limited by the current state of RIS hardware technology. Consequently, there has also been a push toward the development of more advanced RIS platforms to provide greater functionality and ease of deployment.

Although the traditional concept of a smart radio environment was built around the idea of zero- or near-zero-power consumption passive RIS deployments, active RIS platforms have begun to garner some interest due to the potential for extended coverage and supporting SWIPT to energy-constrained network nodes (G. Chen et al., 2022). Hybrid RISs are able to reap the benefits of passive and active implementations by deploying a mix of active and passive elements to provide both extended signal coverage and higher beamforming gain (Kang et al., 2023).

A new category of RIS implementations has also recently gained traction within the research community. In contrast to conventional reflection-only RIS platforms, simultaneously transmitting and reflecting intelligent omni-surfaces (STAR-IOMs) achieve higher tuning degrees of freedom and full 360° coverage (Xu et al., 2022). In addition, STAR-IOMs can improve indoor and outdoor sensing and localization capabilities when there are severe signal blockages. Similar to conventional RIS platforms, STAR-IOMs can be implemented either using planar antenna arrays or metasurfaces, with the required reconfiguration achieved either via embedded switching elements or the use of tunable materials. In addition, the amplitude coefficients of constituent unit elements can be adjusted to enable operation in either transmit mode, reflection mode, or a simultaneous transmit and receive mode.

Concurrent to these investigations is research into new materials for implementing conformal hardware surfaces for manipulating the EM environment. New developments in the realization of metasurfaces using flexible and non-conventional media (Walia et al., 2015) are expected to lead to a novel class of conformal RIS platforms that can be more seamlessly integrated into indoor and outdoor wireless environments (Jia et al., 2024).

To complement the programmable functionalities provided by RIS platforms, static passive smart skins are also proposed as conformal structures that could be embedded within an operational environment to create a more lenient propagation environment (Oliveri et al., 2022). They are realized using thin metasurfaces, whose EM responses are fixed by design. Tuning elements are not embedded in them, and so they are not reconfigurable. The idea of deploying smart skins in addition to RISs, however, is to provide a basic level of enhancement of the radio environment, such as providing a fixed signal reflection toward an area without LOS coverage (Bilotti et al., 2024).

An alternative vision for smart radio environments is based on the manipulation of trapped surface waves rather than radio wave propagation in free space (Wong et al., 2020). These surface waves are non-radiating and propagate at the interface between two different media, for instance, along the surface of a metasurface in free space. This alternative propagation mechanism creates opportunities for novel RIS implementations and wavefront manipulations that can be used to realize smart radio environments.

11.4.3 Real-Time Radio Environment Awareness and Channel Estimation

The key to a smart radio environment, as distinct from one that is only tunable, is that there is real-time, comprehensive awareness of the dynamic wireless environment in which wireless devices are deployed. Optimization of the radio environment to service high-level operational functionalities requires an accurate determination of the state of the wireless channel based on EM-realistic channel models. However, as a consequence of their near-passive implementations, conventional RIS platforms have minimal signal-processing capabilities. This constraint presents a formidable challenge for accurately estimating the real-time condition of the dynamic radio channel.

AI approaches, in concert with edge intelligence-based architectures, may provide the required framework to handle the required computational complexity and network intelligence to acquire real-time CSI and orchestrate RIS deployments accordingly. AI-based channel modeling approaches hold key advantages over conventional techniques. In addition to predicting channel characteristics using channel training data collected by RIS sensors, they are able to integrate the complex nonlinear channel characteristics with RIS configuration state information to provide actionable insights (Huang et al., 2022). Specifically, learning mechanisms can be integrated into the control architecture used for coordinating deployed RIS platforms. For instance, machine learning, deep learning, or reinforcement learning approaches can be employed to extract wireless environment data and provide actionable insights for wave manipulation (Di Renzo et al., 2020b). It has also been shown that modeling the radio environment as a back-propagating neural network can provide an effective methodology for configuring deployed RIS platforms (Liaskos et al., 2019).

Although various studies have demonstrated the feasibility of adopting learning methodologies, an important trade-off relates to the evolution of the dynamic environment compared to the speed of model training. Training data ages quicker in a highly dynamic environment. In situations where model training is accomplished centrally, a corresponding communication overhead introduces significant latency to the scheme. In this case, federated learning frameworks are a potential solution, where convolutional neural networks can be trained on node-specific local datasets without involving the central base station. This would lead to a significant decrease in the protocol overhead (Elbir and Coleri, 2021).

11.4.4 Technology Standardization

The progress realized so far in developing enablers for smart radio environments has been engendered by multidisciplinary academic research drawing from the fields of computing, wireless communications, signal processing, physics, and material science. There has also been participation from the industry in this regard. A notable example is the reported progress by the Japanese network operator, NTT DOCOMO, including field trials on metasurface prototypes for enhancing wireless connectivity.[3] Concurrent to these efforts have been various funded projects supporting both fundamental and applied research, with notable outcomes provided in the literature and developed prototypes. Between 2012 and 2022, there were 11 key funded projects dedicated to the in-depth analysis and engineering of smart radio environments (Liu et al., 2022).

One consequence of this widespread research activity is the proliferation of proposed technology solutions for smart radio environments, many of which may not guarantee interoperability, scalability, global adoption, or commercial viability. It is therefore imperative that technical standards and specifications for the implementation and deployment of the required hardware and software enablers for smart radio environments are developed for the mainstreaming of the technology. This requires direct interventions by relevant standards organizations as well as collaborative efforts involving industry stakeholders and academic researchers.

On a positive note, initial steps toward the development of technical standards have commenced in some regions of the world. In Europe, the European Telecommunications Standards Institute has an Industry Specification Group on RISs to, among others, identify specific requirements for RIS-related applications and deployment scenarios.[4] Elsewhere, the China Communications Standards Association kicked off its standardization studies in 2020, focusing on various aspects of RIS-related technology. This work is complemented by the RIS Tech Alliance (RISTA), formed in 2022 to deepen collaboration, exchange, and cooperation toward promoting research, standardization, and industry realization of RIS-related technologies.[5]

At the global level, the International Telecommunications Union (ITU) has identified the RIS as one of the key technologies for the radio interface of envisaged IoT-centric 6G wireless networks (ITU, 2022). Consequently, it is expected that specifications for RIS and related technologies for smart radio environments will feature prominently in emerging 6G standardization efforts.

11.5 CONCLUSION

IoT technologies have become increasingly pervasive. Growth trends suggest a continued exponential increase in the number of connections and a greater variety in the types of connected devices. Wireless access technologies are key enablers of IoT networks. The concept of smart radio environments presents a radical departure from the traditional approach of modeling wireless access channels as intractable entities that should be adapted. Rather, through the use of RIS platforms, radio environments can be transformed into controllable and programmable entities that can be optimized to enhance higher-level network objectives.

A significant amount of progress has been made on RIS-based realizations of smart radio environments, as evidenced by the considerable research output, developed prototypes, and industry whitepapers. However, there are a number of issues that need to be resolved prior to the mainstreaming of this technology. Some of these include the characterization of fundamental performance limits of RIS-supported radio environments based on radio channel models that are consistent with the EM environment, as well as the development of hardware enablers able to achieve theoretical performance limits. AI paradigms for real-time channel learning and control of networked RISs constitute another focus area. Also, the security of installed enablers for smart environments requires special attention since their presence opens up a new vista of attack vectors targeting IoT networks.

With discussions underway on the development of technical specifications for 6G wireless communications networks, next-generation networks are envisaged to be considerably more IoT-centric than previous generations. Consequently, enablers for smart radio environments have begun to receive attention in these standardization efforts. These developments will go a long way in stimulating industry-scale realizations and widespread adoptions for overcoming fundamental limitations posed by wireless access in IoT deployments.

NOTES

1 https://www.visorsurf.eu/.
2 https://iot-analytics.com/number-connected-iot-devices.
3 https://www.agc.com/en/news/detail/1201961_2814.html.
4 https://www.etsi.org/committee/ris.
5 https://www.risalliance.com/en/riswp2023_en/.

REFERENCES

Alexandropoulos, G.C., Crozzoli, M., Phan-Huy, D.-T., Katsanos, K.D., Wymeersch, H., Popovski, P., Ratajczak, P., Bénédic, Y., Hamon, M.-H., Gonzalez, S.H., D'Errico, R., Strinati, E.C., 2022. Smart Wireless Environments Enabled by RISs: Deployment Scenarios and Two Key Challenges. In *2022 Joint European Conference on Networks and Communications & 6G Summit (EuCNC/6G Summit)*, Grenoble, France, 1–6.

Alexandropoulos, G.C., Phan-Huy, D.-T., Katsanos, K.D., Crozzoli, M., Wymeersch, H., Popovski, P., Ratajczak, P., Bénédic, Y., Hamon, M.-H., Gonzalez, S.H., Mursia, P., Rossanese, M., Sciancalepore, V., Gros, J.-B., Terranova, S., Gradoni, G., Di Lorenzo, P., Rahal, M., Denis, B., D'Errico, R., Clemente, A., Strinati, E.C., 2023. RIS-enabled Smart Wireless Environments: Deployment Scenarios, Network Architecture, Bandwidth and Area of Influence. *Journal on Wireless Communications and Networking*, 2023, 103. https://doi.org/10.1186/s13638-023-02295-8.

Bilotti, F., Barbuto, M., Hamzavi-Zarghani, Z., Karamirad, M., Longhi, M., Monti, A., Ramaccia, D., Stefanini, L., Toscano, A., Vellucci, S., 2024. Reconfigurable Intelligent Surfaces as the Key-enabling Technology for Smart Electromagnetic Environments. *Advances in Physics: X*, 9(1), 2299543. https://doi.org/10.1080/23746149.2023.2299543.

Chen, G., Wu, Q., He, C., Chen, W., Tang, J., Jin, S., 2023. Active IRS Aided Multiple Access for Energy-Constrained IoT Systems. *IEEE Transactions on Wireless Communications* 22 (3), 1677–1694, https://doi.org/ 10.1109/TWC.2022.3206332.

Chen, L., Hu, W., Jamieson, K., Chen, X., Fang, D., Gummeson, J., 2021. Pushing the Physical Limits of IoT Devices with Programmable Metasurfaces. In *18th USENIX Symposium on Networked Systems Design and Implementation (NSDI 21)*, USENIX. 425–438. https://www.usenix.org/conference/nsdi21/presentation/chen

Chen, R., Liu, M., Hui, Y., Cheng, N., Li, J., 2022. Reconfigurable Intelligent Surfaces for 6G IoT Wireless Positioning: A Contemporary Survey. *IEEE Internet of Things Journal*, 9, 23570–23582. https://doi.org/10.1109/JIOT.2022.3203890.

Chen, Y., Li, X., Nassar, H., Hu, G., Huang, G., 2018. A Programmable Metasurface for Real Time Control of Broadband Elastic Rays. *Smart Materials and Structures*, 27, 115011. https://doi.org/10.1088/1361-665X/aae27b.

Chu, Z., Hao, W., Xiao, P., Mi, D., Liu, Z., Khalily, M., Kelly, J.R., Feresidis, A.P., 2021. Secrecy Rate Optimization for Intelligent Reflecting Surface Assisted MIMO System. *IEEE Transactions on Information Forensics and Security*, 16, 1655–1669. https://doi.org/10.1109/TIFS.2020.3038994.

Chu, Z., Hao, W., Xiao, P., Shi, J., 2020. Intelligent Reflecting Surface Aided Multi-Antenna Secure Transmission. *IEEE Wireless Communications Letters*, 9, 108–112. https://doi.org/10.1109/LWC.2019.2943559.

Di Renzo, M., Ntontin, K., Song, J., Danufane, F.H., Qian, X., Lazarakis, F., De Rosny, J., Phan-Huy, D.-T., Simeone, O., Zhang, R., Debbah, M., Lerosey, G., Fink, M., Tretyakov, S., Shamai, S., 2020a. Reconfigurable Intelligent Surfaces vs. Relaying: Differences, Similarities, and Performance Comparison. *IEEE Open Journal of the Communications Society*, 1, 798–807. https://doi.org/10.1109/OJCOMS.2020.3002955.

Di Renzo, M., Zappone, A., Debbah, M., Alouini, M.-S., Yuen, C., de Rosny, J., Tretyakov, S., 2020b. Smart Radio Environments Empowered by Reconfigurable Intelligent Surfaces: How it Works, State of Research, and Road Ahead. *IEEE Journal on Selected Areas in Communications, 38(11),* 2450–2525. https://doi.org/10.1109/jsac.2020.3007211. IEEE. USA

Elbir, A.M., Coleri, S., 2021. Federated Learning for Channel Estimation in Conventional and RIS-Assisted Massive MIMO. *IEEE Transactions on Wireless Communications*, *21*(6), 4255–4268. IEEE. USA. https://doi.org/10.1109/twc.2021.3128392.

Feng, K., Li, X., Han, Y., Jin, S., Chen, Y., 2020. Physical Layer Security Enhancement Exploiting Intelligent Reflecting Surface. *IEEE Communications Letters 25 (3)*, 734–38. IEEE. USA. https://doi.org/10.1109/lcomm.2020.3042344.

Glybovski, S.B., Tretyakov, S.A., Belov, P.A., Kivshar, Y.S., Simovski, C.R., 2016. Metasurfaces: From Microwaves to Visible. *Physics Reports*, 634, 1–72. https://doi.org/10.1016/j.physrep.2016.04.004

Hassan, Q. (Ed.), 2018. *Internet of Things A to Z: Technologies and Applications*. John Wiley & Sons.

Huang, J., Wang, C.-X., Sun, Y., Feng, R., Huang, J., Guo, B., Zhong, Z., Cui, T.J., 2022. Reconfigurable Intelligent Surfaces: Channel Characterization and Modeling. *Proceedings of the IEEE* 110 (9), 1290–1311. IEEE. USA. https://doi.org/10.1109/JPROC.2022.3186087.

ITU, 2022. REPORT ITU-R M.2516-0 - Future Technology Trends of Terrestrial International Mobile Telecommunications Systems towards 2030 and beyond (No. ITU-R M.2516-0).

Jia, X., Tan, H., Dong, X., Ye, F., Cui, H., Chen, L., 2024. Dual-Polarized Reconfigurable Manipulation Based on Flexible-Printed Intelligent Reflection Surface. *Photonics*, 11, 69. https://doi.org/10.3390/photonics11010069

Kang, Z., You, C., Zhang, R., 2023. Active-IRS-Aided Wireless Communication: Fundamentals, Designs and Open Issues. *IEEE Wireless Communications* 31 (3): 368–74. IEEE. USA. https://doi.org/10.1109/mwc.003.2300013.

Khawaja, W., Ozdemir, O., Yapici, Y., Erden, F., Guvenc, I., 2020. Coverage Enhancement for NLOS mmWave Links Using Passive Reflectors. *IEEE Open Journal of the Communications Society*, 1, 263–281. https://doi.org/10.1109/OJCOMS.2020.2969751

Kisseleff, S., Martins, W.A., Al-Hraishawi, H., Chatzinotas, S., Ottersten, B., 2020. Reconfigurable Intelligent Surfaces for Smart Cities: Research Challenges and Opportunities. *IEEE Open Journal of the Communications Society*, 1, 1781–1797. https://doi.org/10.1109/OJCOMS.2020.3036839

Liaskos, C., Mamatas, L., Pourdamghani, A., Tsioliaridou, A., Ioannidis, S., Pitsillides, A., Schmid, S., Akyildiz, I.F., 2022. Software-Defined Reconfigurable Intelligent Surfaces: From Theory to End-to-End Implementation. *Proceedings of the IEEE*, 110, 1466–1493. https://doi.org/10.1109/JPROC.2022.3169917

Liaskos, C., Nie, S., Tsioliaridou, A., Pitsillides, A., Ioannidis, S., Akyildiz, I., 2018. Realizing Wireless Communication through Software-defined HyperSurface Environments. In *2018 IEEE 19th International Symposium on World of Wireless, Mobile and Multimedia Networks (WoWMoM)*, Chania, Greece. IEEE. 14–15. https://doi.org/10.1109/wowmom.2018.8449754.

Liaskos, C., Tsioliaridou, A., Nie, S., Pitsillides, A., Ioannidis, S., Akyildiz, I., 2019. An Interpretable Neural Network for Configuring Programmable Wireless Environments. In *2019 IEEE 20th International Workshop on Signal Processing Advances in Wireless Communications (SPAWC)*, Cannes, France, 1–5. IEEE. https://doi.org/10.1109/spawc.2019.8815428.

Liu, R., Wu, Q., Di Renzo, M., Yuan, Y., 2022. A Path to Smart Radio Environments: An Industrial Viewpoint on Reconfigurable Intelligent Surfaces. *IEEE Wireless Communications*, 29, 202–208. https://doi.org/10.1109/MWC.111.2100258

Luo, J., Wang, F., Wang, S., Wang, H., Wang, D., 2021. Reconfigurable Intelligent Surface: Reflection Design Against Passive Eavesdropping. *IEEE Transactions on Wireless Communications*, 20, 3350–3364. https://doi.org/10.1109/TWC.2021.3049312

Oliveri, G., Zardi, F., Rocca, P., Salucci, M., Massa, A., 2022. Building a Smart EM Environment -- AI-Enhanced Aperiodic Micro-Scale Design of Passive EM Skins. *IEEE Transactions on Antennas and Propagation*, 70, 8757–8770. https://doi.org/10.1109/TAP.2022.3151354

Petrou, L., Kossifos, K.M., Antoniades, M.A., Georgiou, J., 2022. The First Family of Application-specific Integrated Circuits for Programmable and Reconfigurable Metasurfaces. *Scientific Reports*, 12, 5826. https://doi.org/10.1038/s41598-022-09772-y

Rana, B., Cho, S.-S., Hong, I.-P., 2023. Review Paper on Hardware of Reconfigurable Intelligent Surfaces. *IEEE Access*, 11, 29614–29634. https://doi.org/10.1109/ACCESS.2023.3261547

Sabella, D., Vaillant, A., Kuure, P., Rauschenbach, U., Giust, F., 2016. Mobile-Edge Computing Architecture: The role of MEC in the Internet of Things. *IEEE Consumer Electronis Magazine*, 5, 84–91. https://doi.org/10.1109/MCE.2016.2590118

Strinati, E.C., Alexandropoulos, G.C., Sciancalepore, V., Di Renzo, M., Wymeersch, H., Phan-huy, D.-T., Crozzoli, M., D'Errico, R., De Carvalho, E., Popovski, P., Di Lorenzo, P., Bastianelli, L., Belouar, M., Mascolo, J.E., Gradoni, G., Phang, S., Lerosey, G., Denis, B., 2021. Wireless Environment as a Service Enabled by Reconfigurable Intelligent Surfaces: The RISE-6G Perspective. In *2021 Joint European Conference on Networks and Communications & 6G Summit (EuCNC/6G Summit)*, Porto, Portugal, 562–567. IEEE. https://doi.org/10.1109/EuCNC/6GSummit51104.2021.9482474.

Tran, N.M., Amri, M.M., Park, J.H., Kim, D.I., Choi, K.W., 2022. Reconfigurable Intelligent Surface-Aided Wireless Power Transfer Systems: Analysis and Implementation. *IEEE Internet of Things Journal*, 9, 21338–21356. https://doi.org/10.1109/JIOT.2022.3179691

Tsilipakos, O., Tasolamprou, A.C., Pitilakis, A., Liu, F., Wang, X., Mirmoosa, M.S., Tzarouchis, D.C., Abadal, S., Taghvaee, H., Liaskos, C., Tsioliaridou, A., Georgiou, J., Cabellos-Aparicio, A., Alarcón, E., Ioannidis, S., Pitsillides, A., Akyildiz, I.F., Kantartzis, N.V., Economou, E.N., Soukoulis, C.M., Kafesaki, M., Tretyakov, S., 2020. Toward Intelligent Metasurfaces: The Progress from Globally Tunable Metasurfaces to Software-Defined Metasurfaces with an Embedded Network of Controllers. *Advanced Optical Materials*, 8, 2000783. https://doi.org/10.1002/adom.202000783

Tsioliaridou, A., Pitilakis, A., Pyrialakos, G.G., Liaskos, C., 2020. The Internet of MetaMaterial Things and their Software Enablers. *ITU Journal on Future and Evolving Technologies*, 1, 55–77. https://doi.org/10.52953/PPWI5546

Vaezi, M., Azari, A., Khosravirad, S.R., Shirvanimoghaddam, M., Azari, M.M., Chasaki, D., Popovski, P., 2021. Cellular, Wide-Area, and Non-Terrestrial IoT: A Survey on 5G Advances and the Road Towards 6G. *IEEE Communications Surveys & Tutorials*, 1117–1174. IEEE. USA. https://doi.org/10.1109/comst.2022.3151028.

Vasic, B., Isic, G., Beccherelli, R., Zografopoulos, D.C., 2020. Tunable Beam Steering at Terahertz Frequencies Using Reconfigurable Metasurfaces Coupled with Liquid Crystals. *IEEE Journal of Selected Topics in Quantum Electronics*, 26, 1–9. https://doi.org/10.1109/JSTQE.2019.2956856

Walia, S., Shah, C.M., Gutruf, P., Nili, H., Chowdhury, D.R., Withayachumnankul, W., Bhaskaran, M., Sriram, S., 2015. Flexible Metasurfaces and Metamaterials: A Review of Materials and Fabrication Processes at Micro- and Nano-Scales. *Applied Physics Reviews*, 2, 011303. https://doi.org/10.1063/1.4913751

Wong, K.-K., Tong, K.-F., Chu, Z., Zhang, Y., 2020. A Vision to Smart Radio Environment: Surface Wave Communication Superhighways. *IEEE Wireless Communications* 28 (1), 112–19. IEEE. USA. https://doi.org/10.1109/mwc.001.2000162.

Xu, J., Liu, Y., Mu, X., Zhou, J.T., Song, L., Poor, H.V., Hanzo, L., 2022. Simultaneously Transmitting and Reflecting Intelligent Omni-Surfaces: Modeling and Implementation. *IEEE Vehicular Technology Magazine*, 17, 46–54. https://doi.org/10.1109/MVT.2022.3157069

You, C., Zheng, B., Mei, W., Zhang, R., 2022. How to Deploy Intelligent Reflecting Surfaces in Wireless Network: BS-Side, User-Side, or Both Sides? *Journal of Communications and Information Networks*, 7, 1–10. https://doi.org/10.23919/JCIN.2022.9745477

Zhang, X.G., Yu, Q., Jiang, W.X., Sun, Y.L., Bai, L., Wang, Q., Qiu, C.-W., Cui, T.J., 2020. Polarization-controlled Dual-programmable Metasurfaces. *Advanced Science*, 7, 1903382.

Zheng, Y., Bi, S., Zhang, Y.J., Quan, Z., Wang, H., 2020. Intelligent Reflecting Surface Enhanced User Cooperation in Wireless Powered Communication Networks. *IEEE Wireless Communication Letters*, 9, 901–905. https://doi.org/10.1109/LWC.2020.2974721

12 RISC-V
A Comprehensive Overview of an Emerging ISA for the AI-IoT Era

Qusay F. Hassan and Assim Sagahyroon

12.1 INTRODUCTION

The number of deployed Internet of Things (IoT) devices is increasing exponentially, ranging in capability from basic sensor nodes that collect data to sophisticated platforms at the edge capable of processing the incoming data and making informed decisions accordingly (Hassan, 2025; Hassan, Khan, and Madani, 2017). Machine learning (ML) techniques are typically used to efficiently and intelligently process the huge volumes of data made available by the millions of these devices. Hence, various IoT-based applications in areas such as healthcare, intelligent transportation systems, smart agriculture, and smart grids are in need of timely and efficient processing solutions.

The deployment of these sensors and edge devices is often done over low-bandwidth connectivity, and hence communication is constrained in terms of both bandwidth and latency with a limited energy budget (Shafique, 2018). The recent trend is to shift computation from datacenters and the cloud to the edge, where intelligent real-time processing can be performed. This paradigm shift created a demand for alternative architectures and a departure from the traditional X86-64 designs, which are suitable for general-purpose computing applications but not for embedded IoT applications, for example. This is mainly due to the unique requirements of IoT and embedded artificial intelligence (AI) applications. To fulfill these requirements, hardware architectures must address constraints such as real-time processing performance, power consumption, and acceleration for AI computing needs. At the forefront of these alternative architectures, ARM and RISC-V are leading the way. The two architectures follow the reduced instruction set computer (RISC) philosophy that seeks simplicity and efficiency, although they differ in many other aspects.

ARM is a proprietary instruction set architecture (ISA) that is used in many embedded systems. It has licensing fees and allows for limited customization to a specific application. Unfortunately, this proprietary nature has prevented small businesses, startups, and researchers from fully participating in the creative process and the development of modern processors.

On the other hand, RISC-V (pronounced "risk-five") is a royalty-free open standard ISA based on the principles of RISC. It is maintained by the RISC-V International organization and is not a proprietary architecture. Over 4,400 RISC-V members from 70 countries contribute and collaborate to define RISC-V open specifications, as well as convene and govern related technical, industry, domain, and special interest groups. International members include companies such as Alibaba, Raspberry Pi, and SiFive that offer or have announced commercial Systems on a Chip (SoCs) that incorporate one or more RISC-V compatible cores. The basic 32-bit integer instruction set (RV32I), required for all implementations, is extendable to a 64-bit instruction set (RV64I) as well as a 128-bit set (RV128I). Additionally, 16-bit instructions (for common instructions) are supported. Because of this flexibility, RISC-V has become a universal ISA capable of supporting different processor sizes, from small embedded systems to high-performance computers. It also supports a wide variety of software stacks and programming languages.

 DOI: 10.1201/9781003506638-15

Motivated by the accelerated adoption of RISC-V and the growth of its ecosystems, this chapter provides the readers with a comprehensive overview of this emerging ISA. The chapter first presents a brief summary of the underlying principles behind complex instruction set computer (CISC) and RISC design approaches, with an explanation supporting the case for RISC architecture. Due to many reasons detailed in the various sections of this chapter, RISC-V is used as an example of a promising futuristic ISA that is poised to lead in the implementations of low- and high-end computing platforms. The chapter does not stop at discussing the technical details of RISC-V alone; it continues to provide the reader with example implementations and introduces key alliances and vendors advocating for the ISA. Moreover, it discusses in great detail the present global initiatives and the current adoption challenges. The chapter concludes by discussing a case study where RISC-V is used in developing solutions that are accelerating the execution and improving the performance of today's social media platforms.

12.2 OVERVIEW OF ISA

ISA is one of the key elements of a processor design. It is an abstract model that defines how software controls the processor.[1] That is, a set of instructions that specify the operations that a processor can execute, such as memory access operations (e.g., load, store*)*, arithmetic operations (e.g., add, subtract), logical operations (e.g., AND, OR), and flow control operations (e.g., jump, return). The ISA defines the instructions' format, as well as the number and data types of operands they can work on. It also defines the memory addressing modes a processor can use and the registers it can access. As processors evolve over time, ISAs adapt by adding new instructions to utilize new hardware elements.

As illustrated in Figure 12.1, ISA acts as an interface between hardware and software. On the hardware side, ISA defines the specifications that processor designers must adhere to. On the software side, it defines how software compilers convert high-level language code, such as Java or C, into executable machine code for the processor. This gives programmers who write code in high-level languages the ability to write their programs without caring about the underlying hardware implementation details.

ISA is implementation-independent; one ISA can have different hardware (microarchitecture) implementations. This allows programs to run on different processors, as long as they implement the same ISA. For example, the Windows operating system and its applications can run on both Intel and AMD central processing units (CPUs) because both of them use the x86-64 ISA, despite their different hardware implementations.

Although there are many ISAs available on the market, two key design philosophies dominate the industry, namely, complex instruction set Architecture computer (CISC) and reduced instruction set architecture computer (RISC), which we will discuss below.

12.2.1 CISC

Before the 1980s, compilers were inefficient. They were slow and generated a large amount of code per program. Memory was also inefficient and very expensive (US $5000/MB in the mid-1970s). Programmers tended to use assembly language to write their programs, as it was more efficient than high-level language compilers. Processors, on the other hand, were improving, and Moore's law was in full force. Several advancements in the transistor industry have enabled processor designers to implement new instructions in ISAs, making assembly language programming easier.[1] Many of these new instructions were complex, designed to perform multiple low-level operations, thereby minimizing the burden on programmers. The goal was to execute more operations with fewer instructions (i.e., less assembly code). This was achieved by adding complex functional units to the processor's circuitry and mapping their functions to more abstract ISA instructions. For example, FSQRT is a complex instruction in the x86 assembly language that calculates the square root of a

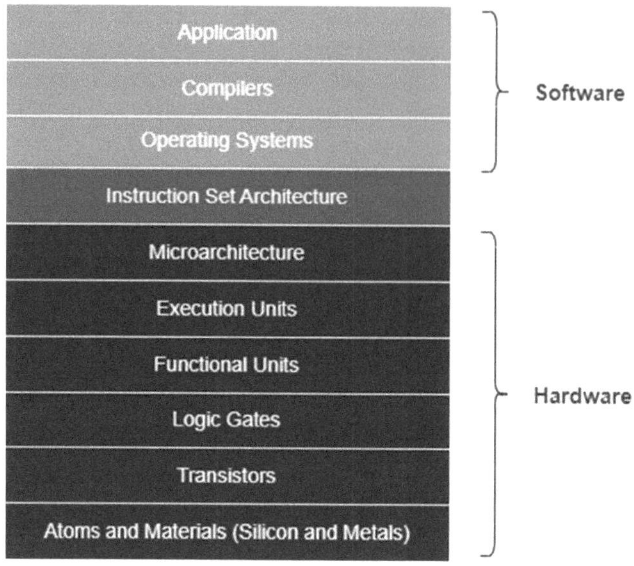

FIGURE 12.1 Computing Abstraction Layers. ISA Acts as an Interface between the Software and Hardware Layers

number – a non-essential function that programmers could perform otherwise by combining several basic arithmetic instructions together. In computer architecture, this model is known as CISC (pronounced "sisk").

Performance is the most important factor in determining the efficiency of a processor, which can be measured by calculating the time it takes to run a program, as presented below (Hennessy and Patterson, 2017):

$$\frac{time}{program} = \frac{instructions}{program} \times \frac{average\ clock\ cycles}{instruction} \times \frac{time}{clock\ cycle} \tag{1}$$

According to this formula, the total number of instructions in a program, the average number of clock cycles required to execute an instruction (i.e., cycles per instruction (CPI)), and the time taken per cycle affect the time needed for a processor to run a program. The last parameter refers to the processor's clock speed, which is dependent on the hardware implementation, leaving ISA designers with a tradeoff between the number and complexity of instructions. CISC aims to optimize the number of instructions per program while compromising the underlying complexity. The following are among the key features of CISC:

- *Number of Instructions:* CISC offers a large number of instructions (in the thousands), some of which have several variations.
- *Clock Cycles:* A CISC instruction is typically complex and takes several clock cycles to execute.
- *Code Size:* CISC attempts to use a small number of instructions per program, resulting in a small code size; hence, less memory is required to run the program.
- *Operand Manipulation:* Operands are entities operated upon by the instruction. In CISC, an operand can be residing in memory and not only in registers. In such scenarios, memory access operations are included in the instruction.
- *Instruction Length:* Because data can be either in memory or in registers, CISC has variable-length instructions. That is, some instructions take more bytes than others. Memory-based instructions are longer than register-based instructions.
- *Pipelining:* Instruction pipelining is a parallelization technique within a single processor that divides the execution unit into several phases that can overlap during execution,

reducing the overall execution time. Because CISC uses variable-length instructions, standard pipelining is difficult to achieve. However, CISC processors often utilize advanced techniques to achieve instruction-level parallelism (e.g., superscalar).

- *Addressing Modes:* This refers to the mechanisms by which memory addresses an operand. CISC ISAs offer a large number of addressing modes.
- *Extensibility:* Variable-length instructions and a wide address space make it easy to add new instructions to the ISA when needed.
- *Chip Design:* A large number of instructions, including complex ones, lead to complicated chip designs.
- *Power Use:* Because of the complex chip designs and enormous number of transistors used, CISC processors tend to consume significant amount of power.

The CISC approach was conceived in the 1970s and has gained significant popularity since then. It is widely used on PCs and servers, with a few implementations in the mobile domain. Examples of CISC ISAs include x86-64, IBM System/360, and VAX. Today, only Intel and AMD make CISC processors, both using x86-64; the x86-64 ISA is very complex and has approximately 1,500 instructions, growing from less than 100 since its inception.

12.2.2 RISC

In the mid-1970s, a study revealed that compilers only use a small portion of an instruction set (Alexander and Wortman, 1975). To validate this, IBM developed an experimental processor, the IBM 801, which used a simple design and instruction set. The results showed that simple processor designs can outperform the commercially available CISC processors. Consequently, researchers and major companies from the field, including Apple, IBM, and Motorola (known as the "AIM alliance"), started developing a new instruction set that is simple and only has the most essential instructions. The objective was to remove unnecessary instructions from the hardware and let high-level language compilers figure out a way to generate code that utilizes these basic instructions. This was possible due to advances in compilers. The new model was coined RISC (Patterson and Ditzel, 1980).

RISC adopts a minimalistic approach where instructions are simple and optimized. Unlike CISC, RISC attempts to optimize CPI and compromise the number of cycles per program (total number of instructions). The following are among the key features of RISC:

- *Number of instructions:* RISC ISAs often have a small number of instructions (10s–100s), leading to simple chip designs.
- *Clock cycles:* A RISC instruction is simple and typically takes one clock cycle to execute.
- *Code size:* Because instructions are simple, programmers must combine more of them to write a program, resulting in a large code size and, in turn, inefficient use of memory and cache.
- *Operand manipulation:* In RISC, only load and store instructions have access to memory, and all arithmetic logic unit (ALU) operations are performed on data in registers. RISC processors tend to have an abundance of registers.
- *Instruction length:* Instructions in RISC have a fixed length.
- *Pipelining:* Because of the fixed-length instructions, efficient machine instruction pipelining is easy to achieve.
- *Addressing modes:* RISC has a few addressing modes.
- *Extensibility:* Adding new instructions is possible, although it is not as flexible as in CISC.
- *Chip design:* A smaller number of instructions in the ISA means simpler chip designs, fewer transistors, and a smaller die[2] area dedicated to the circuitry. This leaves room on the die area to add cache and coprocessors and, in turn, improve performance.
- *Power use:* Because of their simple designs and efficient pipelining, RISC processors tend to be power efficient.

In the late 1980s, RISC gained popularity and was widely used in various applications, including gaming consoles, embedded systems, and even some PCs. Examples of RISC processors include MIPS, SPARC, PowerPC, ARM, and RISC-V. Today, RISC dominates the mobile market, with numerous applications in embedded systems and IoT. It can also be found in PCs and servers; Apple uses ARM processors in new MAC computers, and Amazon uses them in AWS datacenters.

It is worth noting that the term CISC was coined retroactively after the emergence of RISC to differentiate between the old complex model and the new simple one. Today, the differences between CISC and RISC have become blurry. Modern x86-64 CPUs use microprogrammed controls to break CISC instructions into smaller, RISC-like internal operations (i.e., micro-operations) that can be processed more efficiently. On the other hand, ARM, which is a RISC-based ISA has added many complex instructions into its latest designs.

12.3 RISC-V

The history of RISC-V dates back to 2010 when it initially started as a research project at the Parallel Computing Laboratory (Par Lab) at the University of California (UC) Berkeley. RISC-V is the fifth generation of the Berkeley RISC ISA. The objective of the project was to create a new ISA that overcomes the limitations of existing proprietary ISAs. The project team set out to create a simple and efficient ISA that is suitable for research and education purposes and be flexible to accommodate future use cases. A year later, the team released the base integer instruction set, known as "RV32I," which has the most essential instructions to run full-stack software (Waterman et al., 2011). In 2014, the full specifications came out, which continued to grow over the years by adding new extensions (Waterman et al., 2014).

12.3.1 RISC-V International

While working on RISC-V, the project team realized the need for an open standard ISA, especially since there are open standards for almost all other computing areas (Asanović and Patterson, 2014). The team set out to make their new ISA an open standard, and in 2015, the RISC Foundation,[3] a non-profit organization, was founded to serve as a governing body for RISC-V. The goal was to promote the ISA and push its development. In 2020, the RISC-V Foundation changed its name to RISC-V International and moved from the US to Switzerland as a sign of neutrality and out of concern about potential trade restrictions. RISC-V International manages the RISC-V specifications, which are published alongside other resources on GitHub.[4]

The organization started with 23 founding members, and today it has over 4,400 members across 70 countries.[5] Its members come from both industry and non-profit/research institutions. Industry members include Alibaba, Andes, Huawei, IBM, Intel, Google, Meta, Nvidia, Qualcomm, Samsung, and SiFive, to name a few. From non-profit and research institutions, members include lowRISC, the Ministry of Science, Technology, and Innovation in Brazil, ETH Zurich, the Technical University of Munich, the Chinese Academy of Sciences (CAS), IIT Madras, TII in the UAE, and Future University in Egypt, to name a few. The work done by the RISC-V International members has led to the creation of many IP cores[6] and hardware processors, as well as tools that support the ISA.

12.3.2 Why RISC-V?

In 2023, MIT named RISC-V among the breakthrough technologies of 2023.[7] This is primarily due to its unique characteristics and significance to the tech sector. In this section, we address some of the reasons that make RISC-V an attractive ISA.

12.3.2.1 Free and Open Standard

The processor market today is dominated by three key vendors, namely, Intel, AMD, and ARM. Intel and AMD, with their x86-64 ISA, have the largest share of the program counter (PC) and server markets. ARM designs, on the other hand, are in virtually all mobile devices (tablets, smartphones, and wearables). Intel and AMD do not license their ISA; rather, they make chips that implement it. ARM, on the other hand, relies on selling licenses for its ISAs and IP designs to chipmakers and collecting royalties for sold units. ARM has two types of licenses: the *standard license*, which allows the licensee to use a core without changing it, and the *architecture license*, which is extremely expensive but gives the licensee the right to re-implement the ISA and add instructions. Many chipmakers use ARM-based designs, including Apple, Nvidia, Qualcomm, Samsung, and Texas Instruments.

Unlike these proprietary ISAs, RISC-V is an open standard, similar to HTTP, SQL, and XML. Anyone can use the ISA to design processors without paying license fees or royalties (Asanović and Patterson, 2014). However, its implementations do not have to be; implementers are free to make their designs open or closed. This new business model helps lower the barrier for startups and research groups and, in turn, allows for more involvement and innovation. It also prevents vendor lock-in. Table 12.1 provides a brief comparison between the different ISA business models available on the market today.[8]

Additionally, because RISC-V is open, there is transparency. The community can participate in its development to ensure that all design decisions are thoroughly investigated and taken solely for technical reasons, not marketing considerations. Transparency also leads to improved trust and security.

12.3.2.2 Simple

RISC-V is a clean, well-engineered ISA. Developers are only required to implement the base integer instruction set, which contains 47 unique instructions and is capable of running a full software stack, i.e., a basic operating system and simple computations. Developers can optionally implement extensions for more complex applications. This separation and minimalistic approach lead to simpler designs, smaller die areas, improved power and cost efficiency, fewer architectural bugs, and faster design and verification processes.

12.3.2.3 Modular

Commercial ISAs follow an accumulative approach, in which a new version of the ISA adds new instructions without removing anything from previous versions in order to maintain backward compatibility with legacy software. This does not exclude unwise design decisions from the past. In other words, to design a new processor, developers must implement both the new and old ISA instructions. RISC-V, on the other hand, is relatively new and is built in a modular fashion from the ground up. It provides a set of independent modules, known as extensions, from which developers can implement only the ones that meet their application needs. Besides the base integer instruction sets, RISC-V comes with a broad range of optional standard extensions that are well-suited for various use cases.

12.3.2.4 Extensible

In addition to the standard extensions, RISC-V allows implementers to define their own extensions and instructions. This is unique to RISC-V, as legacy ISAs do not offer this. The extensibility RISC-V offers allows for innovation, as implementers are not limited to what comes with the ISA. This

Table 12.1 Comparison between Available ISA Business Models

ISA	Vendors?	License?	Add Custom Instructions?	Commercial IP Cores?	Open-Source IP Cores?
x86-64	Yes, only Intel and AMD	No	No	No	No
ARM	Yes, many	Yes, expensive	No, mostly	Yes, only by ARM	No
RISC-V	Yes, many	Yes, free	Yes	Many	Many

makes the ISA scalable and well-suited for a wide range of applications – from tiny microcontrollers to datacenter applications.

12.3.2.5 Stable

To guarantee stability and backward compatibility in RISC-V, no changes are allowed to an instruction set once it has been ratified. This applies to both the base instruction sets and standard extensions. Future technical requirements and use cases can be met with new extensions or new instructions in non-ratified extensions. This simplifies the ISA design and frees developers from long testing and verification processes.

12.3.3 Driving Forces

RISC-V is disruptive technology that is gaining momentum in recent years. In this section, we analyze the reasons behind the global interest in it.

12.3.3.1 Proliferation of IoT and AI Deployments

Numerous objects today are smart (i.e., autonomous) and connected to the network. The phenomenon is widely known as the IoT, which goes beyond traditional consumer electronics to include applications in automotive, agriculture, healthcare, urban management, surveillance, and manufacturing (Bara, Brito, and Hassan, 2025; Schoder, 2018). Such objects usually have sensors, network connectivity, some processing/storage capabilities, and even sometimes some AI capabilities. Additionally, in most cases, these smart objects send collected data to servers, whether at the edge or in the cloud, for analysis and storage. Complex AI models on these servers enable users to gain insights into data by identifying trends and patterns (Samaraweera et al., 2025). However, for IoT to work, we need chips with different form factors and capabilities to power all of the devices in this IoT-edge-cloud continuum. And as IoT adoption continues to grow, so will the demand for chips.

Similarly, AI is ubiquitous. In addition to IoT applications, AI powers augmented reality (AR) and virtual reality (VR), the metaverse, voice recognition, and many other applications (Byabazaire et al., 2025). Moreover, major software vendors such as Microsoft, Google, and IBM have plans to employ AI in their products. These emerging applications inherently increase the demand for new advanced chips.

Fortunately, the birth of RISC-V has coincided with the rise of AI and IoT. The growing demand for chips, coupled with the need to reduce energy use and overall cost, makes RISC-V a very attractive option for many chip makers.

12.3.3.2 End of Moore's Law and Dennard Scaling, and Rise of Accelerated Compute

Since the creation of the first IC in 1959, advances in transistors and lithography technologies have led to a sharp increase in the number of transistors packaged in a chip. The first IC had less than ten transistors, whereas an average computer or smartphone chip today has billions of transistors; for example, the recent M2 Ultra chip, used in Mac computers, has approximately 132B transistors.

Gordon E. Moore in the mid-1960s predicted that transistors would continue to miniaturize, enabling us to double their count on a chip every 18–24 months with relatively no cost increase (Moore, 1965). Derived from Moore's prediction, Robert Dennard explained in the early 1970s that as transistors become smaller, they will use less power, keeping power use per chip almost constant despite the performance improvement (Dennard et al., 1974). Both Moore's prediction and Dennard scaling allowed us to achieve leaps in chip performance over the past five decades. However, since the late 1990s, it has become costly and physically difficult to miniaturize transistors further (Moore, 2003). Likewise, by 2007, Dennard scaling had come to an end, and it was no longer feasible to increase chip performance without significant power use and heat dissipation, i.e., "power wall" (Bose, 2011).

Then came multicores (and later multiprocessors) as a new way to improve performance (Valero-Lara et al., 2016). These maintain relatively static clock speeds and split the computational workload for parallel processing. Although these architectures have significantly increased performance (Hassan, 2016), they proved to be inefficient in general-purpose computing, as the serial part of a program will determine the overall performance (Amdahl, 2013). Also, their low power efficiency and high heat dissipation led to issues such as dark silicon (Esmaeilzadeh et al., 2011), where parts of the chip remain inactive to prevent overheating, limiting the overall performance gain.

These issues have pushed the field to seek novel ways to improve performance. Among the techniques that have emerged lately are domain-specific architectures (DSAs). The idea is to customize computer architectures that tackle specific domain problems in order to improve performance (Hennessy and Patterson, 2019). DSAs rely on processing specific algorithms in a massively parallel fashion, using, for example, many cores/hardware threads or large two-dimensional systolic multipliers. The industry has seen an increasing adoption of DSAs, primarily graphical processing units (GPUs), due to performance gains from their inherent hardware parallelism (Fei et al., 2016). GPUs are replacing CPU (software)-accelerated compute. Their applications go beyond graphics processing to include other areas such as AI/ML, big data, and other high-performance computing (HPC) applications. Nvidia chips are the most popular example in this category with its CUDA parallel programming platform and application programming interface (API), which offer a great level of application adaptability (i.e., programmability).

However, despite this adaptability and significant performance gain, general-purpose computing on GPUs has its drawbacks. Such implementations tend to be power hungry, and they do not consider the unique characteristics and requirements of the different application domains. This inspired tech giants like Google and Meta to build more specialized DSAs with better performance per watt ratios for their specific use cases. Popular examples of these include data processing units for data-centric workloads; neural processing units (NPUs) for deep learning workloads; tensor processing units for ML workloads; and single instruction, multiple data (SIMD) coprocessors for 3D graphics and audio/video processing. xPUs, or simply accelerators, is a generic term used to collectively refer to these different architectures. Novel architectures today use heterogeneous accelerated compute.

RISC-V, with its versatility, space and power efficiency, and the huge opcode space reserved for future use cases (something that is missing in legacy ISAs), is well suited for implementing many of these specialized architectures.

12.3.3.3 SoC Market Growth and ISA Fragmentation

The continuous technological advancement and hardware miniaturization have allowed us to move from large discrete components like those used before the 1990s to more compact integrated solutions like SoCs. A SoC, in simple terms, is a chip that has all or most of the components of a computing system (Risset, 2011). Over the past two decades, the proliferation of mobile devices and embedded systems has driven growth in the SoC market. Recent reports estimated the global SoC market at US $154.8B in 2022, with forecasts to reach US $327.8B by 2032.[9] Emerging technologies such as 5G/6G, AI, autonomous/electric vehicles, edge computing, wearables, robotics, and the Industrial Internet of Things (IIoT) are expected to significantly contribute to the growth of the SoC market in the future.

A typical SoC, like those used in virtually all smartphones today, contains heterogeneous architectures, including both digital and analog subsystems. It would have a CPU, GPU, NPU, power management processor, security processors, radio and audio digital signal processors (DSPs), and many other components. Most of these components require their own ISAs and different software stacks to run. In many cases, these units are built by different vendors using various proprietary instruction sets, leading to a large number of ISAs running on a single SoC. Managing all these ISAs and their software stacks is neither simple nor cheap.

RISC-V, on the other hand, with its standard base instruction set, modular architecture, and availability of various address spaces (i.e., 32/64/128-bits), makes it easy for chip makers to build and run all of these components without running into compatibility issues or long verification cycles.

12.3.3.4 Geopolitical Tension

Today, chips are essential to virtually all industries. Any disruption to a nation's access to chips can have major social and economic impacts. During the COVID-19 shutdown, disruptions in the global supply chains caused chip shortages, which had a huge impact on many industries. The automotive industry, for example, suffered significantly; vehicle production stopped, and used vehicle prices surged globally. This incident gave us a sense of the potential consequences of disrupting access to chips – something we previously took for granted. In response to these chip shortages (and as a way to counter China), the US issued the CHIPS and Science Act, authorizing a US $52.7B fund to support domestic semiconductor research, development, and manufacturing.[10]

Although the pandemic is over now, we are still under threat. The war in Ukraine, the rising tension between the US and China, and the ever-increasing hostilities in the Middle East are putting the global supply chains at severe risk. Moreover, the recent setback of globalization and the global shift away from free-market policies will not only have drastic economic impacts, but they will also threaten the technological sovereignty of many nations (Klementina and Benjamin, 2024). As we enter a new era, many nations are attempting to achieve some level of technological autonomy. Among the areas receiving special attention is the advancement of domestic chip industries in many of those nations.

As we will discuss later, various governments have committed themselves to open-source hardware and software with a specific focus on RISC-V. Among the top adopting nations are Brazil, China, India, Japan, South Korea, Taiwan, and the EU.

12.3.3.5 Nvidia-ARM Merger

In 2020, Nvidia attempted to acquire ARM for US $40B. Although the Federal Trade Commission blocked the deal in 2022,[11] the event left the chip industry uncertain about ARM's future. It was a wake-up call for many chip companies, making them realize that they rely on a single vendor for many of the processors they use in their products. This led many to seek alternatives, with RISC-V emerging as a viable option due to its openness and growing ecosystem. In fact, several hardware vendors, including Nvidia itself, have been using RISC-V for in-house applications for years prior to any plans regarding the Nvidia-ARM deal.

12.3.3.6 New ISA

The industry has not developed a new major ISA for decades, and the progress in legacy ISAs has virtually stalled. There might be an emotional factor behind the global interest in RISC-V, especially among students, researchers, and open-source communities. Experts are also drawn to the development of emerging and disruptive technologies, and RISC-V, as a clean-slate ISA, provides them with this opportunity.

12.3.4 Technical Specifications

In its basic form, RISC-V has 32 integer registers and uses a load-and-store architecture, i.e., instructions address registers only except for memory access instructions. Before looking at the RV32I instruction set, the processor's internal general-purpose register set needs to be defined. Table 12.2 lists these registers along with their abbreviations and descriptions.

As depicted in the table, the RV32I has 32 general-purpose registers from x0 to x31, plus a PC. Except for x0, which always returns a zero when read, all are general-purpose registers. For x1 to x31, the specification does not define the values of these registers after reset, and these may be written and read using instructions. The application binary interface (ABI) column gives the name for each register. All the registers are 32-bit wide, as we are looking at the 32-bit specification. If this were for the 64-bit specification (RV64I), then these would all be 64 bits, but they would still have the same names. The processor's instructions utilize this register set to perform the various operations that are part of the instruction set.

TABLE 12.2
RISC-V Registers Set

Register	ABI Name	Description
x0	zero	Hardwired zero
x1	ra	Return address
x2	sp	Stack pointer
x3	gp	Global pointer
x4	tp	Thread pointer
x5	t0	Temporary 0
x6	t1	Temporary 1
x7	t2	Temporary 2
x8	s0/fp	Saved register0/frame pointer
x9	s1	Saved register 1
x10	a0	Function argument/return value 0
x11	a1	Function argument/return value 1
x12	a2	Function argument 2
x13	a3	Function argument 3
x14	a4	Function argument 4
x15	a5	Function argument 5
x16	a6	Function argument 6
x17	a7	Function argument 7
x18	s2	Saved register 2
x19	s3	Saved register 3
x20	s4	Saved register 4
x21	s5	Saved register 5
x22	s6	Saved register 6
x23	s7	Saved register 7
x24	s8	Saved register 8
x25	s9	Saved register 9
x26	s10	Saved register 10
x27	s11	Saved register 11
x28	t3	Temporary 3
X29	t4	Temporary 4
X30	t5	Temporary 5
X31	t6	Temporary 6

12.3.4.1 Instruction Format

The underlying design principles followed by the RISC-V designers are (Patterson and Hennessy, 2017):

- Simplicity favors regularity.
- Make the common case fast.
- Smaller is faster.
- Good design necessitates good compromises.

Table 12.3 lists the six base instruction formats in the RISC-V ISA, as well as their usage. All of them consist of 32-bit instructions (Waterman et al., 2011).

Instructions are divided into different fields, as shown in Figure 12.2, that include a different combination of the following: opcode, rs1, rs2, rd, funct3, funct7, and imm, which stands for immediate. The opcode is used to identify the type of instruction used. rs1 and rs2 are used for the source registers, while rd is the destination register. funct3 and funct7 are 3 bits and 7 bits, respectively,

TABLE 12.3
RISC-V Instruction Format

Instruction format	Usage
R-type	Operations between registers
I-type	Immediate and load
S-type	Storing
U-type	Instructions with long immediate values
B-type	Conditional branch operations
J-type	Unconditional jump operations

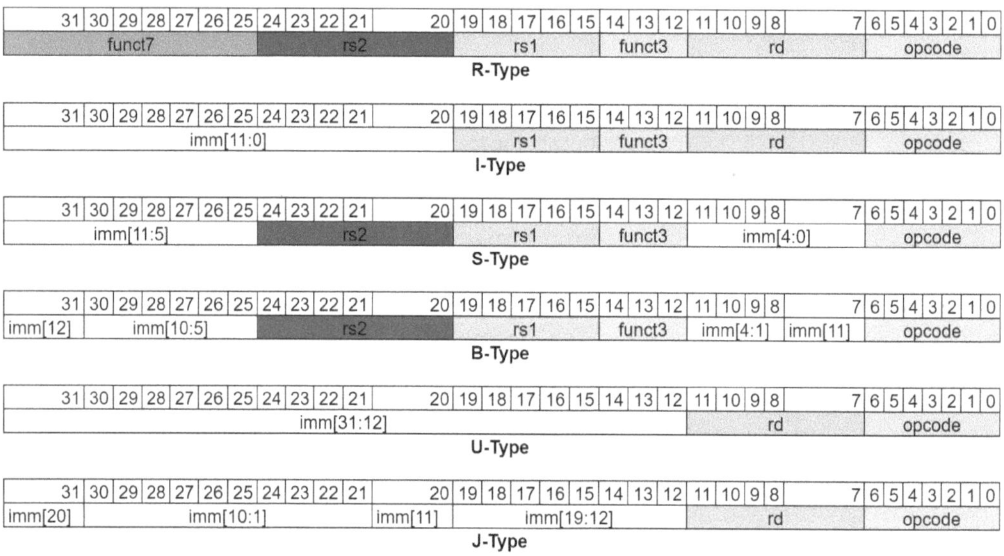

FIGURE 12.2 RISC-V Base Instruction Formats

and they are used along with the opcode to determine the type of operation to be performed. The immediate field stores a constant that is used as part of the operands while executing the instruction.

There are four other instruction formats, and they have the exact same sorts of fields as the R- and I-type formats, but formatted differently or, in the case of immediate fields, a different number and range of bits is used. The S-Type format is used for memory store instructions (load instructions use the I-type format). The main difference here is that the immediate value is split up into different bits. The B-type and J-type formats are for the PC updating instructions (branch and jump). The U-type is used for a couple of special data manipulation instructions to load a 20-bit immediate value into the upper 20 bits of a register, used for forming addresses within a register. Thus, only a destination register index (rd) and the 20-bit immediate number are required after the opcode. No attempt is made to align these immediate bits with the other formats.

The RISC-V designers attempted to keep formats that have a particular field, such as **rd,** in the same bit positions (7 to 11) for all of the different types. This makes decoding easier and faster.

12.3.4.2 Base Integer Instruction Set

The base integer instruction set consists of a set of instructions that are used to perform different operations, such as computational, control, load, store, and memory ordering instructions. Implementing the base integer instruction is mandatory. The base integer instruction set implements minimal operations to keep it small. For example, multiplication and division operations are not included in the base integer instruction set of RISC-V, which are covered by the M extension, which will be further explained later. The sets have different lengths. There is RV32I, RV64I, and RV128I. The name starts with the keyword RV, which stands for RISC-V, followed by the number of bits that represent the address space used and followed by a letter such as I to represent an integer. They are all instruction sets that are 32, 64, and 128 bits wide, respectively. Other instruction sets are available, such as RV32E and RV64E, which are reduced versions of the RV32I and RV64I, respectively. They are designed for the microcontrollers used in embedded systems. The only change is that they have reduced the number of integer registers to 16 from 32 registers. The removal of the upper 16 registers saves around 25% of the core area and power consumption in the corresponding core (Cheng, 2022). Table 12.4 summarizes the categories of the base instruction set, RV32I (Raveendran, 2016).

Furthermore, RISC-V has extensions that increase the capabilities of the ISA. Extensions will be further discussed in the section below.

12.3.4.3 Extensions

RISC-V ISA consists of basic instruction sets and modular ISA extensions. Furthermore, it supports domain-specific custom extensions. Extensions are used to increase the capabilities of the ISA. Extensions are divided into two types: *standard* and *custom*. Standard extensions are supported by the RISC-V International Association, while custom extensions allow users to define their own to cover their specific application needs. Custom extensions are designed to enhance areas in multiple sectors, such as AI, IoT, graphic computing, and many more fields. It is also expected that RISC-V extensions can contribute to areas such as cloud computing, in-network computing, and many more fields. RISC-V ISA and its related subsets and extensions can be categorized, as shown in Figure 12.3.

The non-privileged ISA includes the basic integer instruction set and several non-privileged ISA extensions. Some examples of these extensions are summarized below (Cui, Li and Wei, 2023):

- The *M-extension* is for signed and unsigned integer multiplication and division between two integer registers.
- The *A-extension* is also referred to as the atomic instruction set. It is used to perform read, modify, and write memory atomically. Atomic extension targets multiprocessor and multithreaded environments as it provides synchronization across multiple harts (hardware threads).

TABLE 12.4
RISC-V Base Instruction Set (RV32I) Categories

Category	Instructions
Branch instructions	BEQ, BNE, BLT, BGE, BLTU, BGEU
Load instructions	LB, LH, LW, LBU, LHU
Store instructions	SB, SH, SW
Arithmetic and logic unit (ALU) instructions	ADDI, LUI, SLTI, SLTIU, XORI, ORI, AND, ANDI, SLLI, SRLI, SRAI, ADD, SUB, SLL, SLT, SLTU, XOR, SRL, SRA, OR
Control status register	CSRRW, CSRRS, CSRRC, CSRRWI, CSRRSI, CSRRCI
Miscellaneous	AUIPC, JAL, JALR, FENCE, FENCE.I, ECALL, EBREAK

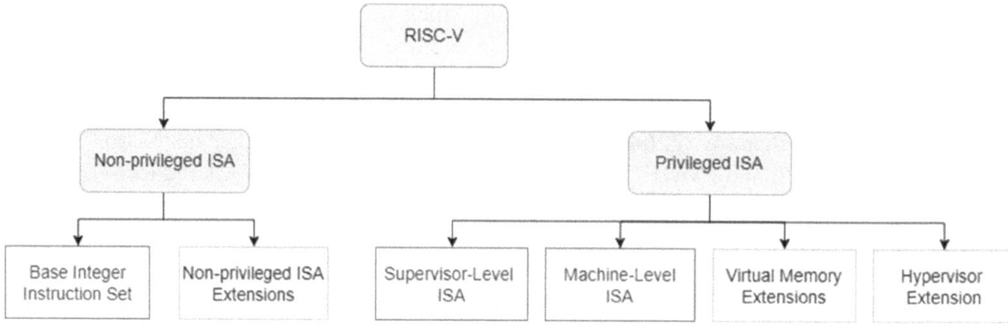

FIGURE 12.3 RISC-V Instruction Set Categories

- The *F-extension* is used for single-precision floating-point operations that follow the IEEE 754–2008 arithmetic standard.
- The *D-extension* is for double-precision floating-point operation as well as to increase the 32 floating-point registers to 64 and hence depends on the F-extension.
- The *Q-extension* is the quad-precision binary floating-point ISA extension, which extends the register to 128-bit. The D-extension is required for the Q-extension.
- The *C-extension* is used for compressing commonly used instructions by replacing 32-bit instructions with 16-bit instructions and can be used to enhance any base ISA.
- The *B-extension* is for bit manipulation, which includes "bit counts, shifts, rotations, insertions, extractions, permutations, etc."
- The *P-extension* is the SIMD instructions extension for efficient parallel data computing for audio, voice, and image processing.
- The *V-extension* is for adding the vector operation instructions.
- The *Zicsr-extension* is for the CSR and the instructions to read-modify-write them atomically.
- The *Zifencei-extension* is for the fetch instruction, which is just one, i.e., Fence. This instruction offers explicit synchronization between writing and fetching on the same hardware thread.
- The *Ztso-extension* is the total store ordering extension.

On the other hand, the privileged ISA extensions include the virtual memory extension and the hypervisor extension to the machine-level ISA and the supervisor-level ISA. The hypervisor extension, also referred to as the H-extension, provides hardware support for efficient virtualization. Its goal is to allow the operating system to run under Types 1, 2, and hybrid hypervisors, as well as provide recursive virtualization. The H-extension also transforms the supervisor mode into a hypervisor-extended mode. It also introduces the virtual supervisor mode, also known as the VS-mode, for the guest operating systems to run and the virtual user mode, also known as the VU mode, for the user applications to run. The H-extension also provides extra Control and Status Registers (CSRs) and hypervisor instructions. It also increases the machine CSRs for the guest virtual memory control. The virtual memory extensions used are Sv32, Sv39, Sv48, and Sv57. Each of the extensions mentioned has a number that corresponds to the number of bits in the virtual address space. The translation between virtual and physical addresses happens through multi-level page tables. Examples of custom extensions for different fields, including IoT, AI, and communication, are shown in Table 12.5 (Cui, Li and Wei, 2023).

12.3.4.4 Profiles

The inherent flexibility found in the way the RISC-V ISA was perceived and implemented might lead to numerous possible ISA choices. The use of profiles simplifies this by specifying a smaller common

TABLE 12.5
RISC-V Custom Extensions in Different Fields and Applications

Domain	Possible Applications	Summary Description
AI	Deep neural network (DNN) acceleration	DNN weights compression to speed up inference
	Convolutional neural network (CNN) acceleration	Convolution operation instruction extension to reduce computing time
	Graph convolutional network (GCN) acceleration	Feature vector aggregation and combination instruction extensions to speed up the GCN calculations
	Transformer acceleration	Improving the performance efficiency of activation and Softmax instruction extensions
Communication	Signal processing	Edge Q has more than 50 custom instructions and extensions designed for baseband signal processing such as FFT, complex modulation, equalization, matrix decomposition, etc.
	Error correction coding	Error correction codes related instruction extensions to perform absolute value, addition, subtraction, multiplication, minimum, etc.
	Radio resource management	Tanh and sigmoid instruction extensions to speed up recurrent neural network (RNN) calculations in RNN-based 5G radio resource management
Graphics computing	Single instruction multiple thread (SIMT) execution	SIMT instruction extensions for the GPU vortex
	Graphics rendering	Graphics rendering instruction extension to boost performance while reducing the power consumption
HPC	Global address space	Extended base global address space (xBGAS) extension
	Remote atomic	Remote atomic extension (RAE)
IoT	Floating-point computation	Introducing lower precision floating-point formats
	Digital signal processing (DSP)	DSP extension that includes dot-product and shuffle instructions to improve efficiency Wireless DSP algorithms: Fast Fourier Transform (FFT) computation, gain control
Post-quantum (PQ) cryptography	Module learning with error (LWE) cryptography	Cooley–Tukey (CT) Butterfly and Gentleman–Sande (GS) to accelerate Number Theoretic Transform (NTT), and Barrett/reduce 32 reduction to accelerate modular reductions
	Ring-LWE cryptography	Four instructions proposed: SHA256, polynomial multiplication, Chien search, and modulo reduction
	Code-based cryptography	Galois field instruction extension to accelerate Classic McEliece cryptography
Security	Memory attack protection	Memory protection during hardware operations
	Control-flow integrity	Bratter for forward control flow integrity
	Control-flow attack protection	Morpheus II for control-flow attack protection that resists attacks by code pointers and code encryption
Virtualization	Delegated virtualization	Delegated virtualization extension to reduce virtualization overhead

set of ISA choices that capture the most value for users, with an interface built around a well-defined set of extensions designed to assist software developers and support a certain set of applications.

In essence, profiles is a set of a base ISA, a group of mandatory extensions, and a group of optional extensions. The profiles are documented in a specific format. It starts with the keyword RV. It is then followed by either A for application or I for integer (base) unprivileged instructions. Followed by the release year, then the privilege mode, then the base length. Managing extensions individually can be difficult. Profiles make it easy for implementers to cover specific use cases without having to implement each of the desired extensions individually. The intention is to define a shared set of compatible extensions.

For the base integer set, the RVI20U profile does not consist of any mandatory extension but contains an initial set of extensions and has two variants, which are the 32-bit and the 64-bit. There are two ratified profiles (RVA20 and RVA22) for the application profile, as well as a new iteration, RVA23. The application profiles are 64-bit, with a long list of mandatory extensions and a small list of options. Profiles are also divided into two categories, which are the major and minor versions. This division was created to balance stability and improvement.[12]

12.3.4.5 Memory Overview

A RISC-V hart has a single byte-addressable address space of 2^{XLEN} bytes for all memory accesses, where XLEN represents the width of an integer register in bits (either 32 or 64). A *word* of memory is defined as 32 bits (4 bytes). Correspondingly, a *halfword* is 16 bits (2 bytes), a *doubleword* is 64 bits (8 bytes), and a *quadword* is 128 bits (16 bytes). The memory address space is circular so that the byte at address $2^{XLEN} - 1$ is adjacent to the byte at address zero. RISC-V supports both little-endian and big-endian formats, with little-endian being the most common implementation.

It uses load and store instructions, using register operands to access memory.

There are different addressing modes, and the most common are: *Base+Offset* addressing, *Immediate* addressing, and *Register-Indirect* addressing. The base plus offset calculates the address by adding the base register value to an offset value. Immediate addressing is when the address is directly specified as the immediate value. Finally, the Register-Indirect is based on an address stored in a register. This mode is useful for memory locations to be determined at runtime, such as pointers or dynamic memory (Patterson and Hennessy, 2017).

RISC-V supports both virtual and physical memory. Virtual memory is used for memory isolation, memory management, and address space extension. Memory isolation enhances security by providing a separate address space for each process. Memory management helps reduce memory fragmentation and improve performance. Systems with limited physical memory benefit from the address space extension that the virtual memory provides. The virtual memory is managed using a multi-level page table mechanism. The translation is performed using a series of table lookups. The number of levels is based on the address space size and the page size. The larger address spaces and smaller page sizes necessitate a higher number of levels. The translation between the virtual and physical memory in the RISC-V architecture is managed by the memory management unit. The privileged architecture (Supervisor Mode, Machine Mode) provides control over memory protection and mapping.

Cache design in RISC-V follows the general established guiding principles that are common in most architectures, with the possibility of customization to a specific ISA because of its open architecture. Processors typically use a 2-Level Cache, taking into consideration design constraints, such as performance and power. In multi-core environments, RISC-V supports cache coherence protocols, such as MESI, which ensures that all processor cores have a consistent view of memory.

12.3.4.6 Exceptions and Interrupts

Exceptions and interrupts are meant to handle unexpected events. Exceptions typically refer to unexpected changes in the control flow regardless of whether the cause is internal or external, and the term interrupt is typically used to refer to externally caused events. Both events are asynchronous in nature. Exceptions occur during run-time, such as when the processor tries to execute an undefined instruction or a memory access violation. Interrupts, on the other hand, occur independent of the program being executed and are due to requests from external devices, such as an I/O device, requesting attention. The operating system will then take the appropriate action. In RISC-V-related terminologies, the term trap refers to the transfer of control to a trap handler, and this is due to the occurrence of an unexpected or exceptional condition within a RISC thread.

Traps encountered by a hart at runtime can have four different scenarios[13]:

- *Contained trap:* The trap is visible to and handled by software running inside the execution environment. The control is transferred to a supervisor mode handler running on the same hart.
- *Requested trap:* The trap is a synchronous exception that is an explicit call to the execution environment requesting an action on behalf of the software inside the execution environment. A system call is one example.
- *Invisible trap:* The trap is handled transparently by the execution environment, and execution resumes normally after the trap is handled. Examples include emulating missing instructions, handling non-resident page faults in a demand-paged virtual memory system, or handling device interrupts for a different job in a multiuser environment.
- *Fatal trap:* The trap represents a fatal failure and causes the execution environment to terminate execution. Examples include failing a virtual memory page protection check or allowing a watchdog timer to expire.

12.3.5 Ecosystem

For any ISA to thrive, it needs a community that supports it. Developing a new processor requires a broad and rich software ecosystem that covers the entire development process (Mezger et al., 2022). It also needs means to realize generated designs on hardware. In the list below, we discuss some of the environments that support RISC-V:

- *Implementations:* As we will discuss later, there are numerous implementations that use RISC-V designs. These range from processor cores (e.g., Ascalon and XuanTie C910) to complete chips and accelerators (e.g., ET-SoC-1 and Veyron V2). There are also single-board computers that allow users to develop and test RISC-V code on affordable hardware (e.g., Allwinner D1 and HiFive Premier P550).
- *Toolchain:* These are essential tools for developing programs that can run efficiently on RISC-V hardware, which include assemblers, compilers, linkers, and debuggers. Such tools allow developers to write a program in high-level languages, such as C/C++, compile it and generate optimized machine code for RISC-V hardware, combine generated object files into an executable file, inspect variables, and monitor code execution. Examples include compilers such as Chisel,[13] LLVM/CLANG, and GCC, as well as supporting tools, such as Binutils, Glibc, GDB, and Newlib.
- *Services:* These include software development kits (SDKs) and integrated development environments (IDEs), such as RiscFree; code analysis and optimization platforms, such as Siemens EDA (formerly known as Mentor Graphics); and security and IP protection for embedded systems, such as Emproof Nyx. In addition, some cloud environments support RISC-V, such as CloudV and Scaleway, which offer cloud-based RISC-V environments that allow programmers to write/test code for RISC-V without having to own the necessary hardware.
- *Hypervisors:* Various operating system-level virtualization solutions support RISC-V, such as KVM, Salus, Xvisor, ZAYA, Kubernetes, and Docker.
- *Bootloaders:* Computing systems need bootloaders (firmware) to initiate operating systems. Fortunately, several bootloaders support RISC-V, including the Berkeley boot loader, CoreBoot, OpenSBI, and Bao Project.

- *Operating systems:* Numerous operating systems support RISC-V, including both real-time operating systems (RTOS) and general operating systems. Examples of RTOS that support RISC-V include AliOS, Eclipse ThreadX (formerly known as Azure RTOS), embOS, FreeRTOS, RIOT, RT-Thread, and Zephyr. In addition, there is support for RISC-V in some Linux kernel distributions, such as seL4 and Little Kernel, which are widely used in embedded systems and IoT. For general-purpose computers, common operating systems such as FreeBSD, Haiku, and OpenBSD, as well as major Linux distributions, such as Debian, Fedora, OpenSuse, and Ubuntu, support RISC-V.
- *Emulators and simulators:* There are various emulators and simulators that support RISC-V, such as EmulsiV, Jupiter, OVPsim, QEMU, and Spike. Such tools enable programmers to test their code in RISC-V-like (software) environments without the need for physical RISC-V hardware. They also help designers verify their RISC-V hardware designs.
- *Runtimes:* Several language runtimes support RISC-V, including Dart, Go, OpenJDK, PHP, Python, Rust, and.NET. There are also runtime libraries for specific use cases, such as cloud native/edge applications, network applications, packet processing applications, and storage applications. These can be found in frameworks such as the WasmEdge Runtime, DPDK, and SPDK, respectively.
- *Libraries:* There are a variety of libraries that support RISC-V and cover a wide range of use cases. This includes security libraries such as OpenSSL and wolfSSL; AI/ML libraries such as PyTorch and TensorFlow; media codecs such as FFmpeg, Speex, and FLAC; computer vision libraries such as OpenCV; and parallel computing APIs such as OpenMP.
- *Applications:* Several user applications, databases, and web hosting software environments support RISC-V, including LibreOffice, MySQL, WordPress, Redis, Apache, TLS, SSH, and ONNX.

12.3.6 Processors

According to the RISC-V International organization, there are currently over 13B RISC-V cores available on the market,[14] which vary in their designs and targeted use cases. Their implementations also take many forms, ranging from soft-core implementations to hard-core implementations. Soft-core implementations on programmable logic devices, e.g., FPGAs (field-programmable gate arrays), are affordable and reprogrammable. On the other hand, hard-core implementations, e.g., ASICs (application-specific integrated circuits) and SoCs, are more performant and power efficient.

RISC-V is being used in deeply embedded systems (e.g., programmable alternatives for finite state machines), microcontrollers, co-processors (e.g., DSPs and accelerators), and applications processors. As we will discuss below, the ISA can be found in in-house, open-source, and commercial (closed-source) implementations.

12.3.6.1 In-house Implementations

Various companies have used RISC-V cores internally and in their products, as we discuss below:

- *Nvidia:* In 2016, Nvidia decided to replace its proprietary microcontroller, known as Falcon, with a RISC-V-based microcontroller. Falcon was used in over 15 designs of Nvidia's GPUs and shipped in over 3B chips.[15] The new core, known as NV-RISCV, is a high-performance core that implements the RV64IM_Sdef ISA and has a 5-stage dual-issue in-order pipeline.[16]
- *Western Digital:* In 2019, Western Digital decided to abandon its proprietary processors and rely on RISC-V cores in its solid-state drive controllers. As a result, Western Digital developed SweRV, a 32-bit superscalar core with a 9-stage, dual-issue, in-order pipeline (Wheeler, 2019). The company later contributed SweRV to the CHIPS Alliance, which currently maintains it under the name VeeR (Very Efficient & Elegant RISC-V). SweRV has three variances, i.e., EH1, EH2, and EL2.

- *Seagate:* In 2020, Seagate announced that it had developed two in-house RISC-V-based cores for its hard disk drive (HDD) controllers.[17] Examples of Seagate products that use RISC-V include Mozaic 3+, a 30 TB HDD that targets datacenters. Mozaic 3+ implemented its controller on a 12 nm chip, which is smaller (vs. 28 nm) and offers improved reliability, customizability, and security, as well as 3x performance over its predecessors.
- *Qualcomm:* Qualcomm chips can be found in many Android mobile devices, most of which use ARM cores. Vendor lock-ins and recent lawsuits with ARM have pushed the company to look for alternatives to ARM designs. Qualcomm is a founding member of RISC-V International, and since 2019, it started using RISC-V cores in its products, including the Snapdragon SoCs. To date, it has shipped almost 1B chips with RISC-V cores – although not as main processors. Qualcomm is also working with Google to build next-generation Wear OS solutions using a RISC-V-based Snapdragon Wear platform.[18]
- *Google:* Starting with Pixel 3, Google separated the Trusted Execution Environment, an area on the chip used to generate and store security keys, from the main chip to enable a higher level of security. Google first developed the Titan M using proprietary technology, but then, in 2021, it developed a new generation known as the Titan M2 using RISC-V. The chip was first introduced in Pixel 6. It has its own memory, cache, and cryptographic accelerator, thereby eliminating the need to share encryption keys with the main processor. Google is a founding member of RISC-V International.

12.3.6.2 Open-Source Implementations

As previously stated, RISC-V is an open standard, not an open-source processor. However, over 100 open-source RISC-V cores, processors, and SoCs are available for research, educational, and industrial purposes.[19] RISC-V open-source cores come from both research and industry, and they vary in their designs, capabilities, and targeted application domains (Dörflinger et al., 2021), as highlighted in the few examples below:

- *BOOM:* An out-of-order, industry-grade, Linux-capable core developed, originally for educational purposes, by UC Berkeley (Celio et al., 2015). BOOM is written in Chisel and is based on the RV64GC ISA. There are three generations of this core, with Sonicboom representing the third generation (Zhao et al., 2020). The core can be implemented on both ASICs and FPGAs.
- *BlackParrot:* A Linux-capable, cache-coherent, multicore processor, developed by the University of Washington and Boston University (Petrisko et al., 2020). BlackParrot is written in SystemVerilog and is based on the RV64GC ISA, with a variance that supports the RV64GCBSU_Zfencei_Zicsr_Zcbo ISA. It can be used in lightweight accelerators and as standalone Linux cores. BlackParrot is scalable and can support up to 16-core (per chip) configurations.
- *CV32E40P:* Formerly known as RI5CY, CV32E40P is a single-issue core with a 4-stage in-order pipeline (Gautschi et al., 2017). CV32E40P is based on the RI5CY core, which was developed by the Parallel Ultra-Low-Power (PULP) Platform[20] and is currently maintained by the OpenHW Group. CV32E40P is written in SystemVerilog and implements RV32IM[F|Zfinx]C instructions and the PULP custom extension. CV32E40P is designed for ultra-low-power signal processing applications and is suitable for IoT endpoint devices.
- *Ibex:* Formerly known as "Zero-risky," it is a single-issue core with a 2-stage in-order pipeline (Schiavone et al., 2017). Ibex was also developed by the PULB Platform and is currently maintained by lowRISC.[21] Ibex is written in SystemVerilog, supports RV32IMCB/RV32EMCB, and is highly parameterizable, allowing programmers to configure it in various ways for different use cases. Ibex is used in Microsoft CHERIoT, a security platform for low-cost embedded systems (Amar et al., 2023).

- *SCR1:* An industry-grade and silicon-proven microcontroller core with a 2–4-stage pipeline.[22] The processor is developed by Syntacore. SCR1, written in SystemVerilog, implements the RV32I or RV32E instruction sets and can optionally use the M and C extensions. Only the M-mode privilege level is used. SCR1 is optimized for power and area and targets embedded systems and IoT.
- *XiangShan:* An industry-grade multi-core processor with an 11-stage, out-of-order, multi-issue superscalar pipeline. XiangShan is developed by the Institute of Computing Technology (ICT), the CAS, and the Peng Cheng Laboratory (Xu et al., 2022; Wang et al., 2023). There are three generations of XiangShan, all written in Chisel, namely, Yanqihu (based on RV64GC), Nanhu (based on RV64GCBK and achieves over 40% performance improvement over Yanqihu), and Kunminghu (still under development and will add the vector and hypervisor extensions). XiangShan is considered one of the highest-performance open-source RISC-V cores in the world.[23]
- *XuanTie C910:* An industry-grade, Linux-capable multicore (up to 4 cores per cluster) processor with a 12-stage, out-of-order, multi-issue superscalar pipeline (Chen et al., 2020). The processor is developed by T-Head Semiconductors, from Alibaba. It is based on the RV64GCV ISA and implements the XuanTie Instruction Extension. C910 is used in the T-head RVB-ICE development boards, which support Android and Debian operating systems. Moreover, FPGA-based implementations have been deployed in the Alibaba Cloud datacenters.

12.3.6.3 Commercial Implementations

There are thousands of closed-source, commercial RISC-V implementations offered by both industry leaders and startups. In the list below, we introduce some of these implementations:

- *Ascalon:* Developed by Tensorrent, Ascalon is a high-performance, out-of-order, superscalar, 8-wide decode processor.[24] The processor implements the RV64ACDHFMV instruction sets and targets automotive, wireless communication, client computing devices, servers, and datacenters. Ascalon is offered as a standalone IP processor or on a cluster (up to 8 cores). It is also used in other Tensorent products, such as the Wormhole PCIe (Peripheral Component Interconnect express) board accelerator.
- *AX45MP:* Developed by Andes, AX45MP is a high-performance multicore processor (up to 8 cores) with an in-order, dual-issue, 8-stage, superscalar pipeline.[25] The processor implements the RV64GCP instruction set and targets automotive, high-performance applications, video and image processing, ML acceleration, and network controllers.
- *E20:* Developed by SiFive, E20 is a simple core with a 2-stage, single-issue, in-order pipeline.[26] The core only supports M-mode operations and uses the RV32IMC ISA. E20 is considered one of the smallest, most power- and area-efficient RISC-V cores available. It targets microcontroller applications such as analog/mixed-signal processing and IoT.
- *ET-SoC-1:* A SoC developed by Esperanto for large-language model inferencing and massively parallel workloads (e.g., DSP applications), both at the edge and on the cloud.[27] The chip has 1,088 64-bit low-power, single-threaded, general-purpose RISC-V cores with an in-order pipeline known as *ET-Minion* cores. Each ET-Minion core has a custom tensor/vector unit optimized for ML inference applications. Additionally, the chip has four high-performance 64-bit RISC-V cores with quad-issue out-of-order execution, known as *ET-Maxion* cores, for resource management, data movement, and the host operating system.
- *I8500 Series:* A high-performance multicore, multithreaded processor with a 3-wide, 9-stage, in-order pipeline.[28] The processor is developed by MIPS. It implements RV64GHZ, Bitmanip Zba and Zbb, and CMO instruction sets. The I8500 Series is suitable for datacenter, IIoT, and automotive applications.
- *NOEL-V:* A space-grade, high-performance, fault- and radiation-tolerant processor with an advanced dual-issue in-order pipeline.[29] NOEL-V is developed by Frontgrade Gaisler, a prominent provider of aerospace embedded systems.[30] NOEL-V implements the

RV32IMAFDBC[H]/RV32IMAFDBC[H] instructions and is offered as a synthesizable VHDL model. An open-source version of NOEL-V is also available.

12.3.7 Power Consumption

Power consumption has become a significant design parameter, particularly when related to the energy demands of personal computing devices. Researchers are constantly searching for new design techniques and methods to mitigate its effects.

ARM has been a major player in embedded systems applications and System-on-Chips (SoCs). However, in recent years, RISC-V-based products have become more accessible as more companies and organizations adopt and support RISC-V. There has been a growing demand for alternative hardware platforms that implement different ISAs with superior performance per watt (Elsadek and Tawfik, 2021). In this context, both ARM and RISC-V are noteworthy since they have advantages over traditional architectures such as X86. RISC-V has the added value of being an open specification, providing a needed opportunity for innovation and creativity in designing processor cores that are energy efficient.

In this section, we first discuss the factors that influence processors' power consumption, and we follow that by presenting example cases of RISC-V implementations where power consumption and energy efficiency were primary design objectives.

12.3.7.1 Factors Affecting Power Consumption

Power consumption in complementary metal-oxide-semiconductor (CMOS) circuits, the technology of choice for portable electronic design, can be categorized into static and dynamic power.

Static power is mainly determined by the type of transistors and process technology. The static power consumption, or leakage power, is caused by factors, such as MOS subthreshold and gate leakage currents, that are present in any active circuit, independently of the circuit clock rates or scenarios. Scaling down the supply voltage has played a major role in adding to the static power consumption of circuits in this era of deep-submicron. This is because the threshold voltage must be scaled down to avoid affecting small geometry device delays. However, this reduction in the threshold voltage by the same factor as the supply voltage in small geometry MOSFETS usually results in a noticeable increase in standby current and, subsequently, an increase in static power dissipation.

Dynamic power consumption is created by circuit activity (i.e., circuit nodes switching values) and depends mainly on a specific usage scenario, clock rates, and I/O activity. The sources of the dynamic power consumption are short-circuit current and switched capacitance. Short-circuit current is a small percentage of the total power consumption, and so far, no way has been found to reduce this value without compromising the performance. Switched capacitance is the primary source of the dynamic power consumption; therefore, the dynamic power consumption can be defined as:

$$P_{\text{dynamic}} = aCV^2f \tag{2}$$

where α is the switching activity, C is the physical capacitance, V is the supply voltage, and f is the clock frequency (Weste and Harris, 2011).

Therefore, from the above equation, power consumption reduction can be achieved by reducing the supply voltage, reducing the output capacitance, reducing the clock frequency, and reducing the switching activity.

Designers use various techniques at different levels of abstraction to try and decrease the effect of each parameter above or a combination of them. The values of switching activity and capacitance are determined by the low-level system design. The reduction of both depends to a great extent on the designers' experience and creativity. For example, reducing the switching activity might require a careful redesign of circuit nodes that are exhibiting high activity, balancing the various logic paths, and selecting the appropriate logic style. On the other hand, the capacitive load can be optimized by using techniques, such as technology scaling, appropriate logic family selection, and transistor resizing (Wiltgen et al., 2013).

To reduce the effect of the supply voltage and clock frequency, a dynamic power management technique called dynamic voltage and frequency scaling (DVFS) is used.

The main idea behind this approach is to downscale the CPU performance when it is not fully utilized by decreasing the voltage and frequency of the CPU. In the ideal case, this should result in a cubic reduction of the dynamic power consumption. DVFS is supported by most modern CPUs, including mobile, desktop, and server systems.

12.3.7.2 Power in RISC-V Implementations

The fact that RISC-V distinguishes itself as an open specification provides a great opportunity for innovation and creativity in designing processor cores. Cores can be designed to target different application domains, from high-performance needs to lower-power applications. In the case of resource-constrained electronic devices targeting IoT applications, the set-up would typically be battery-dependent; hence, this limitation on energy supply requires the use of processors that are energy efficient to allow for longer battery life. For companies with the end-user in mind, reducing power while optimizing performance is very important. It can mean the difference, for example, between watching an entire video on a smartphone or only a segment of it; in the medical field, too, the energy efficiency of implanted devices with built-in sensors is critical.

Companies and academic institutions that provided open-source core implementations include SiFive, Qualcomm, T-head, and research groups at MIT and UC Berkeley, among others (Dörflinger et al., 2021). No doubt, the emerging and different implementations of RISC-V processors underline the success of its ISA specification.

Comparisons between these implementations are not straightforward because results are reported for varying technologies and configurations. It is not easy to say which implementation fits best for a system with given requirements, such as processing performance or power consumption.

In general, RISC-V platforms allow for various low-power modes, enabling fine-grained control over the power consumption of the processor. These modes include idle, sleep, and deep sleep. Depending, different components of the system can be selectively powered down or put into a low-power state when they are not active. By utilizing low-power modes intelligently, developers can achieve substantial power savings without compromising system performance.

Low-power cores are typically characterized by a lower pipeline depth and 32-bit ISA usage. A few examples of cores designed with low power consumption in mind targeting resource-constrained devices, include SCR1, Ibex, and E20 (Elsadek and Tawfik, 2021):

There are a number of designs targeting IoT applications that are based on these cores. For example, Mr. Wolf (Pullini et al., 2019), which is an energy-efficient system on chip (SoC) for IoT edge processing, uses the Ibex core as the controller; the core has an efficient data transfer subsystem for communication with peripherals.

12.3.8 Security

One of the arguments made in favor of RISC-V openness is that the best way to build a secure platform is to make the entire design open and available for public scrutiny, thereby allowing the entire community to work on securing it. However, the risks of open-source platforms cannot be overlooked; established commercial architectures, such as those of ARM and Intel, have a longer and proven track record and a plethora of techniques and solutions for security-related issues and otherwise.

Nicole Fern, a senior security analyst at Riscure, warns that "there's also a risk of threats being intentionally introduced. "Because anyone can publish an open-source RISC-V implementation, there are more opportunities for bad actors to publish designs with known vulnerabilities or hardware Trojans," Fern said. "For open-source projects with multiple contributors, there is opportunity for untrusted entities to try and push vulnerable patches or updates into otherwise trustworthy open-source projects."[31]

Security threats include hardware Trojans (HTs) that cause hardware vulnerabilities, Side Channels attacks, and the recently disclosed speculation-based timing attacks Meltdown and Spectre, plus a few others (Marchese, 2020). A modification or undesired alteration to existing circuit elements done in order to apply malicious activity is called HT. HTs possess harmful effects, including changing the functionality of a circuit, reducing its reliability, or leaking valuable information (Cassano et al., 2022). A side channel attack is defined as a technique of stealing information indirectly by exploiting unintended information leakage, such as side-channel signals generated during computation. These channel signals allow for the measurement and analysis of physical parameters, such as supply current, execution time, and electromagnetic emission that, in turn, can be utilized to compromise the system security. Spectre and Meltdown are two security deficiencies that can provide access to personal data for hackers and can potentially affect the performance of Linux, Windows, and MAC devices (Abu-Ghazaleh, Ponomarev, and Evtyushkin, 2019).

Over the last few years, many general security and memory safety techniques have shifted from being software-based to hardware, such as in ARM Trustzone and Intel CET. Similar approaches have been followed in RISC-V as well.

The RISC-V architecture utilizes three privilege modes, namely, Machine Mode (M-mode), Supervisor Mode (S-mode), and User Mode (U-mode). A combination of these modes is feasible as well, for example, (M,U) for embedded systems with security. The privilege level is used to provide differentiated protection for different components of the software stack, which lays the foundation for the security of the RISC-V platform. M-mode is the highest privilege level. It is the only mandatory privilege level of the RISC-V hardware platform. Code running in M mode is usually inherently trustworthy because it has low-level access to the machine implementation. M-mode can be used to manage the secure execution environment on RISC-V. Many RISC-V implementations support U-mode to protect the rest of the system from application code. S-mode can be added to provide isolation between a supervisor-level operating system and the Supervisor Execution Environment (Lu, 2021).

Each privilege level has a core set of privileged ISA extensions, as well as optional extensions and variants. For example, the M-mode supports an optional standard extension for physical memory protection (PMP). The privilege mechanism also allows for flexible routing of traps (e.g., supervisor call or timer interrupt) to different privilege levels, which is an important security enabler of RISC-V.

For memory, RISC-V offers a practical solution to address security concerns by using the PMP unit within the RISC-V architecture, which plays a critical role in enhancing its security by controlling access to physical memory addresses. It is a programmable hardware component that allows the definition of multiple memory regions, each governed by its own access policy per processor core. This feature guarantees that software running on a given processor core abides by its specified read, write, and execute permissions depending on the execution mode of the core. The M-mode of privilege has access to the entire machine after reset; it controls the PMPs and can dynamically swap PMP settings to run different security contexts on a hart.

The case for continued research efforts intended to close the doors on security threats within a RISC-V environment is made clear by a recent work that discussed microarchitectural attacks on RISC-V CPUs (Gerlach et al., 2023). Two commercially available off-the-shelf 64-bit RISC-V (hardware) CPUs used in most RISC-V systems running a full-fledged commodity Linux system were analyzed. Specifically, the microarchitectural attack surface analysis was performed on the SiFive U74 and the XuanTie C906, which are two of the first COTS 64-bit RISC-V CPUs. During the course of this analysis, three new attack techniques based on RISC-V-specific design and implementation decisions are identified.

Researchers demonstrated that known attacks, such as Flush+Reload and Prime+ Probe, are applicable to these CPUs. Furthermore, it is demonstrated that keys from AES and RSA implementations can be leaked, KASLR can be broken, and file names from inaccessible write-only folders can be identified.

As SoCs continue to grow in complexity and application scope, the importance of robust security mechanisms cannot be overstated. It is clear that hardening an ISA and the microarchitecture should be included in CPU designs to reduce the attack surface and ensure the security of future devices. Researchers continue to explore the design, implementation, verification, and evaluation of microarchitecture security features, hardware-assisted security techniques, and secure executions around the RISC-V ISA.

12.3.9 Key Alliances and Vendors

Many industry, technical, and regional alliances support RISC-V.[32] There are also numerous vendors that design and implement RISC-V processors.

From a long list of many alliances, we highlight the following three:

- *Common Hardware for Interfaces, Processors, and Systems (CHIPS) Alliance:* An organization that develops and hosts high-quality open-source hardware (e.g., IP cores and interconnect IP) and software (e.g., design and verification tools) for FPGAs and ASICs.[33] Members include industry leaders such as AMD, Antmicro, Google, Infineon, Intel, Microsoft, Meta, Micron, Nvidia, and SiFive, as well as research institutions such as UC Berkeley, the Institute of Software, the Chinese Academy of Sciences (ISCAS), Standford Engineering, and Yale University. The CHIPS Alliance manages several RISC-V projects, including OmniXtend, a cache coherence framework for RISC-V; RISCV-DV, an open-source verification tool for RISC-V designs; RocketChip, a SoC generator that instantiates the RISC-V Rocket Core and related items (Asanović et al., 2016); and the VeeR cores. It also manages the development of the Chisel language.
- *OpenHW Group:* Similar to the CHIPS Alliance, the OpenHW Group is an organization that develops and hosts high-quality open-source hardware and related software and verification tools.[34] Members include industry leaders such as Alibaba, Amazon, Ashling, Bluespec, Beagle Board, Codasip, Red Hat, and Siemens, as well as research institutions such as the Barcelona Supercomputing Center (BSC), the Beijing Institute of Open Source Chip (BOSC), ETH Zurich, ISCAS, Technion, and the University of Bologna. Among the projects the OpenHW Group manages are several cores developed by the PULP Platform, CVA5 (formerly known as TAIGA; Matthews and Shannon, 2017), and CVE2 (based on the Ibex core).
- *RISC-V Software Ecosystem (RISE):* The project is managed by the Linux Foundation, through which several members are working together in collaboration with RISC-V International on developing high-quality, open-source software and toolchains for RISC-V.[35] The key market segments of the project include automotive, consumer electronics, datacenter, and mobile. Members include industry leaders such as Alibaba, Andes, Google, Intel, Nvidia, Qualcomm, Red Hat, Samsung, and SiFive, as well as research institutions such as BOSC and ISCAS.

From the many vendors that develop RISC-V-based designs, we highlight the following:

- *Alibaba:* A major Chinese technology company and a founding member of the RISE project. Alibaba is one of the key adopters of RISC-V and has built several RISC-V-based cores under the XuanTie 9 Series. The series has three families of processors designed for various workloads and use cases. The C Series is for edge and AI applications. The E Series targets low-energy microcontroller units and IoT. The R Series is well-suited for high-performance real-time systems, including IIoT devices such as control systems and vehicle mount terminals.
- *Andes Technologies:* A Taiwanese semiconductor company and one of the main suppliers of RISC cores for embedded systems. Andes is a founding member of RISC-V International. It offers a wide range of high-performance, power-efficient 32-bit and 64-bit,

in-order and out-of-order RISC-V cores for various applications. The cores vary in complexity and features, ranging from RTOS to Linux support, vector processing support, and DSP and packed SIMD instructions support.

- *Codasip:* A European startup founded in the Czech Republic and headquartered in Germany. Codasip is a founding member of RISC-V International. The company offers two families of IP cores with different word lengths, arithmetic units, and pipeline depths, namely, the Embedded Processors and the Application Processors families. The Embedded Processors family includes the L11 and L31 cores, which target sensors, IoT, AI, wireless devices, cameras, and wearables. The Application Processors family, which includes the A70 and A730 cores, is well-suited for AI, edge IoT, and sensor fusion. Codasip allows its clients to fully customize their cores by offering baseline designs (in its proprietary HDL, The Codasip Architectural Language (CodAL)[36]), which can be modified according to application needs.
- *Esperanto Technologies:* An American company specializing in developing high-performance and energy-efficient solutions for AI and HPC applications based on RISC-V. Esperanto is a founding member of RISC-V International. The company is known for its ET-SoC-1 chip, which is offered as IP and also shipped on its PCIe cards and servers. An Esperanto server can be configured to have either 8 or 16 ET-SoC-1 chips. Additionally, Esperanto provides software development tools for optimizing trained ML models for parallel processing on its chips, with support for major ML frameworks, including TensorFlow and PyTorch.
- *MIPS:* A pioneer of RISC processors, with a history that dates back to the 1980s. MIPS is considered one of ARM's main competitors. Its cores are widely used in automotive systems, embedded systems, network systems, and datacenters. MIPS offers two models of its RISC-V processor, eVocore, namely, the P8700 Series and the I8500 Series.
- *Renesas:* A Japanese semiconductor company and a leading provider of microcontrollers. The company specializes in providing RISC-V solutions for embedded systems, IoT devices, and IoT edge applications. Among these solutions is RZ/Five, a general-purpose 64-bit, 1GHz microprocessor that uses the Andes AX45MP RISC-V core. RZ/Five is ideal for IoT gateways, such as home security systems and solar inverters.[37]
- *SiFive:* An American company founded in 2015 by the team that created RISC-V. SiFive, a founding member of RISC-V International and a leading provider of RISC-V IP processors, was the first company to build RISC-V chips. SiFive has four families of processors that cover a broad range of applications, from IoT to high-performance applications. The Essential Family is intended for IoT and embedded systems. The Automotive Family offers a variety of processors for automotive systems. The Intelligent Family is for AI and vector processing applications. Finally, the Performance Family, which offers various high-performance processors for mobile, edge, and datacenter applications. SiFive also offers several development boards.
- *Syntacore:* A founding member of RISC-V International, which offers two classes of RISC-V cores, namely, the Application Class and the Microcontroller Class. The Application Class offers several Linux-capable, energy-efficient cores for IoT, storage devices, network devices, automotive, and industrial control systems. The Microcontroller Class is for embedded systems and small devices.

12.3.10 Global Initiatives

Since the emergence of RISC-V, several nations, including the US,[38] have shown interest in it, with plans to use it one way or another. As we discuss below, we see most activities taking place in the EU *and several countries from the BRICS.*[39]

12.3.10.1 The European Union (EU)

The EU identified the lack of European hardware and the heavy reliance on imported chips, which are mainly fabricated in the US and Asia, as a threat to its autonomy. The EU also raised concerns that imported chips could have backdoors that facilitate espionage. The EU currently contributes less than 10% of the global semiconductor supply, which reached 1 trillion chips made globally in 2020. The EU, however, aims to change that and reach a 20% market share by 2030. Through the European Chip Act, the EU is committing €43B to advance its chip industry. One of the key objectives is to develop an ecosystem that is based on open technologies. RISC-V was identified as an essential element of this open ecosystem, and, in turn, the EU is devoting several projects to support it, both directly and indirectly. For instance, the EU Commission announced in 2022 its plans to release €270 million to fund projects that develop RISC-V hardware and software.[40]

The EU is taking several initiatives to achieve chip self-sufficiency, among which is the European Processor Initiative (EPI),[41] a project to build European energy-efficient processors for exascale supercomputers and scientific applications (Kovač, 2022). RISC-V is one of the core technologies being used in building these processors. Thirty-nine research and academic institutions, along with industry leaders from ten European countries, currently manage the project. Research and academic members include BSC, ETH Zurich, Fraunhofer, and the University of Bologna, whereas companies include Atos, BMW Group, GENCI, and Infineon. EPI-funded projects have led to the development of several advanced chips, including the European Accelerator (EPAC), an AI accelerator that uses RISC-V for vector processing.[42] We also mentioned earlier the PULP Platform (developed several low-power open-source RISC-V cores, e.g., RI5CY, Ariane, and Zero-risky), which is a partially funded project by the EPI.

Another key initiative is the Important Project of Common European Interest (IPCEI),[43] under which the EU Commission has approved €8.1B in public funding for building next-generation European chips for research and industrial applications. Currently, 56 companies and 68 research and development (R&D) projects receive funding from the IPCEI. Openchip,[44] a BSC spin-off founded with the goal of building high-performance, power-efficient accelerators using RISC-V for AI and HPC applications, is among the fund recipients.

Similarly, eProcessor is another important EU-funded project (Alvarez, 2023), which aims to build a European open full-stack ecosystem (both hardware and software) based on RISC-V for HPC, AI, data analytics, and bioinformatics applications.[45] The project is managed by the European High Performance Computing Joint Undertaking (EuroHPC JU) and led by BSC.[46] Both research institutions and industry players are participating in this project. Participants from the research community include Chalmers Tekniska Högskola, Universität Bielefeld, and Università degli Studi di Roma, whereas industry participants include Cortus, Extoll, and Thales SA.

Another key project is TRISTAN (stands for **T**ogether for **RIS**c-V **T**echnology and **A**pplicatio**N**s),[47] a research project that aims at developing and commercializing the European RISC-V ecosystem. The project's main goal is to make it simple to build, integrate, and secure embedded systems based on RISC-V by offering a library of industry-grade building blocks necessary for developing SoCs for various application domains. The project covers both electronic design automation (EDA) tools and the full software stack – we may consider TRISTAN as a counterpart to the RISE project (discussed earlier). A diverse consortium of 46 members from 12 countries, including 2 non-EU states (i.e., Israel and Turkey), currently manages the project. Members include leading companies, small and medium-sized enterprises, and research and academic groups working with RISC-V. Industry members include Antmicro, Bosch, Codasip, Infineon, Nokia, and Siemens, whereas research and academic members include ETH Zurich, Fraunhofer, Graz University of Technology, and the Technical University of Munich.

In addition to state-funded projects, several European startups were founded specifically to support RISC-V. For instance, Quintauris[48] is a German-based startup that aims to offer compatible RISC-V-based products, reference architectures, and standards for building solutions for various domains. The company was founded by Bosch, Infineon, Nordic, NXP, and Qualcomm, with an initial focus on automotive, and plans to expand to cover IoT and mobile.

12.3.10.2 China

Over the past few years, the US has imposed several sanctions on the sale of advanced chips, lithography technologies, and EDA tools to China. The list of sanctioned items includes the sale of various Intel, Qualcomm, Nvidia, and other advanced chips to Huawei, as well as banning foreign foundries that use American technology, including Taiwan Semiconductor Manufacturing Company (TSMC), from manufacturing Huawei's chips. China recognizes the significance of the West's control over the chip industry. Intel and AMD are American companies, and ARM is British (now owned by Softbank, a Japanese company). Moreover, most of the chip foundries are located in countries allied with the US. These factors have pushed China to focus more on its chip industry and stay behind its leading foundries.

China has plans to become a global high-tech superpower, shifting its focus from mass production of cheap and low-quality products to advanced technologies for all major industries. To accomplish this, China has implemented many policies over the past two decades (Chen and Naughton, 2016), including the "Made in China 2025" (MIC2025) industrial policy, which was launched as a ten-year plan in 2015 (Levine, 2020; Chen et al., 2024). MIC2025, inspired by Germany's "Industry 4.0" (Zhang et al., 2016; Li, 2018), focuses on enabling "indigenous innovation" by providing government funds, venture capital, and tax exemptions to Chinese technology firms with the aim of improving technological entrepreneurship. The policy clearly states the importance of both designing and manufacturing domestic advanced chips, with a specific focus on AI, HPC, and IoT.

Following the US and EU steps, China has allocated large funds to advance its chip industry. For example, the third round of the National Integrated Circuit Industry Investment Fund (also known as the "Big Fund") managed to collect US $47B from several Chinese banks and government agencies.[49] These funds are helping the Chinese chip industry advance, and although it is still behind its American counterpart, it has achieved considerable progress over the past few years. For instance, China's top (and partially state-owned) foundry, Semiconductor Manufacturing International Corporation, has recently become the third-largest chip foundry in the world, after TSMC and Samsung.[50] In addition, with Huawei's latest chip, the Ascend 910C, China is closing the gap with US top AI chips.The performance of the Ascend 910C is comparable to Nvidia's H100 GPU, a product that has been banned by the U.S. government from being sold to China.

China is a major player in the RISC-V arena, having invested more than US $50 million from 2018 to 2023 and published over 1,060 patents in 2023 alone, some of which have military applications.[51] Moreover, funding for the top ten Chinese RISC-V startups has exceeded US $1.1B. China views RISC-V as a viable option that could help it mitigate some of the impacts of the US sanctions and improve its technological autonomy (Cheung, 2023). This explains why many Chinese entities (over 70) are members of the RISC-V International organization, including premier members such as Alibaba, Huawei, Tencent, and ZTE. China is capitalizing on the global demand for RISC-V processors, with over 300 tech companies involved in designing, manufacturing, and using them. Prominent examples, besides those we just mentioned, include Nuclei, StarFive, and VeriSilicon. In fact, in 2022, China accounted for 50% of the global supply of RISC-V processors (totaled 1B).

China has devoted several initiatives to advance the development and use of RISC-V. This includes the involvement of the ICT in founding the RISC-V Foundation with the goal of staying close to the ISA development. Furthermore, China has created several institutes, including BOSC, the Tsinghua-Berkeley Shenzhen Institute (TBSI), and the RISC-V International Open Source Laboratory (RIOS Lab), which is part of TBSI, all with the goal of connecting industry, academia, and government and integrating resources. These, along with many other research institutes in China, including the CAS and ISCAS, are involved in developing and using RISC-V. In 2018, China established the China RISC-V Alliance (CRVA) with three goals in mind[52]: the delivery of silicon-proven open-source RISC-V-based IP and SoCs within 3–5 years; the development of open-source SoCs using open-source EDA tools within 5–7 years; and to automatically build open-source hardware using open-source EDA tools within 10–15 years.

China's investment in RISC-V has led to the creation of many RISC-V processors, including some of the popular open-source cores such as XuanTie E902, E906, C906, and C910 from Alibaba, XiangShan from CAS (in collaboration with Alibaba, Tencent, and ZTE), and Hummingbird E203 from Nuclei. China has been successful not only at designing and building efficient RISC-V processors but also using them in both research and commercial applications. For instance, researchers from the Information Engineering University in Zhengzhou used a Hummingbird E203 core to make an energy-efficient AES cryptographic accelerator that can speed up data processing by 700% with only a nearly 2% increase in hardware resources (Xiaolei et al., 2023). The accelerator is well-suited for IoT applications. Furthermore, China was able to deploy the first commercial RISC-V cluster in the cloud,[53] which was implemented by Alibaba (using Sophgo's SG2042 SoCs[54]) for Shandong University for teaching and research purposes as well as for commercial cloud computing instances. China has also used RISC-V in locally made Linux-based single-board computers (e.g., Allwinner D1) and PCs.[55]

12.3.10.3 India

India, as an emerging power, is also trying to achieve technological autonomy (Chandrasekhar, 2023). Digital transformation is taking place across the board in India, with a special focus on 5G, AI, and IoT/IIoT (Ghising, 2025). A few years ago, India started manufacturing mobile devices, and today it is trying to enter the chip industry with the goal of achieving self-sufficiency and becoming a major semiconductor manufacturing hub that competes with China, South Korea, and Taiwan (Cyrill and Kapur, 2024; Rao and Cyrill, 2024).

India has previously struggled to establish a robust semiconductor industry. However, it has recently dedicated several initiatives to fix this problem, including the India Semiconductor Mission, "Digital India," and "Make in India." Through these initiatives, India is providing incentives to invest in the domestic semiconductor industry. So far, India has authorized the construction of four semiconductor foundries, which will bring in US $18.15B in investment and have a combined production capacity of 70 million chips per day (Cyrill and Kapur, 2024). The government is also evaluating several proposals from tech leaders, both local and international, for other semiconductor foundries. Interested companies include AMD, Foxconn, Kaynes, Larsen & Toubro, Micron, Qualcomm, Tata, and Tower Semiconductor.

India has also signed agreements with several countries, including Japan, Singapore, the EU, and the US, with the aim of facilitating collaboration in the field of semiconductors and securing access to chips. Its recent agreements with the US, for example, are part of an effort to expand the strategic partnership between the two countries.[56] The partnership will help India establish a new semiconductor foundry that targets advanced sensing, communication and telecommunications, defense electronics, and clean energy applications. Teams from both countries will collaborate on designing and producing many hardware solutions for both defense and industrial uses.

In regards to RISC-V, India has shown interest in the ISA since its birth. The Indian Institute of Technology (IIT) in Madras is one of the founding members of the RISC-V Foundation. The country's plans to develop homegrown processors led to the launch of the Digital India RISC-V (DIR-V) Program in 2022, recognizing RISC-V as its "national ISA."[57] Using RISC-V, the RISE (Reconfigurable Intelligent Systems Engineering) group at IIT Madras created the Shakti processor as part of their plan to create open-source, industry-grade processors and a supporting software ecosystem (Gala et al., 2016, 2018). There are several models of the Shakti processor, namely, E-Class, C-Class, I-Class, M-Class, S-Class, H-Class, T-Class, and F-Class, which target the entire computing spectrum, from microcontrollers for IoT/IIoT to multicore clusters for datacenter applications. Furthermore, Shakti comes with a software stack that includes an SDK and IDE that allow implementers to program the processors. Although not much information about commercial products that use Shakti is currently available, the processor is expected to be used in many of the chips produced by these new foundries.

12.3.11 Adoption Challenges

A recent report by the SHD Group estimated the market revenues for RISC-V SoCs at US $6.1B in 2023, up from US $1.6B in 2022, and projected them to reach US $92.7B by 2030 (Wawrzyniak, 2024). It also highlighted the sale of 1,260.8M SoC units in 2023, up from 536.4M units in 2022, and projected it to reach 16.181B units by 2030. The report indicates the widespread use of RISC-V across various sectors, highlighting the potential for growth the ISA has. However, despite these promising estimates, like any emerging technology, RISC-V's adoption faces several challenges, as discussed below.

12.3.11.1 Switching Costs

Adopting an ISA is expensive. This goes beyond the ISA's licensing fees to include toolchains and software stacks, training staff, designing and fabricating the chips, and building relationships with foundries and other service providers. Mastering one ISA's ecosystem also takes time, which adds to the total cost. Additionally, reporting code between ISAs is time-consuming and requires significant effort, adding to the overall cost. Although RSIC-V is royalty-free, this does not exempt adopters from the other costs, making it difficult for adopters of one ISA, especially startups and small businesses, to give up on existing investments for a new ISA. However, with the growing demand for RISC-V cores and chips, we expect to see new adopters over time, some of whom might even support multiple ISAs.

12.3.11.2 ISA Fragmentation

Even though RISC-V is flexible and allows implementers to use custom extensions, there are worries that this could potentially lead to ISA fragmentation and vendor lock-in. Such fragmentation could cause compatibility issues between the different implementations and with the underlying software stacks. The RISC-V International organization tries to alleviate this problem by receiving feedback from its members and implementing frequently requested features in future standard extensions. Similarly, popular open-source projects only accept ratified/frozen extensions. Also, creating a proprietary software stack can be very expensive – even for nation-states. We believe all of these factors minimize the potential for ISA fragmentation.

12.3.11.3 Immature Ecosystem

Despite the significant advancement of the RISC-V ecosystem over the past three to four years, it still lags behind legacy ISAs in terms of the available options and features . However, with more support from startups and industry leaders, this should gradually improve. We highlighted in our discussion earlier some of the projects dedicated to enriching the RISC-V ecosystem (e.g., the CHIPS Alliance, RISE, TRISTIAN, and OpenHW Group).

12.3.11.4 Security

With the widespread adoption of the ISA, RISC-V-based implementations will come under attack. The proliferation of IoT has increased the attack surface (Sha et al., 2018). IoT is vulnerable by nature because it lacks the compute and power resources required to handle advanced security capabilities (Tiloca and Raza, 2018). Many of these tiny devices will contain RISC-V designs, putting them to the test. Therefore, implementers must think about security right from the start, i.e., security by design (Benzel et al., 2012), and ensure that their designs have comprehensive security features against cyberattacks.

Some might think that RISC-V, being an open ISA with many open-source implementations, is more vulnerable to security threats.[58] However, as we learned from open-source software many years ago, security by obscurity proved to be less effective when compared to open-source implementations backed by a large community of experts who continuously analyze the code and fix found vulnerabilities (Witten, Landwehr, and Caloyannides, 2001).

12.3.11.5 Power Consumption

One of ARM's key strengths over other architectures is its power efficiency, thanks to the enormous R&D work it does in this area. While there are many excellent and power-efficient RISC-V designs, they are still inferior to ARM designs in terms of performance per watt ratios (Suárez, Almeida, and Blanco, 2024). However, as the community grows, we expect more R&D work done to address this issue and potentially close the gap.

12.3.11.6 Export Controls

One of RISC-V's key advantages is that it follows a non-discriminatory model with its adopters, allowing countries that are not on favorable terms with the US to use it freely. However, there are now discussions about the possibility of restricting access to RISC-V in such countries. The way of applying such restrictions remains unclear. Experts believe that if the US uses RISC-V as a weapon against its foes, this could limit the ISA's potential and eventually lead to fragmentation when such countries decide to develop their own versions of the ISA and ecosystems.

12.3.11.7 Lack of Expertise

Although RISC-V has been around for a decade, many people still lack a thorough understanding of it.[59] The popularity of other more mature ISAs (e.g., ARM, MIPS, and x86-64) partly explains this. The lack of exposure to RISC-V by computer engineers, as well as the scarcity of related courses in college programs, makes it difficult for adopters to find skilled workers who can use the ISA. Additionally, because of its simplicity, skeptics often mistakenly believe it is only suitable for embedded systems and small devices. However, this perception has started to change with the growing interest in RISC-V over the past few years.

Several events and venues now cover RISC-V, such as the RISC-V Summit,[60] which is held annually in various parts of the world. Similarly, some college programs have started covering the ISA.[61] Furthermore, there are a few textbooks that explain the ISA, making it easier for educators to introduce it to new tech generations and for self-learners to understand it better (Patterson and Hennessy, 2017; Patterson and Waterman, 2017; Harris and Harris, 2021). This chapter is also an attempt to familiarize more people with RISC-V.

12.4 CASE STUDY: META

To efficiently serve the millions of users of its social media platforms, Meta was forced to seek novel solutions for two of its widely used computation engines, namely, video transcoding and deep learning recommendation models (DLRMs). As the size of the data and the complexity of the algorithms used increased, existing accelerators and other available CPU-based solutions became inefficient.

As a result, Meta identified three key quality metrics to maximize their return on investment, namely, high performance, power efficiency, and low latency. In addition, Meta needed the new accelerators to be programmable, resilient, and flexible in handling various workloads.[62] The development teams also identified three architectural requirements for the new cores, namely, support for 64-bit addressing, support for vector and SIMD processing, and the ability to add custom instructions. Support for large memory spaces is necessary to handle large amounts of data, and vector/SIMD processing is essential for processing complex tasks such as AI models. Similarly, the ability to add custom instructions is necessary to define common operations such as matrix multiplication, non-linear functions, and data manipulations.

A few years ago, Meta evaluated RISC-V and concluded that its openness would allow the company to build their accelerators in-house. This gives the company access to their real workloads, allowing its teams to analyze performance and optimize their designs accordingly. This not only gives Meta control over their innovations but also the RISC-V's flexibility makes it easy to adjust designs in response to new business requirements.

12.4.1 Accelerators

With the widespread use of AI and the potential of virtual reality AR/VR, Meta foresees a huge growth in the size and complexity of its workloads. In response, the company is developing new rack systems as well as custom-made domain-specific chips for its applications.[63] Using the RISC-V ISA, Meta has developed two accelerators so far: the Meta Scalable Video Processor (MSVP) and the Meta Training and Inference Accelerator (MTIA), both implemented in ASICs.

12.4.1.1 MSVP

Videos represent a high percentage of the content being shared and viewed on Meta's platforms. On Facebook, for example, people spend 50% of their time watching videos, leading to over 4B views per day. These views take place on a variety of devices with different computation capabilities, such as computers, smartphones, and TVs. Additionally, access to these videos is done over different network bandwidths. This means that in order to provide the best user experience, each uploaded video needs to be transcoded into several bit rates using different formats, resolutions, and quality metrics.

Video transcoding involves a compute-intensive pipeline that is made of several stages, namely, decoding, preprocessing, encoding, and quality metrics. At each stage, several complex operations take place before moving on to the next stage. Video transcoding is usually done using software libraries such as libx264 and libVPX. However, software transcoders are slow and power-inefficient. Processing videos at Meta's scale requires an efficient solution that can produce the highest quality possible in the shortest time and with the least amount of energy. Efficient compression is also crucial to reduce data usage and storage. To meet these requirements, Meta developed MSVP, which is a chip specialized in video transcoding (Reddy et al., 2023).

The chip uses a RISC-V core, which implements the video transcoding process. The core IP implements the RV64IMACD extensions. A PCIe Gen4 interface is used to connect the host with the chip core. Communication between the host and the core takes place during the transcoding process, in which bitstreams are moved from the host memory to the local memory. After processing these bitstreams, they are stored back in the host memory. The chip has large cache memory and is connected to 8 GB of external LPDDR (Low-Power Double Data Rate) 5 memory. The chip, along with the LPDDR5 memory modules, are mounted on an M.2 board. Figure 12.4 depicts the main components of the MSVP SoC.

Meta uses its new rack systems to deploy MSVP (Haken et al., 2021). Testing the chip against software libraries, it has achieved 9x throughput for H.264 videos over the libx264 library and 50x throughput for VP9 videos over the libVPX library (on speed 2). At peak, the chip consumes ~10 watts, allowing for single input and multiple outputs of 1B pixels/second. The entire board consumes ~12 watts, saving 50% of the energy used by the software transcoders. Meta deployed the chip in their datacenters, which now handle 100% of user video uploads. MSVP now handles 85% of the work CPUs used to do.

12.4.1.2 MTIA

AI and ML sit at the center of Meta's operations. The company uses DLRMs for tasks such as ad ranking, feeds, content understanding, and recommendations to improve the user experience. However, with the increasing complexity of these models and the sheer amount of data they use, Meta needed an efficient way to process their AI/ML models. As a result, the company developed MTIA.

MTIA is an inference accelerator made of 64 processing elements (PEs) organized in an 8x8 grid (Firoozshahian et al., 2023). Each PE has two RISC-V cores: a scalar core for handling data movement and a vector core for computation. These cores implement the RV64IMC extensions, where each PE has 128 KB of local static random-access memory (SRAM) for data reuse and improved performance. There is also a RISC-V core for control and 128 MB of on-chip SRAM (placed at the four edges of the grid), offering low latency and high bandwidth for frequently accessed instructions and data. Moreover, the chip has access to 64 GB of off-chip LPDDR5 memory. The chip has a memory bandwidth of 400 GB/s per PE, 800 GB/s of local memory, and 176 GB/s of off-chip LPDDR5 memory. Figure 12.5 illustrates the high-level architecture of the MTIA chip.

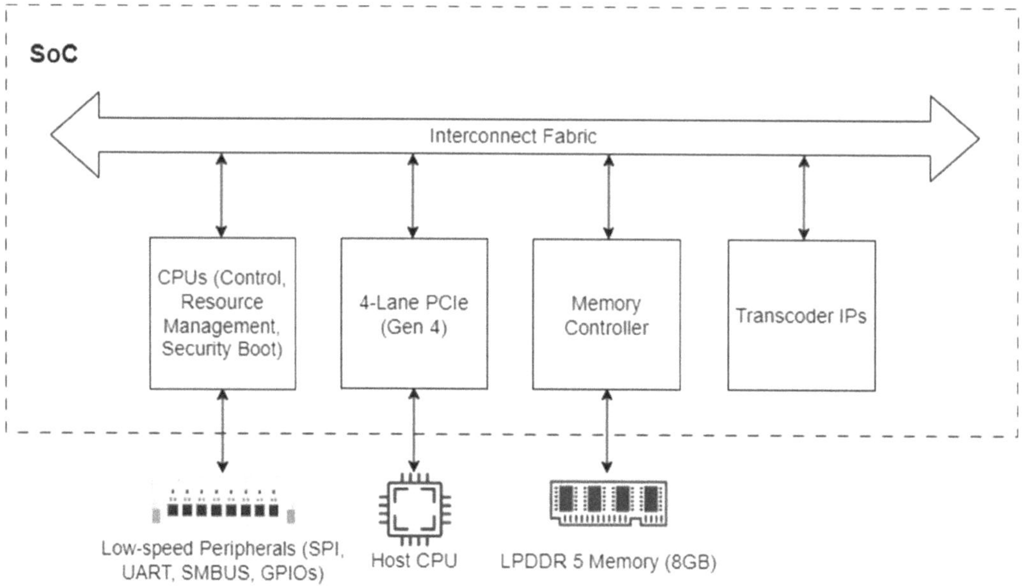

FIGURE 12.4 MSVP SoC Main Components (Reddy et al., 2023)

The chip is built using TSMC 7 nm technology and has a clock frequency of 800 MHz, providing 102.4 TOPS/s at INT8 precision and 51.2 TFLOPS/s at FP16 precision.[64] The chip has a thermal design power (TDP) of 25 watts. It is mounted on a dual M.2 board, which consumes ~35 watts during operation. The board is then connected to the host CPU using 8 links of PCIe Gen4 for a total of 16 GB/s bandwidth/host. A server has 12 boards, which are connected to the host and to each other via PCIe switches. This allows the accelerators to communicate with each other in a peer-to-peer fashion without involving the host CPU. The chip offers a high level of parallelism, allowing developers to either dedicate the whole chip to run one complex operation or split it into parts to run several smaller operations simultaneously.

MTIA also comes with a software stack that allows developers to efficiently write programs that can run on it. PyTorch, initially developed by Meta, is fully integrated with the chip, offering developers simple APIs for efficient code writing. With PyTorch 2.0, developers can write programs for the chip in a similar way to writing code for CPUs and GPUs. MTIA also allows developers to define compute kernels using PyTorch, C/C++, and KNYFE.[65] The LLVM compiler is used to generate executables optimized for MTIA's cores.

Meta's data shows that the chip has achieved a 2x improvement in performance per watt over commercially available GPUs. This improvement, however, was limited to low-complexity models, as MTIA was not fully optimized for medium- and high-complexity models on the software stack. Meta also explained that the slow development of memory and interconnect technologies is limiting their accelerators' potential.

With MTIA's success, its development team decided to build a new generation of the chip, and while writing this chapter, MTIA v2 was released (Maddury, Kansal, and Wu, 2024). In the new design, the team decided to double the compute power and memory bandwidth of the first chip, which resulted in a 3.5x improvement in dense compute performance and a 7x improvement in sparse compute performance.[66]

The new chip is built using TSMC 5nm technology and has a clock frequency of 1.35 GHz, providing 708 TOPS/s at INT8 (sparsity) precision, 354 TOPS/s at INT8, 354 TFLOPS/s at FP16 (sparsity) precision, and 177 TFLOPS/s at FP16 precision. Each PE has 384 KB of local SRAM, 256 MB of on-chip memory, and 128 GB of off-chip LPDDR5 memory. Memory bandwidth has increased in the new generation to 1 TB/s per PE, 2.7 TB/s for on-chip memory, and 204.8 GB/s for off-chip LPDDR5 memory. The new chip has a TDP of 90 watts, achieving a 1.5x performance per watt improvement over MTIA v1. The new chips are housed on boards where each board has two chips. The board has 8 links of PCIe Gen5 for a total of 32 GB/s bandwidth/host. Meta deployed the new accelerators in its datacenters using its custom-built rack system, where each rack can have up to 72 accelerators, split over three chassis, each hosting 12 boards.

The new chip also supports PyTorch 2.0, making code written for it compatible with the previous chip. This allowed Meta to deploy the new chip in 16 regions and complete the migration from the old chip in nine months. The new chip also supports Triton (Tillet, Kung, and Cox, 2019), an open-source, Python-like language for parallel programming that allows developers to write efficient ML compute kernels for GPUs without the closeness or complexity of the commonly used CUDA platform. It also allows developers to write code for both MTIA accelerators and GPUs seamlessly.

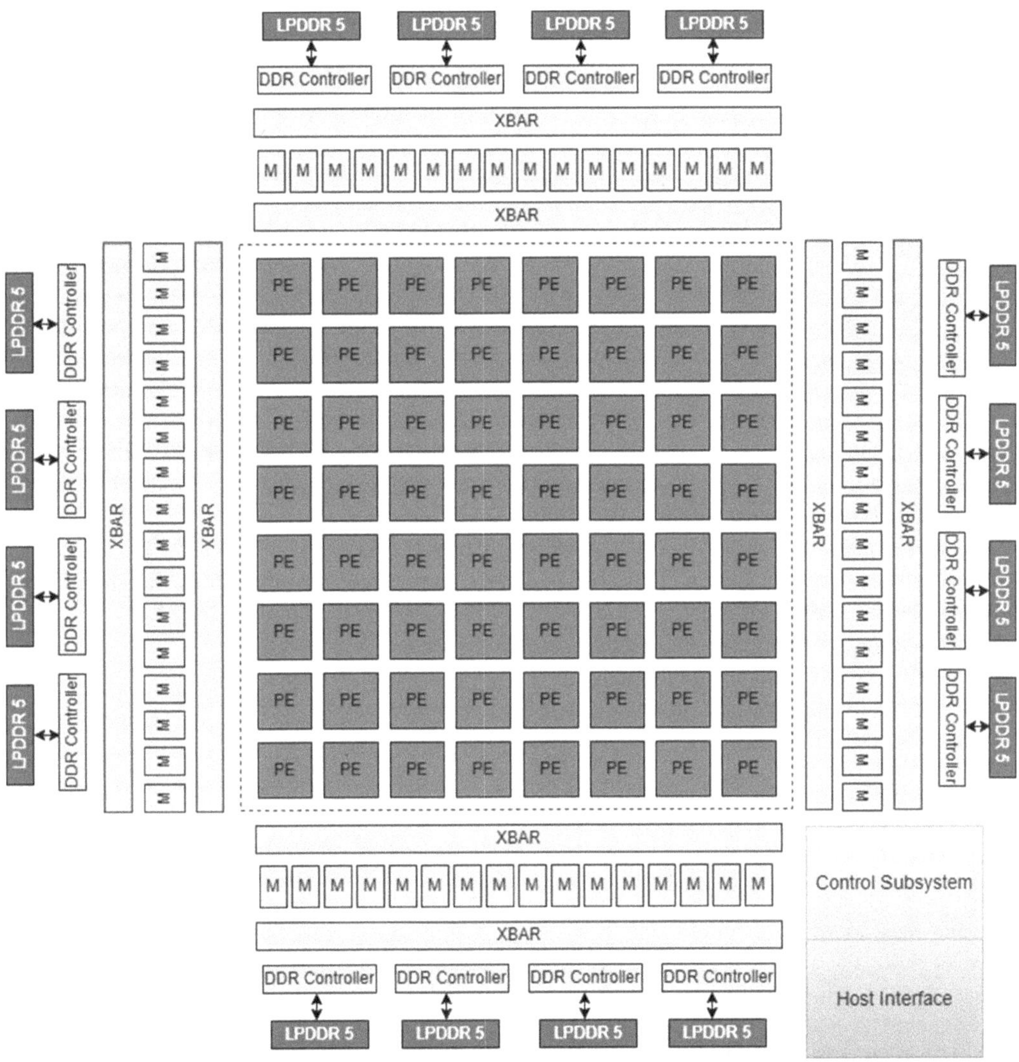

FIGURE 12.5 High-Level Architecture of MTIA Chip (Firoozshahian et al., 2023)

12.4.2 Challenges

Although the development teams at Meta believe that the RISC-V ecosystem has been evolving significantly over the past few years, they have identified a few areas that still need some work.

On the hardware and architecture level, the teams believe that there should be ways to allow adopters to add custom instructions seamlessly. This includes tools that can turn high-level descriptions of the custom instructions into synthesizable RTL (register transfer level) designs, simulators, and software tools. Moreover, the teams are encouraging the RISC-V international organization to accelerate the ratification process of important extensions, such as the vector extension, which is necessary for AI and HPC applications. The teams also think there should be a ratified extension for matrix operations. Another issue Meta highlighted was the lack of interoperability between RISC-V cores from various IP vendors. Finally, the development teams think there is a lack of sophisticated debug and trace environments comparable to something like CoreSight (from ARM).

On the software level, the teams identified the need to support large memory models for AI and HPC applications. Furthermore, they believe that support for LLVM, while improving, is still lagging. Finally, the teams think there should be support for major software libraries, including linear algebra functions and signal-processing kernels.

In fact, these challenges seem to be common among different RISC-V adopters regardless of the application domain, as we discussed earlier.

12.4.3 Future Plans

With the success of MSVP and MTIA, Meta is planning to develop new generations of these chips for improved performance and new use cases. The MSVP development team will continue to work on improving the chip's performance and video quality. Support for new encodes (e.g., AV1), advanced preprocessing methods (e.g., noise reduction and image enhancement), and postprocessing methods (e.g., artifact removal and super resolution) are planned for the next-generation chip. Furthermore, Meta intends to offload the majority of video transcoding to MSVP, limiting the use of software libraries to special cases. The company believes the chip will have other use cases in the future, including generative AI, live streams, and metaverse applications. As for MTIA, the company is planning to go beyond AI inference to support AI training and generative AI workloads in future generations.

We should note that while Meta is developing its own chips for its datacenter applications, it is one of the main users of Nvidia GPUs. The company announced recently that by the end of 2024, there will be 350,000 H100 graphics cards from Nvidia in its datacenters.[69] What Meta seems to be doing is meeting its demanding workloads with cutting-edge commercial GPUs while simultaneously developing its own chips for the long run. Developing RISC-V-based chips that are on par with such GPUs require significant time and a extensive R&D efforts. We believe that each approach will have its own pros and cons and a set of use cases where it best fits.

12.5 CONCLUSION

For decades, proprietary ISAs, mainly X86-64 and ARM, have dominated the chip industry. The former is not licensable, limiting chip makers to the latter, which incurs very high royalty and licensing fees, as well as limited customizability, posing significant challenges for new entrants. However, with the emergence of RISC-V as a free and open standard ISA, this is expected to change. RISC-V has gained momentum over the past few years, as it opens the door for new players to enter the chip industry and create innovative solutions.

In this chapter, we attempted to provide a holistic overview of RISC-V. We began by discussing key ISA principles, including CISC and RISC, to lay the foundations. Then, we provided a detailed discussion of RISC-V, covering its technical aspects, popular implementations, catalysts driving its adoption, key alliances and vendors, initiatives that support it, and the challenges it faces. In addition, a detailed case study on how Meta uses RISC-V in building their homegrown chips for video transcoding and AI workloads was provided. Although some skeptics argue that RISC-V is immature and only suitable for embedded systems and IoT, Meta's decision to use it for AI and HPC applications serves as a strong endorsement of its outstanding efficiency and capabilities. It also emphasizes how domain-specific hardware is shaping the computing industry in the emerging AI-IoT era.

Relatively speaking, RISC-V deployment efforts have just started. However, given the increasing demand for chips, partially driven by the extensive deployment of AI and IoT, it is estimated to be in 25% of computing systems by 2030. RISC-V's openness and neutrality are already allowing several nations to advance their chip industries and improve their technological sovereignty.

ACKNOWLEDGMENT

We are grateful to the assistance of Mohamed Elmohandes for contributing to the initial draft of Section 3.4 (Technical Specifications). We are also grateful to Ahmed Emad ElDean Khalifa, Axel Kloth, and Dongming Peng for their valuable reviews and comments on earlier drafts, which helped improve this chapter.

NOTES

1 It is worth noting that each assembly instruction has a matching machine-code instruction, i.e., a one-to-one relationship. This is not the case with high-level language statements, where one instructions is often converted into several machine-code instructions
2 A die is a piece of semiconductor material, usually silicon, on which the circuit is fabricated.
3 https://riscv.org/.
4 https://github.com/riscv.
5 The founding members of the RISC-V Foundation were Andes, Antmicro, Bluespec, CEVA, Codasip, Cortus, Esperanto, Espressif, ETH Zurich, Google, IBM, ICT, IIT Madras, Lattice, lowRISC, Microchip, MIT (Csail), Qualcomm, Rambus, Rumble, SiFive, Syntacore, and Technolution. A complete list of all members can be found at: https://riscv.org/members/.
6 An IP refers to all information (data, software code, design blueprints, etc.) necessary to make a processor. An IP core is a reusable block of logic or circuit layout design. IP cores can be either soft cores (i.e., synthesizeable RTL in a hardware description language such as Chisel or Verilog) or hard cores (i.e., low-level transistor layouts). Designers can sell, license, open-source, or internally use their IP cores.
7 https://www.technologyreview.com/2023/01/24/1064974/riscv-chip-design-10-breakthough-technologies-2023/.
8 https://www.youtube.com/watch?v=HYvDKxYzp68.
9 https://marketresearch.biz/report/system-on-chip-soc-market/.
10 https://www.whitehouse.gov/briefing-room/statements-releases/2022/08/09/fact-sheet-chips-and-science-act-will-lower-costs-create-jobs-strengthen-supply-chains-and-counter-china/.
11 https://nvidianews.nvidia.com/news/nvidia-to-acquire-arm-for-40-billion-creating-worlds-premier-computing-company-for-the-age-of-ai.
12 https://fprox.substack.com/p/risc-v-profiles.
13 Chisel is an open-source hardware definition language (HDL) based on Scala (Bachrach et al., 2012) created by the RISC-V team at UC Berkeley with the goal of providing a great level of abstraction for hardware designs. Chisel is highly parameterizable and supports both object-oriented and functional programming (Schoeberl, 2019).
14 https://riscv.org/blog/2023/12/the-top-10-risc-v-milestones-highlights-from-2023/#:~:text=You%20could%20point%20to%20the,getting%20bigger%20and%20more%20diverse.
15 https://riscv.org/wp-content/uploads/2017/05/Tue1345pm-NVIDIA-Sijstermans.pdf.
16 Terms such as n-stage, dual-issue, in-order and out-of-order pipeline are explained in the book's glossary.
17 https://www.seagate.com/em/en/news/news-archive/seagate-designs-risc-v-cores-to-power-data-mobility-and-trustworthiness-master-pr/.
18 https://www.qualcomm.com/news/releases/2023/10/qualcomm-to-bring-risc-v-based-wearable-platform-to--wear-os-by-.

19 https://github.com/riscvarchive/riscv-cores-list.
The PULP Platform is an ongoing partnership between ETH Zurich, a founding member of RISC-V International, and the University of Bologna to develop open-source, high-performance, power-efficient processors.
20 A UK-based not-for-profit engineering company that develops and promotes open-source hardware designs and tools. Partners and contributors include ETH Zurich, Google, Rivos, Segate, and WesternDigital, to name a few. https://lowrisc.org/.
21 https://syntacore.com/products/scr1.
22 https://docs.xiangshan.cc/zh-cn/latest/.
23 https://tenstorrent.com/ip/ascalon.
24 https://www.andestech.com/en/products-solutions/andescore-processors/riscv-ax45mp/.
25 https://static.dev.sifive.com/SiFive-E20-Manual-v1p0.pdf.
26 https://www.esperanto.ai/products/.
27 https://mips.com/products/hardware/i8500/.
28 https://www.gaisler.com/index.php/products/processors/noel-v.
29 It is worth pointing out that RISC-V has been gaining traction in the space domain with numerous studies and projects dedicated to this area (Di Mascio et al., 2019a, 2019b; Andersson, 2020; Aranda et al., 2020). For instance, NASA is planning to use RISC-V in its future space missions: https://www.eejournal.com/article/nasa-recruits-microchip-sifive-and-risc-v-to-develop-12-core-processor-soc-for-autonomous-space-missions/. Likewise, Europe is dedicating significant efforts to creating a RISC-V ecosystem for space applications (Furano et al., 2022).
30 https://semiengineering.com/how-secure-are-risc-v-chips/.
31 A full list of RISC-V alliances can be found at: https://riscv.org/alliances/#.
32 https://www.chipsalliance.org/.
33 https://www.openhwgroup.org/.
34 https://riseproject.dev/.
35 https://codasip.com/2021/02/26/what-is-codal/.
36 https://www.renesas.com/us/en/about/press-room/renesas-pioneers-risc-v-technology-rzfive-general-purpose-mpus-based-64-bit-risc-v-cpu-core.
37 The US government's Defense Advanced Research Projects Agency (DARPA) was one of the early advocates of open-source IP, specifically using RISC-V, for government electronic systems: https://riscv.org/wp-content/uploads/2017/12/Wed-1042-RISCV-Open-Source-LintonSalmon.pdf, https://www.theregister.com/2021/04/07/sifive_darpa_chips/ and https://www.darpa.mil/program/ssith. Moreover, several projects under the DARPA Common Heterogeneous Integration and IP Reuse Strategies (CHIPS) program have used RISC-V, e.g., https://www.youtube.com/watch?v=wHfRdC6-vwk.
38 BRICS stands for Brazil, Russia, India, China, and South Africa, an intergovernmental body initially comprising these countries with the goal of improving their economies and technical prowess. Iran, Egypt, Ethiopia, and the United Arab Emirates later joined BRICS. While this section primarily focuses on China and India within the BRICS, it is important to note that other members of the group are also involved in RISC-V projects. For example, Brazil recently became a premier member of RISC International and signed an agreement between Eldorado Research Institute and BSC to exchange experiences and help Eldorado build a matrix multiplication accelerator for HPC applications using RISC-V. Russia also has plans to use the ISA to build domestic computers for the government.
39 https://www.hpcwire.com/2022/12/16/europe-to-dish-out-e270-million-to-build-risc-v-hardware-and-software/.
40 https://www.european-processor-initiative.eu/.
41 https://www.european-processor-initiative.eu/accelerator/.
42 https://ec.europa.eu/commission/presscorner/detail/en/IP_23_3087.
43 https://openchip.com/.
44 https://eurohpc-ju.europa.eu/research-innovation/our-projects/eprocessor_en.
45 Barcelona Supercomputing Center (BSC), Spain, is a leading European supercomputing center and pioneer in RISC-V chip development.
46 https://tristan-project.eu/.
47 https://www.quintauris.eu/.
48 https://www.theregister.com/2024/05/28/china_semiconductor_big_fund_3/.
49 https://www.cnbc.com/2024/05/23/chinas-smic-is-now-worlds-third-largest-chip-foundry-counterpoint.html.

50 https://www.reuters.com/technology/china-bets-open-source-chips-us-export-controls-mount-2024-02-05/.
51 https://riscv.org/blog/2023/08/risc-v-expanding-in-china/.
52 https://www.hpcwire.com/2023/11/08/china-deploys-massive-risc-v-server-in-commercial-cloud/#:~:tex t=China's%20Plan%20for%20RISC%2DV&text=The%20Chinese%20Academy%20of%20Sciences,which%20was%20announced%20in%202021.
53 https://en.sophgo.com/sophon-u/product/introduce/sg2042.html.
54 China recently deployed 10,000 indigenous PCs powered by Loongson's 3A5000 CPU, which use RISC-V, in schools: https://www.theregister.com/2024/04/05/china_schools_riscv_pc/.
55 https://www.whitehouse.gov/briefing-room/statements-releases/2024/09/21/joint-fact-sheet-the-united-states-and-india-continue-to-expand-comprehensive-and-global-strategic-partnership/.
56 https://pib.gov.in/PressReleasePage.aspx?PRID=1820621.
57 https://semiengineering.com/how-secure-are-risc-v-chips/.
58 https://riscv.org/blog/2023/03/top-ten-fallacies-about-risc-v/.
59 https://riscv.org/events-all/.
60 Among the popular education programs devoted to introducing RISC-V in schools is the "One Student, One Chip" program in China, which teaches students the practical aspects of designing and taping out RISC-V chips.
61 Keynote: Building Data Center Scale SoC's using RISC-V at Meta, https://www.youtube.com/watch?v=ad9XXHEqilE.
62 https://about.fb.com/news/2023/05/metas-infrastructure-for-ai/.
63 Unlike CPUs, which are designed for general-purpose use, an ASIC is a form of chip that is tailored for a specific use.
64 According to Meta, KNYFE is a new domain-specific language that generates optimized low-level C++ kernel code for high-level description of ML operators. Not much information about KNYFE is currently available except this post, https://ai.meta.com/blog/meta-training-inference-accelerator-AI-MTIA/.
65 In machine learning, sparsity refers to a matrix that includes many zeros or irrelevant values that do not affect a calculation.
66 https://www.cnbc.com/2024/01/18/mark-zuckerberg-indicates-meta-is-spending-billions-on-nvidia-ai-chips.html.

REFERENCES

Abu-Ghazaleh, N., Ponomarev D., and Evtyushkin, D. 2019. How the Spectre and Meltdown Hacks Really Worked. *IEEE Spectrum*, 56(3), pp. 42–49. https://doi.org/10.1109/MSPEC.2019.8651934.

Alexander, W.G., and Wortman, D.B. 1975. Static and Dynamic Characteristics of XPL Programs. *Computer*, 8(11), pp. 41–46. https://doi.org/10.1109/c-m.1975.218804.

Alvarez, L., Ruiz, A., Bigas-Soldevilla, A., Kuroedov, P., Gonzalez, A., Mahale, H., Bustamante, N., Aguilera, A., Minervini, F., Salamero, J., Palomar, O., Papaefstathiou, V., Psathakis, A., Dimou, N., Giaourtas, M., Mastorakis, I., Ieronymakis, G., Matzouranis, G.M., Flouris, V., Kossifidis, N., Marazakis, M., Goel, B., Manivannan, M., Ejaz, A., Strikos, P., Vázquez, M., Sourdis, I., Trancoso, P., Stenström, P., Hagemeyer, J., Tigges, L., Kucza, N., Philippe, J.M., and Papaefstathiou, I. 2023. EProcessor: European, Extendable, Energy-Efficient, Extreme-Scale, Extensible, Processor Ecosystem. In *Proceedings of the 20th ACM International Conference on Computing Frontiers (CF '23)* (pp. 309–314). New York, NY: Association for Computing Machinery. https://doi.org/10.1145/3587135.3592178.

Amar, S., Chisnall, D., Chen, T., Filardo, N.W., Laurie, B., Liu, K., Norton, R., Moore, S.W., Tao, Y., Watson, R.N., and Xia, H. 2023. CHERIoT: Complete Memory Safety for Embedded Devices. In *Proceedings of the 56th Annual IEEE/ACM International Symposium on Microarchitecture,* Association for Computing Machinery. New York. United States (pp. 641–653). https://doi.org/10.1145/3613424.3614266.

Amdahl, G.M. 2013. Computer Architecture and Amdahl's Law. *Computer*, 46(12), pp. 38–46. https://doi.org/10.1109/mc.2013.418.

Andersson, J. 2020. Development of a NOEL-V RISC-V SoC Targeting Space Applications. In *2020 50th Annual IEEE/IFIP International Conference on Dependable Systems and Networks Workshops (DSN-W)* (pp. 66–67). Valencia, Spain: IEEE Computer Society. https://doi.org/10.1109/dsn-w50199.2020.00020.

Aranda, L.A., Wessman, N.J., Santos, L., Sánchez-Macián, A., Andersson, J., Weigand, R., and Maestro, J.A. 2020. Analysis of the Critical Bits of a RISC-V Processor Implemented in an SRAM-based FPGA for Space Applications. *Electronics*, 9(1), p.175. https://doi.org/10.3390/electronics9010175.

Asanović, K., Avizienis, R., Bachrach, J., Beamer, S., Biancolin, D., Celio, C., Cook, H., Dabbelt, D., Hauser, J., Izraelevitz, A., Karandikar, S., Keller, B., Kim, D., Koenig, J., Lee, Y., Love, E., Maas, M., Magyar, A., Mao, H., Moreto, M., Ou, A., Patterson, D., Richards, B., Scmidt, C., Twigg, S., Vo, H., and Waterman, A. 2016. The Rocket Chip Generator. EECS Department, University of California, Berkeley, Tech. Rep. UCB/EECS-2016–17, 4, pp.6–2. Available at: https://www2.eecs.berkeley.edu/Pubs/TechRpts/2016/EECS-2016-17.pdf.

Asanović, K., and Patterson, D.A. 2014. Instruction Sets Should Be Free: The Case for Risc-v. EECS Department, University of California, Berkeley, Tech. Rep. UCB/EECS-2014–146. Available at: https://people.eecs.berkeley.edu/~krste/papers/EECS-2014-146.pdf. Bachrach, J., Vo, H., Richards, B., Lee, Y., Waterman, A., Avižienis, R., Wawrzynek, J., and Asanović, K. 2012. Chisel: Constructing Hardware in a Scala Embedded Language. In *Proceedings of the 49th Annual Design Automation Conference,* Association for Computing Machinery. New York. United States (pp. 1216–1225). https://doi.org/10.1145/2228360.22285 84.

Baras, K., Brito, L.M., and Hassan, Q.F. 2025. Internet of Things: Architectures, Technologies, Applications, and Challenges. In Hassan, Q.F. (ed.), *Advances in the Internet of Things Challenges, Solutions, and Emerging Technologies.* Chapman and Hall/CRC. DOI 10.1201/9781003506638-23.

Benzel, T.V., Irvine, C.E., Levin, T.E., Bhaskara, G., Nguyen, T.D., and Clark, P.C. 2012. *Design Principles for Security.* Monterey, CA: Naval Postgraduate School. Available at: https://apps.dtic.mil/sti/pdfs/ADA437854.pdf.

Bose, P. 2011. Power Wall. In Padua, D. (ed.), *Encyclopedia of Parallel Computing.* Boston, MA: Springer (pp. 1593–1608). https://doi.org/10.1007/978-0-387-09766-4_499.

Byabazaire, J., Tirana, J., Chouliaras, A., Koutsos, V., Aslanidis, T., Panagiotidis, I., and Chatzopoulos, D. 2025. Deep Learning and the Internet of Things: Applications, Challenges, and Opportunities. In Hassan, Q.F. (ed.), *Internet of Things A to Z: Technologies and Applications*, Second Editions. In Press. Wiley and Sons.

Cassano, L., Masico, S., Palumbo, A., Menicucci, A., Furano, G., Bianvhi, G., and Ottavi, M. 2022. Is RISC-V Ready for Space? A Security Perspective. In *Proceedings of the IEEE International Symposium on Defect and Fault Tolerance in VLSI and Nanotechnology Systems (DFT)* (pp. 1–6). Austin, TX: IEEE. https://doi.org/10.1109/DFT56152.2022.9962352.

Celio, C., Patterson, D.A., and Asanovic, K. 2015. The Berkeley Out-of-order Machine (Boom): An Industry-Competitive, Synthesizable, Parameterized Risc-v processor. EECS Department, University of California, Berkeley, Tech. Rep. UCB/EECS-2015–167. Available at: https://www2.eecs.berkeley.edu/Pubs/TechRpts/2015/EECS-2015-167.pdf.

Chandrasekhar, R. 2023. Catalyzing the Indian Innovation Ecosystem – The RISC-V Way. *The Economic Times.* Available at: https://economictimes.indiatimes.com/tech/catalysts/ettech-opinion-catalysing-the-indian-innovation-ecosystem-the-risc-v-way/articleshow/105482322.cms?utm_source=contentofinterest&utm_medium=text&utm_campaign=cppst.

Chen, C., Xiang, X., Liu, C., Shang, Y., Guo, R., Liu, D., Lu, Y., Hao, Z., Luo, J., Chen, Z., Li, C., Pu, Y., Meng, J., Yan, X., Xie, Y., and Qi, X. 2020. Xuantie-910: A Commercial Multi-core 12-stage Pipeline Out-of-Order 64-bit High Performance RISC-V Processor with Vector Extension: Industrial Product. In *2020 ACM/IEEE 47th Annual International Symposium on Computer Architecture (ISCA)* (pp. 52–64). Valencia, Spain: IEEE. https://doi.org/10.1109/isca45697.2020.00016.

Chen, K., Meng, Q., Sun, Y., and Wan, Q. 2024. How Does Industrial Policy Experimentation Influence Innovation Performance? A Case of Made in China 2025. *Humanities and Social Sciences Communications*, 11(1), pp. 1–17. https://doi.org/10.1057/s41599-023-02497-x.

Chen, L., and Naughton, B. 2016. An Institutionalized Policy-making Mechanism: China's Return to Techno-industrial Policy. *Research Policy*, 45(10), pp. 2138–2152. https://doi.org/10.1016/j.respol.2016.09.014.

Cheng, Y.H., Huang, L.B., Cui, Y.J., Ma, S., Wang, Y.W., and Sui, B.C. 2022. RV16: An Ultra-Low-Cost Embedded RISC-V Processor Core. *Journal of Computer Science and Technology*, 37, pp. 1307–1319. https://doi.org/10.1007/s11390-022-0910-x.

Cheung, S. 2023. Examining China's Grand Strategy for RISC-V. *China Brief*, 23(23). Available at: https://jamestown.org/program/examining-chinas-grand-strategy-for-risc-v/.

Cui, E., Li, T., and Wei, Q. 2023. RISC-V Instruction Set Architecture Extensions: A Survey. *IEEE Access*, 11, pp. 24696–24711. https://doi.org/10.1109/ACCESS.2023.3246491.

Cyrill, M., and Kapur, Y. 2024. India's Semiconductor Sector: Tracking Government Support and Investment Trends. *India Briefing.* Available at: https://www.india-briefing.com/news/setting-up-a-semiconductor-fabrication-plant-in-india-what-foreign-investors-should-know-22009.html/.

Dennard, R.H., Gaensslen, F.H., Yu, H.N., Rideout, V.L., Bassous, E., and LeBlanc, A.R. 1974. Design of Ion-implanted MOSFET's with Very Small Physical Dimensions. *IEEE Journal of Solid-State Circuits*, 9(5), pp. 256–268. https://doi.org/10.1109/JSSC.1974.1050511.

DiMascio, S., Menicucci, A., Furano, G., Monteleone, C., and Ottavi, M. 2019a. The Case for RISC-V in Space. In *Applications in Electronics Pervading Industry, Environment and Society*. ApplePies 2018 (pp. 319–325). Cham: Springer International Publishing. https://doi.org/10.1007/978-3-030-11973-7_37.

DiMascio, S., Menicucci, A., Gill, E., Furano, G., and Monteleone, C. 2019b. Leveraging the Openness and Modularity of RISC-V in Space. *Journal of Aerospace Information Systems*, 16(11), pp. 454–472. https://doi.org/10.2514/1.i010735.

Dörflinger, A., Albers, M., Kleinbeck, B., Guan, Y., Michalik, H., Klink, R., Blochwitz, C., Nechi, A., and Berekovic, M. 2021. A Comparative Survey of Open-source Application-class RISC-V Processor Implementations. In *Proceedings of the 18th ACM International Conference on Computing Frontiers,* Association for Computing Machinery. New York. United States. (pp. 12–20). https://doi.org/10.1145/3457388.3458657.

Elsadek, I., and Tawfik, E. 2021. RISC-V Resource-Constrained Cores: A Survey and Energy Comparison. In *2021 19th IEEE International New Circuits and Systems Conference (NEWCAS)* (pp. 1–5). Toulon, France: IEEE. https://doi.org/10.1109/NEWCAS50681.2021.9462781.

Esmaeilzadeh, H., Blem, E., St. Amant, R., Sankaralingam, K., and Burger, D. 2011. June. Dark Silicon and the End of Multicore Scaling. *Proceedings of the 38th Annual International Symposium on Computer Architecture*, 32(3), pp. 365–376. https://doi.org/10.1109/mm.2012.17.

Fei, X., Li, K., Yang, W., and Li, K. 2016. CPU-GPU Computing: Overview, Optimization, and Applications. In Hassan, Q.F. (ed.), *Innovative Research and Applications in Next-Generation High Performance Computing*, IGI Global. Hershey, Pennsylvania, USA (pp. 159–193). https://doi.org/10.4018/978-1-5225-0287-6.ch007.

Firoozshahian, A., Coburn, J., Levenstein, R., Nattoji, R., Kamath, A., Wu, O., Grewal, G., Aepala, H., Jakka, B., Dreyer, B., Hutchin, A., Diril, U., Nair, K., Aredestani, E.K., Schatz, M., Hao, Y., Komuravelli, R., Ho, K.H., Abu Asal, S., Shajrawi, J., Quinn, K., Sreedhara, N., Kansal, P., Wei, W., Jayaraman, D., Cheng, L., Chopda, P., Wang, E., Bikumandla, A., Sengottuvel, A.K., Thottempudi, K., Narasimha, A., Dodds, B., Gao, C., Zhang, J., Al-Sanabani, M., Zehtabioskuie, A., Fix, J., Yu, H., Li, R., Gondkar, K., Montgomery, J., Tsai, M., Dwarakapuram, S., Desai, S., Avidan, N., Ramani, P., Narayanan, K. S., Mathews, A., Gopal, S., Naumov, M., Rao, V., Noru, K., Reddy, H., Venkatapuram, P., and Bjorlin, A. 2023. MTIA: First Generation Silicon Targeting Meta's Recommendation Systems. In *Proceedings of the 50th Annual International Symposium on Computer Architecture,* Association for Computing Machinery. New York. United States (pp. 1–13). https://doi.org/10.1145/3579371.3589348.

Furano, G., Di Mascio, S., Menicucci, A., and Monteleone, C. 2022. A European Roadmap to Leverage RISC-V in Space Applications. In *2022 IEEE Aerospace Conference (AERO)* (pp. 1–7). Big Sky, MT: IEEE. https://doi.org/10.1109/aero53065.2022.9843361.

Gala, N., Madhusudan, G. S., George, P., Sahoo, A., Menon, A., and Kamakoti, V. 2018. SHAKTI: An Open-Source Processor Ecosystem. *Advanced Computing and Communications*. Available at: https://www.researchgate.net/publication/330577797_SHAKTI_An_Open-Source_Processor_Ecosystem.

Gala, N., Menon, A., Bodduna, R., Madhusudan, G.S., and Kamakoti, V. 2016. SHAKTI Processors: An Open-source Hardware Initiative. In *2016 29th International Conference on VLSI Design and 2016 15th International Conference on Embedded Systems (VLSID)* (pp. 7–8). Kolkata, India: IEEE Computer Society. https://doi.org/10.1109/vlsid.2016.130.

Gautschi, M., Schiavone, P.D., Traber, A., Loi, I., Pullini, A., Rossi, D., Flamand, E., Gürkaynak, F.K., and Benini, L. 2017. Near-threshold RISC-V Core with DSP Extensions for Scalable IoT Endpoint Devices. *IEEE Transactions on Very Large Scale Integration (VLSI) Systems*, 25(10), pp. 2700–2713. https://doi.org/10.1109/tvlsi.2017.2654506.

Gerlach, L., Weber, D., Zhang, R., and Schwartz, M. 2023. A Security RISC: Microarchitectural Attacks on Hardware RISC-V CPUs. In *Proceedings of the IEEE Symposium on Security and Privacy (SP)* (pp. 2321–2338). San Francisco, CA: IEEE. https://doi.org/10.1109/SP46215.2023.10179399.

Ghising, A. K. 2025. India and the Internet of Things: Challenges and Opportunities for an Emerging Power. In Hassan, Q. (ed.) Advances in the Internet of Things Challenges, Solutions, and Emerging Technologies. CRC Press. https://doi.org/10.1201/9781003506638-17.

Haken, M., Clow, J., Li, Y., Wei, B., Chong, D., Thoon, K.Y., Vemuri, K., Westhauser, T., Shetty, P., Chan, A., Wu, H., Mehta, D., and Liao, S. 2021. Yosemite V3: Facebook Multi-Node Server Platform Design Specification. *Open Compute Project*. Available at: https://www.opencompute.org/documents/ocp-yosemite-v3-platform-design-specification-1v16-pdf.

Harris, S., and Harris, D. 2021. Digital Design and RISC-V Computer Architecture Textbook. In *2021 ACM/IEEE Workshop on Computer Architecture Education (WCAE)* (pp. 1–5). Raleigh, NC: IEEE. https://doi.org/10.1109/wcae53984.2021.9707615.

Hassan, Q.F. ed. 2016. *Innovative Research and Applications in Next-Generation High Performance Computing*. Hershey, PA: IGI Global. https://doi.org/10.4018/978-1-5225-0287-6.

Hassan, Q.F. ed. 2025. *Internet of Things A to Z: Technologies and Applications*. Second Edition. John Wiley & Sons, New Jersey. In press.

Hassan, Q.F., Khan, A.R., and Madani, S.A. eds. 2017. *Internet of Things: Challenges, Advances, and Applications*. Chapman and Hall/CRC, New York. https://doi.org/10.4324/9781315155005.

Hennessy, J., and Patterson, D.A. 2019. A New Golden Age for Computer Architecture. *Communications of the ACM*, 62(2), pp. 48–60. https://doi.org/10.1145/3282307.

Hennessy, J.L., and Patterson, D.A. 2017. *Computer Architecture: A Quantitative Approach*. Sixth Edition. Elsevier, Amsterdam.

Klementina, K., and Benjamin, B. 2024. *The US-China Tech War and Its Implications for Transatlantic Relations*. Budapest, Hungary: Pazmany Peter Catholic University. https://doi.org/10.13140/RG.2.2.15792.16645.

Kovač, M., Denis, J.M., Notton, P., Walter, E., Dutoit, D., Badstuebner, F., Stilkerich, S., Feldmann, C., Dinechin, B., Stevens, R. Magugliani, F., Chaves, R., Knezović, J., Hofman, D., Brčić., M., Vukušić, K., Krajna, A., Dragić, L., Piljić, I., Kovač, M., Malnar, B., and Duspara, A. 2022. European Processor Initiative: Europe's Approach to Exascale Computing. In Olivier Terzo and Jan Martinovič (eds) *HPC, Big Data, and AI Convergence Towards Exascale* (pp. 273–290). CRC Press, Boca Raton, Florida.

Levine, D.A. 2020. Made in China 2025: China's Strategy for Becoming a Global High-Tech Superpower and Its Implications for the U.S. Economy, National Security, and Free Trade. *Journal of Strategic Security*, 13(3), 1–16. https://doi.org/10.5038/1944-0472.13.3.1833.

NewYork, NY: Association for Computing Machinery. https://doi.org/10.1145/3505253.3505257.

Li, L. 2018. China's Manufacturing Locus in 2025: With a Comparison of "Made-in-China 2025" and "Industry 4.0". *Technological Forecasting and Social Change*, 135, pp. 66–74. https://doi.org/10.1016/j.techfore.2017.05.028.

Lu, T. 2021. *A Survey on Risc-v Security: Hardware and Architecture*. Cornell University Technology. Report, arXiv:2107.04175v1. https://doi.org/10.48550/arXiv.2107.04175.

Maddury, M., Kansal, P., and Wu, O. 2024. Next Gen MTIA -Recommendation Inference Accelerator. In *2024 IEEE Hot Chips 36 Symposium (HCS)* (pp. 1–27). Stanford, CA: IEEE. https://doi.org/10.1109/HCS61935.2024.10665192.

Marchese, S. 2020. Security Vulnerabilities and Hardware Trojans in RISC-V Processors. OneSpin Solutions. Available at: https://www.onespin.com/fileadmin/user_upload/OneSpin_ElectronicsWorld_Security_vulnerabilities_and_hardware_Trojans_in_RISC-V_processors.pdf.

Matthews, E., and Shannon, L. 2017, September. TAIGA: A New RISC-V Soft-Processor Framework Enabling High Performance CPU Architectural Features. In *2017, the 27th International Conference on Field Programmable Logic and Applications (FPL)* (pp. 1–4). Ghent, Belgium: IEEE. https://doi.org/10.23919/fpl.2017.8056766.

Mezger, B.W., Santos, D.A., Dilillo, L., Zeferino, C.A., and Melo, D.R. 2022. A Survey of the RISC-V Architecture Software Support. *IEEE Access*, 10, pp. 51394–51411. https://doi.org/10.1109/access.2022.3174125.

Moore, G. 2003. No Exponential Is Forever: But 'forever' Can Be Delayed! In *Proceedings of the IEEE International Solid-State Circuits* (pp. 20–23). San Francisco, CA: IEEE https://doi.org/10.1109/isscc.2003.1234194.

Moore, G. 2021. *Cramming More Components onto Integrated Circuits (1965)*. MIT Press, San Francisco, CA. https://doi.org/10.7551/mitpress/12274.003.0027.

Patterson, D., and Waterman, A. 2017. *The RISC-V Reader: An Open Architecture Atlas*. Strawberry Canyon, Berkeley, CA.

Patterson, D.A., and Ditzel, D.R. 1980. The Case for the Reduced Instruction Set Computer. *ACM SIGARCH Computer Architecture News*, 8(6), pp. 25–33. https://doi.org/10.1145/641914.641917.

Patterson, D.A., and Hennessy, J.L. 2017. *Computer Organization and Design RISC-V Edition: The Hardware Software Interface*. Morgan Kaufmann, Burlington, Massachusetts.

Petrisko, D., Gilani, F., Wyse, M., Jung, D.C., Davidson, S., Gao, P., Zhao, C., Azad, Z., Canakci, S., Veluri, B., and Guarino, T. 2020. BlackParrot: An Agile Open-source RISC-V Multicore for Accelerator SoCs. *IEEE Micro*, 40(4), pp. 93–102. https://doi.org/10.1109/mm.2020.2996145.

Pullini, A., Rossi, D., Loi I., Tagliavini, G., and Benini, L. 2019. Mr. Wolf: An Energy-Precision Scalable Parallel Ultra Low Power SoC for IoT Edge Processing. *IEEE Journal of Solid-State Circuits*, 54(7), pp. 1970–1981. https://doi.org/10.1109/JSSC.2019.2912307.

Rao, A., and Cyrill, M. 2024. India's Emerging Semiconductor Ecosystem: Key Players. *India Briefing*. Available at: https://www.india-briefing.com/news/india-emerging-semiconductor-ecosystem-key-players-34506.html/.

Raveendran, A., Patil, V., Selvakumar, D., and Desalphine, V. 2016. A RISC-V Instruction Set Processor-Micro-Architecture Design and Analysis. In *Proceedings of International Conference on VLSI Systems, Architectures, Technology and Applications (VLSI-SATA)* (pp. 1–7). Bengaluru, India: IEEE. https://doi.org/10.1109/VLSI-SATA.2016.7593047.

Reddy, H.M., Chen, Y., Lan, J., Katsavounidis, I., Anandharengan, B., Lalgudi, H.G., Alaparthi, S., Hua, G., Chuang, H.C., Wu, P.H., Lei, Z., Mastro, A., Petersen, C., Chaudhari, G., Prakash, P., Regunathan, S., and Grandhi, D. 2023. Efficient Video Processing at Scale Using MSVP. In *Applications of Digital Image Processing XLVI* (Vol. 12674, pp. 320–335). San Diego, CA: SPIE. https://doi.org/10.1117/12.2685875.

Risset, T. 2011. SoC (System on Chip). In Padua, D. (ed.), *Encyclopedia of Parallel Computing* (pp. 1837–1842). Boston, MA: Springer. https://doi.org/10.1007/978-0-387-09766-4_5.

Samaraweera, C., Bhattarai, A., Peng, D., and Sharif, H. 2025. Artificial Intelligence and the Internet of Things: Algorithms, Applications, and Challenges. In Hassan, Q.F. (ed.), *Internet of Things A to Z: Technologies and Applications*, Second Editions. In Press. John Wiley and Sons, Hoboken, New Jersey.

Schiavone, P.D., Conti, F., Rossi, D., Gautschi, M., Pullini, A., Flamand, E., and Benini, L. 2017, September. Slow and Steady Wins the Race? A Comparison of Ultra-low-power RISC-V Cores for Internet-of-Things Applications. In *2017 27th International Symposium on Power and Timing Modeling, Optimization and Simulation (PATMOS)* (pp. 1–8). Thessaloniki, Greece: IEEE. https://doi.org/10.1109/patmos.2017.8106976.

Schoder, D. 2018. Introduction to the Internet of Things. In Hassan, Q.F. (ed.), *Internet of Things A to Z: Technologies and Applications* (pp. 1–50). Wiley, Hoboken, New Jersey. https://doi.org/10.1002/9781119456735.ch1.

Schoeberl, M. 2019. *Digital Design with Chisel*. Kindle Direct Publishing, Seattle, Washington.

Sha, K., Wei, W., Yang, T.A., Wang, Z., and Shi, W. 2018. On Security Challenges and Open Issues in Internet of Things. *Future Generation Computer Systems*, 83, pp. 326–337. https://doi.org/10.1016/j.future.2018.01.059.

Shafique, M., Theocharides, T., Bouganis, C., Hanif, M., Kahlid, F., Hafiz, R., and Rehman, S. 2018. An Overview of Next-generation Architectures for Machine Learning: Roadmap, Opportunities and Challenges in the IoT Era. In *2018 Design, Automation & Test in Europe Conference & Exhibition (DATE)* (pp. 827–832). Dresden, Germany: IEEE. https://doi.org/10.23919/DATE.2018.8342120.

Suárez, D., Almeida, F., and Blanco, V. 2024. Comprehensive Analysis of Energy Efficiency and Performance of ARM and RISC-V SoCs. *The Journal of Supercomputing*, 80, pp. 12771–12789. https://doi.org/10.1007/s11227-024-05946-9.

Tillet, P., Kung, H.T., and Cox, D. 2019, June. Triton: An Intermediate Language and Compiler for Tiled Neural Network Computations. In *Proceedings of the 3rd ACM SIGPLAN International Workshop on Machine Learning and Programming Languages*, Association for Computing Machinery. New York, United States (pp. 10–19). https://doi.org/10.1145/3315508.3329973.

Tiloca, M., and Raza, S. 2018. Security Mechanisms and Technologies for Constrained IoT Devices. In Hassan, Q.F. (ed.), *Internet of Things A to Z: Technologies and Applications* (pp. 219–254). Hoboken, NJ: Wiley and Sons. https://doi.org/10.1002/9781119456735.ch8.

Valero-Lara, P., Paz-Gallardo, A., Foster, E.L., Prieto-Matías, M., Pinelli, A., and Jansson, J. 2016. Multicore and Manycore: Hybrid Computing Architectures and Applications. In Hassan, Q.F. (ed.), *Innovative Research and Applications in Next-Generation High Performance Computing* (pp. 107–158). IGI Global. https://doi.org/10.4018/978-1-5225-0287-6.ch006.

Wang, K., Xu, Y., Yu, Z., Tang, D., Chen, G., Chen, X., Gou, L., Hu, X., Jin, Y., Li, Q., Li, X., Zhang, C., Zhang, F., Zhang, L., Zou, J., Cai, Y., Zun, D., Zusong, L., Jiye, Z., Wei, H., Sun, N., and Bao, Y. 2023. XiangShan Open-source High Performance RISC-V Processor Design and Implementation. *Journal of Computer Research and Development*, 60(3), pp. 476–493. https://doi.org/10.7544/issn1000-1239.202221036.

Waterman, A., Lee, Y., Patterson, D., and Asanović, K. 2011. The RISC-V Instruction Set Manual, Volume I: Base User-Level ISA. Available at: https://www2.eecs.berkeley.edu/Pubs/TechRpts/2011/EECS-2011-62.pdf.

Waterman, A., Lee, Y., Patterson, D., and Asanović, K. 2014. The RISC-V Instruction Set Manual. Volume I: User-Level ISA', version 2. Available at: https://isrc.iscas.ac.cn/gitlab/mirrors/github.com/riscv_riscv-isa-manual/-/raw/3b4c695efcb7933ca9b647913792ccb04db02553/release/riscv-spec-v2.0.pdf.

Wawrzyniak, R. 2024. RISC-V Market Analysis Report; Application Forecasts in a Heterogeneous World. The SHD Group. Available at: https://theshdgroup.com/2023risc-v_download/.

Weste, N., Harris, D., and Harris, D. 2011. *CMOS VLSI Design: A Circuits and Systems Perspective*, Second Edition. Boston, MA: Addison-Wesley.

Wheeler, B. 2019. WD Rolls Its Own RISC-V Core. The Linley Group Microprocessor Report, 3. Available at: https://documents.westerndigital.com/content/dam/doc-library/en_us/assets/public/western-digital/collateral/analyst-report/WD_Rolls_Its_Own_RISC-V_Core.pdf.

Wiltgen, A., Escobar K., Reis, A., and Ribas, R. 2013. Power Consumption Analysis in Static CMOS Gates. In *2013 26th Symposium on Integrated Circuits and Systems Design (SBCCI)* (pp. 1–6). Curitiba, Brazil: IEEE. https://doi.org/10.1109/SBCCI.2013.6644863.

Witten, B., Landwehr, C., and Caloyannides, M. 2001. Does Open Source Improve System Security? *IEEE Software*, 18(5), pp. 57–61. https://doi.org/10.1109/52.951496.

Xiaolei, Z., Zibin, D., Pengfei, G., and Yang, L. 2023. Design and Verification of AES Cryptographic Acceleration Engine Based on RISC-V. *Application of Electronic Technique*, 49(2), pp. 39–44. National Computer System Engineering Research Institute of China. https://doi.org/10.16157/j.issn.0258-7998.223046.

Xu, Y., Yu, Z., Tang, D., Chen, G., Chen, L., Gou, L., Jin, Y., Li, Q., Li, X., Li, Z., Lin, J., Liu, T., Liu, Z., Tan, J., Wang, H., Wang, K., Zhang, C., Zhang, F., Zhang, L., Zhang, Z., Zhao, Y., Zhou, Y., Zhou, Y., Zou, J., Cai, Y., Huan, D., Li, Z., Zhao, J., Chen, Z., He., W., Quan, Q., Liu, X., Wang, S., Shi, K., Sun, N., and Bao, Y. 2022. Towards Developing High Performance RISC-V Processors Using Agile Methodology. In *2022 55th IEEE/ACM International Symposium on Microarchitecture (MICRO)* (pp. 1178–1199). IEEE. https://doi.org/10.1109/mm.2023.3273562.

Zhang, X., Peek, W.A., Pikas, B., and Lee, T. 2016. The Transformation and Upgrading of the Chinese Manufacturing Industry: Based on "German Industry 4.0". *Journal of Applied Business and Economics*, 18(5), pp. 97–105. Available at: https://www.proquest.com/docview/1855298156?sourcetype=Scholarly%20 Journals.

Zhao, J., Korpan, B., Gonzalez, A., and Asanovic, K. 2020. Sonicboom: The 3rd Generation Berkeley Out-of-Order Machine. In *Fourth Workshop on Computer Architecture Research with RISC-V* (Vol. 5, pp. 1–7). Available at: https://people.eecs.berkeley.edu/~krste/papers/SonicBOOM-CARRV2020.pdf.

Part IV

Case Study

13 India and the Internet of Things

Challenges and Opportunities for an Emerging Power

Aditya Kant Ghising

13.1 INTRODUCTION

The discourse on the advent of globalization, Industry 4.0 and Industry 5.0, driven by the Internet of Things (IoT), has been profoundly characterized by issues pertaining to the enhancement of connectivity and technological development among nation-states. While it is generally agreed that there has been significant progress in some areas in this regard, there do remain certain bottlenecks in the proper channelization and distribution of the shared benefits accrued from such progress. It has been established that, through the use of sensors and gateways, IoT systems gather data and transmit it to centralized platforms. These, in turn, are programmed to collect, handle, store, and examine the data to produce insightful information that may be applied to boost business and process operational effectiveness. With the use of IoT technologies and insights, it is considerably easier to make decisions and take actions that lead to an improvement in productivity and quality of life, particularly in developing countries.

IoT, for example, can assist agronomists and farm owners in measuring important performance metrics and keeping an eye on livestock and agricultural sites. It also provides access to new resources for engaging in economic activity, facilitating economic exchange and consumption, and increasing productivity (Kshetri, 2023). Moreover, applying the latest IoT technologies in agriculture practice can fundamentally change the traditional ways of farming in every aspect. This paves the way for a new "precision agriculture" pattern, which is a new technology that will have a profound effect on the lives of billions of people (Zhang et al., 2018).

This chapter assesses the ongoing digital transformation in India with a focus on IoT integration as well as its extensions into the policy domain in the neighborhood. The critically acclaimed writings by modern intellectuals, such as Yuval N. Harari, James Barrat, Alec Ross, and others, point toward the urgency of the subject of IoT in the twenty-first century. Many of such concerns have the potential to shape the emerging political praxis of the developing world. It also gains relevance in the emerging context of including technological advancements in the foreign policy formulation of nation-states in their international relations. This technological diplomacy or "techplomacy" is slated to play a major role in determining the success of bilateral and multilateral engagements in the near future. Japan, Germany, the USA, and, most recently, China, have all set notable examples in this regard.

In this direction, there are ongoing efforts in India aimed at the introduction of curriculum pertaining to IoT along with artificial intelligence (AI) and data analytics. In this regard, although private institutions have been more effective in the implementation of such courses, government institutions have the advantage of being able to educate millions more than private firms can, simply from an economic standpoint. These efforts are aside from the ongoing focus on digitalization under the leadership of Prime Minister Narendra Modi.

DOI: 10.1201/9781003506638-17

The effects of the IoT revolution may also be felt in the way India engages with her neighbors. For instance, the RuPay[1] payment method (the global card payment network of India) under the wider "Digital India" initiative is emerging as an important element of this eco-system. Other emerging elements in India's ongoing IoT revolution are smart cities, data-driven healthcare systems, and smart factories. From an industry standpoint, TechSagar[2] is another effort by the Indian government toward the adoption of IoT. It is a portal that lists the business and research capabilities of various entities from the IT industry, start-ups, academia, and research and development (R&D) institutes.

India is among a handful of countries supporting the global EV30@30 campaign, which targets to have at least 30% new electric vehicle sales by 2030 (Mandhanya et al., 2023). The achievement of such objectives depends on the employment of technologies associated with IoT, as evidenced by the government's focus on similar initiatives in recent years. Currently, Faster Adoption and Manufacturing of (Hybrid and) Electric Vehicles (FAME) is India's flagship scheme for promoting electric mobility. Under the electric vehicle incentive, the government offers various types of financial incentives to make electric vehicles more affordable. These include incentives related to actual purchases, road tax exemptions, income tax benefits, scrapping incentives, discounts offered on the interest rate while applying for loans, and coupons. Additionally, the Indian Government's Department of Heavy Industries has sanctioned 2,636 electric vehicle charging stations in 62 cities across 24 States and Union Territories under FAME India scheme phase II.[3]

Additionally, for emerging economies like India, the declining cost of IoT hardware in recent years has provided a significant boost to governmental efforts in implementing IoT solutions under various flagship schemes. Globally, this has led to a reduction in the cost of creating and implementing IoT solutions to solve social and economic issues that the bottom 4 billion (B4B) community faces. As a result, the IoT is being quickly adopted by nations with high rates of poverty. In theory, the availability of smartphones and enhanced cellular mobile networks, such as 5G technology, contributes to the quick adoption of IoT. In 2022, during the India Mobile Congress, Airtel and Jio, two of India's prime telecom operators, announced their plans to introduce 5G technology in India. However, its real-world accessibility in India remains limited, with many phones constantly switching between 4G and 5G networks.

These aspects of IoT integration in the governance model of India are analyzed in various sections of this chapter. This chapter attempts to analyze the recent trends in IoT adaptation in India, along with its growing relevance in India's diplomacy. Some of the major ongoing initiatives of the Indian government in this regard are discussed. However, as a post-colonial society, India experiences several obstacles both at the societal as well as policy levels in the proper implementation of IoT. These have also been addressed in this chapter.

13.2 INDIA'S ROAD TO IOT ADOPTION

Although IoT is not a construct that has appeared suddenly or without precursors (Schoder, 2018), its presence in India's overall growth trajectory has been a recent phenomenon owing to the largely agriculture-based policies of various governments until the early 1990s. The post-Cold War reforms of liberalization, privatization, and globalization paved the way for corporations engaged in IoT to enter the Indian market in the twenty-first century. This was backed by a new industrial policy that led to the launch of a price liberalization scheme along with an important reduction in external tariffs, thereby dismantling the old system. As a post-colonial society, the need for technology-driven solutions was realized much later than in most other parts of the world.

13.2.1 Projects and Schemes

In 2011, a scheme backed by the Government of India for the creation of a National Optical Fiber Network was initiated with the objective of connecting all the Gram Panchayats[4] of the country with high-speed Internet. However, this initiative has been criticized for severely lagging behind

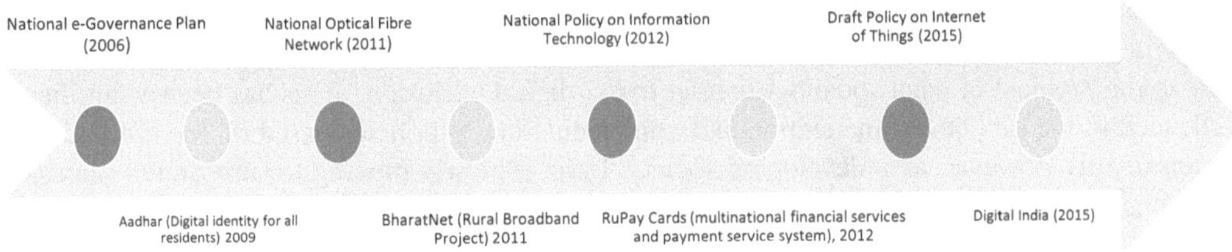

FIGURE 13.1 Timeline of the Major Milestones in the Evolution of India's IoT Adoption

its original timeline. By 2018, under another initiative known as "BharatNet" (launched in 2011), some progress was observable. By then, out of 1.24 lakh (or 124,000) Gram Panchayats, 1.08 lakh (or 108,000) were service ready. Additionally, the National Information Infrastructure (NII) was tasked with ensuring the integration of the networks and cloud infrastructure with the goal of providing high-speed connectivity to various government departments up to the panchayat (village) level. The NII comprises networks such as the State Wide Area Network, National Knowledge Network, BharatNet, Government User Network, and the "MeghRaj" Cloud, India's national cloud initiative.[5] Some of the major milestones in India's IoT adoption have been highlighted in Figure 13.1.

In 2014, Indian Prime Minister Narendra Modi declared his intention of creating 100 smart cities across the nation.[6] He defined a smart city as one that has the fundamental infrastructure needed to provide a respectable standard of living as well as a clean, sustainable environment through the use of smart solutions. His concept of smart solutions revolves around three areas, which comprise the following:

1. Electronic dissemination of information between local governments and citizens to automatically "push" advisories and enable rapid citizen feedback.
2. Improved energy management through the use of automated metering controls and green building development.
3. Advanced analytics and intelligent traffic management systems to support efficient use of the urban road infrastructure network.

In 2015, the Government of India formulated a draft IoT policy[7] with a vision to develop a connected and smart IoT-based system for India's economy, society, environment, and global needs. This policy launched a Smart City project by allocating INR 7,060 crores (approximately 844,620,000 USD) for the same. In continuation of this endeavor, the launch of the "Digital India" initiative aims to transform Indian society into a digitally empowered society and boost the IoT industry. The proposed smart cities shall consist of smart homes, smart parking, smart phone detection, smart transportation, smart roads, and smart lighting. However, owing to certain constraints and bottlenecks, which are discussed in Section 2.2, most of the goals set out in India's IoT policy have not been achieved. On the other hand, it is also true that IoT as an industry has been growing in India under separate government schemes dedicated to individual sectors. The Digital India initiative is an example of this.

13.2.2 Bottlenecks and Government Solutions

Although these goals have been set out by the administration, there are many areas within India that, for several reasons, struggle to get access to basic Internet services. A report published by the World Economic Forum (WEF) in 2016 highlighted this glaring digital divide within India. Titled "The Global Competitiveness Report: 2015–2016," it showed that in terms of macroeconomic stability, India ranked 91st out of 139 countries covered by the WEF (Schwab, 2015). On the basis of technology readiness, it was found to be the least digitally connected country in the world (ranked

120th). More alarmingly, the report depicted that the use of Internet services in India is limited to the privileged class and less than one in five Indians has access to the Internet on a daily basis. This leads to the problem of what sociologists refer to as "digital exclusion." This has been a significant bottleneck in the proper channelization and implementation of policies based on the use and promotion of IoT. However, as a developing society, India is slowly moving toward digital inclusion through government initiatives that aim to address the problems associated with digital exclusion, some of which are as follows:

1. The Pradhan Mantri Gramin Digital Saksharta Abhiyan (PMGDISHA), which literally translates to "Prime Minister Rural Digital Literacy Campaign." This was launched in 2017 and aims to provide 6 crore (60 million) rural households with access to digital devices, Internet connectivity, and digital literacy training.
2. Internet Saathi Program, which was launched in 2015 by Google India and Tata Trusts with the aim of facilitating digital literacy among rural Indian women.
3. Optical Fiber Network, which is a project aimed at ensuring broadband connectivity to over two lakh (200,000) gram panchayats of India.

Additionally, there are ongoing efforts on the government's part to address the digital divide within India. For instance, the Ministry of Electronics and Information Technology (MeitY) has approved ₹ 436.87 Crore (approximately USD 52 million) for the implementation of the project titled "FutureSkill Prime," which aims to create a reskilling/upskilling ecosystem in ten emerging and futuristic technologies, including IoT.[8] In this regard, the National Institute of Electronics and Information Technology, Aurangabad, has been approved as the co-lead resource center in IoT.

Another bottleneck to the IoT adoption in India is the lack of indigenous chips required for realizing tailored and cost-effective IoT solutions. The fact that India is not self-reliant when it comes to hardware (chips) is a serious impediment to its vision of embarking on the IoT journey, be it for development projects or for its international diplomacy. In this regard, the potential of India's indigenous engineering institutions, with their vast supply of young talent, could play a vital role in innovation. With this in mind, India's Ministry of Electronics and IT flagged off a nationwide roadshow on the Digital India RISC-V (DIR-V) program in November 2023 to improve the domestic semiconductor industry. This program is playing a pivotal role in realizing the "vision of creating and adopting open-source technologies laid out by the Prime Minister."[9] The aim of this project is to enable India to create microchips necessary for building its devices, many of which are to be used in IoT and industrial IoT, thereby realizing its digital transformation plans using RISC-V, an open standard instruction set architecture (ISA; Hassan and Sagahyroon, 2025). This is crucial to avoid any future chip shortages, especially with the rising tension between China and Taiwan, which might affect the supply chain since the latter produces approximately 60% of the world's semiconductors.

As a by-product of the DIR-V Program, SHAKTI started as an academic initiative in 2014. It is an open-source processor development ecosystem initiative by the Reconfigurable Intelligent Systems Engineering group at the Indian Institute of Technology (IIT) Madras (Gala et al., 2018). The initiative focuses on the production of open-source processors, complete System on Chips, development boards, and supporting software platforms.

13.3 EMERGING TRENDS IN INDIA'S GOVERNMENT-CORPORATE EFFORTS AT IOT ADOPTION

Indian Prime Minister Narendra Modi's idea of smart cities and "Information and Communications Technology (ICT) for smart solutions" is not exclusive to India. In this regard, similar initiatives and strategies that have been successful in similar conditions elsewhere need to be understood and adjusted as per the societal setup in India.

ICT is being used by many emerging smart cities globally to provide new methods of monitoring pollution in the environment, managing traffic and city transportation systems, as well as distributing energy. Building automation technology is also making major advancements in physical security, space heating and cooling, and lighting controls within smart cities. *Statista* highlights India's IoT potential going forward, with projections showing that revenue in the IoT market is set to reach USD 33.34 billion by 2024.[10] Furthermore, this revenue is anticipated to exhibit an annual growth rate of 15.83% from 2024 to 2028.

13.3.1 Ongoing Efforts at IoT Integration

The gradual emergence of environmental consciousness, urbanization, and the swift advancement of inventive and technological advancements have all contributed to the demand for smart cities. These elements have made it necessary to reconsider urban management practices and have prompted the creation of strategies for carrying them out effectively.

It has been observed that five key segments are contributing to the overall IoT market demand in India, which comprise large enterprises, global capability centers (GCCs), small and medium businesses (SMBs), consumers, and the government. At present, there are over 1,000 large enterprises in India that are public-listed, and the IoT adoption rate in these enterprises stands at around 35%. Supplementing this is the presence of over 1,400 GCCs, which are known to be engaged in rapidly testing, implementing, and scaling IoT technologies and use cases, further boosting the IoT segment in India. Furthermore, there are over 75 million SMBs in India, which makes it the second-largest base of SMBs in the world. Many of these businesses today rely on some form of advanced digital technology solution for driving their operations, which include AI, cloud computing, and IoT.

A significant boost to the government's promotion of IoT in India came in the form of an announcement for the creation of the Centre for Excellence. Prime Minister Narendra Modi announced this in July 2015 as part of the "Digital India" initiative to jumpstart the IoT ecosystem based on India's IT strengths. The main objective of the center, as put forward by the prime minister, is to create innovative applications and foster domain capability by harnessing the innovative nature of the country's emerging startup community and combining this with the experience of corporate players. The hope is that such initiatives will address many of the bottlenecks in streamlining India's overall approach to IoT in the near future, with major implications for many sectors. In this regard, the other objectives of the Center for Excellence are to cultivate industry-capable talents within India by developing neutral and interoperable as well as multi-technology stack laboratory facilities, promoting start-up communities, and creating an entrepreneurial ecosystem for the implementation of IoT. Five startups, namely, Wireless Controls, Uncanny Vision, LightMetrics, ThingsCloud, and SAAR Microsystems, have been chosen to be a part of India's first Centre for Excellence in IoT (CoE IoT), under the segment that focuses on sensors in devices that communicate and share data using the Internet.

13.3.2 Achievements in IoT Adoption

Overall, the program has performed rather well since its inception and its objectives and scope have grown in the sense that it has fostered the creation of strong start-ups. The CoE IoT, in the city of Bangalore, is emerging as a well-recognized contributor to the start-up ecosystem in India, not only by innovators but also by industry players. The center has the potential to house 40 startups and has been jointly set up by the software industry group National Association of Software and Service Companies (NASSCOM) in collaboration with the Department of Electronics and Information Technology (DEITY) and the Education and Research Network (ERNET) of India. It allows start-ups to not only build prototypes and test them but also focus on creating solutions for sectors, such as agriculture, automotive, telecom, healthcare, and consumer goods.

Overall, as a rising economy, India stands to benefit from the IoT integration given the rapidly increasing usage and dependence of various sectors on the Internet. To harness the benefits of this technology, however, strong and resilient policies regulating it need to be in place, given that the number of Internet users globally as well as within India seems to be on the rise.

13.4 IMPACT OF IOT ON INDIA'S NEIGHBORHOOD POLICY

The emerging transformations in IoT technologies provide ample opportunities for South Asian countries to transition from the existing economic models based on low value-added services to a products-and-innovation ecosystem. However, this would depend on the implementation of policies pushing for more aggressive investments and regional collaborations in the R&D of IoT. With about 25% of the global population, there is a promising market for innovative goods and services in the region. India has been regarded as a rising power when it comes to digitalization by global entities ranging from the European Union to the RAND[11] Corporation of the USA. The Indian government's ongoing efforts in this domain are indicative of this narrative.

Given the fact that IoT-driven solutions bring more discussions on sustainable use of resources, planning, development, employment patterns, etc., they are seen as the next frontier in human evolution. Within the South Asian region, India plays a pivotal role owing not only to its historical, strategic, and socio-economic roots but also to the fact that it is seen by its smaller neighbors[12] as a balancer to China's increasing footprint in the region. From a strategic perspective, this role of balancing China brings with it the logical corollary that IoT technology is a key political and economic resource of the future. Additionally, with China now working toward enhancing a new digital economy under a grand foreign policy priority through the Digital Silk Road initiative, the future of India's policy calculations inherently features an increased focus on digitalization and technology.

13.4.1 Potential Areas of Collaboration

Although still at its nascent stage, IoT technologies in the region have the potential to change the existing dynamics of bilateral relations between India and its neighbors. Based on the existing consensus in academic circles, South Asia carries the potential to become a hub of innovation in the Fourth Industrial Revolution (4IR). This is premised on, among other factors, the region's young population's access to knowledge of emerging global technological developments as well as opportunities created by the growth of 4IR technologies. This is supplemented by a large and educated labor market.

The ongoing thrust on IoT technologies presents many opportunities in tackling some, if not all, of the prevalent large-scale systemic challenges in South Asia. One major hindrance in tapping this potential has been the lack of understanding of these technologies along with the fear regarding their disruptive potential, which gains particular relevance in a tradition-rooted society. Nevertheless, it can be argued that in the light of some crucial developments on the policy front, as highlighted in Section 2, tech diplomacy is increasingly becoming a driving force for India's worldview, through which it strives to capture the potential of the nexus between technological innovation and geo-economic development. For instance, IoT-based financial collaborations, which aim at providing better services related to payments across borders, have emerged as one of India's prime economic policies when it comes to the neighborhood. RuPay (short for Rupee Payment), a payment service system, launched in 2012 by the Government of India with the stated goal of a "cashless India" (meaning all financial transactions within India would be digital) in the near future. As a part of its ongoing evolution, a strategic partnership agreement for Domestic Card Scheme implementation with Al Etihad Payments of the UAE was made in 2023. Although RuPay started off as a pan-India payment initiative by the government of India, it has quickly transitioned into the sphere of bilateral relations. In 2023, the National Payments Corporation of India – International Payments Ltd. (NIPL), the wholly owned subsidiary and international arm of the National Payments

Corporation of India (NPCI), announced the significant milestone of issuing over 10,000 RuPay cards in Bhutan with a partner bank of the Royal Monetary Authority of Bhutan, Bhutan National Bank Ltd (BNBL). A joint press release by India's NPCI and Bhutan's BNBL highlights that the BNBL-RuPay debit cards are accepted at more than 265,994 ATMs and 7.9 million Point of Sale (PoS) terminals deployed by the member banks for cash withdrawals, balance inquiries, and purchases in India.

This makes RuPay the first-of-its-kind global card payment network from India. Additionally, RuPay (Phase II), which is a cross-border payment initiative between India and Bhutan, was launched in November 2020. In a political landscape shaped by a state's ability to develop and employ its soft-power[13] potential, RuPay has emerged as India's way of navigating its relations with its neighbors in South Asia. This is a region that is witnessing an increasing Chinese economic and technological footprint, both categorized as soft-power strategies in international relations. The RuPay initiative between India and Bhutan was aimed at promoting the use of low-cost digital payment solutions, advancing digital financial inclusion, and allowing Bhutanese citizens to withdraw cash and make payments through all ATM and PoS terminals in India. Currently, Bhutan National Bank is the only bank in Bhutan that issues RuPay cards.[14] There are also agreements in place with SBI Bank Limited (Nepal) and Mercury Payments Services (UAE), which allow Indian travelers to pay via their RuPay cards at select merchant locations in Nepal and the UAE, respectively. Apart from these countries, Mauritius and Singapore accept RuPay as a payment method.

The rising footprint of China in India's immediate neighborhood has been a catalyst in the formulation and implementation of IoT-based policies as a part of its foreign policy agenda. In the discourse on the rise of China in the region, one of the lessons has been the co-relationship between R&D in hitherto underdeveloped areas. Its persistent and sustained investment in R&D enabled the Chinese government to craft IoT policies based on local requirements, thereby providing new models of rural development in Asia that have been applauded in many parts of the region. Similarly, in order to enable progress toward an innovation-powered economy, it would serve the governments in South Asia well to invest a larger percentage of their GDP back into the R&D ecosystem. As a positive move in this direction, in 2024, India and Sri Lanka agreed to leverage India's digital public infrastructure in line with Sri Lanka's needs for the effective delivery of citizen-centric services.[15]

Another potential area of collaboration is checking illegal immigration across the long borders, which India shares with its South Asian neighbors. Monitoring of sensitive borders in the region through IoT-enabled devices could not only lead to more precision in tracking illegal immigration but also open new avenues for collaboration among India and her neighbors in the sector. Technologies that foster de-territorialized border controls have been tested on some of the borders within Europe under the Smart Border initiative. Such successes could be replicated in South Asia as policies based on the use of IoT become more common in the region.

13.4.2 Public–Private Synergy

In the case of IoT adoption, at least so far, the practical aspects (including the implementation) of government initiatives largely hinge on support from the private sector. It would not be possible to stay updated on the latest cutting-edge technologies or the pursuit of a more digitally connected society without this support. In line with this assessment, one may observe the increasing reliance on private firms for IoT integration in India. The government-owned telecommunications company, Bharat Sanchar Nigam Limited (BSNL), which used to be the only major telecommunications company in India, has been largely sidelined with the emergence of companies such as Reliance Jio, Vodafone, and Airtel. These and similar companies have become the backbone of the government's Digital India initiative, highlighting the emphasis on public-private synergy in the sector.

Additionally, most of the major initiatives shaping the emerging landscape of IoT in India, including promoting grassroots literacy about the subject as well as conducting employment-based activities, are being undertaken by private entities through privately funded institutions. Cognizant, a major IT company based in the USA, as well as its Indian branch, Cognizant Technology Solutions India Pvt Ltd, has been one of the leaders in this domain, providing new employment avenues in the IoT domain. Other such innovators in India are Accenture, Infosys, Tata Consultancy Services, Tech Mahindra, and HCLTech. Cognizant has been rated among the top IoT service providers by HFS Research Limited.[16]

The efficiency of private corporations may in part be due to the limited constraints when it comes to operability as compared to government agencies. The latter have to abide by the norms and values of the Indian constitution and work toward addressing other concerns related to a myriad of issues concerning citizens' rights, justice, liberty, equality, and electoral politics.

Combining both private and government enterprises and their initiatives on IoT integration pits India directly against China and its technological prowess in the region in the form of a new soft-power strategy. Beijing's growing collaboration with India's neighbors has created a sense of unease in New Delhi (Baruah, 2018). One key arena in this regard could be the emerging space programs of India and China, which would, in all likelihood, be affected by advancements in IoT technology in the years to come.

Much of the ongoing research on the use of IoT focuses on out-of-the-box solutions to local problems, due to the fact that most traditional solutions have been tried and tested, with limited success in improving the lives of citizens at large. The increasing usage of robots in surgical tasks could also be seen in this regard. Apollo Hospitals Enterprise Limited, an Indian multinational healthcare group, has conducted more than 10,000 robot-assisted surgeries since 2011, with a 400% increase in such procedures between 2021 and 2023.[17] This trend is likely to continue based on its success rates over the years and yet, only a decade ago, this would have been something unimaginable in Indian society, which has deep roots in alternative and more traditional forms of medicine.

13.4.3 Extra-Regional Developments

India and the USA have agreed on Critical and Emerging Technology (iCET) on the sidelines of the Quad meeting in Tokyo, held in May 2022. This initiative talks about supporting the enhancement of a semiconductor ecosystem, advancing cooperation between India and the USA on 5G and 6G development, and incorporating Open RAN (Open Radio Access Network) technology in India. In the backdrop of limited success in other areas, these technologies could become a part of India's regional diplomacy going forward. Earlier that year, India's defense public-sector undertaking Bharat Electronics Limited (BEL) entered into a contract with Hyperion Global Group, an infrastructure telecommunication distribution company based in the USA, to develop, manufacture, and supply IoT devices worth $73 million for the US market.[18]

As a welcome step in this trajectory, as announced in March 2022, the Government of India seeks to fund a new area office for the International Telecommunication Union (ITU) with the aim of fostering digital technology uptake, which is in line with the United Nations sustainable development goals, across South Asia.[19] The perceived impact of such developments on India's external relations within South Asia could be felt in both bilateral as well as multilateral engagements going forward. Regional initiatives such as the Bay of Bengal Initiative for Multi-Sectoral Technical and Economic Cooperation (BIMSTEC), the Bangladesh-Bhutan-India-Nepal (BBIN) network, and the SAARC Preferential Trading Agreement (SAPTA) could be re-evaluated under the new IoT-based reforms in India, thereby leading to new perspectives on growth and development in the region.

The post-COVID world order has only heightened the need for such collaborations. The role of governments and nodal agencies becomes paramount in such endeavors since they can spur innovation along with providing a strong support system for micro, small, and medium-sized enterprises through policies aimed at import tax subsidies on equipment as well as infrastructure subsidies for setting up advanced manufacturing facilities. A World Bank report titled "South Asia's Digital Opportunity: Accelerating Growth, Transforming Lives"[20] shows that while some countries are benefiting from expanded access to services and markets, new jobs, innovation, reduced transaction costs, and a better quality of life due to increased digitalization, its full game-changing potential remains untapped (World Bank, 2022).

13.5 CHALLENGES AND PROSPECTS OF IOT INTEGRATION

Globally, the challenges to IoT technologies are similar, although there is a clear indication that success rates vary according to the level of technological development existing in a given society. As an emerging technology, IoT has been implemented successfully in some developed parts of the world, which serves as the template for technological advancement elsewhere. However, merely attempting to imitate the successes of more technologically advanced societies would neither be feasible nor sustainable. Therefore, there is a need to discuss some of these challenges.

13.5.1 Tradition vs Technology

As highlighted in a 2017 report by the Global Commission on Internet Governance, new technologies and projects are seen as a threat to traditional wireless telecommunications service providers.[21] These include China Unicom, Nippon, Telefonica, and Telegraph, among many others, that, in turn, are pressurizing their respective governments as well as multilateral organizations to intervene to protect their interests. A majority of their current earnings are from their wireless (voice and data) services. With increasing exposure to new technologies, traditional service providers have been facing some challenges. Based on a report by Ericsson,[22] repeated and increasing interactions of consumers with digital providers such as Amazon, Netflix, and Apple are setting consumer expectations for their interactions with other companies, including telecom service providers. Their expectations of and interactions with traditional companies in this domain will only be heightened with the eventual societal penetration of AI. The report provides some key insights into some of the challenges that traditional telecom service providers will have to address in order to not only utilize the benefits of IoT but also to exist alongside the new providers.

13.5.2 Data Privacy

From an Indian standpoint, one of the major hindrances to IoT integration is the general concern with data privacy among individuals as well as enterprises, especially in the large rural sector. In this regard, although the government of India introduced the Digital Personal Data Protection Act (DPDP Act)[23] in 2023, the purpose of which is to protect the personal data of individuals, the act has been met with strong resistance on the grounds of making personal information more transparent for the sake of governmental and private interests. The lack of strong data protection laws such as the European Union's General Data Protection Regulation is therefore a challenge that is not easy to address in the case of India. Whereas the EU's regulation is regarded as the toughest law in the world on privacy and security, in India, the vagaries of electoral politics based on more traditional concerns, such as caste and religion, prove to be a significant hindrance in the formulation and implementation of laws related to data privacy. This is also amplified by the general unawareness about the significance of data privacy among the masses. Additionally, although the introduction of IoT-backed surveillance measures has been known to contribute toward crime reduction, civil society organizations in India have often expressed their concerns regarding the encouragement of surveillance techniques to track down political rivals.

13.5.3 Skill Gap

The challenges to successful IoT integration are compounded by the existing skill gap in the domain, which necessitates the government's active involvement in developing skills in hardware and software engineering as well as data analytics and cybersecurity. Re-skilling and training a

TABLE 13.1
Indian AI Research Institutes and Universities

Institute/University Name	Centre/Group
Indian Institute of Science (IISc), Bengaluru	The Artificial Intelligence Group
Indian Institute of Technology (IIT) Kharagpur	The Center for Excellence in Artificial Intelligence
Vellore Institute of Technology (VIT) University, Bhopal	Artificial Intelligence Division
IIT Bombay	The Center for Machine Intelligence and Data Science
IIT Madras	The Robert Bosch Center for Data Science and Artificial Intelligence
IIT Delhi	Yardi School of Artificial Intelligence
IIT Kanpur	Centre of Innovation and Translational Research
Amrita Vishwa Vidyapeetham Kollam	Atal Innovation Mission established Incubation (Amrita Technology Business Incubator) for AI and Deep Tech innovations

Source: State of the Education Report for India 2022: Artificial Intelligence in Education.

large workforce have emerged as a major bottleneck in this regard. It is also true that most societies look at IoT and AI as disruptive technologies, which place emphasis on machines taking over day-to-day activities, thereby leading to increased unemployment for humans.

The concerns emanate from the idea that IoT will "automate many routine activities, allowing machines to make decisions without human interaction" (Lewis, 2016). The Inmarsat Research Program report *Industrial IoT in the Time of Covid-19* also shows that skill shortages are hindering IoT innovation, "inhibiting the adoption of the technology and the effectiveness of IoT deployments across businesses."[24] India is the world's most populous country, with a large share of youth as a percentage of the population. Moreover, as per the recent findings of the International Labor Organization, its youth account for almost 83% of the unemployed workforce.[25] This comes at a time when there are multiple policies aimed at IoT integration and should lend urgency to the skilling and re-skilling efforts of the government with the goal of substantially addressing unemployment.

In a developing country, the real-life usage of IoT would depend on the existence of a knowledge base that is aware of its potential. In order to create this knowledge base with regard to IoT and as an acknowledgment of the demands of the new workforce, courses on AI have been introduced in the curriculum of most educational institutions in India.[26] This is in addition to the already existing (and successful) computer science and engineering institutions in almost every state in the country. Table 13.1 shows some of the prominent Indian institutions engaged in imparting quality education in AI (Zleshe and Bhagat, 2022).

13.5.4 Lack of Uniformity

An observable irritant in the proper implementation of IoT-based solutions in the country is the lack of uniform standards and protocols in the IoT ecosystem, which has significantly hindered the process of seamless integration and interoperability. On a practical level, it has been observed that IoT devices from different manufacturers may not be compatible, thereby leading to fragmentation and complexity. This makes it essential to work toward the establishment of industry-wide standards and the promotion of open-source frameworks in order to encourage collaboration for a thriving IoT ecosystem in India.

For the government and citizens alike, implementing and managing IoT solutions can often be complex and demanding due to the involvement of a multitude of interconnected devices, networks, sensors, and software components, all requiring seamless coordination. This makes it crucial for businesses to possess the necessary technical expertise and resources to handle the intricacies of IoT projects.

To further enhance this potential, there is a need to focus on setting up technology transfer offices at colleges and universities where the primary objective would be to commercialize innovation. Additionally, it could serve as a platform for communication among scientists working on breakthrough research, entrepreneurs engaged in building companies involving new technologies, and policy experts who create laws related to the adoption of these technologies in society. With close to 44,000 colleges and universities in India, such initiatives could significantly enhance the emerging IoT landscape in the country while also addressing the digital divide.

13.5.5 Industry 5.0 and Its Prospects

Perhaps as a reaction to the perception of Industry 4.0, the possibility of moving toward Industry 5.0 is being discussed by experts in the domain. Although the subject is rather understudied at the moment, its focus is on how its stakeholders, which include the government, could leverage this agenda in order to "materialize its sustainability values" (Ghobakhloo et al., 2023). This emerging discourse could generate new ideas about the sustainable integration of mechanization and sustainability, largely owing to its emphasis on human and societal concerns. The possibilities of synergizing human and mechanical collaborations toward achieving modern solutions to economic and societal problems are complemented by its emphasis on preserving employment for humans as well as contributing toward more sustainable forms of growth.

As with the Industry 4.0 agenda, proper channels of sensitization at various levels with regard to Industry 5.0 need to be worked out, focusing on the pros and cons of IoT and the new types of employment avenues it can generate. Its impact on daily use applications needs to be highlighted more effectively. This would go a long way in creating a knowledge base for the proper integration of IoT into societal policy-making.

13.6 CONCLUSION

Technology, or more specifically, advancements therein, are a double-edged sword. In traditional and post-colonial societies such as India, policies that rely on technology have often failed to grasp the mindset of the voters. However, although the proper integration of IoT into the digital economy has been a profound challenge for policymakers, it may also be seen as an opportunity and incentive, especially for the private sector to invest in India. It is by virtue of India's potential for IoT that many multinational corporations have placed their bets on homegrown companies such as *Reliance Jio* and *Flipkart,* which have helped accelerate the growth of digital commerce in India. For its part, the Indian government is highly focused on strengthening global economic cooperation in the realm of technology. This is mainly due to the remarkable rise in the presence of global technology businesses within the country and the associated inflow of foreign direct investments, mainly into the information technology sector.

Given the fact that IoT is a rather new entrant in the already volatile landscape of Indian politics, many of its existing models may need to be adjusted to suit the vast majority of citizens who are either unaware of the benefits or are not in a position to have access to it. So far, the achievements on this front have been noteworthy, with prospects for further growth. However, in order to realize the full potential of IoT in India, these developments would need to focus on the regional initiatives for implementing similar policies and work toward expanding the reach and role of the stakeholders in the region. It would serve the governments of South Asian countries well to make the digital skills agenda a key part of their national strategies. The ongoing revolution in the manufacturing industry, based on IoT, has transformed traditional factories into smart and connected powerhouses. These are programmed to significantly increase productivity and reduce costs while maintaining optimal manufacturing conditions. This shift should have many takers in the developing world as well. India's ongoing efforts at adapting to these changes represent a duality of a post-colonial society as well as a modern emerging economy. Although not quite unique to India, this duality does lead to new perceptions of the discourse on development, technology, and IoT.

It must also be borne in mind that while arguing for the implementation of IoT solutions in governance and manufacturing in India or South Asia in general, one should not imagine the same kind and intensity of IoT-based activities that one may observe in the developed parts of the world. In all likelihood, technologies that have started to become obsolete in the Western world would be replicated (with or without adjustments) elsewhere. This chapter was an attempt at understanding the emerging landscape of IoT in India, its potential, prospects, and challenges in contributing toward India's economic and societal development, as well as its relevance to India's neighborhood policy.

NOTES

1 Shah, I., and Sethia, T., "All about RuPay, India's Payments Network," *The Economic Times*, August 2, 2021. Available at https://bfsi.economictimes.indiatimes.com/news/financial-services/all-about-rupay-indias-payments-network/84969331.
2 See: Kumari, S., "TechSagar – Repository of India's Cybertech Capabilities," Data Security Council of India, February 17, 2020. Available at https://www.dsci.in/blogs/techsagar-repository-of-indias-cybertech-capabilities/#:~:text=TechSagar%20aims%20to%20provide%20the,analysing%20capabilities%20for%20different%20purposes.
3 "In Phase-II to Fame India Scheme 2636 EV Charging Stations Sanctioned," Press Information Bureau, Ministry of Heavy Industries & Public Enterprises, Government of India, January 3, 2020. Available at https://pib.gov.in/newsite/PrintRelease.aspx?relid=197263.
4 A "Gram Panchayat" or "Panchayat" refers to an institution of self-government for the rural areas in India, constituted under article 243B of the Indian Constitution. For more details, see: "What Is a Panchayat?" Available at https://www.pria.org/panchayathub/panchayat_text_view.php.
5 Strategy for New India @ 75 (Policy Document), National Institution for Transforming India (NITI-Aayog), Government of India, November 2018. Available at https://www.niti.gov.in/sites/default/files/2019-01/Strategy_for_New_India_2.pdf.
6 "PM Narendra Modi's Pet 100 Smart Cities Project Gets Cabinet's Approval," *The Economic Times*, May 11, 2015. Available at https://economictimes.indiatimes.com/news/economy/infrastructure/pm-narendra-modis-pet-100-smart-cities-project-gets-cabinets-approval/articleshow/47094719.cms?from=mdr.
7 Draft Policy on Internet of Things, 2015, Department of Electronics & Information Technology (DeitY), Ministry of Communication and Information Technology, Government of India. Available at https://www.iea-4e.org/wp-content/uploads/publications/2015/12/07l_-_India_-_IoT_Strategy.pdf.
8 Annual Report: 2019–20, Ministry of Electronics & Information Technology, Government of India. Available at https://www.meity.gov.in/static/uploads/2024/02/33-1.pdf.
9 See: "Minister Rajeev Chandrasekhar Flags Off the Nationwide Roadshow on Digital India RISC-V (DIR-V) Program," Ministry of Electronics and IT, Government of India, November 17, 2023. Available at https://pib.gov.in/PressReleaseIframePage.aspx?PRID=1977576.
10 "Internet of Things: India," *Statista: Market Insights*. Available at https://www.statista.com/outlook/tmo/internet-of-things/india#analyst-opinion.
11 The RAND Corporation is an American nonprofit global policy think tank, research institute, and public sector consulting firm.
12 These include Nepal, Bhutan, Bangladesh, Sri Lanka, Maldives, Pakistan, and Afghanistan.
13 Soft power refers to a state's economic and cultural power, which it may use to achieve a national interest objective in the global political arena. Military power is regarded as "hard power" in this case.
14 "Press Release: RuPay Card Issuance Crosses 10,000 Milestone in Bhutan," NPCI International & Bhutan National Bank Limited, July 24, 2023. Available at https://www.npci.org.in/PDF/npci/press-releases/2023/Press-Release-RuPay-Card-Issuance-Crosses-10,000-Milestone-in-Bhutan.pdf.
15 "India to assist Sri Lanka in rolling out Digital Public Infrastructure, extending use of UPI payments system," *India News Network* (online), December 17, 2024. Available at https://www.indianewsnetwork.com/en/20241217/india-to-assist-sri-lanka-in-rolling-out-digital-public-infrastructure-extending-use-of-upi-payments-system.
16 "Cognizant Rated among Top IoT Service Providers by HfS Report," *The Hindu: Business Line* (online), January 22, 2018. Available at https://www.thehindubusinessline.com/info-tech/cognizant-rated-among-top-iot-service-providers-by-hfs-report/article64342295.ece.

17 "400 per cent Increase in Robotic-assisted Surgeries in across Apollo Hospitals in Last 2 Years: Healthcare Group," *The Economic Times* (online), March 15, 2023. Available at https://economictimes.indiatimes.com/industry/healthcare/biotech/healthcare/400-per-cent-increase-in-robotic-assisted-surgeries-in-across-apollo-hospitals-in-last-2-years-healthcare-group/articleshow/98671831.cms?from=mdr.

18 Kumar, C., "BEL Inks Multi-million Dollar Deal to Make IoT Devices for US Market," *The Times of India* (online), January 27, 2022. Available at https://timesofindia.indiatimes.com/india/bel-inks-multi-million-dollar-deal-to-make-iot-devices-for-us-market/articleshow/89157476.cms.

19 "India Supports ITU South Asia Presence," Press Release: 03 March 2022, International Telecommunication Union, The United Nations. Available at https://www.itu.int/hub/2022/03/india-supports-new-itu-area-office/.

20 World Bank. 2022. South Asia's Digital Opportunity: Accelerating Growth, Transforming Lives. © Washington, DC: World Bank. Available at https://openknowledge.worldbank.org/entities/publication/f39ee0a0-b01e-5c27-ab5b-53600d8d4052.

21 See: "The Shifting Geopolitics of Internet Access: From Broadband and Net Neutrality to Zero-rating", *GCIG Research: Volume 6* (online), Global Commission on Internet Governance, July 26, 2017. Available at https://www.cigionline.org/publications/shifting-geopolitics-internet-access-broadband-and-net-neutrality-zero-rating/

22 See: "The Zero Touch Customer Experience: Uncovering the Future of Consumer Interactions with Telecom Service Providers," *Insight Report*, Ericsson Consumer & Industry Lab, May 2018. Available at https://www.ericsson.com/4ac64d/assets/local/reports-papers/consumerlab/reports/2018/the_zero_touch_consumer_experience_consumerlab_rgb_screen.pdf.

23 The Digital Personal Data Protection Act: 2023, Ministry of Law and Justice, Government of India. Available at https://www.meity.gov.in/static/uploads/2024/06/2bf1f0e9f04e6fb4f8fef35e82c42aa5.pdf.

24 See: "Skills Shortages Remain the Top Barrier to Industrial IoT Adoption, Inmarsat Research Reveals," *Inmarsat* (online), March 15, 2022. Available at https://www.inmarsat.com/en/news/latest-news/enterprise/2022/skills-shortage-iot-adoption-research.html.

25 India Employment Report 2024: Youth employment, education and skills, Institute for Human Development, International Labour Organization, March 2024. Available at https://www.ilo.org/wcmsp5/groups/public/---asia/---ro-bangkok/---sro-new_delhi/documents/publication/wcms_921154.pdf.

26 See: "Steps Taken by the Government to Use Artificial Intelligence for Education Transformation," Press Release, Ministry of Education, Government of India, March 15, 2021. Available at https://www.pib.gov.in/PressReleasePage.aspx?PRID=1704878.

REFERENCES

Baruah, D. M., 2018. India's Answer to the Belt and Road: A Road Map for South Asia. *Carnegie Endowment for International* Peace, 21 August. Available at: https://carnegieendowment.org/research/2018/08/india s-answer-to-the-belt-and-road-a-road-map-for-south-asia?lang=en.

Gala, N., Madhusudan, G. S., George, P., Sahoo, A., Menon, A., and Kamakoti, V., 2018. SHAKTI: An Open-Source Processor Ecosystem. *Advanced Computing and Communications*. Available at: https://www.researchgate.net/publication/330577797_SHAKTI_An_Open-Source_Processor_Ecosystem.

Ghobakhloo, M., Iranmanesh, M., Foroughi, B., Tirkolaee, E. B., Asadi, S., and Amran, A., 2023. Industry 5.0 Implications for Inclusive Sustainable Manufacturing: An Evidence-Knowledge-Based Strategic Roadmap. *Journal of Cleaner Production*, 417, p. 138023. https://doi.org/10.1016/j.jclepro.2023.138023.

Hassan, Q. F., and Sagahyroon, A., 2025. RISC-V: A Comprehensive Overview of an Emerging ISA for the AI-IoT Era. In Q.F. Hassan (ed.), *Advances in Internet of Things: Challenges, Solutions, and Emerging Technologies*. https://doi.org/10.1201/9781003506638-15

Kshetri, N., 2023. *Fourth Revolution and the Bottom Four Billion: Making Technologies Work for the Poor.* Ann Arbor: University of Michigan Press. https://doi.org/10.3998/mpub.12205632.

Kumari, S., 2020. *Techsagar-Repository of India's Cybertech Capabilities.* Available at: https://www.dsci.in/blogs/techsagar-repository-of-indias-cybertech-capabilities/#:~:text=TechSagar%20%E2%80%93%20India's%20Cybertech%20Repository%2Cdesigned,Indian%20Industry%2C%20Academia%20%26%20Research%3B.

Lewis, J. A., 2016. *Managing Risk for the Internet of Things.* Washington, DC: Center for Strategic and International Studies. Available at: https://csis-website-prod.s3.amazonaws.com/s3fs-public/legacy_files/files/publication/160217_Lewis_ManagingRiskIoT_Web_Redated.pdf.

Mandhanya, Y., Masih, J., Mathur, M., and Mishra, S., 2023. India Moving Ahead with Sustainable Transportation – A Case Study of Consumer Perception towards E-Charging Stations in India via Social Media Analysis. In A. Hannoon and A. Mahmood (eds.), *Artificial Intelligence, Internet of Things, and Society 5.0.* Switzerland: Springer, pp. 523–532. https://doi.org/10.1007/978-3-031-43300-9_43.

Schoder, D., 2018. Introduction to the Internet of Things. In Q.F. Hassan (ed.), *Internet of Things A to Z: Technologies and Applications.* Princeton, NJ: IEEE Press, pp. 3–50. https://doi.org/10.1002/9781119456735.ch1.

Schwab, K., 2015. *The Global Competitiveness Report: 2015–2016.* Geneva: World Economic Forum. Available at: https://www.weforum.org/publications/global-competitiveness-report-2015/.

Shah, I., and Sethia, T., 2021. *All about RuPay, India's Payments Network.* Available at: https://bfsi.economictimes.indiatimes.com/news/financial-services/all-about-rupay-indias-payments-network/84969331.

The World Bank, 2022. *South Asia's Digital Opportunity: Accelerating Growth, Transforming Lives.* Washington, DC: The World Bank. Available at: https://openknowledge.worldbank.org/server/api/core/bitstreams/778a3dc7-9da1-5722-a128-50b6b2aa38d7/content.

Zhang, L., Dabipi, I. K., and Brown, W. L., 2018. Internet of Things Applications for Agriculture. In Q.F. Hassan (ed.), *Internet of Things A to Z.* Princeton, NJ: IEEE Press, pp. 507–528. https://doi.org/10.1002/9781119456735.ch18.

Zlesche, S., and Bhagat, K. K., 2022. *State of the Education Report for India 2022.* New Delhi: UNESCO New Delhi Cluster Office. https://doi.org/10.54675/WBOV9023.

Glossary

5-Tuple: A group of five values, i.e., source IP address, destination IP address, source port number, destination port number, and protocol. It uniquely identifies a network connection or traffic flow.

Aadhaar: A 12-digit individual identification number that serves as proof of identity and proof of address for residents of India.

Abstract Syntax Tree: A tree data structure that represents the source code structure. Often, it is used by compilers and interpreters as an intermediate representation for source code analysis.

Access Point: A networking hardware device through which other devices connect to a wired or wireless network.

Accuracy: In machine learning, accuracy refers to the proportion of correct predictions made by a model, which is calculated as the sum of true positives and true negatives divided by the total number of predictions.

AD8232: An electrocardiogram (ECG) measurement device with an analog ECG output signal.

Advanced Persistent Threat Group: Well-resourced cybercriminal groups engaged in sophisticated malicious cyber activity with the aim of prolonged network/system intrusion.

Adversarial Machine Learning: A technique designed to trick a model into making incorrect inferences by using adversarial samples. Adversarial samples are created by making minimal alterations to legitimate data. They look almost the same as the original data but can cause the model to produce incorrect results.

Adversary: An attacker or hacker who attacks or tampers with a computer system. An adversary can be a human being or an organization.

AI Spread-out: The novel process and architectural models of distributing the AI learning and execution algorithms throughout the diverse IoT systems. They are adaptively developed so that not only the AI learning schemes are coordinated among the nodes in IoT networks but also the execution schemes of the AI model are partitioned among the diverse computing platforms in the IoT based on that learning.

Amazonification: The concept of operating to the standards set by Amazon.

Analog-to-Digital Converter: A converter that converts an analog signal to a digital representation integrated into various microcontrollers to read analog sensor output.

Anomaly Detection: The process of detecting abnormally appearing behaviors, events, or agents in an information security processing system. Examples include malicious software agents, emergent healthcare situations, or computer network intrusions using specific intelligent algorithms, such as machine learning and AI. Anomaly detection is widely used in IoT applications, such as smart grids, healthcare, precision agriculture, and smart transportation.

Application Programming Interface (API): An interface that enables software applications to communicate with one another to exchange data, functionalities, and features.

Application-Specific Integrated Circuits (ASICs): An ASIC is a form of chip tailored for a specific use (e.g., machine-learning applications). ASICs are different from CPUs, which are designed for general-purpose use.

Assembly Language: A low-level programming language that uses mnemonics (i.e., human-friendly abbreviations of instructions) such as LDA for load and STA for store to represent machine instructions. Before assembly language, programmers wrote their code in binary (i.e., sequences of 0s and 1s).

Asset: A major application, general support system, high-impact program, physical plant, mission-critical system, personnel, equipment, or a logically related group of systems.

Asymmetric Encryption: A type of encryption that uses a pair of keys, a public key and a private key.

Attack Vector: An attack vector, or a threat vector, is a way for attackers to enter a network or a system.

Autoregressive Integrated Moving Average (ARIMA): A statistical forecasting model used in time-series analysis that combines autoregressive, integrated differencing (integration), and moving average components for predicting future patterns based on past observations.

Backend Data Sharing: A pattern that is used when there is a need to analyze combined data from several sources.

Behavior Learning: A method to train a model that captures users' daily routines from their behavior sequences in the past. It is useful for downstream tasks like behavior prediction and abnormal behavior detection.

Blockchain: A decentralized record-keeping system based on the consensus of the nodes in a secure distributed network. In this system, information is stored in a chain-like structure where each block of information is assigned a unique cryptographic value called a hash value.

Bluetooth: A short-range wireless communication standard used for data exchange between devices. Typical distances are below 10 meters, and the operating frequency lies between 2.40 and 2.48 GHz. It was initially standardized as IEEE 802.15.1.

Botnet: A group of computing devices (e.g., personal computers, IoT devices, and smartphones) that are infected by malware and controlled by an attacker. A botnet can conduct malicious activities, e.g., DDoS attacks.

Byzantine Fault Tolerance: A consensus algorithm designed to tolerate Byzantine faults, where nodes in the network reach an agreement despite some nodes failing or behaving maliciously.

Callback: In programming, a callback is a function passed into another function as an argument, which is then invoked inside the outer function to execute an action.

Capacitive Soil Moisture Sensor: A capacitance-based sensor used to measure the volumetric moisture content in the soil.

Central Processing Unit (CPU): People call it the computer brain. A CPU is responsible for executing all operations that take place on a computing system. CPUs are commonly referred to as processors or microprocessors.

Channel Estimation: In wireless communication, channel estimation is the process of gauging how signals propagate from a transmitter to a receiver, bearing in mind the impact of the environment on such a signal.

Channel State Information: Refers to the known properties of a communication link. In wireless communication, it denotes the known effect of the wireless environment on signals propagating within it, where such information is acquired through channel estimation.

Chips: In the context of semiconductors, a chip is a small, thin, flat wafer of a semiconductor material, often silicon, on which the circuit is fabricated. Chips are also known as microchips and computer chips.

Cloud Computing: A centralized infrastructure that consists of a pool of computing resources (hardware and software).

Cloud-Edge Continuum: A hierarchical computing system; at its top is centralized cloud computing and at its bottom are multiple edge computing entities. Moreover, it allows for seamless integration and orchestration across these distributed environments.

Compiler: A program that converts code written in a high-level language, such as C/C++, into executable machine instructions for the processor.

Complex Instruction Set Computer (CISC): A computer architecture that uses complex instructions to manage and handle the processor's tasks. In CISC, one ISA instruction typically handles several internal tasks and takes several cycles to execute. x86-64, VAX, and IBM System/360 are popular examples of CISC ISAs.

Consensus Mechanism: A protocol used in blockchain systems to achieve agreement on a single data value among distributed processes or systems.

Constrained Devices: Computing devices with limited resources (i.e., computing, memory, and power capacity). Usually, such devices are IoT, mobile devices, wearables, etc.

Convolutional Neural Network (CNN): A special type of feed-forward neural network that learns feature engineering by itself. It is a deep-learning network architecture that learns directly from data.

Cryptographic Hash Function: A function that takes an input and returns a fixed-size string of bytes.

Cyber-Physical System (CPS): An integrated system that consists of physical components, computers, and communication networks, wherein physical conditions and activities are monitored by sensors, processed by computers, and transmitted by communication networks.

Decentralized AI: A system for applications that support AI operations (e.g., model inference, model training, data processing) implemented in a distributed manner, where input data is stored in a decentralized manner.

Decision Boundary: A hyperplane in the feature space that separates data samples of different classes.

Deep Learning: A subfield of machine learning, where "deep" refers to successive layers of the neural network.

Deep Packet Inspection: An advanced packet filtering technique that not only analyzes headers from OSI L2 to L4 but also examines the application layer data.

Dennard Scaling: A theory by Robert Dennard that dates back to the early 1970s, in which he explained that as transistors get smaller, they will use less power, keeping power use per chip almost constant despite the performance improvement.

Device Fingerprinting: The process of identifying devices through their device features or unique device identifiers.

Device-to-Cloud Communication: This pattern of communication occurs when a device captures data from its environment and uploads it to an application service provider. When the device maker and app service provider are the same, integrating other devices may be hard, even though communication is IP-based.

Device-to-Device Communication: A communication pattern that is applied when two devices communicate directly, normally using a wireless network.

Device-to-Gateway Communication Pattern: A communication pattern that may be used when the system contains non-IP devices, when support for legacy devices is needed, or when additional security functionality must be implemented.

Digital Divide: The uneven distribution of information and communication technologies in society.

Digitalization: The process of converting something to digital form.

Directed Acyclic Graph (DAG): A data structure used in some blockchain systems where transactions are not grouped into blocks but rather organized in a graph.

Distributed Denial-of-Service (DDoS): A category of cyberattacks that generates a massive amount of fabricated connection requests or malicious traffic to exhaust the capacity of a target server. As a result, the victim server may become temporarily unavailable for legitimate users.

Distributed Learning: In machine learning, a learning procedure distributes a training task across multiple processing devices, and the final learned model is obtained by combining all the learning results achieved by the devices in the network.

Domain-Specific Architectures: Also known as accelerators, they are computer architectures tailored to tackle specific domain problems to improve performance. They rely on processing specialized algorithms in a massively parallel way. Examples include GPUs, DPUs, NPUs, and TPUs.

E-governance: The employment of information and communication technologies such as the Internet at various levels of the government, public sector, and beyond for the purpose of enhancing governance.

Eavesdropper: An entity that obtains sensitive information by intercepting and listening to network traffic without legitimate authorization.

ECG: An electrocardiogram is the measure of heart electric activity over time.

Edge Computing: Similar to cloud computing, edge computing is also composed of multiple computing resources, but they can be less powerful than cloud computing and are usually geographically distributed and closer to the end nodes (the clients).

emWave Pro: A PPG-based commercial device that accesses the R-R interval sequence from the change in blood volume of the tissues' microvascular bed.

Encryption: The process of converting human-readable plaintext into ciphertext to enhance the safety of data.

ESP32: A microcontroller with integrated Wi-Fi used in IoT applications to read and log data to a remote server.

Extra Trees (ET) Algorithm: In Random Forest's analogy, this model does not consider all features at each split point of the decision trees.

F1 Score: A performance metric that calculates the harmonic mean of the precision and recall of a classification model. The F1 score can be between 0 and 1, where 0 indicates poor performance and 1 indicates perfect performance.

Feature Engineering: The process of transforming raw data into a better format or representation for the neural networks to learn from, resulting in improved efficiency.

Federated Learning: A type of collaborative machine learning that engages multiple entities, often referred to as clients. These clients work together to train a model while their data remains decentralized (i.e., they do not share the raw data format with another entity)

Field Programmable Gate Arrays (FPGAs): A type of semiconductor that is programmable. Unlike ASICs, which are not reprogrammable after manufacturing, implementers can reprogram FPGAs to meet changing application requirements.

Gateway: A networking device used to connect networks with dissimilar protocol sets. It operates by translating communications between the different protocols in use.

General Data Protection Regulation (GDPR): Europe's new data privacy and security law, which includes new requirements for organizations around the world.

GeoSense IoT Framework: An intelligent IoT-based framework for landslide prediction and monitoring purposes.

Global Capability Centers: Offshore units of multinational corporations that operate across the globe.

Global Positioning System (GPS): A system that detects the location of an object on Earth's surface.

Global South: The poorest and least industrialized countries, located mainly in the southern hemisphere.

Global System for Mobile Communications (GSM): An international standard for second-generation (2G) digital cellular networks that facilitate voice and data transmission. It uses SIM cards, supports global roaming and offers SMS and Internet access services.

gNodeB: In 5G terminology, a node in the cellular network that provides connectivity between user equipment and the evolved packet core. It is the 5G radio base station that connects devices, such as mobile phones, to the 5G core network.

Gradient Boosting (GB) Algorithm: An ML model that includes diverse weak learners.

Gram Panchayat: The local self-government organization in India.

Heart rate variability (HRV): The variability in the time interval between R peaks in successive heartbeats.

Human-factor integration: The process of applying human factors within engineering programs.

Human-machine collaboration: Coordinated interaction of humans and machines with a common goal.

In-Order Execution: In computer engineering, in-order execution means that a processor fetches and executes program instructions in the same order generated by the compiler. In-order processors are simple and do not consume much power, as they do not require complex hardware resources for dependency checking and instruction scheduling. However, program execution may stall while waiting for an instruction to be executed. In-order processors are widely used in low-end and low-power embedded systems that do not require high performance.

Industry 4.0: Smart and connected production systems that are designed to sense, predict, and interact with the physical world so as to make decisions that support production in real-time.

Industry 5.0: The term refers to people working with robots and smart machines. It is about robots helping humans work faster by leveraging advanced technologies such as the IoT and big data analytics.

InfluxDB: An open-source time-series database on the cloud widely used for remote logging of real-time sensor data.

Instruction Pipelining: A parallelization technique that divides instructions into a series of steps that can be performed simultaneously, with the aim of keeping the processor busy and increasing throughput.

Instruction Set Architecture (ISA): An abstract model that defines how software controls the processor.

Intellectual Property (IP): In the context of computer chips, an IP refers to all information (data, software code, design blueprints, etc.) necessary to make a chip.

Internet of Things (IoT): A computing model made of uniquely identifiable (physical and virtual) objects that are capable of capturing their context (via sensors) and of transmitting and/or receiving data over the network and, in the case of actuators, are capable of changing their own state or the state of their surroundings with no or very little human intervention.

IoT-MediGuard: A smart system designed to monitor the temperature and humidity of blood, medicines, and vaccines during transportation to ensure they do not go bad.

IoTMaxHRV: IoTMaxHRV is a proposed ECG measuring device with remote connectivity that can be utilized for remote HRV assessment.

IP Core: A reusable block of logic or circuit layout design. IP cores can be either soft cores (i.e., synthesizable RTL in a hardware description language like Chisel or Verilog) or hard cores (i.e., low-level transistor layouts). Designers can sell, license, open-source, or internally use their IP cores.

Landslide Monitoring System: A novel ML-based system that measures slope movements as well as weather via IoT sensors deployed at the monitored locations.

Latency: The time it takes for a transaction to complete.

Localization: The process of estimating the position and orientation of a target object within an environment.

Long-Term Evolution (LTE): Also known as 4G LTE, it is the fourth-generation wireless communication standard for broadband communications for mobile devices and data terminals.

LoRa: A proprietary wireless communication technique based on modulation with chirp pulses that provides low-power, long-range transmissions. It is a component of a low-power networking protocol to enable the connection of battery-powered devices to the Internet.

Machine Learning (ML): A subfield of artificial intelligence (AI) that enables machines to learn from data, e.g., training the neural networks, without explicit instructions to perform tasks.

Machine-Learning Model: A statistical model or program that can find patterns or make decisions from a previously unseen dataset without being explicitly programmed.

Machine-Type Communications: A broad range of communication technologies that facilitate information exchange between sensors, actuators, and devices without human intervention.

Merkle Tree: A tree structure in which each leaf node is a hash of a block of data.

Meta Training and Inference Accelerators (MTIA): A family of custom-made (ASIC) chips developed internally by Meta for its AI workloads.

Meta's Scalable Video Processor (MSVP): A custom-made (ASIC) chip developed internally by Meta for video transcoding.

Metasurface: An artificial sheet material of sub-wavelength thickness with an array of patterns along its horizontal dimension whose sizes are small compared to the operating wavelength.

Micro-Electro-Mechanical Systems (MEMS) Sensor: A low-power electronic chip embedded with tiny sensors and actuators (mechanical parts) to monitor diverse physical properties (e.g., humidity, movement, pressure). MEMS are widely used in IoT applications.

Microcontroller: A chip that contains a CPU, memory, and I/O. Microcontrollers are often optimized for low-power and ultra-low-power applications such as IoT.

Microwave: An electromagnetic wave with a frequency in the range from 300 MHz to 300 GHz.

Model Checking: An automatic verification technique that validates whether a system meets certain desired properties by exhaustively searching the state space.

Moore's Law: Not a real law, but rather an economic observation made by Gordon E. Moore in the mid-1960s, where he predicted that transistors would continue to miniaturize, enabling chipmakers to double their count on a chip every 18–24 months with relatively no cost increase. Transistor count is one of the key factors that define a chip's performance.

Multilateral Organizations: A multilateral organization comprises three or more nations to work on issues of common interest and is typically established through a treaty, a resolution, or an agreement between the participating member states.

Multiple-Issue Pipeline: It means the processor can move multiple instructions from one pipeline stage to the next per clock cycle. For example, in a dual-issue pipeline, the processor can handle two instructions per clock cycle, whereas in a quad-issue pipeline, it can handle four instructions per clock cycle. Multiple-issue pipelines are used to improve instruction-level parallelism and increase throughput.

MOESI (Modified, Owned, Exclusive, Shared, Invalid): A cache coherency protocol that ensures data consistency in multicore environments.

N-Stage Pipeline: It refers to the number of stages a processor's pipeline has, indicating the number of stages necessary to process an instruction. For example, in a 3-stage pipeline, the process is broken into fetch, decode, and execution, whereas a 5-stage pipeline has fetch, decode, execution, memory access, and writeback. The number of desired pipeline stages mainly depends on the targeted applications and inherent constraints (e.g., design complexity and power consumption).

Narrow-Band IoT (NB-IoT): A wireless IoT protocol that employs low-power wide-area network technology.

Network Address Translation (NAT) Gateway: A gateway that converts private IP addresses in private networks into public network IP addresses. NAT gateways protect devices in private networks from attacks from public networks.

Network Intrusion Detection System (NIDS): A system that monitors network traffic with the goal of detecting intrusions and malicious behaviors.

Out-of-Order Execution: A parallelism technique used in high-performance processors to increase throughput, where program instructions are executed in a different order than the compiler generated them. Instructions are executed according to the availability of input data and execution units. Out-of-order processors consume a lot of power as they require complex hardware resources for dependency checking and instruction scheduling.

Phishing Attacks: Phishing attacks are fraudulent emails, text messages, phone calls, or websites that are designed to trick users into downloading malware, sharing sensitive information or personal data (e.g., social security and credit card numbers, bank account numbers, and login credentials), or taking other actions that expose themselves or their organizations to cybercrime.

Physical Objects: All kinds of everyday objects that are present in our environments and that can contain sensors, actuators and communication capabilities. Virtual objects exist in the information sphere and can be stored, accessed, and processed.

PPG: Photoplethysmography is an optical method to measure the blood volume change in the cluster of tissues under the skin over time.

Programmable Data Plane: An advanced technology that offers programmability on the data plane so that users can implement their desired packet processing algorithms conveniently and quickly. Programmable data planes provide high packet processing speeds. However, they typically do not support complex arithmetic operations, such as floating-point operations, to maintain this speed.

Python: A high-level, procedural programming language with various existing libraries. Python is widely used in data science applications.

R-R Intervals Sequence: The interval between the successive R peaks in ECG data obtained from raw ECG and PPG devices.

Radio Frequency Fingerprinting (RFF): A unique feature associated with wireless transmitters that identifies the transmitted Radio Frequency (RF) signal by the analog components of the RF transceivers, such as the oscillators, mixers, filters, amplifiers, and antennas.

Random Forest (RF) Model: An ensemble of multiple decision trees, where each decision tree gives its result and overall output is obtained by either averaging the results in regression or voting in classification.

Ransomware: A common and dangerous type of malware. It works by locking up or encrypting your files so you can no longer access them.

Ransomware-as-a-service (RaaS) Model: A business model between ransomware operators and affiliates in which affiliates pay to launch ransomware attacks developed by operators. Think of RaaS as a variation of the software-as-a-service (SaaS) business model.

Recall: Also known as the true positive rate, which measures the percentage of true positives identified by a machine-learning model out of the actual positives present in the evaluated dataset.

Reconfigurable Intelligent Surface (RIS): A programmable surface structure that can be used to control the propagation of electromagnetic waves.

Reduced Instruction Set Computer (RISC): A computer architecture that relies on simple instructions to manage and handle the processor's tasks. The term "reduced" in RISC refers not to the number of instructions but rather to the complexity of tasks an instruction can perform. ARM and RISC-V are popular examples of RISC ISAs.

Research and Development (R&D): Refers to "research" (observation, analysis, and experimentation) and "development" (creation of ideas, theories, and practices).

RISC-V: A royalty-free, open standard ISA based on the RISC principles. RISC-V was originally developed by UC Berkeley for research and education purposes, but today it has many commercial implementations in a wide range of applications. Although RISC-V is free and open, designs and implementations based on the ISA are not necessarily free or open.

RISC-V International: The governing body of RISC-V, which is responsible for promoting and expanding the ISA and its ecosystem. The organization's members contribute and collaborate to define the ISA's open specifications, as well as convene and govern related technical, industry, and special interest groups.

Risk Mitigation: Refers to identifying, assessing, and reducing threats.

Root Mean Square Error (RMSE): The square root of the average difference between two measurements.

Second of Century (SOC) Counter: The GPS timestamp, which records the count of seconds from the coordinated universal time at midnight (00:00:00) on January 1, 1970, to the current second.

Sensing: Using radio waves to detect and estimate characteristics of targeted objects within an environment.

Sequence-Learning Algorithm: A machine-learning algorithm that mines useful contextual information from a large number of user behavior sequences. It aims to capture user habits accurately.

Serial Peripheral Interface (SPI): A communication protocol used by microcontrollers to communicate with peripheral devices such as sensors and actuators.

SigFox: A proprietary low-power wide-area network technology that allows battery-powered devices to connect to the Internet.

Single-Issue Pipeline: It means the processor can move a single instruction from one stage of the pipeline to the next per clock cycle.

Smart Contracts: Self-executing contracts with the terms of agreement directly written into lines of code. Smart contracts commonly run on blockchains to execute predefined clauses (e.g., digital exchanges).

Smart Device: A device that not only performs its basic functions but also achieves some intelligence and automation by communicating with other connected devices or a cloud server.

Smart Home: Refers to a housing environment with Internet-enabled smart devices, such as smart meters, smart locks, and surveillance cameras. Typically, a smart hub connects the smart devices in a smart home. Users can monitor and control devices in a smart home remotely through user interfaces on smartphones or tablets. Smart homes may also support hands-free control of some devices via voice assistants.

Software-as-a-service (SaaS): A way of delivering applications over the Internet—as a service. SaaS is a common model of cloud computing.

Software-Defined Networking (SDN): A networking approach that provisions, automates, manages, and programs networks through software.

Split Computing: The implementation of ML applications (model inference and model training) with the help of the cloud-edge continuum. The procedure involves segmenting the task into smaller ones that are offloaded onto the cloud or edge for processing.

Split Learning: A collaborative distributed learning technique that allows resource-constrained devices to train large neural network models on the cloud.

Spoofing Attack: A type of cyberattack in which an attacker pretends to be a trusted user by using forged data to gain unauthorized access to a system.

Stakeholder: A person, group, or organization with a vested interest or stake in the decision-making and activities of a business, organization, or project.

Stale Block: In blockchain terminology, a stale block is a block that was successfully mined but not included in the main blockchain.

Steady-State Signal: Refers to the part of the signal that remains after the transient phases.

Symbolic Executor: A program analysis technique that investigates which input triggers the execution of each part of a program. More specifically, it uses symbolic values as input rather than concrete values.

Synergetic Inference: The support of inference tasks using split computing. The client and the cloud collaborate to implement inference model requests.

System on a Chip (SoC): A microchip that contains all or most of the components of a computing system. SoCs are widely used in embedded systems and mobile devices today.

The Third World: Countries with less developed economies and industries.

Threat Actor: Threat actors, also known as cyberthreat actors or malicious actors, are individuals or groups that intentionally cause harm to digital devices or systems.

Threat Model: A threat model is a structured representation of all the information that affects the security of an application. In essence, it is a view of the application and its environment through the lens of security.

Throughput: The number of transactions a system can process in a given amount of time.

Toolchain: In computer engineering, a toolchain is a set of tools essential for developing programs that run efficiently on a specific hardware architecture. Such tools include assemblers, compilers, linkers, and debuggers.

Traffic Morphing: A technique that reshapes traffic using methods like padding packets and changing packet timing. It changes the original patterns of traffic and, therefore, is an effective measure to defend against fingerprinting attacks.

Transient: The term refers to the segment of signals generated during the power-up or power-down phases of a transmitter. It is a sudden change at the beginning or end of steady-state signals.

Trigger-Action Platform (TAP): A web-based system where users can define conditional rules to control the behavior of smart devices. It helps users conveniently realize desired automation among connected devices.

Unified Payments Interface: An Indian payment system that powers multiple bank accounts into a single mobile application (for any participating bank), merging several banking features, seamless fund routing, and merchant payments under one umbrella.

Unmanned Aerial Vehicle (UAV): An aircraft or aerial robot, also known as a drone, that is guided either autonomously or by remote control without a human pilot on board.

Virtual Private Network (VPN): A technique to secure communication between entities. It uses encryption and encapsulation to protect data transmission, which can make communication endpoints appear as if they are located on the same local network.

Watchdog Timer (WDT): A hardware timer used to detect and recover a microcontroller from malfunction.

Index

Note: **Bold** page numbers refer to tables and *italic* page numbers refer to figures.

FSC
www.fsc.org
MIX
Papier | Fördert gute Waldnutzung
FSC® C083411

Kolibri 360 GmbH
Ferdinand-Jühlke-Straße 7
99095 Erfurt, Deutschland
produktsicherheit@kolibri360.de